D0899189

WITHDRAWN
UTSA LIBRARIES

Pointlike Structures Inside and Outside Hadrons

THE SUBNUCLEAR SERIES

Series Editor: **ANTONINO ZICHICHI**
European Physical Society
Geneva, Switzerland

1. 1963 STRONG, ELECTROMAGNETIC, AND WEAK INTERACTIONS
2. 1964 SYMMETRIES IN ELEMENTARY PARTICLE PHYSICS
3. 1965 RECENT DEVELOPMENTS IN PARTICLE SYMMETRIES
4. 1966 STRONG AND WEAK INTERACTIONS
5. 1967 HADRONS AND THEIR INTERACTIONS
6. 1968 THEORY AND PHENOMENOLOGY IN PARTICLE PHYSICS
7. 1969 SUBNUCLEAR PHENOMENA
8. 1970 ELEMENTARY PROCESSES AT HIGH ENERGY
9. 1971 PROPERTIES OF THE FUNDAMENTAL INTERACTIONS
10. 1972 HIGHLIGHTS IN PARTICLE PHYSICS
11. 1973 LAWS OF HADRONIC MATTER
12. 1974 LEPTON AND HADRON STRUCTURE
13. 1975 NEW PHENOMENA IN SUBNUCLEAR PHYSICS
14. 1976 UNDERSTANDING THE FUNDAMENTAL CONSTITUENTS OF MATTER
15. 1977 THE WHYS OF SUBNUCLEAR PHYSICS
16. 1978 THE NEW ASPECTS OF SUBNUCLEAR PHYSICS
17. 1979 POINTLIKE STRUCTURES INSIDE AND OUTSIDE HADRONS

Volume 1 was published by W. A. Benjamin, Inc., New York; 2-8 and 11-12 by Academic Press, New York and London; 9-10 by Editrice Compositori, Bologna; 13-17 by Plenum Press, New York and London.

Pointlike Structures Inside and Outside Hadrons

Edited by
Antonino Zichichi
European Physical Society
Geneva, Switzerland

PLENUM PRESS · NEW YORK AND LONDON

Library of Congress Cataloging in Publication Data

International School of Subnuclear Physics, 17th, Erice, Italy, 1979.
Pointlike structures inside and outside hadrons.

(The Subnuclear series; 17)
Includes index.
1. Hadrons—Congresses. I. Zichichi, Antonino. II. Italy. Ministero della pubblica istruzione. III. Title. IV. Title: Subnuclear physics. V. Series: Subnuclear series; 17.
QC793.5.H322I56 1979 539.7'216 80-25632
ISBN 0-306-40568-7

Proceedings of the Seventeenth International School of Subnuclear Physics, held in Erice, Trapani, Sicily, Italy, July 31–August 11, 1979.

© 1982 Plenum Press, New York
A Division of Plenum Publishing Corporation
233 Spring Street, New York, N. Y. 10013

All rights reserved

No part of this book may be reproduced, stored in a retrieval system, or transmitted, in any form or by any means, electronic, mechanical, photocopying, microfilming, recording, or otherwise, without written permission from the publisher

Printed in the United States of America

PREFACE

In August 1979 a group of 94 physicists from 60 laboratories in 21 countries met in Erice to attend the 17th Course of the International School of Subnuclear Physics.

The countries represented at the School were: Argentina, Australia, Austria, Belgium, Canada, Federal Republic of Germany, France, Hungary, India, Italy, Japan, Korea, the Netherlands, Norway, Poland, Rumania, Spain, Sweden, the United Kingdom, the United States of America, and Yugoslavia.

The School was sponsored by the Italian Ministry of Public Education (MPI), the Italian Ministry of Scientific and Technological Research (MRST), the Sicilian Regional Government, and the Weizman Institute of Science.

In the theoretical sessions, as expected, Sidney Coleman's (1/N) lectures provided a new masterpiece in his Erice series. André Martin lectured on the theory of new particles, R.L. Jaffe on the Bag Model and Francis E. Low on quark status at low energy. In the more specialized section we were able to reflect on the problem of asymptotic changes in gauge theory, troubles experienced with instantons and finally on the fate of false vacua, thanks to A. Patrascioiu, R. Petronzio, and H. Kleinert.

A quite exceptional event this year was the special QCD session which I leave the reader to enjoy without further introduction.

On the experimental front, e^+e^- physics, neutral weak currents and their relation with charged currents, deep inelastic phenomena and the status of the diquark structure in hadronic physics were the hottest problems and were discussed by five outstanding experts: S.C.C. Ting, E. Lohrmann, F. Dydak, J.J. Aubert, and R.T. Van de Walle. The opening lecture of Victor F. Weisskopf was the inspiring introduction to all courses, and the lecture on his life as a physicist by H.B.G. Casimir a fitting end to what I believe all the participants found to be a most stimulating occasion.

I hope the reader will enjoy the book as much as the students enjoyed the School including one of the most attractive aspects: the informal, free, and unconventional discussion sessions. Thanks to the efforts of the scientific secretaries, these discussions have been reproduced in a form reflecting as closely as possible the real happenings.

At various stages of my work I have enjoyed the collaboration of many friends in Erice, Bologna, and Geneva, whose contributions have been extremely important for the School and are highly appreciated.

Antonino Zichichi

CONTENTS

OPENING LECTURE

THEORETICAL LECTURES

REVIEW LECTURES

PERSONAL IMPRESSIONS OF RECENT TRENDS IN PARTICLE PHYSICS

Victor F. Weisskopf

M.I.T., Cambridge, MA 01239, USA and

CERN, Geneva, Switzerland

There are three discoveries made in the last ten years that have deeply influenced our understanding of particle physics.

1) Deep inelastic scattering experiments with electrons and neutrinos as proof for the actual existence of quarks within hadrons.

2) The discovery of two more quark flavours and one more heavy electron.

3) The increasing significance of Yang-Mills field theories.

The first group of discoveries put the quarks definitely on the map. It showed that small ($< 10^{-16}$ cm) charged units indeed exist within the hadrons, with spin ½ and with not too strong interactions at large momentum transfers.

The discoveries of more flavours (c and b quarks) and of the τ electron were to some extent gratifying and to some extent disquieting. The existence of the c quark was a confirmation of theoretical predictions by Bjorken, Glashow and collaborators, but the discovery of the b quark opened up the possibility of an unending series of flavours with increasing masses. Certainly, five is not infinity, and six flavours have indeed been proposed in order to explain the violation of CP conservation; but what if there is a seventh flavour? Does the beginning proliferation of flavours indicate an internal structure of the quark, a new spectroscopy, a higher rung of the quantum ladder? Does the third electron (the τ particle) indicate an extended spectrum of electrons coming from some internal structure? Two remarks have to be made here: one is this: A new spectrum of internal excitations would require the appearance of states with $J > \frac{1}{2}$. Only the discovery of a particle with $J = 3/2, 5/2$, etc. would be a clear indication of a new spectroscopy. Second: We know from deep-inelastic scattering that the size of the quark is less than several 10^{-16} cm. Hence, true excitations of an internal dynamics should be expected at energies $> 1/R$ which is < 20 GeV, much higher than the masses of the new quarks and electrons. Thus, if the newly discovered particles are part of a spectrum of an internal structure, they can only be the fine structure of the ground state.

It is worth recalling that Nature in the whole Universe consists only of the u and d quark, the ordinary electron and neutrino. There is not enough energy available to excite the higher flavours. Possible exceptions are neutron stars and the earliest stages of the Universe. The question arises what is the role of those other short-lived particles. Why are they here? As Rabi has asked "Who ordered them?"

It may be significant, however, that for each quark pair of charge 2/3 and 1/3, there is one electron: the ordinary electron gas with the u-d pair, the muon with the c-s pair and the τ with a t-b pair, of which the t is not yet discovered. Does this indicate something or is it accidental?

We now come to the Yang-Mills theories. The present views of weak and strong interactions make use of non-Abelian field theories. In a non-Abelian field theory the field itself is a carrier of charge.

The charge is not bound to the fermions that are the sources of the field. The field itself is a source. Therefore, there are direct interactions between the field quanta, such as gluons or intermediate bosons.

Let us first discuss QCD. Here the field quanta (gluons) remain massless. The fact that the charge (colour) is exchanged between the quarks and the gluon field is the basic reason for asymptotic freedom. The effective charge is not tied to the quark but spread over the adjacent field. Thus high momentum transfers (small distances) "see" only part of the charge. In addition we believe (no proof yet; see below) that, at large distances, the interaction between quarks becomes infinite so that they cannot exist as free particles. These circumstances lead to the concept of a "running coupling constant", depending on the amount of momentum transfer Q^2. It goes to zero for high Q^2 and to infinity for low Q^2. We are considering here the effective coupling constant g, not g_0, that appears in the original Lagrangian. The latter one is fixed; the running coupling constant g is the result of taking into account gluon-quark and gluon-gluon interactions plus applying the renormalization methods.

If, for the moment, we consider only u and d quarks which, probably, have no or negligible mass, we face a theory without any length entering the equations, since the coupling constant g_0 is dimensionless. Still a length appears in the following way: there must necessarily be a momentum transfer Q_1 for which the effective coupling g is of order unity. This value Q_1 determines a mass Q_1 and a length Q_1^{-1}. For higher Q, one can use perturbation approach, for lower Q one cannot. Since the coupling becomes quite large for $Q < Q_1$, the corresponding bound states (hadrons) are of the size Q_1^{-1} and the hadron masses are $\sim Q_1$. Obviously Q_1 must be of the order of a good fraction of one GeV. In other words, we have to choose g_0 such that $g(Q^2 \sim 1 \text{ GeV}) \sim 1$. (Actually, this is a somewhat simplified description of the situation since the non-renormalized coupling constant goes to zero. But it illustrates the logical connections.)

In QED the situation is quite different. There, the effective coupling constant increases with Q^2 because, at high Q, the vacuum polarization ceases to shield the electric charge. What we understand by "charge e" is the fully shielded charge at large distances. That value is finite for $Q = 0$ and equal to $(137)^{\frac{1}{2}}$ but it increases as $\log (Q/m)$ for high Q.

Let me say a few words on how one can, perhaps, describe the situation*) in QCD for small Q. All this is tentative since we do not have a reasonable theoretical approach for the strong coupling situation at $Q < Q_1$. Let us look at the vacuum in QED and in QCD. In the first case it is full of field fluctuations and virtual pairs. The photons and pairs "present" in the vacuum are virtual because their energy is positive. We call such a vacuum a "simple" vacuum. Let us be sure that the energy ε of an electron positron pair always is positive. For the sake of simplicity, we neglect the masses. Then the energy consists of two parts, a kinetic term $\sim r^{-1}$, where r is the distance between partners, and an attractive energy $-(e^2/r)$. Clearly $\varepsilon \sim r^{-1} (1 - e^2)$ will always be positive, since e^2_{eff} is small and increases only logarithmically with $Q \sim r^{-1}$. In QCD things are different. The energy ε of a quark pair or of a gluon pair can become negative if $Q \sim r^{-1} < Q_1$; then $g^2 > 1$ and $\varepsilon \sim r^{-1}(1 - g^2)$ becomes negative! The true vacuum, therefore, should consist of real (not virtual) gluon and quark pairs or "balls" of a size $\sim Q_1^{-1}$. The very big ones, $r >> Q^{-1}$, seem to

*) The following ideas were suggested to me by S. Coleman and K. Johnson. I take the responsibility for the formulation.

have very large negative energy, but their phase space is very small; hence we expect a finite average size of those balls and a finite negative energy density of the true vacuum. The true vacuum is liquid-like, with gluon- and quark-balls of size Q_1^{-1} forming and transforming.

Now we make a daring hypothesis. The true vacuum expels all gluo-electric field lines in a similar way as a superconductor expels magnetic field lines. Then, an assembly of real quarks (sources of gluo-electric field lines) would have to form a bubble around it in the true vacuum. In that bubble the true vacuum cannot exist (because of the presence of gluo-electric fields) and the bubble will be filled with a "simple" vacuum. After all, in a small region (smaller than Q_1^{-1}) the effective coupling constant is smaller than unity and, therefore, it does not pay to form real gluon or quark pairs ($\varepsilon = r^{-1}(1 - g^2) > 0$). Quarks can only exist within a simple vacuum and are caught in the bubble. The energy density of the simple vacuum is zero; it is higher than the true vacuum. Thus, the forming of a bubble in the true vacuum costs energy proportional to the bubble volume. All this is identical with the assumptions of the bag model. If it is true, we have a QCD argument in support of the bag model.

T.D. Lee has presented a way that perhaps makes plausible the expulsion of gluo-electric fields from the true vacuum. It goes as follows: A change of the effective coupling constant can also be expressed by a changing dielectric constant $\kappa(Q^2)$. Let us assume arbitrarily that $\kappa(Q_1^2) = 1$, and call g_1 the coupling constant for $Q = Q_1$ (it is per definition equal or near unity). Then $g^2(Q^2) = g_1^2/\kappa(Q^2)$. Therefore at large distances ($Q^2 \to 0$), κ must go to zero, in order to describe the fact (it may be a fact - we are not sure) that $g^2(Q^2)$ goes to infinity. So, the true vacuum at large ought to behave as a medium with a gluo-dielectric constant going to zero. In ordinary media made of atoms, the dielectric constant is always

$\kappa < 1$; that means effective charges are smaller than true charges. This is because the little dipoles in the medium turn their negative ends to the (positive) true charge, thus shielding it. When $\kappa < 1$, it is as if the little dipoles would turn their positive ends towards the true charge, thus antishielding (increasing) its effectiveness. (Of course, real dipoles would not do that; there must be another yet unclear mechanism. But we know, after all, that gluonic charges indeed are antishielded at distances $> Q_1^{-1}$). Now, the electric energy density is $\vec{D} \cdot \vec{E} = \vec{D}^2/\kappa$. It becomes infinite for $\kappa \to 0$, since $\vec{D}$ is fixed by the charges. Thus electric fields will not penetrate into this medium, because it costs infinite energy.

The discussion so far applies only to QCD with u and d quarks. The appearance of quarks with non-negligible masses, such as the other quarks, represents a great difficulty for QCD because it contains no mechanism to provide masses to the quarks. Only the masses of the non-strange or non-charmed hadrons (that part which does not come from the intrinsic quark masses) are understandable. They are derived from the kinetic energies of the quarks, confined to a volume of the size Q_1^{-1}, and from contributions of the spin-spin interactions between quarks that are caused by the gluo-magnetic field. They raise the masses when the spins are parallel and lower them for anti-parallel spin.

In QED there may be some hope that the strong effective coupling at high Q gives rise to structures that may explain the masses of the diverse electrons. This hope does not exist in QCD because of asymptotic freedom. The masses of the higher flavours must be produced by another mechanism, such as a coupling to a Higgs-type field.

This brings me to the other Yang-Mills field theory: the

---

*) Most references to a theory by names are intrinsically unfair, except in the case of Einstein. There are many more people than those two who have contributed to it. Perhaps one should call it Q.WE.D. By now, the name W-S has become established.

Weinberg-Salam*) theory of electro-weak interactions. I have less to say about this theory, not because I think less of it, but because it triumphs and shortcomings are well known and less controversial. The triumphs consist of a successful unification of weak and electromagnetic forces. This is seen most clearly in the fact that the neutral current effects of the weak interaction are different in character from the charged current effects: they are not purely (V-A) couplings, that is, they do not have maximal parity violations. It comes from the most astonishing, but experimentally well-established, fact that nature chooses to mix the neutral part of the weak interactions (the I_3 component of an isotopic triplet of intermediate bosons) with the electromagnetic interactions. The mixing angle is the famous Weinberg angle. This shows clearly that there are four "components" of weak-electric field: W^+, W^-, Z^0 and γ, the latter two being the two orthogonal mixtures between W^3 and an isotopic scalar representing the "Ur-electromagnetism" (U(1) group) before mixing. What I call "shortcomings" is the necessity of introducing a new field, the Higgs field, which is responsible for the masses of the participating fermions and bosons, and for the above-mentioned mixing. It is necessary to provide masses by a coupling with a field whose vacuum expectation value does not vanish; with finite masses ab initio, the theory would not be renormalizable. The Higgs coupling contains as many arbitrary coupling constants as there are masses. This is a rather awkward way to "explain" the existence of masses and their magnitudes. It is possible, of course, that those Higgs particles really exist. Then the Higgs coupling _is_ Nature's way to make masses. I believe that Nature should be more inventive, but experiments may prove me wrong.

Experiments have verified a great deal of the predictions of the W-S theory, in particular, in respect to the detailed properties of the neutral current events. The deservedly famous SLAC experiment about the parity non-conserving scattering of electrons by nucleons is an outstanding example. The fact that the Weinberg angle comes out to be the same in all experiments, certainly is a strong support of the theory. However, we still have no experimental evidence for the existence of intermediate bosons, to say nothing of Higgs bosons. In a few years, facilities will be available with enough energy to produce them. Woe to the theory if they do not show up!

We have indicated before that a Yang-Mills type of field theory leads to asymptotic freedom and infinite binding at low momentum transfer. This is the case (most probably) with QCD. In the W-S theory of electro-weak interactions it is not so. Things do not blow up at low Q^2 because the particles get masses from the Higgs field. Thus the "true" vacuum does not form. (In the electromagnetic part, the bosons are still massless, but they do not carry charge.) At high momentum transfer, it is again the Higgs coupling (being not of the Yang-Mills type) which prevents asymptotic freedom.

Although the last decade has given us many more insights into the world of particles, some of the great questions are still open. We do not even know whether QCD makes sense at low momentum transfer. There is the question of the origin of the masses of the higher quarks, the question of the nature of quark flavours and of heavy electrons (is there a limit or is there an internal structure?), the question of the unification of electro-weak and strong forces, and the question of the uniqueness of the electric charge e, all of which are still completely unexplained. The fractional charges of the quarks make the last problem even more mysterious.

There are, of course, a number of tentative efforts to get at some of the unsolved questions. The studies of supersymmetries and of grand unification schemes are examples. So far, these studies have not yet yielded solid results. The uncertainty of the number of flavours and heavy electrons makes it hard to invent supersymmetries that contain the right number. The present grand unification schemes are forced to make simplistic assumptions such as that no essentially new phenomena will be found up to the incredibly high energies where supposedly the three interactions merge. Past experience shows that this is not very probable.

Nobody can predict, however, what the experiments will tell us and what new ideas will emerge. There is a Danish proverb: Predicting is difficult, especially if it concerns the future. One thing, however, seems to be sure: Ten years from now, the picture will be very different and much richer, perhaps even more profound.

1/N*

Sidney Coleman**

Stanford Linear Accelerator Center
Stanford University
Stanford, California 94305

1. INTRODUCTION

More variables usually means greater complexity, but not always. There exist families of field theories with symmetry group SO(N) (or SU(N)) that become simpler as N becomes larger. More precisely, the solutions to these theories possess an expansion in powers of 1/N. This expansion is the subject of these lectures.

There are two reasons to study the 1/N expansion.

(1) It can be used to analyse model field theories. This is important. Most of us have a good intuition for the phenomena of classical mechanics. We were not born with this intuition; we developed it toiling over problems involving rigid spheres that roll without slipping and similar extreme but instructive simplifications of reality. One reason we have such a poor intuition for the phenomena of quantum field theory is that there are so few simple examples; essentially all we have to play with is

---

*Work supported in part by the Department of Energy under contract DE-AC03-76SF00515 and by the National Science Foundation under Grant No. PHY77-22864.

**Permanent Address: Department of Physics, Harvard University.

perturbation theory and a handful of soluble models. The 1/N expansion enables us to enlarge this set.

In Section 2 I develop the 1/N expansion for ϕ^4 theory and apply it to two two-dimensional models with similar combinatoric structures, the Gross-Neveu model and the CP^{N-1} model. These models display (in the leading 1/N approximation) such interesting phenomena as asymptotic freedom, dynamical symmetry breaking, dimensional transmutation, and non-perturbative confinement; they are worth studying.

(2) It is possible that the 1/N expansion, with N the number of colors, might fruitfully be applied to quantum chromodynamics. In the real world, N is 3, so an expansion in powers of 1/N may not seem like such a good idea. This objection is without force, as is shown by the following wisecrack by Ed Witten:

$$\frac{e^2}{4\pi} = \frac{1}{137} \Longleftrightarrow e = .30$$

Of course, this does not show that the 1/N expansion in QCD will necessarily be as good an approximation as perturbation theory in QED, but it does show that there is no reason to reject it a priori.

Unfortunately, it is not possible to make a decisive test of the approximation, because no one knows how to compute even the first term in the expansion in closed form. However, it is possible to argue that this first term, whatever its detailed form, has many properties that are also shared by the real world, and which are otherwise underived from field theory. These include the saturation of scattering amplitudes by an infinite number of narrow resonances, the essential feature of dual-resonance models. I discuss these matters in Section 3.

Although united here, these two classes of applications have very different standings. The work on model field theories is modest but solid, a permanent part of our knowledge. In contrast, the work on chromodynamics is ambitious but conjectural. It is possible that it will lead to great breakthroughs; it is possible that it will fizzle out, like so many hopeful programs before it.

I should warn you that these lectures are introductory rather than encyclopedic. Much more could be said about every topic I discuss. I have not gone into more depth in part because of lack of time and in part because of lack of competence. I am not an expert in these matters; one reason I decided to lecture on them this summer was to force myself to learn them.

Much of what I do know I have learned from conversations with Roman Jackiw, Hugh Osborne, Howard Schnitzer, Gerard 't Hooft, Ken Wilson, and Edward Witten. My debt to Witten is enormous; most of the second half of Section 3 is plagiarism of his ideas.

[Note to the reader: If you are only interested in chromodynamic applications, I suggest you read just the first five paragraphs of Section 2, and then proceed directly to Section 3.]

2. VECTOR REPRESENTATIONS, OR, SOLUBLE MODELS

2.1. ϕ^4 Theory (Half-way)[1]

I will begin the development of the 1/N expansion with a theory that is (I hope) familiar to you, the O(N) version of ϕ^4 theory. I warn you in advance that I will stop the discussion half-way, after I have worked out all the combinatorics but before I have evaluated any Feynman integrals. This is not because the details of the model are not interesting and instructive (they are, especially in the Nambu-Goldstone mode), but because I am using this theory only as a warm-up, and want to get on to the even more interesting and instructive Gross-Neveu and CP^{N-1} models.

The dynamical variables of the theory are a set of N scalar fields, ϕ^a, a = 1 ... N, with dynamics defined by the Lagrange density,

$$\mathscr{L} = \frac{1}{2}\partial_\mu \phi^a \partial^\mu \phi^a - \frac{1}{2}\mu_0^2 \phi^a \phi^a - \frac{1}{8}\lambda_0 \left(\phi^a \phi^a\right)^2 \quad , \tag{2.1}$$

where the sum on repeated indices is implied. Since I am going to stop the investigation before evaluating any Feynman integrals, I might as well keep the dimension of space-time arbitrary (but less than or equal to four).

To get an idea of what is going on, I have written down in Fig. 1 the first few diagrams (in ordinary perturbation theory) for the scattering of two mesons of type a into two mesons of type b ($a \neq b$). The first diagram displayed, the Born term, is $O(\lambda_0)$. The second diagram is $O(\lambda_0^2 N)$, because there are N possible choices for the internal index, c. The third diagram, in contrast, is only $O(\lambda_0^2)$; the internal indices are fixed and there is no sum to do.

The explicit factor of N in the second diagram makes the large-N limit seem nonsensical, but this is easily rectified. All we need do is define

$$g_0 \equiv \lambda_0 N \quad , \tag{2.2}$$

and declare that we wish to study the limit of large-N with fixed g_0 (not fixed λ_0). The first diagram is now $O(g_0/N)$; as we shall see, this is the leading non-trivial order in $1/N$. The second diagram is $O(g_0^2/N)$, the same order in $1/N$. The third diagram is $O(g_0^2/N^2)$, next order in $1/N$ and negligible compared to the two

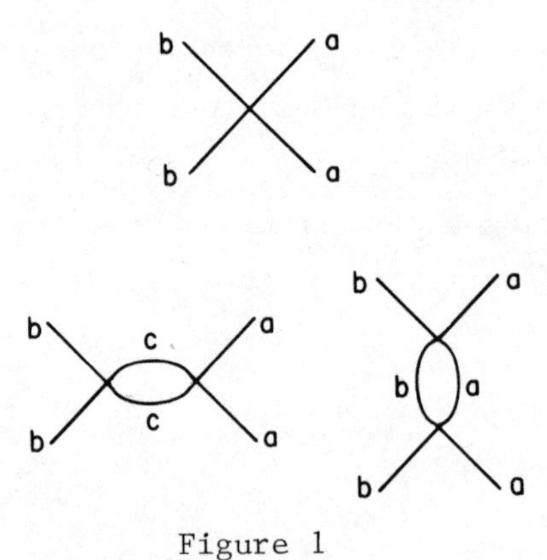

Figure 1

preceding diagrams in the large-N limit.

This is the first step in constructing the 1/N expansion. We must decide what parameters to hold fixed as N becomes large. If we make the wrong choice, we can obtain either a trivial theory (only the Born term survives) or one without a 1/N expansion (there are graphs proportional to positive powers of N). Of course, we have not yet shown that the second possibility does not occur in the theory at hand. However, there are clearly an infinite number of graphs proportional to 1/N, times various powers of g_0; two of them are shown in Fig. 2. (To keep the graphs from being hopeless jumbles, I have left out the index labels; I hope you can figure out where they go.)

To keep all these diagrams straight, and to show that there are no diagrams proportional to positive powers of N, is a combinatoric challenge. We can simplify life considerably by introducing an auxiliary field, σ, and altering the Lagrange density:

$$\mathscr{L} \rightarrow \mathscr{L} + \frac{1}{2}\,\frac{N}{g_0}\left(\sigma - \frac{1}{2}\,\frac{g_0}{N}\,\phi^a\phi^a\right)^2 \quad . \tag{2.3}$$

This added term has no effects on the dynamics of the theory. This is easy to see from the viewpoint of functional integration. The functional integral over σ is a trivial Gaussian integral; its only effect is to multiply the generating functional of the theory by an irrelevant constant. It is also easy to see from the viewpoint of canonical quantization. The Euler-Lagrange equation for σ is

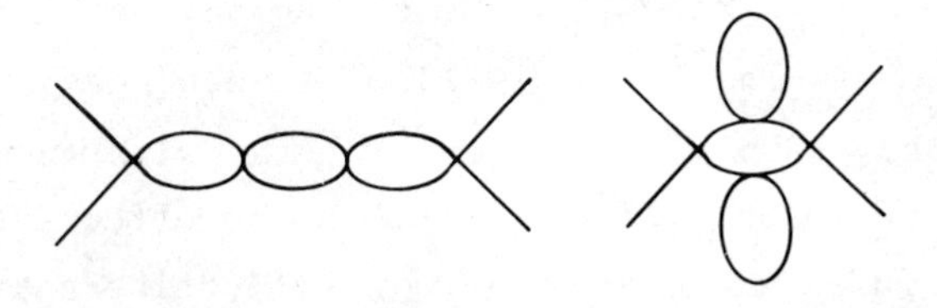

Figure 2

$$\sigma = \frac{1}{2} \frac{g_0}{N} \phi^a \phi^a \quad . \tag{2.4}$$

This involves no time derivatives; it is not a true equation of motion, but an equation of constraint, like the Euler-Lagrange equation for the fourth component of a massive vector field. When we construct the Hamiltonian, σ must be eliminated from the Lagrangian, using Eq. (2.4); this cancels the added term.

However, although the dynamics defined by our new Lagrangian are the same as those defined by the old one, the Feynman rules are different. By elementary algebra

$$\mathscr{L} = \frac{1}{2} \partial_\mu \phi^a \partial^\mu \phi^a - \frac{1}{2} \mu_0^2 \phi^a \phi^a$$
$$+ \frac{1}{2} \frac{N}{g_0} \sigma^2 - \frac{1}{2} \sigma \phi^a \phi^a \quad . \tag{2.5}$$

Thus, in the new formalism, the only non-trivial interaction is the $\phi\phi\sigma$ coupling. All factors of 1/N come from the σ propagator (ig_0/N). Every line on a closed ϕ loop must always carry the same index, and this index must always be summed over; thus, we need not write explicit indices on ϕ loops, and every closed ϕ loop always gives a factor of N.

Figure 3 shows the graphs of Fig. 1 in our new formalism. (The dashed line is the σ propagator.) Counting powers of 1/N is now much easier than before, but things can be made easier yet. Let us imagine analyzing a general Feynman graph as follows: First, let us strip away all the ϕ lines that end on external lines, that is to say, that are not part of closed loops. This yields a graph that has only external σ lines. Second, let us do all the momentum integrals over the closed ϕ loops. Every ϕ loop thus becomes a (non-local) interaction between the σ fields that terminate on that loop. We thus generate a graph with only σ lines; it can be thought of as a graph in an effective field theory whose Feynman rules are derived from an effective action, $S_{eff}(\sigma)$.

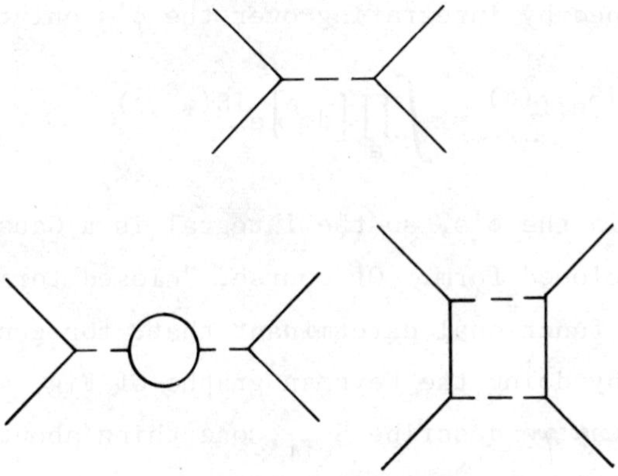

Figure 3

There are two ways of describing S_{eff}, in terms of Feynman graphs or in terms of functional integrals. The description in terms of graphs is shown in Fig. 4. The first graph gives the term linear in σ; the second graph gives a term quadratic in σ, which must be added to the third graph, the quadratic term already present in Eq. (2.5); the fourth graph gives the cubic term; etc. In terms of functional integrals, the quantum theory is defined by integrating the exponential of iS, the classical action, over all configurations of all fields in the theory. The effective

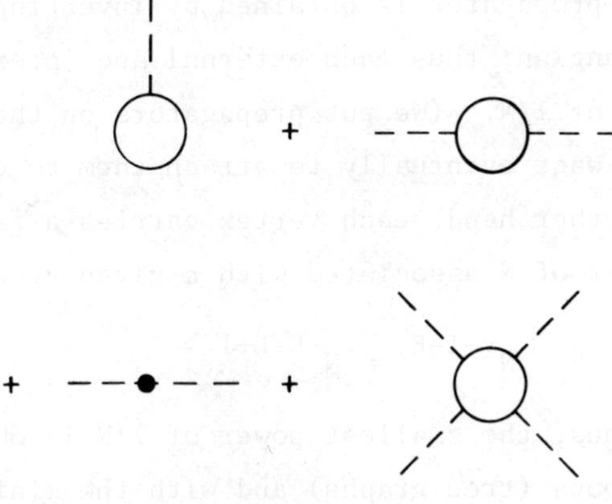

Figure 4

action is obtained by integrating over the ϕ's only:

$$e^{iS_{eff}(\sigma)} = \int \prod_a \left[d\phi^a\right] e^{iS(\phi^a,\sigma)} \quad . \tag{2.6}$$

S is quadratic in the ϕ's, so the integral is a Gaussian one, and can be done in closed form. Of course, "closed form" is a hoax; the answer is a functional determinant that, for general σ, can be evaluated only by doing the Feynman graphs of Fig. 4.

Whichever way we describe S_{eff}, one thing about it is obvious; every term in it is proportional to N:

$$S_{eff}(\sigma,N) = NS_{eff}(\sigma,1) \quad . \tag{2.7}$$

This makes counting powers of N very easy. Consider a graph in our effective field theory with E external lines, I internal lines, V vertices, and L independent loop integrations. These quantities are not independent. For a connected graph,

$$L = I - V + 1 \quad . \tag{2.8}$$

That is to say, we have one integration momentum for each internal line, but we also have one delta-function for each vertex; each delta-function cancels one momentum, except for one delta-function that is left over for overall momentum conservation. The power of N associated with a graph can be expressed in terms of these quantities. The propagator is obtained by inverting the quadratic part of the Lagrangian; thus each external and internal line carries a factor of 1/N. (We put propagators on the external σ lines because we want eventually to attach them to external ϕ lines.) On the other hand, each vertex carries a factor of N. Thus the net power of N associated with a given graph is

$$N^{V-I-E} = N^{-E-L+1} \tag{2.9}$$

by Eq. (2.8). Thus, the smallest power of 1/N is obtained from graphs with no loops (tree graphs) and with the minimum number of external lines required to connect the external ϕ lines. In the case we began by studying, meson-meson scattering, two external σ lines are required, and thus the leading power is 1/N.

This is no great surprise, of course; we hardly needed all this formalism to get this piddling result. However, we are almost in a position to compute meson-meson scattering to O(1/N) in closed form. I say "almost" because S_{eff} has the awkward feature of containing a term linear in ϕ. In the presence of such a term, there are an infinite number of tree graphs with two external lines; all one has to do is build a tree graph of arbitrary complexity, and then terminate all but two of its external lines on linear vertices. The cure for this problem is well known. We define a new, shifted field,

$$\sigma' \equiv \sigma - \sigma_0 \quad , \tag{2.10}$$

where σ_0 is a constant chosen such that $\sigma = \sigma_0$ is a stationary point of S_{eff},

$$\left.\frac{\delta S_{eff}}{\delta \sigma'}\right|_{\sigma'=0} = 0 \quad . \tag{2.11}$$

(I will shortly show that σ_0 exists.) In terms of σ', there are no linear vertices.

The easiest way to construct $S_{eff}(\sigma')$ is to express $\mathscr{L}$ in terms of σ'. From Eq. (2.5),

$$\mathscr{L} = \frac{1}{2}\partial_\mu\phi^a\partial^\mu\phi^a - \frac{1}{2}\mu_1^2\phi^a\phi^a$$

$$+ \frac{1}{2}\frac{N}{g_0}\sigma'^2 - \frac{1}{2}\sigma'\phi^a\phi^a + \frac{N}{g_0}\sigma_0\sigma' \tag{2.12}$$

plus an irrelevant constant, where

$$\mu_1^2 \equiv \mu_0^2 + \sigma_0 \quad . \tag{2.13}$$

μ_1 is the ϕ mass, to leading (zeroth) order in 1/N; it will be convenient to use it as an independent parameter of the theory instead of μ_0. The graphical construction of $S_{eff}(\sigma')$ is now the same as that shown in Fig. 4, with two exceptions: (1) The internal ϕ lines now carry a mass μ_1 rather than μ_0. (2) There is an additional linear vertex, coming from the last term in Eq. (2.12). We use this to cancel the linear vertex from the first graph in Fig. 4, thus at one stroke fixing σ_0 and eliminating if from all future computations.

We are now in a position to compute whatever we want. For example, let me sketch out the computation of ϕ-ϕ scattering, to leading order. There are only three graphs that can contribute, shown in Fig. 5. I have put a shaded blob on the σ' propagator to remind you that it is not just ig_0/N, but the full propagator obtained from inverting the quadratic term in S_{eff}, the sum of the second and third graphs in Fig. 4. In momentum space,

$$D^{-1}(p) = -N\left[ig_0^{-1} + \int \frac{d^dk}{(2\pi)^d} \frac{1}{(k^2 - \mu_1^2 + i\varepsilon)} \frac{1}{\left([p+k]^2 - \mu_1^2 + i\varepsilon\right)}\right] . \tag{2.14}$$

where d is the number of space-time dimensions. Note that for $d = 4$, the integral is logarithmically divergent, but that its divergence can be absorbed in the bare coupling constant, g_0.

Of course, this is just the beginning. We could evaluate this integral, study the properties of the scattering amplitude it defines, investigate the interesting case in which we choose μ_1^2 to be negative (spontaneous symmetry breakdown?), worry about higher-order corrections, etc. However, as I warned you at the

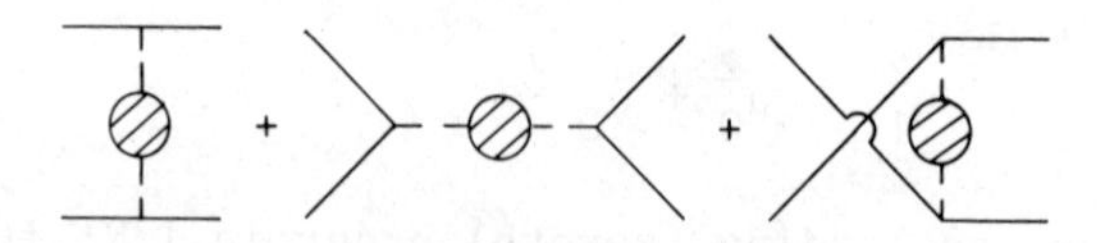

Figure 5

beginning, I am going to stop the discussion of ϕ^4 theory half-way, and go on to investigate other models. If you want to find out more about this theory, you will have to go to the literature[1] (or work it out yourself -- it is not that hard).

2.2. The Gross-Neveu Model[2]

The Gross-Neveu model is a renormalizable field theory that admits a 1/N expansion and displays asymptotic freedom, dynamical symmetry breakdown, and dimensional transmutation. This is the good news; the bad news is that it is a field theory in two space-time dimensions.

The fundamental dynamical variables of the theory are a set of Dirac fields, ψ^a, $a = 1 \ldots N$. In two dimensions, Dirac fields have only two components and the Dirac matrices are 2×2 matrices. In standard representation,

$$\gamma^0 = \sigma_z \, , \qquad \gamma^1 = i\sigma_y \, , \qquad \gamma_5 = \gamma^0\gamma^1 = \sigma_x \, , \tag{2.15}$$

where the σ's are the Pauli spin matrices. In all other ways, conventions are the same as in four dimensions.

The model is defined by

$$\mathscr{L} = \bar{\psi}^a i\partial_\mu \gamma^\mu \psi^a + \frac{g_0}{N}\left(\bar{\psi}^a \psi^a\right)^2 \quad . \tag{2.16}$$

We see from the first term that ψ^a has dimensions of $(\text{length})^{-1/2}$, so g_0 is dimensionless and the interaction should be renormalizable. A mass term is excluded by the discrete chiral symmetry,

$$\psi^a \to \gamma_5 \psi^a \, , \qquad \bar{\psi}^a \to -\bar{\psi}^a \gamma_5 \quad . \tag{2.17}$$

As we shall see, it is this symmetry (not the continuous U(N) symmetry) which suffers spontaneous breakdown.

The construction of the 1/N approximation is a duplicate of that for ϕ^4 theory. First we add a term to $\mathscr{L}$ that has no effect on the physics, involving an auxiliary field, σ,

$$\mathscr{L} \to \mathscr{L} - \frac{N}{2g_0}\left(\sigma - \frac{g_0}{N}\bar{\psi}^a\psi^a\right)^2$$

$$= \bar{\psi}^a i\partial_\mu \gamma^\mu \psi^a - \frac{N}{2g_0}\sigma^2 + \sigma\bar{\psi}^a\psi^a \quad . \qquad (2.18)$$

Next, we integrate over ψ loops to obtain $S_{eff}(\sigma)$. This is shown diagrammatically in Fig. 6. We only have even powers of σ because the trace of an odd number of Dirac matrices vanishes (alternatively, because σ changes sign under the discrete symmetry (2.17)).

At first glance, it would seem that there is no need for the third step in the analysis, shifting the σ field; S_{eff} is even in σ, and thus $\sigma = 0$ is automatically a stationary point. But this is begging the question; if we are interested in spontaneous breakdown of the discrete symmetry, the issue is precisely whether there are stationary points other than $\sigma = 0$.

Fortunately, to settle this issue we do not need to compute S_{eff} for general σ, merely for constant σ. In this case, we can put the universe in a box of spatial extent L and temporal extent T, and define

$$-V(\sigma) = \lim_{L,T\to\infty} S_{eff}(\sigma)/LT \quad . \qquad (2.19)$$

Each stationary point of V is a possible starting point for a 1/N expansion, defines a possible vacuum state of the theory to leading order in 1/N. The energy densities of these vacua are easily

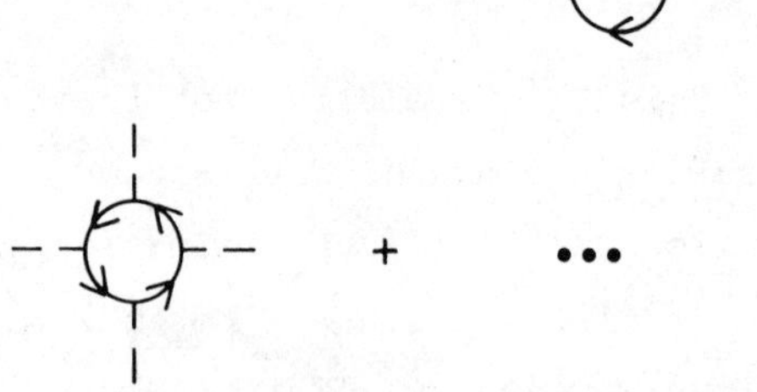

Figure 6

computed. If we denote a vacuum energy density by $\mathscr{E}$, and if we denoted by Σ the sum of all connected vacuum-to-vacuum Feynman graphs, then a general formula of time-dependent perturbation theory states that

$$-i\mathscr{E} = \lim_{L,T\to\infty} \Sigma/LT \quad . \qquad (2.20)$$

To lowest (minus first) order in 1/N, Σ is given by the sum of all connected tree graphs with no external lines. This set consists of precisely one graph, with only one vertex, the term in S_{eff} that contains no powers of the shifted field. That is to say, V at the stationary point is the vacuum energy density (to leading order); if there are several stationary points, only those of minimum V are true vacua.[3]

Diagramatically, V is given by the sum of Feynman diagrams in Fig. 6, with all external lines carrying zero two-momentum, and with the momentum-conserving delta-functions left off. (These give the factor of LT.) The summation of these graphs has been done countless times in the literature. I will bore you by doing it once more:

$$-iV = -i\frac{N}{2g_0}\sigma^2 - \sum_{n=1}^{\infty}\frac{N}{2n}\,\mathrm{Tr}\int\frac{d^2p}{(2\pi)^2}\left(\frac{-\not p\sigma}{p^2+i\varepsilon}\right)^{2n} \quad . \qquad (2.21)$$

The terms in this series have the following origins: (a) From Eq. (2.18), every vertex carries a factor of i and every propagator a factor of $i\not p/(p^2+i\varepsilon)$. (b) There is an N for the N Fermi fields and a (-1) for the Fermi loop. (c) Cyclic permutation of the external lines recreates the same graph; thus the 1/2n! in Dyson's formula is incompletely cancelled and we have a left-over factor of 1/2n. It is trivial to do the trace, sum the series, and rotate the integration to Euclidean two-momentum, p_E. We thus obtain

$$V = N\left[\frac{\sigma^2}{2g_0} - \int\frac{d^2p_E}{(2\pi)^2}\,\ell n\left(1+\frac{\sigma^2}{p_E^2}\right)\right] \quad . \qquad (2.22)$$

The momentum integral is ultraviolet divergent; we cut it off by restricting the integral to $p_E^2 \leq \Lambda^2$, with Λ some large number. We find

$$V = N\left[\frac{\sigma^2}{2g_0} + \frac{1}{4\pi}\sigma^2\left(\ln\frac{\sigma^2}{\Lambda^2} - 1\right)\right] \quad . \tag{2.23}$$

We wish to rewrite this in terms of a (conveniently defined) renormalized coupling constant, g. I will pick an arbitrary renormalization mass, M, and define g by

$$\frac{1}{g} \equiv N^{-1}\left.\frac{d^2V}{d\sigma^2}\right|_M = \frac{1}{g_0} + \frac{1}{2\pi}\ln\frac{M^2}{\Lambda^2} + \frac{1}{\pi} \quad . \tag{2.24}$$

Note that g is $g_0 + O(g_0^2)$, as a good renormalized coupling constant should be. If you do not like my choice of M and want to use another, M', you are free to do so. Your coupling constant is connected to mine by

$$\frac{1}{g'} = \frac{1}{g} + \frac{1}{2\pi}\ln\frac{M'^2}{M^2} \quad . \tag{2.25}$$

V is now given by

$$V = N\left[\frac{\sigma^2}{2g} + \frac{1}{4\pi}\sigma^2\left(\ln\frac{\sigma^2}{M^2} - 3\right)\right] \quad . \tag{2.26}$$

Two of the announced properties of the model are now manifest. Firstly, it is renormalizable, at least in the order to which we are working; Eq. (2.26) is totally free of cutoff-dependence. Secondly, the theory is asymptotically free. This can be seen in the usual two equivalent ways: (1) In Eq. (2.24), if we hold g and M fixed, and let Λ go to infinity, g_0 goes to zero. In a fixed theory, the bare coupling constant vanishes for infinite cutoff. (2) In Eq. (2.25), if we hold g and M fixed, and let M' go to infinity, g' goes to zero. In a fixed theory, the renormalized coupling constant vanishes for infinite renormalization mass.

We can now search for spontaneous symmetry breakdown.

$$\frac{dV}{d\sigma} = N\left[\frac{\sigma}{g} + \frac{\sigma}{2\pi}\left(\ln\frac{\sigma^2}{M^2} - 2\right)\right] \quad . \tag{2.27}$$

This vanishes at

$$\sigma^2 = \sigma_0^2 \equiv M^2 \exp\left(2 - \frac{2\pi}{g}\right) \quad . \tag{2.28}$$

At this point

$$V = -N\sigma_0^2/4\pi \quad . \tag{2.29}$$

This is negative, that is to say, less than V(0). The third announced property of the model is now manifest. The discrete chiral symmetry suffers spontaneous breakdown, and the massless fermions acquire a mass. To leading order, this mass is just σ_0.

I stated earlier that one could choose the renormalization mass, M, arbitrarily; a change in M could always be compensated for by an appropriate change in g. I will now use this freedom to choose M to be σ_0. By Eq. (2.28) this fixes g:

$$g = \pi \quad . \tag{2.30}$$

This is the fourth announced property, dimensional transmutation. We began with a theory that apparently depended on only one continuous parameter, g_0. We have arrived at a theory that depends on only one continuous parameter, σ_0. The surprise is that we began with a dimensionless parameter, on which we would expect observable quantities to depend in a complicated way, while we arrived at a dimensionful parameter, on which observable quantities must depend in a trivial way, given by dimensional analysis.

Of course, dimensional transmutation is an inevitable feature of any renormalizable field theory depending only on a single dimensionless coupling constant. Renormalization trades the single bare coupling constant, g_0, for the pair (g,M), the renormalized coupling constant and the renormalization point. But this is a redundant pair; we still have only a one-parameter theory;

the (g,M) plane is the union of curves such that any two points on the same curve define the same theory. One way of parametrizing these curves is by that value of M at which they pass through some fixed value of g, like 1/2 or π. Our one-parameter family of theories are now labeled by a single parameter, and it is a mass. Nevertheless, even though we expect dimensional transmutation to occur in very general circumstances, it is still pleasant to have a model in which we can explicitly see it happening.

2.3. The CP^{N-1} Model[4]

Like the Gross-Neveu model, the CP^{N-1} model is a two-dimensional renormalizable field theory which displays dimensional transmutation. Also like the Gross-Neveu model, the theory contains a set of particles that are massless in perturbation theory but which acquire a mass in the leading 1/N approximation. However, this is not due to spontaneous symmetry breakdown, but to its reverse. The particles are massless because they are the Goldstone bosons of a spontaneously broken symmetry, and they acquire a mass because the symmetry is dynamically restored. This should be no surprise. In two dimensions there can be no spontaneous breakdown of a symmetry associated with a local conserved current; if the 1/N approximation had not predicted symmetry restoration, we would have known it was a lie.

More interestingly, the massive particles are confined; there is a linear potential between particle and antiparticle which prevents the components of a pair from being separated indefinitely. Of course, a linear potential is not as difficult to achieve in two space-time dimensions as in four; on a line, the classical electric force between oppositely charged particles is independent of distance. However, in this case the linear potential arises in a theory without any fundamental gauge fields; this is astonishing.

The CP^{N-1} model is a generalization of the nonlinear sigma model. I will first remind you of this latter theory and then go on to describe the generalization.

The linear sigma model is a theory of N scalar fields, assembled into an N-vector, $\vec{\phi}$, with dynamics defined by

$$\mathscr{L} = \frac{1}{2}\partial_\mu \vec{\phi}\cdot\partial^\mu\vec{\phi} - \frac{\lambda}{8}\left(\vec{\phi}\cdot\vec{\phi} - a^2\right)^2 \quad , \tag{2.31}$$

where λ and a are positive numbers. This theory is SO(N)-invariant, but, at least in perturbation theory, the symmetry spontaneously breaks down to SO(N-1); the ground states of the theory are constant fields lying on the (N-1)-dimensional sphere,

$$\vec{\phi}\cdot\vec{\phi} = a^2 \quad . \tag{2.32}$$

The nonlinear sigma model is the formal limit of this theory as λ goes to infinity. The fields in general, not just in their ground state, are restricted to obey Eq. (2.32). The Lagrange density then simplifies to

$$\mathscr{L} = \frac{1}{2}\partial_\mu\vec{\phi}\cdot\partial^\mu\vec{\phi} \quad . \tag{2.33}$$

Of course, this simple form does not mean that the dynamics is simple. The N components of $\vec{\phi}$ are not independent, and Eq. (2.33) in fact describes a highly complicated nonlinear theory, as would be manifest were we to write $\vec{\phi}$ as a function of N-1 independent variables (say, angles on the sphere). The nonlinear model can be thought of as a stripped-down version of the linear model, with only the Goldstone bosons retained.

The role of the parameter a can be clarified by rescaling the fields,

$$\vec{\phi} \to a\vec{\phi} \quad . \tag{2.34}$$

Under this transformation, the constraint becomes

$$\vec{\phi}\cdot\vec{\phi} = 1 \quad , \tag{2.35}$$

while

$$\mathscr{L} = \frac{a^2}{2} \partial_\mu \vec{\phi} \cdot \partial^\mu \vec{\phi} \quad . \tag{2.36}$$

From this we see that 1/a is a coupling constant. (See the discussion of powers of N following Eq. (2.7).) This makes sense; if a were infinite, we would not be able to tell the difference between the sphere defined by Eq. (2.32) and ordinary flat space, for which Eq. (2.33) would define a free field theory.

I emphasize that the passage to the nonlinear model is purely a formal limit. For example, the linear model is renormalizable in four dimensions or less, while the nonlinear model is renormalizable only in two dimensions (where scalar fields are dimensionless) or less. We have thrown away some important physics (at least at short distances) by throwing away the non-Goldstone modes. Nevertheless, once we have the nonlinear model, we can certainly study it as a theory in its own right. Indeed, I could have constructed the model directly, as a field theory where the field variables lie in a nonlinear space; I chose to build it from the linear model only for reasons of pedagogy.

The CP^{N-1} model can likewise be constructed directly as a field theory in a nonlinear space. However, again for reasons of pedagogy, I will obtain it as the formal limit of a linear theory.

The linear theory is a theory of N^2-1 scalar fields, assembled into an $N \times N$ traceless Hermitian matrix, ϕ, with dynamics defined by

$$\mathscr{L} = \frac{1}{2} \mathrm{Tr}\, \partial_\mu \phi \, \partial^\mu \phi - \lambda \, \mathrm{Tr}\, P(\phi) \quad , \tag{2.37}$$

where λ is a positive number and P is some polynomial in ϕ. This theory is invariant under SU(N):

$$\phi \rightarrow U \phi U^+ \quad , \qquad U \in SU(N) \quad . \tag{2.38}$$

It is possible to choose P such that the minima of TrP are matrices with N-1 equal eigenvalues and one unequal eigenvalue; SU(N) then breaks down spontaneously to U(N-1). In equations, the ground states of the theory are constant fields of the form

$$\phi = g_0^{-1}\left[N^{\frac{1}{2}} zz^+ - N^{-\frac{1}{2}} I\right] \qquad (2.39)$$

where z is an N-dimensional column vector of unit length,

$$z^+ z = 1 \quad , \qquad (2.40)$$

and g_0 is some parameter derived from $P(\phi)$. I will assume that P has N dependence such that g_0 remains fixed as N goes to infinity. As we shall see shortly, this is necessary to get a 1/N expansion.

The CP^{N-1} model is the formal limit of this theory as λ goes to infinity. The fields in general, not just in their ground state, are restricted to obey Eq. (2.39). The Lagrange density then simplifies to

$$\mathscr{L} = \frac{1}{2} \mathrm{Tr}\, \partial_\mu \phi\, \partial^\mu \phi$$

$$= \left(N/g_0^2\right)\left(\partial_\mu z^+ \partial^\mu z - j_\mu j^\mu\right) \quad , \qquad (2.41)$$

where

$$j_\mu \equiv (2i)^{-1}\left[z^+ \partial_\mu z - \left(\partial_\mu z^+\right) z\right] \quad . \qquad (2.42)$$

It is convenient to rescale z,

$$z \to g_0 N^{-\frac{1}{2}} z \quad . \qquad (2.43)$$

The Lagrange density then becomes

$$\mathscr{L} = \partial_\mu z^+ \partial^\mu z - g_0^2 N^{-1} j_\mu j^\mu \quad , \qquad (2.44)$$

while the constraint equation is

$$z^+ z = N/g_0^2 \quad . \qquad (2.45)$$

We see that we have (in perturbation theory) a theory of massless particles with short-range interactions between them. The theory is slightly more complex than the nonlinear sigma model; there are not only interactions induced by the constraint but also explicit interactions in the Lagrange density.

It is interesting to count the number of particles in the theory. At first glance, it looks like we have N complex fields with one real constraint, yielding 2N-1 real fields. On the other hand, if we count Goldstone bosons, we would expect the number of real fields to be

$$\dim SU(N) - \dim U(N-1) = N^2 - 1 - (N-1)^2 = 2N-2 \quad . \tag{2.46}$$

The second count is the correct one. The first count ignored the fact that the transformation

$$z \to e^{i\theta} z \quad ; \tag{2.47}$$

does nothing to ϕ; the overall phase of z is not a dynamical variable and should not have been counted. The manifold in which our fields lie is the set of complex N-vectors with fixed length, <u>and</u> with N-vectors differing only by a multiplicative phase factor identified. This is complex projective N-1 space, CP^{N-1}.

I will now solve the model to leading order in 1/N, in two-space-time dimensions, where it is renormalizable. The first step is standard, eliminating the quartic interaction by introducing an auxiliary field. Since the quartic interaction is of the form vector times vector, the auxiliary field must be a vector field. Thus, we change $\mathcal{L}$ by

$$\begin{aligned} \mathcal{L} &\to \mathcal{L} + g_0^2 N^{-1}\left(j_\mu + g_0^{-2} N A_\mu\right)^2 \\ &= \partial_\mu z^+ \partial^\mu z + 2j^\mu A_\mu + g_0^{-2} N A_\mu A^\mu \quad . \end{aligned} \tag{2.48}$$

Using the constraint, Eq. (2.45), this can be rewritten in the amusing form,

$$\mathcal{L} = \left(\partial_\mu - iA_\mu\right) z^+ \left(\partial_\mu + iA_\mu\right) z \quad . \tag{2.49}$$

This looks like a piece of a gauge field theory, with the gauge transformation of the fields, Eq. (2.47), cancelled by the gauge transformation of the vector potential,

$$A_\mu \rightarrow A_\mu - \partial_\mu \theta \quad . \tag{2.50}$$

Of course, the "gauge invariance" is a hoax, just a reflection of the fact that we are describing the theory in terms of highly redundant variables.[5]

The next step is to get the constraint into the Lagrange density. We do this with another auxiliary field, σ,

$$\mathscr{L} = \left(\partial_\mu - iA_\mu\right)z^+\left(\partial_\mu + iA_\mu\right)z - \sigma\left[z^+z - g_0^{-2}N\right] \quad . \tag{2.51}$$

The new field is a Lagrange multiplier; its Euler-Lagrange equation is the constraint. Equivalently, performing the functional integral over σ yields a delta-function at each point of space-time which enforces the constraint.

We now have a Lagrange density that is quadratic in z, so we can proceed as before to integrate out the internal z loops and obtain an effective action, a functional of σ and A_μ. The first few terms in the graphical expansion of S_{eff} are shown in Fig. 7; the directed lines are z's, the dashed lines σ's, and the wiggly lines A's. We see that we have pure σ terms (the first two lines), pure A terms (the third line) and mixed terms (the fourth line); however, all the terms are proportional to N, just as before.

Also just as before, to eliminate the linear terms, we must shift to a stationary point of $V(\sigma)$. The computation of V is essentially a rerun of that for the Gross-Neveu model. The relevant graphs are those in the first two lines of Fig. 7; the only difference in the computation is that the Fermi minus sign is missing and that we now have σ where before we had σ^2. Thus,

$$V = -N\left[\frac{\sigma}{g_0^2} + \frac{\sigma}{4\pi}\left(\ln\frac{\sigma}{\Lambda^2} - 1\right)\right] \quad , \tag{2.52}$$

where Λ is the cutoff. To renormalize this, I pick an arbitrary renormalization mass, M, and define the renormalized coupling

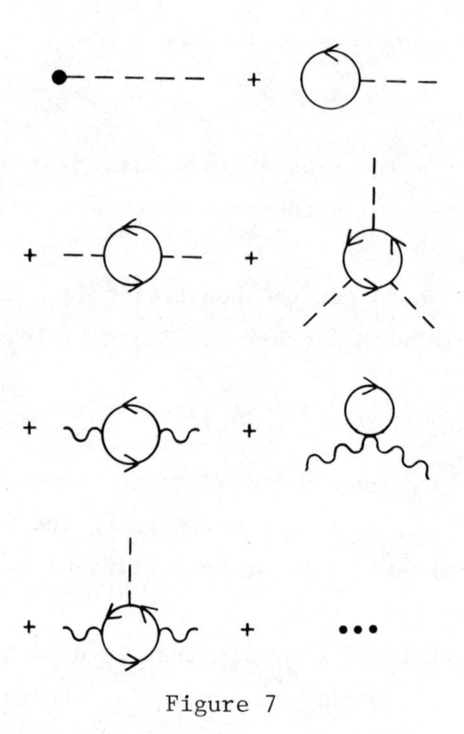

Figure 7

constant, g, by

$$\frac{1}{g^2} \equiv -N^{-1} \left.\frac{dV}{d\sigma}\right|_{M^2} = \frac{1}{g_0^2} + \frac{1}{4\pi} \ln \frac{M^2}{\Lambda^2} \quad . \tag{2.53}$$

Note that g is $g_0 + O(g_0^2)$, as it should be. We can now rewrite V as

$$V = -N\left[\frac{\sigma}{g^2} + \frac{\sigma}{4\pi}\left(\ln \frac{\sigma}{M^2} - 1\right)\right] \quad . \tag{2.54}$$

As before, the theory is renormalizable; all reference to the cutoff has disappeared. As before, it is asymptotically free; for fixed g and M, g_0 vanishes as Λ goes to infinity.

We can now search for stationary points.

$$\frac{dV}{d\sigma} = -N\left[\frac{1}{g^2} + \frac{1}{4\pi}\left(\ln \frac{\sigma}{M^2}\right)\right] \quad . \tag{2.55}$$

This has a unique zero, at

$$\sigma = \sigma_0 \equiv M^2 \exp\left(-4\pi/g^2\right) \quad . \tag{2.56}$$

In terms of σ_0,

$$V = -\frac{N\sigma}{4\pi}\left(\ell n \frac{\sigma}{\sigma_0} - 1\right) \quad . \tag{2.57}$$

Once again, dimensional transmutation has occured; the dimensionless coupling constant, g, has disappeared from the theory, to be replaced by the dimensionful parameter σ_0.

From Eq. (2.51), σ_0 is the squared mass of the z-particles, to lowest (zeroth) order in 1/N. Something remarkable has happened. We started out, in our linear model, with a set of fields transforming according to the adjoint representation of SU(N). By sending λ to infinity, we replaced these by a set of Goldstone bosons which transformed nonlinearly under the action of the group. These have now turned into an ordinary set of massive mesons, transforming linearly, according to the fundamental representation of the group. We are used to making mesons out of quarks; here we have made (bosonic) quarks out of mesons, fundamental representations out of adjoint ones.

Something even more remarkable happens when we study the long-range force between a pair of z's. By the same arguments as were given in ϕ^4 theory, to leading order the force is given by graphs like those shown in Fig. 5. To compute these graphs, we need the σ-σ propagator, the σ-A_μ propagator, and the A_μ-A_ν propagator. By current conservation and Lorentz invariance, the σ-A_μ propagator vanishes. At zero momentum transfer, the σ-σ propagator is given by

$$D_{\sigma\sigma}(0) = -i\left[\left.\frac{d^2V}{d\sigma^2}\right|_{\sigma_0}\right]^{-1} = \frac{4\pi i}{N} \quad . \tag{2.58}$$

This is not infinite; thus σ exchange gives no long-range force.

The A_μ-A_ν propagator is a different story. The term in S_{eff} quadratic in A is obtained from the graphs on the third line of Section 7. These are just the standard second-order photon self-energy graphs; to compute their sum is a trivial exercise in Feynman-graph technology. The answer, in momentum space, is

$$-\frac{iN}{4\pi}\left[g_{\mu\nu}p^2 - p^\mu p^\nu\right]\int_0^1 \frac{dx(1-2x)^2}{\sigma_0^2 - p^2 x(1-x) - i\varepsilon} \quad . \tag{2.59}$$

If we are only interested in long-range forces, that is to say, in small momenta, we may neglect the p^2 in the integrand. We thus obtain

$$-\frac{iN}{12\pi\sigma_0^2}\left[g_{\mu\nu}p^2 - p_\mu p_\nu\right] \quad . \tag{2.60}$$

This corresponds to a term in the effective action of the form

$$S_{eff} = -\frac{N}{48\pi\sigma_0^2}\int d^2x \left(\partial_\mu A_\nu - \partial_\nu A_\mu\right)^2 \quad . \tag{2.61}$$

Aside from a trivial normalization, this is the action for the free electromagnetic field. This is the most astonishing feature of the model; a genuine gauge field has been dynamically generated, produced as a result of radiative corrections in a theory that perturbatively has only short-range interactions. The gauge field now produces a linear potential that confines the z's. Not only does the theory have (bosonic) quarks, it has confined quarks.

There is much more that can be said about the CP^{N-1} model. For example, the classical theory admits instantons (just as in chromodynamics), and there is no infrared cutoff on instanton sizes (again as in chromodynamics). Thus the model can be used as a laboratory for instanton physics (= arena for bloody controversies). Unfortunately, I do not have the time to go into any of this here, and must once again refer you to the literature.[6]

3. ADJOINT REPRESENTATIONS, OR, CHROMODYNAMICS

3.1. The Double-Line Representation and the Dominance of Planar Graphs[7]

The 1/N expansion is vastly more difficult for SU(N) gauge theories than for any of the theories of Section 2. The source of the difficulty has nothing to do with the traditional problems of chromodynamics, the intricacies of gauge invariance or the uncontrollable infrared divergences. It is just that we have to deal with fields that transform according to the adjoint representation rather than the vector representation, objects that carry two group indices rather than one. ϕ^4 theory is just as difficult if ϕ is in the adjoint representation.

The dynamical variables of the theory are a set of Dirac fields, ψ^a, and a set of gauge fields, $A^a_{\mu b}$, where a and b run from 1 to N. The Dirac fields can be thought of as elements of an SU(N) column vector, the gauge fields as those of a traceless Hermitian matrix,

$$A^a_{\mu b} = A^{b\dagger}_{\mu a} \quad , \qquad A^a_{\mu a} = 0 \quad . \tag{3.1}$$

From the gauge fields we define

$$F^a_{\mu\nu b} = \partial_\mu A^a_{\nu b} + iA^a_{\mu c} A^c_{\nu b} - (\mu \leftrightarrow \nu) \quad . \tag{3.2}$$

The dynamics of the theory is defined by

$$\mathscr{L} = \frac{N}{g^2}\left[-\frac{1}{4} F^a_{\mu\nu b} F^{\mu\nu b}_{a} + \bar{\psi}_a \left(i\partial_\mu + A^a_{\mu b}\right)\gamma^\mu \psi^b - m\bar{\psi}_a \psi^a\right] . \tag{3.3}$$

where g and m are real numbers. For $N = 3$, this is the Lagrange density of quantum chromodynamics; the Dirac fields are quarks and the gauge fields gluons. I will retain this nomenclature for general N.

Remarks: (1) I have put the coupling constant in front of the total Lagrangian. Of course, by rescaling the fields, as we did in Section 2.3, we can remove it from the quadratic terms in Eq. (3.3) and put it in its conventional position, a factor of $g/\sqrt{N}$ multiplying the cubic terms and one of g^2/N multiplying the quartic terms. (2) Equation (3.3) is incomplete; I have left out gauge-fixing terms, ghost couplings, renormalization counterterms, and the possibility of more than one flavor. I have done this to keep my equations as simple as possible. Practically all of my analysis will be purely combinatoric, hardly dependent at all on the detailed form of the interactions; thus, the extension to include all these neglected effects will be trivial. (3) I have taken advantage of my knowledge of how things are going to turn out to put the factor of N in the right place from the very beginning. As we shall see shortly, it is the theory defined by Eq. (3.3) that admits a non-trivial 1/N expansion, and not, for example, the one with an N^2 in place of the N.

To take proper account of factors of 1/N, we must keep proper track of the indices within a Feynman graph. Let us begin our analysis with the propagators. The quark propagator is

$$\overline{\psi^a(x)\,\bar{\psi}_b(y)} = \delta^a_b\, S(x-y) \quad , \tag{3.4}$$

where S is the propagator for a single Dirac field. Thus there is no trouble following indices along a quark line; the index at the beginning is the same as the index at the end. The gluon propagator is

$$\overline{A^a_{\mu b}(x)\, A^c_{\nu d}(y)} = \left(\delta^a_d\,\delta^c_b - \frac{1}{N}\,\delta^a_b\,\delta^c_d\right) D_{\mu\nu}(x-y) \quad , \tag{3.5}$$

where $D_{\mu\nu}$ is the propagator for a single gauge field. The term proportional to 1/N is there because the gluon field is traceless; it would not be present if our gauge group were U(N) rather than SU(N). However, precisely because this term is proportional to 1/N, we can drop it, even for SU(N), if we are only interested in the leading order in 1/N. (Of course, we must remember we have dropped it if we want to compute subleading orders. See Appendix C.) There is now no problem following indices along a gluon line; the index pair at the beginning is the same as the index pair at the end. As far as the index structure goes, a gluon propagates like a quark-antiquark pair.

This observation is at the root of the ingenious double-line representation of 't Hooft. This is an alternative way of drawing Feynman graphs in which we draw one line for each index rather than one line for each virtual particle. Thus a quark propagator is represented by a single index line, because a quark carries only one index, but a gluon propagator is represented by two index lines, a double line, because a gluon carries two indices. Figure 8 is a translation dictionary from the old single-line representation on the left to the new double-line representation on the right. The figure shows the translations of propagators, vertices, and a typical vacuum-to-vacuum graph. The great advantage of the double-line representation is immediately obvious: We do not have to clutter our graphs with little letters to show where the indices go; to follow the indices all we have to do is follow the arrows. Phrased in another way, to each double-line graph there corresponds a single-line graph with indices assigned to the lines. Thus, if there is more than one way of assigning indices to the lines in a given single-line graph, there will be more than one double-line graph associated with it.

I will now show that the power of 1/N carried by a double-line graph is determined by certain topological properties of the graph. For simplicity, I will begin by restricting myself to vacuum-to-vacuum graphs, graphs with no external lines. I will later extend the analysis as needed.

Because the graph has no external lines, every index line must close to make an index loop. Let us imagine each index loop to be the perimeter of a polygon. The double-line graph can then be read as a prescription for fitting together these polygons.

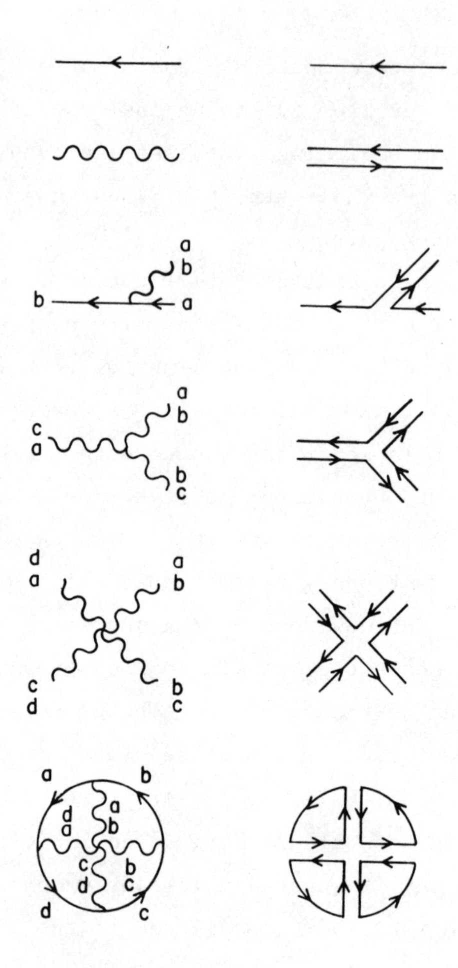

Figure 8

To be more precise, we identify one edge of one polygon with one edge of another if they both lie on the same double line (gluon propagator). In this way, we construct a two-dimensional surface.

We can give an orientation to each polygon by the direction of the arrows around its perimeter and the right-hand rule. Because the two halves of a double line are always oppositely directed, this orientation is consistent as we travel about the surface; we have constructed an oriented surface. Thus we can get spheres or toruses, but not Klein bottles. (If we were doing the parallel analysis for SO(N) rather than SU(N), quark and antiquark would transform equivalently, our lines would not carry arrows and we could get Klein bottles.)

It is easy to count the power of N associated with this surface. Let the surface have V vertices, E edges, and F faces. Every vertex is an interaction vertex of a Feynman graph, and carries a factor of N, by Eq. (3.3). Every edge is a propagator, either quark or gluon, and carries a factor of 1/N, again by Eq. (3.3). Every face is an index loop, and thus yields a factor of N when we sum over all possible values of the index. Thus the graph is proportional to

$$N^{F-E+V} \equiv N^{\chi} \quad . \tag{3.6}$$

χ is the Euler characteristic. It is a famous topological invariant, and can be computed in quite another way. Every two-dimensional oriented surface is topologically equivalent to a sphere with some number of holes cut out of it and some number of handles stuck on to it. For example, a torus is a sphere with one handle; a disc is a sphere with one hole; a cylinder (without end caps) is a sphere with two holes; a loving cup is a sphere with one hole and two handles; etc. Let H be the number of handles and B (for boundary) be the number of holes. Then the Euler characteristic is given by

$$\chi = 2 - 2H - B \quad . \tag{3.7}$$

(If you are not familiar with the Euler characteristic, a quick and dirty proof of this formula is given in Appendix A.)

Thus the leading connected vacuum-to-vacuum graphs are proportional to N^2, and the associated surface has the topology of a sphere. What does this mean in terms of our original single-line graphs? The boundary of a hole is a loop of unpaired index lines, that is to say, a quark loop; thus, the leading graphs involve only gluons. (This is not a deep result. It takes just as many powers of the coupling constant to make a quark pair as to make a gluon pair, but there are N times more gluons than quarks to sum over.)

Let us remove one randomly selected face from our spherical surface and project the remainder of the surface onto a plane. We thus obtain a planar graph. If we collapse the double lines to single lines, the graph remains planar. Conversely, given a planar graph made up only of gluon lines, we can always associate a double-line graph of the desired type with it. A planar graph divides the portion of the plane it occupies into regions. All we need do is draw a clockwise index line just inside the boundary of each of these regions and a counterclockwise index line just outside the boundary of the whole graph.

This analysis will shortly become important to us, so I summarize it in the following: First Result - The leading connected vacuum-to-vacuum graphs are of order N^2. They are planar graphs made up only of gluons.

We might be interested in the leading vacuum-to-vacuum graphs that have a nontrivial dependence on quark parameters, like quark masses or the number of flavors. These graphs must involve at least one quark loop; we see from Eq. (3.7) that the leading graphs have only one quark loop (= one hole) and no handles. Thus one of these graphs defines a spherical surface with one face removed. We can project this onto a plane, just as we did before. The only difference from the preceding case is that the outer boundary of the resultant planar graph is the perimeter of the hole, the quark loop.

Thus we obtain: Second Result - The leading connected vacuum-to-vacuum graphs with quark lines are of order N. They are planar graphs with only one quark loop; the loop forms the boundary of the graph. Thus the first graph in Fig. 9 is leading, but the second is not, even though it is planar.

As we shall see immediately, a very large amount of meson phenomenology is implicit in these two results.

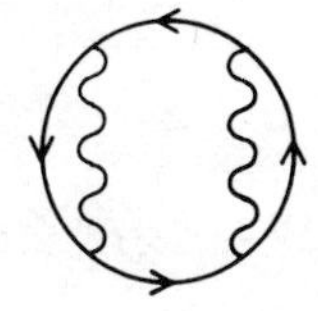
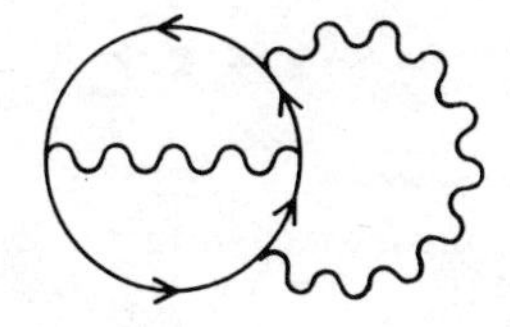

Figure 9

3.2. Topology and Phenomenology

The naive quark model deals with mesons (particles made of a quark and an antiquark) and baryons (particles made of three quarks). Somewhat more sophisticated models worry about glueballs (particles made of gluons) and exotics, in particular exotic mesons (particles made of more than one quark-antiquark pair). Because it takes N quarks to make a color singlet, baryons are a special problem for the 1/N expansion, and I will postpone their study to Section 3.4. However, we do have all the machinery needed to study mesons, glueballs, and exotic mesons.

Our method will be to study the states made by applying certain gauge-invariant local operators to the vacuum. We will restrict ourselves to monomials in ψ, $\bar{\psi}$, $F_{\mu\nu}$, and their covariant derivatives, and, further, to monomials that cannot be written as the product of two gauge-invariant monomials of lower degree. (Practically everything I say will be valid even for non-local non-polynomial gauge-invariant operators, like $\bar{\psi}_a(x)\, U^a_b\, \psi^b(y)$, where U is the ordered exponential integral of the gauge fields over the line from x to y; as far as counting powers of N goes, the key point is that the operator cannot be decomposed into a product of gauge-invariant operators.)

To began with, let us study quark bilinears, operators involving one ψ and one $\bar{\psi}$. Let $B_1 \ldots B_n$ be a string of such bilinears, each at some point, and let $\langle B_1 \ldots B_n \rangle_C$ be the connected Green's function for this string. If we modify the action of our theory by an additional term,

$$S \to S + N \sum_i b_i B_i \quad , \tag{3.8}$$

where the b's are numbers, and if W is the sum of connected vacuum-to-vacuum graphs, then

$$\langle B_1 \ldots B_n \rangle_C = (iN)^{-n} \left. \frac{\partial^n W}{\partial b_1 \ldots \partial b_n} \right|_{b_i=0} . \tag{3.9}$$

The reason for doing things this way is that all the analysis of Section 3.1 applies to the action (3.8) without a word of alteration; every interaction vertex carries a factor of N, and, in the double-line representation, every interaction vertex becomes a vertex of a polyhedron. (Note that this last point would not be true if one of the B's was a product of two gauge-invariant monomials of lower degree. In that case we would get two polyhedral vertices from a single interaction vertex.)

Thus, from Result 2 of Section 3.1, we immediately know that the leading graphs are planar graphs with one quark loop, the boundary of the graph. Of course, all the bilinears must appear as insertions on the quark loop. Figure 10 shows a leading graph for a three-bilinear Green's function; the bilinears are indicated by crosses. (We see that B_3 is linear in gluon fields as well as bilinear in quark fields.) We also know from Result 2 that the

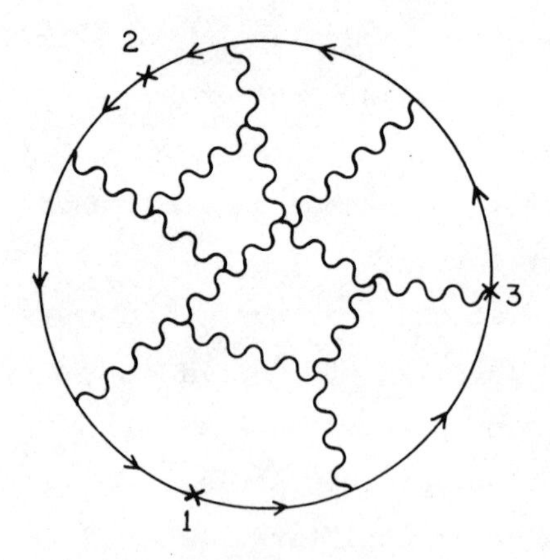

Figure 10

contribution of these graphs to W is proportional to N; thus, to leading order,

$$\langle B_1 \ldots B_n \rangle_C \propto N^{(1-n)} \quad . \qquad (3.10)$$

We will also have use for gauge-invariant local operators made up exclusively of gauge fields. I will denote such operators by G_i. The Green's function for a mixed string of B's and G's can be studied by the same method as before. Once again the leading graphs are planar graphs with only one quark loop, the boundary of the graph. Of course, the G's can appear as insertions anywhere in the graph. Again, the contribution to W is proportional to N; thus, to leading order

$$\langle B_1 \ldots B_n G_1 \ldots G_m \rangle_C \propto N^{(1-n-m)} \quad . \qquad (3.11)$$

The situation is slightly different for a string made up of G's alone. Here Result 1 of Section 3.1 applies; the leading graphs are planar graphs with no quark loops, and their contribution to W is proportional to N^2. Thus, to leading order,

$$\langle G_1 \ldots G_m \rangle_C \propto N^{(2-m)} \quad . \qquad (3.12)$$

We can derive much interesting physics from these equations if we make one assumption, that chromodynamics confines for arbitrarily large N, that all states made by applying strings of gauge-invariant operators to the vacuum are states composed of SU(N)-singlet particles. I stress this is pure assumption. Our main reason for believing in confinement is that we can see quarks within hadrons but we cannot liberate them; this is an experimental reason, not a theoretical one, and experiment exists only for $N = 3$. Nevertheless, the assumption is not unreasonable. To the small extent to which we do have a theoretical understanding of confinement (for example, from strong-coupling lattice gauge theories), there does not seem to be anything special about small N. But the best reason for assuming large-N chromodynamics confines is that, if it does not confine, it bears no resemblance to reality, and the 1/N expansion is hopeless. Thus, we might as well make the assumption and see where it leads us.

I will define a meson to be a one-particle state made by applying a quark bilinear to the vacuum. From Eq. (3.10), our bilinears are not properly normalized to create mesons with N-independent amplitudes. Therefore, we renormalize them, and define

$$B'_i = N^{\frac{1}{2}} B_i \quad . \tag{3.13}$$

For these,

$$\langle B'_1 \ldots B'_n \rangle \propto N^{(2-n)/2} \quad . \tag{3.14}$$

Meson scattering amplitudes are obtained from these Green's functions by the reduction formula; thus a scattering amplitude with n legs is proportional to $N^{(2-n)/2}$. For large N, mesons interact weakly; $N^{-\frac{1}{2}}$ sets the scale of meson interactions just as e sets the scale of the interactions of electrons and photons.

Indeed, the parallel is exact. In quantum electrodynamics, in lowest non-vanishing order in perturbation theory, the tree approximation, an n-field Green's function is proportional to $e^{(n-2)}$. As far as dependence on N is concerned, it is as if our bilinears were linear functions of fundamental fields in some field theory with coupling constant proportional to $N^{-\frac{1}{2}}$, as if the leading order in 1/N were the tree approximation in this theory.

Of course, the tree approximation is characterized by more than just its dependence on the coupling constant. Green's functions in the tree approximation have very simple analytic structures in momentum space; their only singularities are poles. I will now argue that Eq. (3.14) implies that the same holds for our Green's functions.

I will begin with the two-point function. To show that the only singularities are poles is to show that a bilinear applied to the vacuum produces only single-meson states (to leading order). The proof is by contradiction. For example, let us assume a renormalized bilinear produces a pair of singlet particles (A,B), with an amplitude of order unity. We can then construct the sequence of events shown in Fig. 11. This is not a Feynman diagram, but a drawing of events in space-time (time runs upward). Initially a bilinear produces a pair; the components separate until they are each reflected by a bilinear; when they come back to the same point, they are absorbed by yet another bilinear. Every vertex in the figure is obtained from the assumed initial vertex by crossing; thus all the vertices are of order unity and the process produces a physical-region singularity of order unity in the four-bilinear connected Green's function. But this is impossible; Eq. (3.14) tells us that this Green's function is proportional to 1/N. This argument generalizes instantly from a pair to a multi-particle state; all we need to do is declare that each line in the figure represents a cluster of particles.

The argument also generalizes to higher Green's functions. For example, let us consider a four-point function, and let us assume it has a two-particle cut, in a two-bilinear subenergy, in leading (first) order in 1/N. Then there must be a connected amplitude for two bilinears to produce a pair, proportional to

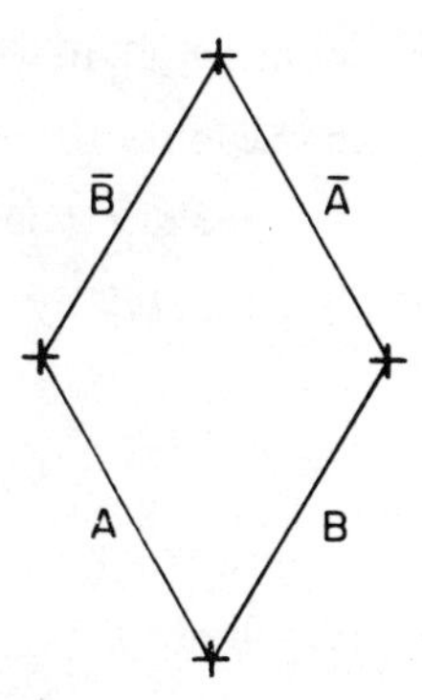

Figure 11

$N^{-\frac{1}{2}}$. If we reproduce the reasoning of the preceding paragraph, with the cross in Fig. 11 now denoting the double bilinear, we deduce that there is a singularity in an eight-bilinear Green's function proportional to $1/N^2$. But this is impossible; this Green's function is proportional to $1/N^3$. And so on.

There is one way in which the large-N theory does not resemble a conventional weakly-coupled field theory; there is an infinite number of mesons. We know this is so because asymptotic freedom is not spoiled in the large-N limit; thus two-point functions must behave logarithmically for large spacelike momenta. This cannot be achieved if the only singularities of these functions are poles and if all poles lie within some bounded region. Thus there must be an infinite number of mesons of ever-increasing mass. We can sharpen this argument a bit: because we can build bilinears of any spin, there must be an infinite number of mesons of each spin. Such an infinite tower of stable mesons makes sense only if all mesonic S-matrix elements vanish in the large-N limit; otherwise, a heavy meson could decay into light ones. As we have seen, they do indeed vanish.

All of our mesonic analysis can be extended trivially to glueballs, particles made by applying gluonic operators to the vacuum. By Eq. (3.12), the G's are already properly normalized and need no renormalization. From the same equation, a glueball scattering amplitude with n legs is proportional to $N^{(2-n)}$. Thus glueballs interact even more weakly than mesons; $N^{-\frac{1}{2}}$ in meson dynamics is replaced by N^{-1} in glueball dynamics.

If glueballs interact more weakly than mesons, they are not mesons. This assertion can be checked by an independent line of argument. From Eq. (3.11),

$$\langle B_1' \ldots B_n' G_1 \ldots G_m \rangle_C \propto N^{(1-m-\frac{1}{2}n)} \quad . \tag{3.15}$$

Thus, glueball-meson mixing vanishes like $N^{-\frac{1}{2}}$ as N becomes infinite. Also, in meson-meson scattering, glueball production is suppressed;

to replace a final-state meson by a glueball costs a factor of $N^{-\frac{1}{2}}$ in amplitude.

Up to now, flavor has been irrelevant to our discussion. This will not be the case for our next (and last) two topics, the validity of Zweig's rule and the nonexistence of exotic mesons. With only one flavor of quark, we would be hard pressed to distinguish Zweig-allowed from Zweig-forbidden processes, or to tell exotic mesons from ordinary ones.

The usual statement of Zweig's rule is that for any mesonic scattering graph, it is impossible to divide the meson legs into two sets unconnected by quark lines. We already have this, in the form of the statement that in leading graphs all bilinears must appear as insertions on a single quark loop. Graphs for Zweig-forbidden processes must involve at least two quark loops, and thus are down in amplitude by at least one factor of 1/N. Thus we have all the usual consequences of the rule: In the limit of strict SU(3), mesons must fall into nonets, an $s\bar{s}$ meson cannot decay into a final state free of strange quarks, etc.

To show the nonexistence of exotics requires a little more work. Exotic mesons, if they existed, would be states created from the vacuum by the application of local gauge-invariant quark quadrilinears. Every such object is the sum of products of local gauge-invariant bilinears. (That is to say, the only way to make an SU(N) scalar is to take the inner product of each quark column vector with an antiquark row vector, perhaps with an interventing SU(N) matrix made of $F_{\mu\nu}$'s and covariant derivatives.) With no loss of generality we can study the states made by a single product,

$$Q(x) = B_1'(x)\, B_2'(x) \quad , \tag{3.16}$$

where, for purposes of this argument, I have restored explicit space-time dependence. For simplicity, I will assume we have chosen the flavors of these operators so that B_1, B_2, and Q all have vanishing vacuum expectation values. Then,

$$\langle Q^+(x)\,Q(y)\rangle = \langle B_1'^+(x)\,B_1'(y)\rangle\langle B_2'^+(x)\,B_2'(y)\rangle$$
$$+ \langle B_1'^+(x)\,B_2'(y)\rangle\langle B_2'^+(x)\,B_1'(y)\rangle$$
$$+ \langle B_2'^+(x)\,B_1'^+(x)\,B_1'(y)\,B_2'(y)\rangle_C \ . \quad (3.17)$$

The first two terms on the right are of order one, while the third is of order 1/N, and thus should be dropped in the large-N limit. But the first two terms simply describe the independent propagation of two mesons from x to y. In the large-N limit, quadrilinears make meson pairs and nothing else.

3.3. The 't Hooft Model[8]

The arguments of the preceding section have been powerful but abstract; it would be nice to have a concrete example in which we could see them at work. Such an example is provided by the 't Hooft model, large-N chromodynamics in two space-time dimensions. The model is almost exactly soluble; the simplest Green's functions can be found in closed form (in an appropriate gauge), and, although the computation of the particle spectrum requires numerical analysis, it is of a sort that can be carried out on a pocket calculator. The model also serves to eliminate a worrisome possibility, that the arguments of Section 3.2 are internally inconsistent, that confinement cannot exist for arbitrarily large N.

Of course, it is no surprise to find confinement in a two-dimensional gauge theory. As a warm-up for the 't Hooft model, let me remind you how confinement occurs in an even simpler theory, two-dimensional quantum electrodynamics. This theory is defined by

$$\mathscr{L} = \frac{1}{2}(F_{01})^2 + \bar{\psi}\left(i\partial_\mu\gamma^\mu - eA_\mu\gamma^\mu - m\right)\psi \quad , \qquad (3.18)$$

where

$$F_{01} = \partial_0 A_1 - \partial_1 A_0 \quad . \qquad (3.19)$$

As always in the analysis of a gauge theory, the first step is to pick a gauge. I will choose axial gauge,

$$A_1 = 0 \quad . \tag{3.20}$$

In this gauge,

$$\mathscr{L} = \frac{1}{2}\left(\partial_1 A_0\right)^2 + \bar{\psi}\left(i\partial_\mu \gamma^\mu - eA_0\gamma^o - m\right)\psi \quad . \tag{3.21}$$

No time derivatives of A appear in this equation; A is not a dynamical variable at all, but a constrained variable, one that must be eliminated from the theory before we can write it in canonical form.

The equation that determines A is

$$\partial_1^2 A_0 = -e\psi^+\psi \equiv -ej^o \quad . \tag{3.22}$$

The general solution of this is

$$A_0(x^0,x^1) = -\frac{e}{2}\int dy^1 \; |x^1-y^1| j^o(x^0,y^1) + Bx^1 + C \quad , \tag{3.23}$$

where B and C are constants. C is irrelevant; it can always be eliminated by a gauge transformation; for simplicity, I will set it to zero. B is relevant; non-zero B corresponds to the existence of a constant background electric field, such as would be caused by classical charges at spatial infinity. In the Abelian case that occupies us at the moment such a background field has a real effect on the physics, and an interesting one. In the non-Abelian case we are heading for, it turns out that the corresponding object has no physical effect whatsoever. This is a fascinating byway, but it is a byway, and I do not want to spend time on it here. Thus, I will assume for this investigation that there is no background field, and set B to zero.

Thus, we can eliminate A_0 and write the Lagrangian as

$$L = L_{0f} + \frac{e^2}{4}\int dx^1\, dy^1\; j_0(x^0,x^1)\,|x^1-y^1|\,j_0(x^0,y^1) \quad , \tag{3.24}$$

where L_{0f} is the free fermion Lagrangian. A linear potential has appeared between charges; confinement is manifest, at least for small coupling, where we can trust perturbation theory.

It will be convenient for our later work to express the current-current interaction in Eq. (3.24) as the effect of exchange of a photon propagator,

$$D_{\mu\nu}(k) = -\frac{i}{2}\delta_{\mu 0}\delta_{\nu 0}\int d^2x\, e^{ik\cdot x}|x^1|\delta(x^o)$$

$$= i\delta_{\mu 0}\delta_{\nu 0}\frac{P}{(k_1)^2} \quad , \tag{3.25}$$

where P is the principal-value symbol,

$$P\frac{1}{z^2} = \frac{1}{2}\left(\frac{1}{z^2 + i\varepsilon} + \frac{1}{z^2 - i\varepsilon}\right) \quad . \tag{3.26}$$

Of course, we could have obtained the momentum-space propagator directly from the Lagrange density, Eq. (3.21), by standard methods; I detoured through position space to justify the somewhat unusual principal-value prescription at the pole.

It requires no work to generalize this to chromodynamics. The nonlinear terms in F_{01} are proportional to the product of A_0 and A_1. Thus they vanish in axial gauge, and with them vanishes one of the characteristic complications of chromodynamics, the self-coupling of the gauge field. Thus the only difference between chromodynamics and electrodynamics is a sprinkling of indices here and there; following the derivation that led to Eq. (3.24), we find

$$L = L_{0f} + \frac{g^2}{N}\int dx'\, dy'\, j^b_{0a}(x^0,x^1)\,|x^1 - y^1|\, j^a_{0b}(x^0,y^1) \quad , \tag{3.27}$$

where

$$j^b_{0a} = \psi^+_a\psi^b - \frac{\delta^b_a}{N}\psi^+_c\psi^c \quad , \tag{3.28}$$

and I have rescaled the fields to put the coupling constant in its conventional location.

The elimination of gluon self-coupling drastically diminishes the number of graphs that contribute in the large-N limit. Figure 12 shows the set of graphs that contribute to a Green's function for two quark bilinears; the shaded blobs represent quark propagators. The simple structure of the graphs arises because a gluon line that connects the upper quark line to the lower quark line forms an impassable barrier. No gluon line can cross it without interaction, because this would violate planarity; no gluon line can cross it with interaction, because there are no interactions. The same simplicity of structure appears in Fig. 13, the equations for the quark self-energy, Σ. The first gluon line to leave the quark must be the last to return, for it forms an impassable barrier.

Things are still not as simple as they could be. For one thing, we still have to keep track of the two components of the quark field; Fig. 13 defines a matrix equation. For another, we are working in a non-covariant gauge, so we do not have the advantages of manifest Lorentz invariance. Both these problems can be eliminated if we switch from axial gauge to light-cone gauge.

Light-cone coordinates are defined by

$$x^{\pm} = \left(x^0 \pm x^1\right)/\sqrt{2} \quad . \tag{3.29}$$

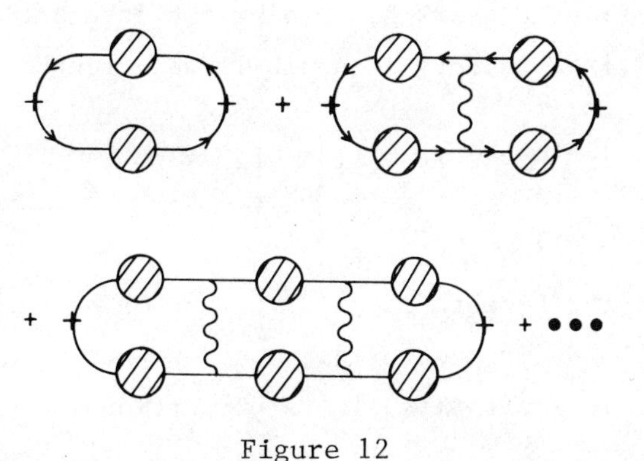

Figure 12

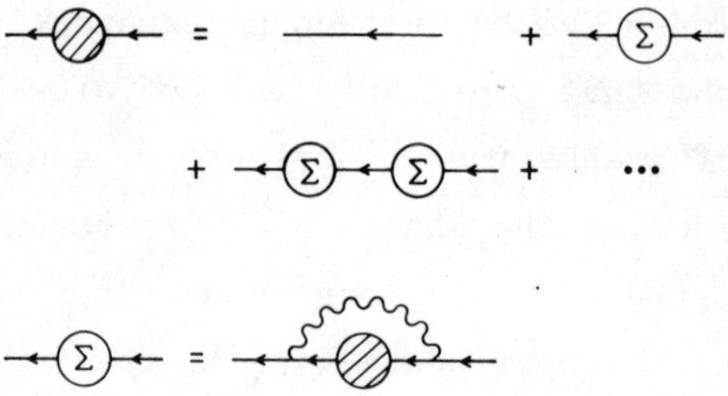

Figure 13

In these coordinates,

$$g^{+-} = g^{-+} = g_{+-} = g_{-+} = 1 \qquad . \qquad (3.30)$$

All other components of the metric tensor vanish. Note that this implies a peculiar Dirac algebra,

$$(\gamma^+)^2 = (\gamma^-)^2 = 0 \quad , \quad \{\gamma^+, \gamma^-\} = 2 \qquad . \qquad (3.31)$$

Light-cone gauge is defined by

$$A_- = A^+ = 0 \qquad . \qquad (3.32)$$

This condition is Lorentz-invariant (but not parity-invariant). A rerun of the derivation of Eq. (3.25) leads to the photon propagator,

$$D_{\mu\nu}(k) = i\delta_{\mu+}\delta_{\nu+} \frac{P}{(k_-)^2} \qquad . \qquad (3.33)$$

At every interaction vertex, only the matrix γ^+ appears. Thus, if we consider a quark line joing two interaction vertices, most of its matrix structure is annihilated; from Eq. (3.31),

$$\gamma^+ \begin{Bmatrix} 1 \\ \gamma^+ \\ \gamma^- \end{Bmatrix} \gamma^+ = 2\gamma^+ \begin{Bmatrix} 0 \\ 0 \\ 1 \end{Bmatrix} \qquad . \qquad (3.34)$$

Hence, all the Lorentz-index structure of our graphs is trivial. The photon propagator has only one non-zero component, and the interaction vertex is always proportional to a single

Dirac matrix, as is the only surviving part of the quark propagator. There is no point in keeping track of these unvarying structures; we might as well drop them and make the substitutions

$$D_{\mu\nu} \to i \frac{P}{(k_-)^2} \quad , \tag{3.35a}$$

$$-\frac{ig\gamma^+}{\sqrt{N}} \to -\frac{2ig}{\sqrt{N}} \quad , \tag{3.35b}$$

and

$$i \frac{p_+\gamma^+ + p_-\gamma^- + m}{2p_+p_- - m^2 + i\varepsilon} \to \frac{ip_-}{2p_+p_- - m^2 + i\varepsilon} \quad . \tag{3.35c}$$

The internal-index structure of our graphs is also trivial, by our earlier analysis; for each graph there is a unique way of distributing the internal indices, and the net effect of this distribution is to cancel the factor of $1/\sqrt{N}$ in Eq. (3.35b).

Thus the first part of Fig. 13, the equation for the quark propagator, $S(p)$, becomes

$$S(p) = \frac{ip_-}{2p_+p_- - m^2 - p_-\Sigma(p) + i\varepsilon} \quad . \tag{3.36}$$

while the second part of Fig. 13, the equation for the quark self-energy, Σ, becomes

$$-i\Sigma = -i4g^2 \int \frac{dk_+dk_-}{(2\pi)^2} S(p-k) \frac{P}{(k_-)^2} \quad . \tag{3.37}$$

If, in this equation, we make the shift of integration variables,

$$p_+ - k_+ \to -k_+ \quad , \tag{3.38}$$

we eliminate all reference to p_+. Σ is a function of p_- only; by Lorentz invariance, it must be a constant multiple of $1/p_-$. Thus,

$$m^2 - p_-\Sigma \equiv M^2 \quad , \tag{3.39}$$

is a constant. To leading order in 1/N, the sole effect of the interaction is to replace the bare quark mass, m, by the "renormalized quark mass," M. (The quotation marks are to remind you that we are working with gauge-dependent entities, entities which do not necessarily have any physical meaning. I will return to this point shortly.) It is now straightforward to evaluate M^2. The details of the computation are in Appendix B; the answer is

$$M^2 = m^2 - (g^2/\pi) \quad . \tag{3.40}$$

Now that we have the quark propagator, we can evaluate the sum of ladder graphs in Fig. 12, and discover the spectrum of meson states. I will only state the answer here (again, the details are in Appendix B). There is a bound state of mass μ for every eigenvalue of the integral equation

$$\mu^2\phi(x) = \left(\frac{M^2}{x} + \frac{M^2}{1-x}\right)\phi(x) - \frac{g^2}{\pi}\int_0^1 dy \, \frac{P}{(x-y)^2}\,\phi(y) \quad , \tag{3.41}$$

where ϕ is a function defined on the interval [0,1] and vanishing at the end points of the interval.

Although it has been derived from field theory, Eq. (3.41) may be read as an equation in particle mechanics; to be more precise, it is a two-particle time-independent light-cone Schrödinger equation.

In the normal Schrödinger formalism, the state of the system is given at fixed x^0, and dynamics is defined by the operator that generates x^0-translations, P_0. This commutes with the generator of x^1-translations, P_1, and thus we can simplify dynamical problems by going to an eigenspace of P_1. In the light-cone Schrödinger formalism, the state of the system is given at fixed x^+, and dynamics is defined by the operator that generates x^+-translations, P_+. This commutes with the generator of x^--translations, P_-, and thus we can simplify dynamical problems by going to an eigenspace of P_-. For example, for a single free particle of mass M,

$$2P_+ = M^2/P_- \quad , \tag{3.42}$$

and dynamics is totally diagonal in a P_- basis, just as it is totally diagonal in the normal formalism in a P_1 basis. An important difference between the two formalisms is that while the spectrum of P_1 is the entire real line, that of P_- is the positive half-line only.

For a two-particle system, it is convenient to work in an eigenspace of total P_- with eigenvalue one. Thus, if we denote the P_- operator for one of the particles by x, that of the other is (1-x). Since each P_- must be positive, x must lie between 0 and 1. For two non-interacting particles of equal mass,

$$2P_+ = \frac{M^2}{x} + \frac{M^2}{1-x} \quad . \tag{3.43}$$

Because P_- is one, the eigenvalues of this operator are the squared masses of the two-particle system. In terms of a momentum-space Schrödinger wave function, the eigenvalue equation is

$$\mu^2\phi(x) = \left(\frac{M^2}{x} + \frac{M^2}{1-x}\right)\phi(x) \quad . \tag{3.44}$$

This is almost Eq. (3.41); all that is missing is the last term. But such a convolution integral in a momentum-space Schrödinger equation is a familiar object; it corresponds to an ordinary potential back in position space, in the case at hand, to a linear potential. Once we strip away the heavy disguise of the light-cone momentum-space formalism, Eq. (3.41) is revealed to be the simplest meson model of all, two quarks interacting through a linear potential.

We would expect such a system to have a purely discrete spectrum. The easiest way to see that this is the case is to reinterpret Eq. (3.41), to think of x as a position operator and the conjugate variable as a momentum operator, p. The operator version of Eq. (3.41) then becomes

$$2P_+ = \frac{M^2}{x} + \frac{M^2}{1-x} + g^2|p| \quad . \tag{3.45}$$

Aside from a trivial multiplicative constant, this is the ordinary Hamiltonian for a mass-zero particle moving in a potential, and restricted to the box [0,1]. It is this last condition that guarantees that the spectrum is purely discrete, that in fact our space of states does not contain any particles that correspond to two free quarks.

We can also use our reinterpretation to get a quantitative idea of the meson spectrum. For a particle moving in a potential and restricted to a box, we would expect the potential to be irrelevant for sufficiently high excited states, and the eigenstates to be those of a free particle in a box,

$$\phi_n = \sin \pi n x \quad , \quad n = 1,2,\ldots \quad , \tag{3.46}$$

with associated eigenvalues

$$\mu_n^2 = g^2 \pi n \quad . \tag{3.47}$$

An easy perturbative calculation shows that this expectation is correct, at least for large n; the corrections to Eq. (3.47) are $O(\log n/n)$.

Of course, Eq. (3.47) is no good for the low-lying mesons. Indeed, since M^2 can be negative for sufficiently large g/m, one might fear that for large g the low-lying spectrum might go crazy. This fear is groundless. From Eq. (3.41) and the identity

$$\int_0^1 \frac{dy\ P}{(x-y)^2} = -\left[\frac{1}{x} + \frac{1}{1-x}\right] \quad , \tag{3.48}$$

it follow that

$$\mu^2 \int_0^1 |\phi|^2 \, dx = m^2 \int_0^1 |\phi|^2 \left(\frac{1}{x} + \frac{1}{1-x}\right) dx$$

$$+ \frac{g^2}{2\pi} \int_0^1 dx \int_0^1 dy \, \frac{|\phi(x) - \phi(y)|^2}{(x-y)^2} \quad . \qquad (3.49)$$

Thus if m^2 is positive, μ^2 is positive, no matter how large g^2 is; tachyonic quarks do not make tachyonic mesons.

To go on requires numerical analysis, so I will stop our investigation here. However, before I leave the model altogether, I would like to make two points.

(1) Everything worked out as we expected it to. Confinement and the large-N limit are not in contradiction.

(2) The structure of the quark propagator tells us nothing about confinement. Here we have a reasonable theory solved in a reasonable approximation; the quark propagator is that of a free particle (sometimes a tachyon); nevertheless, the theory contains no free quarks and no tachyons.

This point needs expansion. Why should the quark propagator be irrelevant? The first answer is that it is a gauge-variant object, and thus not an observable, but this is insufficient; the same might be said about the electron propagator in electrodynamics, and we know the location of the pole here tells us the electron mass, very much an observable quantity. However, it is observable only because it governs the singularity structure of gauge-invariant Green's functions.

For example, Fig. 14 shows a Feynman graph that contributes to a six-current Green's function in electrodyanmics; the dots denote the currents. This graph has a singularity that corresponds to a reading of it as a process going on in space-time (time runs upward): The initial current creates an electron-positron pair, which are then bounced about by widely-separated external fields until they recombine. The location of the corresponding singularity

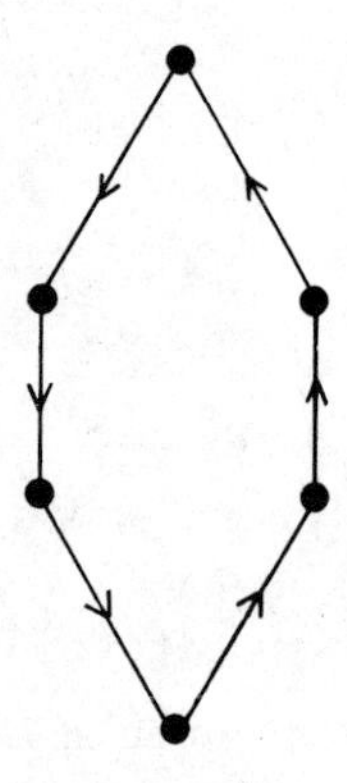

Figure 14

is obviously governed by the electron mass. (This is an idealization of an actual measurement of the mass of a charged particle; the external fields are idealized bending magnets or counters.)

Of course, Fig. 14 is just a lowest-order graph, and we must be sure that higher-order corrections cannot destroy the singularity. We need not worry about propagator corrections, where a photon returns to the same electron line from which it emerged; these merely renormalize the electron mass. Nor need we worry about vertex corrections, where the photon goes from one side of a dot to another; these merely renormalize the strength of the external field. Nor need we worry about corrections where the photon connects electron lines separated by two or more dots; because all the external currents are widely separated, the photon ends are necessarily widely separated, and the electrodynamic interaction is negligible at large distances.

Oh.

We have reached the crux of the matter. The argument that the location of the pole in the quark propagator is an observable quantity rests upon the assumption that there is no confinement. Those who look for confinement in the singularities of the quark propagator are like the man who settled in Casablanca for the waters. They have been misinformed.

3.4. Witten's Theory of Baryons[9]

Baryons present a special problem for the 1/N expansion. The method we used in mesodynamics, the study of fixed Feynman graphs in the large-N limit, cannot be applied here. It takes N quarks (in a totally antisymmetric color state) to make a baryon, and thus we must study graphs with more and more quark lines as N grows larger and larger. This does not mean that there is no hope, that baryons do not obey simple scaling laws. As Witten discovered, the proper procedure is to break the problem into two parts, to first use graphical methods to study n-quark forces in the large N limit, and then to use other methods to study the effects of these forces on an N-body state.

Defining an n-body force is a delicate matter in a quantum field theory. For a two-body force, we must use the Bethe-Salpeter equation; for higher values of n, Faddeev equations. Fortunately, for our purposes, we need worry about none of these niceties. All we want to do is count powers of N, and for this all we need to know is that the n-body interaction kernel is obtained by summing up some family of graphs with n quark lines entering and n quark lines leaving. We can always imagine obtaining such graphs by breaking open n internal quark lines in a vacuum-to-vacuum graph. By result 2 of Section 3.1, the leading vacuum-to-vacuum graphs with internal quark lines are the planar graphs bounded by a single quark loop; these are proportional to N. We want an interaction that will be effective in a totally antisymmetric color state; thus each quark line should carry a different color index, and breaking the quark lines costs us a factor of N^{-n} from lost index sums. Hence, the n-quark interaction is proportional to N^{1-n}.

This completes the first part of the analysis. We must now study the effects of these interactions. From this point on I will assume that the states we are studying are non-relativistic, so we can use ordinary particle mechanics and treat the interactions as ordinary n-body potentials. I am embarassed by the necessity of this assumption. Detailed dynamical assumptions should not be needed just to count powers of 1/N; it should be possible to do the whole analysis in an elegant relativistic formalism. Unfortunately, I have not been able to find such a formalism, so we will just have to plug along with particle mechanics.

We wish to study a bound state made up of a very large number of particles interacting very weakly. This is the traditional domain of the Hartree approximation: Each particle is treated as moving independently of the others in a common potential, which in turn is determined self-consistently from the motions of all the particles. Let me remind you of the justification for this approximation by estimating the sources of error. Firstly, the approximation neglects the fact that as each particle moves, it changes the state of the other particles, and thus the potential which it feels. This effect is proportional to the square of the interaction strength, and is thus a negligible correction to the Hartree potential (directly proportional to the interaction strength) if the interaction is weak. Secondly, the approximation neglects the fact that each particle feels not the potential caused by all the particles, but the potential caused by all the particles but itself. The correction this makes to the Hartree potential is inversely proportional to the number of particles and is negligible if this number is large.

The Hartree approximation leads to many-particle energy eigenfunctions that are products of single-particle wave functions. In our case, the many-particle wave function is antisymmetric in color, and thus symmetric in the remaining quark variables: space, spin, and flavor. Thus, if we factor out the color part of the wave function, the quarks act like identical bosons; in the baryon ground state, all the quarks will be in the same state, the ground state of the Hartree potential.

Let me make this more quantitative. For notational simplicity, I will ignore flavor and spin. The baryon Hamiltonian is then

$$H = \frac{1}{2m}\sum_a |\vec{p}_a^{\,2}| + \frac{1}{2N}\sum_{a\neq b} V^{(2)}(\vec{r}_a,\vec{r}_b)$$

$$+ \frac{1}{6N^2}\sum_{a\neq b\neq c} V^{(3)}(\vec{r}_a,\vec{r}_b,\vec{r}_c) + \ldots \quad , \tag{3.50}$$

where the V's are functions independent of N. The approximate ground state wave function is of the form

$$\psi(\vec{r}_1 \ldots \vec{r}_N) = \prod_a \phi(\vec{r}_a) \quad . \tag{3.51}$$

We find the best choice of ϕ by the variational method. We compute

$$\langle\psi|H|\psi\rangle = N\left[\frac{1}{2m}\int d^3\vec{r}\ |\vec{\nabla}\phi|^2 \right.$$

$$+ \frac{1}{2}\int d^3\vec{r}_1\, d^3\vec{r}_2\, V^{(2)}(\vec{r}_1,\vec{r}_2)|\phi(r_1)\phi(r_2)|^2$$

$$\left. + \frac{1}{6}\int d^3\vec{r}_1\, d^3\vec{r}_2\, d^3\vec{r}_3\ V^{(3)}(\vec{r}_1,\vec{r}_2,\vec{r}_3)|\phi(\vec{r}_1)\phi(\vec{r}_2)\phi(\vec{r}_3)|^2 + \ldots\right] , \tag{3.52}$$

plus terms of O(1), which we neglect. We now minimize this as a function of ϕ, subject to the constraint

$$\int d^3\vec{r}\ |\phi|^2 = 1 \quad . \tag{3.53}$$

We find

$$\left[-\frac{\nabla^2}{2m} + V(\vec{r})\right]\phi = \varepsilon\phi \quad , \tag{3.54}$$

where ε is the Lagrange multiplier associated with the constraint, and V is the Hartree potential,

$$V = \int d^3\vec{r}_1\ V^{(2)}(\vec{r},\vec{r}_1)|\phi(\vec{r}_1)|^2$$

$$+ \frac{1}{2}\int d^3\vec{r}_1\, d^3\vec{r}_2\, V^{(3)}(\vec{r},\vec{r}_1,\vec{r}_2)|\phi(\vec{r}_1)\phi(\vec{r}_2)|^2 + \ldots \quad . \tag{3.55}$$

Equations (3.54) and (3.55) define the Hartree approximation; note that they are independent of N. This is a consequence of the fact that every term in the expression for the total energy, Eq. (3.52), is proportional to N; the n-quark interaction is $O(N^{1-n})$, but there are $O(N^n)$ distinct n-quark clusters. As we shall see, this lucky cancellation of powers of N leads to simple scaling laws for baryon physics.

Let me begin with the static properties of baryons. ϕ is independent of N; thus so is the shape of the ground-state baryon, as measured, for example, by its charge or mass distribution. This is just like the situation for a meson. In contrast to a meson, though, the energy of the baryon grows with N. This is essential to the resolution of what would otherwise be a problem for the approximation, that the ground state is not a momentum eigenstate. Precisely because the Hamiltonian is translationally invariant, any spatial translation of a Hartree ground state is also a Hartree ground state. Thus we can form linear combinations of these states that are momentum eigenstates; more carefully phrased, the problem is not that the energy eigenstates are not momentum eigenstates but that the momentum eigenstates are degenerate in energy. But this is as it should be, if the mass of the states is O(N); in the non-relativistic approximation, for example, the momentum dependence of the energy is proportional to $(\text{momentum})^2/$ mass; this is O(1/N), and should not be seen in leading order.

Low-lying excited baryons are obtained by placing a few quarks in excited energy eigenstates of Eq. (3.54). Note that, to leading order, there is no need to change V; the change in the Hartree potential caused by exciting only a few quarks is an effect of O(1/N). Thus, for example, if ϕ_1 is the first excited state of Eq. (3.54) and ε_1 is the associated eigenvalue, the first excited baryon is given by

$$\psi_1 = \frac{1}{\sqrt{N}} \sum_a \phi_1(\vec{r}_a) \prod_{b\neq a} \phi(\vec{r}_b) \quad . \tag{3.56}$$

Its energy exceeds that of the ground-state baryon by $\varepsilon_1 - \varepsilon$. Thus, although the baryon spectrum begins high, the spacing between successive baryons is O(1), just like the spacing of mesons.

So much for the static properties of baryons. Now let me turn to their interactions with mesons. As explained in Section 3.2, mesons are created and annihilated by quark bilinears, like

$$B = N^{-\frac{1}{2}} \bar{\psi}^a \psi^a \quad . \tag{3.57}$$

Here the ψ's are quark fields, normalized such that they obey canonical commutation relations, that is to say, such that in the non-relativistic approximation they create and annihilate quarks with amplitudes of order unity. Let us consider the matrix element of such a bilinear between two ground-state baryons. Each of the N terms in Eq. (3.57) can annihilate and recreate a quark; thus the baryon-meson-baryon vertex is $O(N/\sqrt{N}) = O(\sqrt{N})$. Of course, because the baryon energy is not changed, the meson must carry energy zero, and cannot be real. However, it can be a virtual meson, for example, one exchanged between baryon and meson in a meson-baryon scattering graph. Because the trilinear meson vertex is $O(1/\sqrt{N})$, this contribution to meson-baryon scattering is O(1).

Indeed, in general the meson-baryon scattering amplitude is O(1). To prove this, let us consider the one-baryon expectation value of the time-ordered product of two quark bilinears. There are two classes of terms that contribute to this expression.
(1) One of the bilinears can annihilate and recreate a quark; the other can do the same. In this case, the two quarks can be of different colors; the sum over colors gives a factor of N^2, and the total contribution is proportional to N. However, since the intermediate state is the ground-state baryon, the bilinears must each carry energy zero, and this contribution vanishes on the meson mass shell. (2) One of the bilinears can annihilate a ground-state quark and replace it with an excited quark; the other can then reverse the process. In this case, all the quarks must be of the

same color, so the sum over colors only gives a factor of N, and the total contribution is O(1). However, this contribution does not vanish on the meson mass shell.

We can also study the matrix element of a bilinear between a ground-state baryon and an excited baryon, like the one given by Eq. (3.56). Here each term in Eq. (3.57) must match up with the corresponding term in Eq. (3.56); thus, as before, we only get a factor of N from the color sum, and the amplitude for mesonic decay of an excited baryon is O(1). By similar reasoning, the amplitude for the process meson + baryon → meson + excited baryon is $O(N^{-\frac{1}{2}})$.

Because the meson-baryon scattering amplitude is O(1), the contribution of continuum states to the absorptive part of the amplitude is of the same order as the contribution of excited baryons. This is in striking contrast to the situation for meson-meson scattering. If, in the real world, meson-baryon scattering is well approximated by a sum over narrow resonances, the explanation of this phenomenon does not lie in the 1/N expansion.

Baryon-baryon scattering has a special feature because the baryon mass increases with N; if one studies scattering at fixed center-of-mass energy and momentum transfer, one soon finds oneself below threshold. The solution is to study scattering at fixed center-of-mass velocity and scattering angle. This implies that momentum transfer grows linearly with N, so it is not profitable to study the scattering process in terms of one-meson exchange, or, indeed, in terms of exchange of any finite number of mesons. The proper strategy is to directly compute the baryon-baryon interaction. When we first worked through the Hartree approximation, we found the interaction of one quark in a baryon with all the rest of the baryon was O(1). For a two-baryon system, by the same reasoning, the interaction of a quark in one baryon with the entirety of the other baryon is O(1). Thus the baryon-baryon interaction is proportional to N, just like the baryon energies, and N factors neatly out of the baryon-baryon scattering equation.

Baryon-baryon scattering, at fixed velocity and scattering angle, is O(1).

Let me summarize our results. The meson spectrum is independent of N, and meson-meson scattering is O(1/N). The baryon spectrum begins at an energy proportional to N, but, after it begins, the spacing of baryons is independent of N. The sizes and shapes of the baryons are also independent of N, as are the amplitudes for meson-baryon scattering and baryon-baryon scattering (at fixed velocities and angles).

As Witten has pointed out, we have heard this tune before. Certain classical field theories admit finite-energy time-independent solutions of the field equations, like the soliton in the sine-Gordon equation and the monopoles that arise in many unified electroweak and grand unified theories.[10] These lumps of energy become particles in the quantum versions of these theories, and, for small coupling, it is possible to study the properties of these particles. They are, word for word, the properties of the baryons enunciated in the preceding paragraph, with 1/N replaced by the small coupling constant (e^2 in the electroweak theories). This is a tantalizing parallelism; it strongly suggests that there should be some way of formulating the 1/N expansion such that the baryons appear directly as lumps. Unfortunately, at this moment, I know of no such formulation.

3.5. The Master Field

It would be good to know the leading term in the 1/N expansion of chromodynamics. We have been able to go far without this knowledge, but with it we could go much farther, and much faster. A direct approach, an attempt to compute and sum all planar graphs, is hopeless; some indirect method is needed. In this section, I will explain an indirect method recently proposed by Witten.[11] Witten's program has not yet been brought to completion. Nevertheless, I think it is worth talking about even in its incomplete state; it may well succeed, and, even if it fails, it involves such novel insights that it may inspire you to discover some other, better method.

I will first describe the method for pure gauge theory, and then go on to explain how to assimilate quarks. I think the clearest way to explain things is by drawing a parallel with the classical limit. Feynman's path integral formula tells us that Green's functions for a quantum theory are obtained by integrating over all possible classical motions. However, as $\hbar$ goes to zero, the measure in function space becomes more and more sharply concentrated about the solution to the classical equations of motion; in the limit of vanishing $\hbar$, all quantities are given by their values at the classical solution. A very similar statement applies to the large-N limit. There is a classical gauge-field configuration, which I will call the master field, such that the large-N limits of all gauge-invariant Green's functions are given by their values at the master field. I emphasize that once things are evaluated at the master field, there are no further steps; in particular, no integrations, functional or other, need to be done.

I will first comment on the method and then give the proof of the existence of the wonder-working master field.

Comments: (1) The master field is a field for the large-N limit of chromodynamics, that is to say, it is a gauge field for gauge group $U(\infty)$. (2) The master field is not unique; because we are interested only in gauge-invariant quantities, any gauge transform of a master field is also a master field. However, this is the end of the non-uniqueness; two gauge-inequivalent fields assign different values to some gauge-invariant quantity, and thus cannot both be master fields. A purist would thus speak not of "the master field" but of "the master orbit of the gauge group." (3) For the classical limit, we not only know that everything is dominated by a single field configuration, we have an algorithm for finding it, solution of the classical equations of motion. This is the missing element in Witten's program. We know that the master field exists but we have no algorithm for finding it. (4) Nevertheless, we can say some things about the master field. We expect large-N Green's functions to be translationally invariant; thus the master field should be translationally invariant also. That is to say, we expect that, in an appropriate gauge, A_μ should be independent of space-time. (Note that this does not mean that the master field is trivial, that $F_{\mu\nu}$ vanishes; the components of A_μ need not commute.) Thus, to find the master field we need only find four matrices. True, these are infinity-by-infinity matrices, so this is not necessarily an easy task. Nevertheless, this is a remarkable reformulation of the problem of summing all planar graphs.

Proof: We wish to show that for large N the measure in function space becomes concentrated on a single orbit of the gauge group. This is equivalent to showing that the probability of finding any gauge-invariant quantity away from its expectation value goes to zero as N goes to infinity. Because all gauge-invariant quantities are sums of products of the G's of Section 3.2, it suffices to prove the proposition for an arbitrary G. Of course, before we begin, we must normalize G such that its expectation value has a large-N limit, that is to say, is of order unity. Thus we define

$$G' = G/N \quad . \tag{3.58}$$

Now let us estimate the probability of G' departing from $\langle G'\rangle$ by computing the variance:

$$\begin{aligned}\langle (G' - \langle G'\rangle)^2\rangle &= \langle G'G'\rangle - \langle G'\rangle\langle G'\rangle \\ &= \langle G'G'\rangle_C \\ &= O(1/N^2) \quad ,\end{aligned} \tag{3.59}$$

by Eq. (3.12). QED.

All of this has been for pure gauge field theory, but the method can readily be extended to the computation of Green's functions involving quark bilinears. For simplicity, let me assume that we are only interested in the bilinear $\bar{\psi}\psi$. We can compute Green's functions for strings of these operators by giving the quarks a space-time dependent mass, m(x), computing the vacuum-to-vacuum amplitude, and then functionally differentiating this with respect to m. We know that in the large-N limit, the dominant vacuum-to-vacuum graphs are those with one quark loop. In a given external gauge field, the sum of all these graphs is given by a famous expression,

$$\mathrm{Tr}\,\ln\left[i\not{\partial} - \not{A} - m\right] \quad . \tag{3.60}$$

But this is a gauge-invariant function of gauge fields only; thus, in the large-N limit, when we integrate over gauge fields, it is given by its value with A_μ replaced by the master field.

You may find these arguments a bit too slick and abstract, and yearn for a concrete example in which one can explicity find the master field. Fortunately, such an example exists. As I have said, the master field has not been found for four-dimensional chromodynamics. It has not even been found for the vastly simpler 't Hooft model, two-dimensional chromodynamics. However, it has been found for zero-dimensional chromodyanmics. This is quite a come-down from field theory; instead of functional integrals over matrix-valued fields we have ordinary integrals over ordinary matrices, and the master field is just a master matrix. Nevertheless, even though the dynamics has been trivialized, the combinatorics retains much of its four-dimensional horror, and the integrals evaluated easily with the master matrix would be nightmares if attempted by the summation of planar graphs. These matrix integrals were first evaluated by this method in a brilliant paper by the Saclay group.[12] I will follow their analysis closely in what follows.

We wish to evaluate integrals of functions of an $N \times N$ Hermitian matrix, H. To begin, we must define integration over H:

$$dH = \prod_{a,b} dH_{ab} \quad , \quad a,b = 1 \ldots N \quad , \tag{3.61}$$

where integration over complex variables is defined in the usual way,

$$dH_{ab}\, dH_{ba} = d(\mathrm{Re}\, H_{ab})\, d(\mathrm{Im}\, H_{ab}) \quad . \tag{3.62}$$

This measure is invariant under zero-dimensional gauge transformations,

$$H \rightarrow U^{+} H U \quad , \tag{3.63}$$

where U is a unitary matrix. We wish to study the zero-dimensional version of the chromodynamic formula for the expectation value of a gauge-invariant operator,

$$\langle \mathrm{Trg} \rangle \equiv \frac{\int dH e^{-S(H)} \mathrm{Tr} g(H)}{\int dH e^{-S(H)}} \quad , \tag{3.64}$$

where g is some function, and

$$S(H) = N \,\mathrm{Tr}\, f(H) \quad , \tag{3.65}$$

for some function f. (Note that the positioning of the factor of N, and thus the combinatoric analysis, is the same as in four-dimensional chromodynamics.)

We want to evaluate (3.64) for large N. We begin by writing H in canonical form,

$$H = U^{+} D U \quad , \tag{3.66}$$

where D is a diagonal matrix, with eigenvalues $\lambda_1 \ldots \lambda_N$. We can rewrite the integration measure in terms of the λ's and U. By gauge invariance, it must be of the form

$$dH = \left(\prod_{a} d\lambda_a\right) h\left(\lambda_1 \ldots \lambda_N\right) dU \tag{3.67}$$

where dU is the invariant measure on U(N) and h is a function we shall find immediately. To determine h, we compare the two sides of Eq. (3.67) in the neighborhood of the identity in U(N). Here,

$$U = 1 + i\varepsilon \quad , \qquad (3.68)$$

where ε is a Hermitian matrix. At this point, Eq. (3.66) becomes

$$H = D - i[\varepsilon, D] \quad , \qquad (3.69)$$

or, written out in components,

$$H_{ab} = \lambda_a \delta_{ab} + i\varepsilon_{ab}(\lambda_a - \lambda_b) \quad , \qquad (3.70)$$

where there is no sum on the repeated indices. Hence,

$$dH = \left(\prod_a d\lambda_a\right) \prod_{a \neq b} i(\lambda_a - \lambda_b)\, d\varepsilon_{ab} \quad . \qquad (3.71)$$

Aside from a possible (but irrelevant) multiplicative constant, the term involving ε is dU. Thus,

$$h = \prod_{a \neq b} (\lambda_a - \lambda_b) \quad . \qquad (3.72)$$

The integration over U(N) factors out of Eq. (3.64). Thus we are left with

$$\langle \mathrm{Tr} g \rangle = \frac{\int \prod_a d\lambda_a \sum_b g(\lambda_b) e^{-S_{eff}}}{\int \prod_a d\lambda_a e^{-S_{eff}}} \quad , \qquad (3.73)$$

where

$$S_{eff} = N \sum_a f(\lambda_a) - \sum_{a \neq b} \ell n |\lambda_a - \lambda_b| \quad . \qquad (3.74)$$

If we count both explicit factors of N and factors of N arising from the number of terms in a sum, we see that both terms in S_{eff} are $O(N^2)$. This can be made more apparent by introducing

$$\rho(\lambda) \equiv \frac{1}{N} \sum_a \delta(\lambda - \lambda_a) \quad . \qquad (3.75)$$

The factor of N has been introduced so ρ obeys an N-independent normalization condition,

$$\int d\lambda\, \rho(\lambda) = 1 \quad . \qquad (3.76)$$

The integral of ρ over any interval gives the fraction of the total number of eigenvalues that lie on that interval; ρ is the fractional density of eigenvalues. For any finite N, ρ is a spiky sum of delta-functions; however, as we shall see shortly, it has a continuous limit as N goes to infinity. In terms of ρ,

$$S_{eff} = N^2\left[\int d\lambda\ \rho(\lambda)f(\lambda) - \int d\lambda\, d\lambda'\ \rho(\lambda)\rho(\lambda')\ \ell n\left|\lambda-\lambda'\right|\right] . \quad (3.77)$$

All factors of N have now been made explicit, and the character of the large-N limit is now clear. The integral is dominated by the ρ which minimizes S_{eff}; to leading order,

$$\langle \mathrm{Tr}g\rangle = N\int d\lambda\ \rho(\lambda)g(\lambda) \quad . \quad (3.78)$$

This can be thought of as Trg(H), where H is a master matrix, a matrix whose density of eigenvalues is given by the minimizing ρ. We can find the minimizing ρ by searching for the stationary points of Eq. (3.77),

$$f(\lambda) - 2\int d\lambda'\ \rho(\lambda')\ \ell n\left|\lambda-\lambda'\right| = \text{constant} \quad , \quad (3.79)$$

where the constant is the Lagrange multiplier associated with the constraint equation, (3.76). It is convenient to eliminate the constant by differentiating with respect to λ. We find

$$f'(\lambda) - 2\int d\lambda'\ \rho(\lambda')\ \frac{P}{\lambda-\lambda'} = 0 \quad . \quad (3.80)$$

This is the equation that must be solved to find the master matrix. There is an important technical point: because ρ is restricted to be positive, Eq. (3.80) holds only within the support of ρ, the region where ρ is non-zero. The easiest way to see this is to enforce positivity by writing ρ as σ^2; $\delta\rho$ is then $2\sigma\delta\sigma$, and deriving Eq. (3.80) outside the support of ρ involves an illegitimate division by zero.

I could go on to solve Eq. (3.80) for special choices of f, but I would prefer to stop here; if you want more, you can find it in the literature.[12] The point has been made. There is nothing wrong with the general arguments; for zero-dimensional chromodynamics, the master field exists, and we have found an algorithm for constructing it, Eq. (3.80). The unsolved problem is to find the appropriate generalization of Eq. (3.80) to four dimensions.

3.6. Restrospect and Prospect

Where are we?

For mesons, things are wonderful. Only an enthusiast who has spent too much time studying dual resonance models and too little time studying reality would claim that the properties we found in Section 3.2 form an accurate portrait of the mesons. They form a caricature. But it is a recognizable caricature; we look upon it and cry, "These are the mesons!" I know of no method other than the 1/N expansion in which the lineaments of the mesons emerge so clearly and unambiguously from chromodynamics.

For the baryons, things are not so good. Witten's theory is an analytical triumph but a phenomenological disaster. It is true that in significant ways baryon phenomenology is qualitatively different from meson phenomenology. To take one famous example, duality plus no exotics works wonderfully for mesons, but leads to contradictions for baryons. Baryons are different from mesons, but not as different as they are in the 1/N expansion. Baryons are not much heavier than mesons, and baryon resonances are not much broader than meson ones. If our picture of the mesons is a good caricature, our picture of the baryons is a bad one.

There are two possibilities. One is that the 1/N expansion is a better approximation for mesons than for baryons. This statement is not as silly as it seems. A reasonable person might agree that a quark and an antiquark is a quark and an antiquark, pretty much the same no matter how many colors there are, while a three-quark baryon is very different from a hundred-quark baryon, no matter how we adjust our coupling constants. The second possibility is that the 1/N expansion is terrible for both mesons and baryons. After all, most of our meson phenomenology was derived only from the dominance of graphs with a single quark loop. Although this is certainly a consequence of the 1/N expansion, it is not inconceivable that it might also be a consequence of some other principle altogether, and thus be valid even though the 1/N expansion is not.

Of course, we would know which of these possibilities is correct, and know much more, if we had an explicit expression for the leading approximation. As Witten has stressed, to seek such an expression is not ridiculously ambitious. One is not searching for a portrait of nature in all its fine shadings, not all phase shifts at all energies, but just for a recognizable caricature of mesodynamics, a table of resonance masses and couplings.

In these lectures I have discussed one attempt to find such an explicit expression. This particular attempt may succeed or fail, but, in any event, I feel future progress in this field rests upon constructing the leading approximation. It is amazing how far we have gone while avoiding this problem, but I do not think we can go much farther without solving it.

APPENDIX A. THE EULER CHARACTERISTIC

Given a surface composed of polygons, with F faces, E edges, and V vertices, the Euler characteristic is defined by

$$\chi = F - E + V \quad . \tag{A.1}$$

In this appendix I give (very sloppy) proofs of two propositions. (1) The Euler characteristic is a topological invariant. (2) If our polygonal surface is topologically equivalent to a sphere with B holes cut out of it and H handles stuck on to it, then

$$\chi = 2 - 2H - B \quad . \tag{A.2}$$

Proof of (1). One can convince oneself that there are only three fundamental ways one can change a polygonal surface continuously. (i) One can distort the surface without changing either F, E, or V. Of course, this does not change χ. (ii) One can shrink an edge to a point. This eliminates one edge, merges two vertices, causing a net loss of one vertex, and does not change the number of faces. Thus χ does not change. The same argument applies to the reverse process. (iii) One can shrink a face to a point. If the face is a polygon with n sides, this procedure eliminates one face, n edges, and n-1 vertices. Again χ does not change. The same argument applies to the reverse process. Processes like conversion of a face to an edge can be obtained as (iii) followed by the reverse of (ii) and do not require independent analysis.

Proof of (2). The argument goes in stages. First we prove the equation for a sphere, then for a sphere with holes, and finally for a sphere with holes and handles. (i) A sphere can be constructed by taking two n-sided polygons and identifying their perimeters. We thus obtain a surface with n edges, n vertices, and two faces, for which

$$\chi = 2 \quad . \tag{A.3}$$

(ii) If we have a polygonal surface that is topologically equivalent to a sphere, we cut holes out of it by removing faces. Thus, each hole reduces χ by one, and

$$\chi = 2 - B \quad . \tag{A.4}$$

(iii) To make a handle, we cut two holes that are both n-sided polygons (reducing χ by two) and then identify the perimeters of the two polygons (reducing both E and V by n and not changing χ at all). Thus,

$$\chi = 2 - 2H - B \quad . \tag{A.5}$$

APPENDIX B. THE 'T HOOFT EQUATIONS

This appendix gives the computations promised in Section 3.3.

I will begin with the quark self-energy. We need a preliminary identity:

$$\int \frac{dx}{x \pm i\varepsilon} = \int dx \left[\frac{P}{x} \mp i\pi\delta(x)\right] = \mp i\pi \quad . \tag{B.1}$$

Hence,

$$\int \frac{dp_+ p_-}{2p_+p_- - a + i\varepsilon} = -i\frac{\pi}{2}\,\mathrm{sgn}\,p_- \quad , \tag{B.2}$$

and Eq. (3.37) becomes

$$\Sigma = \frac{g^2}{2\pi}\int dk_-\ \mathrm{sgn}(p_- - k_-)\ \frac{P}{(k_-)^2}$$

$$= -\frac{g^2}{\pi p_-} \quad . \tag{B.3}$$

I now turn to the eigenvalue equation for meson masses. If the Green's function of Fig. 12 has a meson pole, then standard arguments lead to the Bethe-Salpeter equation shown in Fig. 15. Here all momenta are oriented to the right, and the shaded blob on the right is the matrix element of the time-ordered product of two quark fields between the vacuum and the meson state. If we denote the Fourier transform of this matrix element by ψ, then the Bethe-Salpeter equation is

$$\psi(p,q) = -4g^2 iS(p-q)S(-q)\int \frac{d^2k}{(2\pi)^2}\ \frac{P}{(k_- - q_-)^2}\psi(p,k) \quad . \tag{B.4}$$

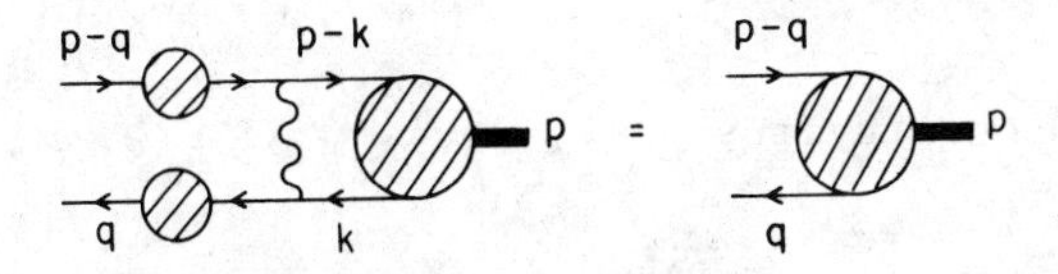

Figure 15

If we define

$$\phi(p,q_-) = \int dq_+ \, \psi(p,q) \quad , \tag{B.5}$$

then

$$\phi(p,q_-) = -\frac{ig^2}{\pi^2}\int dq_+ \, S(p-q)\, S(-q) \int dk_- \, \frac{P}{(k_- - q_-)^2}\, \phi(p,k_-) \ . \tag{B.6}$$

The q_+ integral is an integral over known rational functions and can be done explicitly:

$$\int dq_+ \, S(p-q)\, S(-q) \equiv I(p,q_-)$$

$$= \int dq_+ \frac{1}{2(p-q)_+ - \left[(M^2 - i\varepsilon)/(p-q)_-\right]} \, \frac{1}{2q_+ - \left[(M^2 - i\varepsilon)/q_-\right]} \ . \tag{B.7}$$

If q_- is outside the interval $[0,p_-]$, the two poles of the integrand are on the same side of the real axis, and the integral vanishes. This implies that ϕ also vanishes outside this interval. For q_- within the interval, the integral may be done trivially by closing the contour; the result is

$$I = -\pi i \Big/ \left[2p_+ - M^2/q_- - M^2/(p_- - q_-)\right] \ . \tag{B.8}$$

Equation (B.6) thus becomes

$$\left[2p_+ - M^2/q_- - M^2/(p_- - q_-)\right]\phi(p,q_-) = -\frac{g^2}{\pi}\int_0^{p_-} dk_- \, \frac{P}{(k_- - q_-)^2}\, \phi(p,k_-) \ . \tag{B.9}$$

If we make the substitutions,

$$2p_+ = \mu^2/p_- \quad , \quad q_- = xp_- \quad , \quad k_- = yp_- \quad , \tag{B.10}$$

this becomes Eq. (3.41).

APPENDIX C. U(N) AS AN APPROXIMATION TO SU(N)

In Section 3.1, I dropped the second term in the gluon propagator, Eq. (3.5). To drop this term is to approximate SU(N) gauge theory by U(N) gauge theory; to restore this term is to correct the approximation by introducing a correction gluon, a negative-norm color-singlet gauge meson that cancels the positive-norm color-singlet gauge meson of the U(N) theory. In this appendix I compute the powers of 1/N associated with graphs containing internal correction gluons.

As in Section 3, I will restrict myself to vacuum-to-vacuum graphs; from these, all Green's functions can be obtained by functional differentiation. If all we had in our theory were gauge fields, we would have made no error and we would need no corrections; in a U(N) gauge theory, the U(1) gauge meson is completely decoupled from the SU(N) gauge mesons. Thus we need only consider graphs where all correction gluons terminate on quarks. Because the correction gluon carries no color indices, we can always imagine constructing such a graph by adding correction gluons to graphs without them. How many powers of 1/N do we introduce by this process? For each quark line, we introduce on extra propagator and one extra vertex; the 1/N for the propagator cancels the N from the vertex. However, the correction gluon itself carries 1/N because it is a propagator and an additional 1/N from the explicit factor of 1/N in Eq. (3.5). Thus the net effect of adding a correction gluon is $O(1/N^2)$. This is all we need to know, because we already know how to compute powers of 1/N for graphs without correction gluons.

As an example, let me compute the power of 1/N associated with the leading connected vacuum-to-vacuum graphs containing a correction gluon. We must add the correction gluon to graphs that contain quark loops. The leading graphs are those in which the correction gluon connects two disconnected graphs each of which contains a quark loop; these are $O(N \times N \times 1/N^2) = O(1)$. However, these graphs are trivial; they serve only to exactly cancel the corresponding gluon-exchange graphs in the uncorrected theory. The leading non-trivial graphs are those in which a correction gluon is added to a connected graph containing a quark loop; these are $O(N \times 1/N^2) = O(1/N)$.

NOTES

1. To my knowledge, the first to observe that ϕ^4 theory became simple for large N was K. Wilson, Phys. Rev. D7:2911 (1973). Wilson drew on ideas developed in statistical mechanics, especially Stanley's work on the spherical model [H. E. Stanley, Phys. Rev. 176:718 (1968)]. The auxiliary-field method used here was also developed in statistical mechanics (but not in the context of large-N expansions), by R. L. Statonovich, Doklady Akad. Nauk. S.S.S.R. 115:1097 (1957). The treatment of ϕ^4 theory given here follows that of S. Coleman, R. Jackiw, and H. D. Politzer, Phys. Rev. D10:2491 (1974). Some points left confused in this paper are clarified in L. Abbott, J. Kang, and H. Schnitzer, Phys. Rev. D13:2212 (1976).
2. D. Gross and A. Neveu, Phys. Rev. D10:3235 (1974).
3. This note is for the cognoscenti only; it is written in shorthand. You may know that you can define an effective potential, much like V, in theories with fundamental scalar fields, and that, in such theories, V can be interpreted as an energy density for general arguments, not just stationary points. This is important; for example, it implies that if V is unbounded below the theory is sick, no matter how nicely behaved V is at its stationary points. <u>There is no such interpretation of the effective potential for composite fields.</u> For example, in the case at hand, if we add a source term to the Lagrange density,

$$\mathscr{L} \to \mathscr{L} + J(x)\sigma \quad ,$$

this corresponds to adding a term to the Hamiltonian density,

$$\mathscr{H} \to \mathscr{H} - J\phi^a\phi^a + \frac{g_0}{2N} J^2 \quad .$$

The J^2 term has no analog for a fundamental scalar field, and destroys the standard energy arguments (except at stationary points of V, where it vanishes).

4. The model was devised by H. Eichenherr, Nucl. Phys. B146:215 (1978) and V. Golo and A. Perelomov, Phys. Lett. 79B:112 (1978). The 1/N expansion is worked out in A. D'Adda, M. Lüscher, and P. DiVecchia, Nucl. Phys. B146:63 (1978), and B152:125 (1979).

5. This is a bit too slick. After all, in electrodynamics, gauge invariance could also be thought of as a mere reflection of the presence of a redundant variable, the longitudinal part of the vector potential. The real difference is slightly more subtle. In electrodynamics, it is not possible to eliminate the redundant variables and still have a local theory, as is shown by the Bohm-Aharonov effect; in the case at hand, it is possible to do just this, as is shown by our derivation of the model. As we shall see, this distinction will disappear when we sum up the radiative corrections.

6. E. Witten, Nucl. Phys. B149:285 (1979); A. Jevicki, "Collective Behavior of Instantons in QCD" (IAS preprint); I. Affleck, "Testing the Instanton Method" (Harvard preprint).

7. (This note covers both this subsection and the next.) The large-N expansion for chromodynamics was invented by G. 't Hooft, Nucl. Phys. B72:461 (1974). There are numerous parallels and connections with the topological expansions of S-matrix theory; see G. Veneziano, Nucl. Phys. B117:519 (1976) and G. Chew and C. Rosenzweig, Phys. Rep. 41C:263 (1978).

8. G. 't Hooft, Nucl. Phys. B75:461 (1974). This paper has spawned a large literature, with two branches: (1) Papers which investigate the model in more detail (for example, by computing form factors) and/or use the model to gain insight

into four-dimensional chromodynamics. Some examples: C. G. Callan, N. Coote, and D. J. Groos, Phys. Rev. D13:1649 (1976); M. B. Einhorn, Phys. Rev. D14:3451 (1976); R. Brower, J. Ellis, M. Schmidt, and J. Weis, Nucl. Phys. B128:131, 175 (1977). (2) Papers which attempt to clean up 't Hooft's original derivations and/or to find inconsistencies in the model. To my mind, the paper of this kind that does things best and settles all the problems is I. Bars and M. B. Green, Phys. Rev. D17:537 (1978).

9. E. Witten, Nucl. Phys. B160:57 (1979).
10. For a review of lumps, see my 1975 Erice Lectures (in New Phenomena in Subnuclear Physics, ed. by A. Zichichi, Plenum Publishers, 1977).
11. The proposal was made in a lecture given at Harvard in the Spring of 1979.
12. E. Brezin, C. Itzykson, G. Parisi, and J. B. Zuber, Comm. Math. Phys. 59:35 (1978).

D I S C U S S I O N S

CHAIRMAN: Prof. S. Coleman

Scientific Secretaries: P. Ginsparg, L. Palla and T. Sterling

DISCUSSION 1

- CHRISTOS:

When you computed the effective potential $V(\sigma)$, you summed all one loop diagrams with the external σ lines at zero momentum. Each diagram you summed was infrared divergent, so would not this mean that $V(\sigma)$ is divergent at $\sigma = 0$ and hence one cannot comment about $V(\sigma)$ at $\sigma = 0$?

- COLEMAN:

No. $\sigma = 0$ is no different than any other point. There are two ways of computing the effective potential. The first to compute it as a gaussian integral. According to the standard rules for fermion functional integration, I evaluate

$$\int d\psi\, d\bar{\psi}\; e^{-\int \bar{\psi}\, i(\not{\partial}-\sigma)\psi} = \det i(\not{\partial}-\sigma)$$

where σ is a constant field. By charge conjugation invariance,

$$\det i(\not{\partial}-\sigma) = \det i(\not{\partial}+\sigma) = \det\left[-\partial^2+\sigma^2\right]^{1/2}$$

Finally,

$$\det\left[-\partial^2+\sigma^2\right]^{1/2} = e^{1/2\,\mathrm{tr}\,\ln\left[-\partial^2+\sigma^2\right]} = e^{\int \frac{d^2k}{(2\pi)^2}\ln\left[k^2+\sigma^2\right](2\pi)^2\delta^2(0)}$$

in Euclidean momentum space. The explicit trace over Dirac indices in two dimensions gives a factor of 2 cancelling the 1/2. The $(2\pi)^2\delta^2(0)$ is the usual factor for the volume of space-time. Notice that at no point does an infrared divergence appear in this method of calculation.

Another way of getting the same answer is to sum up an infinite series of Feynman diagrams. This is in fact how we know functional integration is right. We know Feynman diagrams are right and functional integration gives the same answer. Now, the energy momentum conserving δ function can be extracted out before I do the summation since every Feynman graph has a factor of $(2\pi)^2\delta^2(0)$. I know the sum of all Feynman graphs is the exponential of the sum of all connected Feynman graphs, so I do not have to worry about exponentiating. A typical one loop graph has 2n external lines since the trace of an odd number of γ -matrices gives zero.

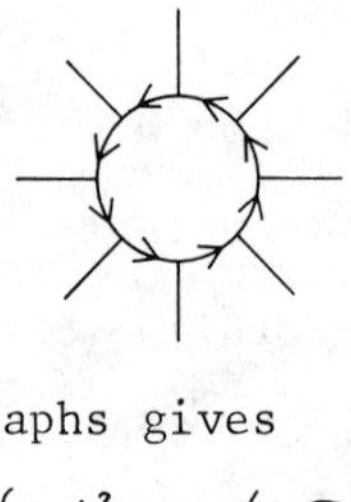

Summing all such 1-loop graphs gives

$$-iV = -\sum_n \mathrm{Tr}\int \frac{d^2p}{(2\pi)^2}\left(\frac{\sigma}{\not p}\right)^{2n}\frac{1}{2n}$$

The combinatoric factor $1/2n$ appears because I have 2n rotations by π/n which leave the graph unchanged and thus the Dyson 1/n! is incompletely cancelled. There is an overall Fermi minus sign and I have suppressed the $i\varepsilon$ which is of course omnipresent.

I find:

$$V = -i\int\frac{d^2p}{(2\pi)^2}\sum_n\left(\frac{\sigma^2}{p^2}\right)^n\frac{1}{n} = -i\int\frac{d^2p}{(2\pi)^2}\ln\left(1-\frac{\sigma^2}{p^2}\right) = \int\frac{d^2p_E}{(2\pi)^2}\ln\left(1+\frac{\sigma^2}{p_E^2}\right)$$

after rotating into Euclidean space.

Aside from an overall additive constant, it is the same formula as before, which is not surprising, since I have calculated the same thing by two correct methods. You might properly point out that each and every term in the summation above is hideously infrared divergent, the divergence becoming worse for larger n. However, the first method shows that the second gives indeed the correct answer although in a sloppy way.

- CHRISTOS:

How do you know that two loop diagrams do not contribute just as much as the sum of one loop diagram?

- COLEMAN:

The two loop graphs are down by a factor of 1/N. I will remind you in general why this is so. Remember for my effective action, the result of taking the ordinary action and integrating out the fermions, I have N times some functional of σ, call it NF(σ). As you quite rightfully say, I still have to sum up graphs with internal σ loops. This hideous theory has 2, 3, 4 ... -point σ interaction terms but it is easy to count powers of N. I have shifted to a stationary point so there is no term linear in σ. The propagator has a 1/N since it is obtained by inverting the quadratic part. Each vertex in this theory has an explicit power of N. Thus I have associated to a general graphs N^{-I+V}, where I = # internal lines and V = # vertices. Now in a general connected graph we know that L, the # loops is equal to the # momentum integrations. This in turn is simply I-V+1 since, although there are naively I integrations, there is also an energy-momentum conserving δ-function at each vertex with one left over for overall energy-momentum conservation. So we have $N^{-I+V} = N^{1-L}$ and we see that counting powers of 1/N is equivalent to counting the number of loops. In a systematic expansion in 1/N I am thus correct to neglect higher loop graphs compared to the 1 loop graphs which I have considered.

- KOH:

The graph

has $\left(\frac{1}{N}\right)^3 \cdot N^2 = \frac{1}{N}$, that is, the same order as the leading bubble expansion. Why is this diagram omitted even in the large N limit?

- COLEMAN:

I talked about these diagrams briefly. For example, the graph

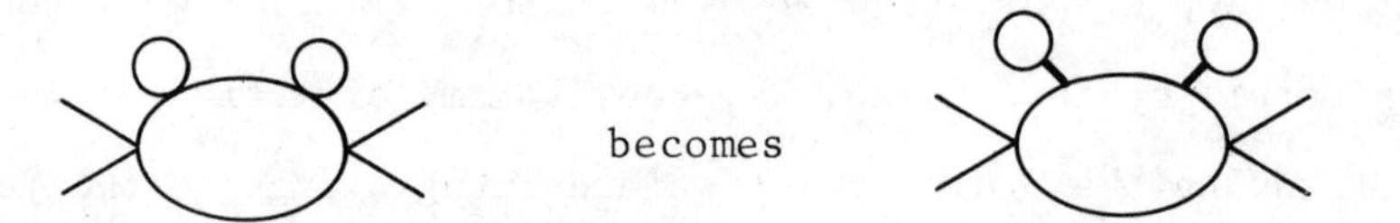

when I expand in both the ϕ and σ fields - the heavy lines indicate the σ field. All diagrams of this kind have an additional loop, a σ tadpole, somewhere. One can always absorb these loops into a renormalization of the mass of the ϕ field. That is, in the σ formalism all such terms can be lumped together with $\frac{N\mu_0^2}{2\lambda_0}\sigma$ to produce a tadpole with an arbitrary assigned value, the physical mass of the ϕ field. In a certain sense the renormalization of this theory is more complicated than that of the Gross-Neveu model where mass renormalization is unnecessary because of the γ_5 invariance.

- OLIENSIS:

Is the mass scale in QCD set by g becoming of order 1 or by $g/\sqrt{N}$ becoming of order 1?

- COLEMAN:

Nobody knows what happens in real QCD. In this case g and $g/\sqrt{N}$ differ by only a factor of 1.7. There are probably several mass scales, for instance the radius of the bag, the mass of the lightest constituent quarks, and the universal slope of Regge trajectories. These are all of the same order of magnitude, but they are by no means all the same thing.

- LOW:

Why not do your integral in three dimensions and thereby obtain a 1/N expansion for that theory?

- COLEMAN:

You want to solve $(\bar{\psi}\psi)^2$ in three dimensions. This is just a one loop computation so the fact that the theory is non-renormalizable would not cause too much trouble. I think Gross and Neveu speculate about whether you can go to higher orders and in fact construct what looks like a renormalizable theory in three dimensions from something that starts out non-renormalizable.

- KLEINERT:

$(\bar{\psi}\psi)^2$ is not defined at all (in three dimensions), but we can instead take an abelian quark gluon theory (that is, electrodynamics wiht massive photons and quarks with several flavors). The whole procedure can be performed just as you did, except that the field σ which was $\bar{\psi}\psi$ in your system becomes $\bar{\psi}(x)\psi(y)$, a bilocal field. This is discussed in my Erice Lectures three years ago.

- COLEMAN:

The question is "What if I start out with this theory and just follow the procedure?" Gross and Neveu say that it looks as though you do not run into the extraordinary divergences you would expect.

- KLEINERT:

But you cannot take the $N \rightarrow \infty$ limit in a theory which is not even defined. That is the problem.

- COLEMAN:

That is the appropriate profound thing to say when one sees the computations have not worked out, but I would like first to see that the computations do not work out.

- KLEINERT:

I do not understand when you talk about "non-renormalizable theories". Are there any?

- COLEMAN:

The question is "Is there a family of quantum theories which have up to now escaped our attention which are consistently defined to all orders in some systematic approximation which in an appropriate classical limit appear as classical field theories of the type we would now call non-renormalizable?" I believe this is an open question.

- GINSPARG:

I do not agree that the relation attributed to Witten, $\frac{e^2}{4\pi} = \frac{1}{137}$ implying $e \sim 1/3$, is quite so frivolous a wisecrack as you suggested. Before one actually understands a theory, it is not clear what the effective expansion parameter is. It might unluckily have turned out to be $4\pi e^2$ in QED and perhaps might luckily turn out to be $1/4\pi N$ in SU(N) QCD, for example.

- COLEMAN:

Absolutely. When I said it was a wisecrack I meant full weight to be given to both syllables. I think it is potentially a very profound remark. At the current state of our knowledge, however, it is a crack, not any sort of theoretical breakthrough, nor piece of dogma on which to base our future investigations.

- JAFFE:

It is $\frac{1}{\sqrt{N}}$ which appears in amplitudes in 4 dimensional QCD and we should therefore compare $\frac{1}{4\pi N}$ rather than $\frac{1}{4\pi N}2$ with α .

DISCUSSION 2

- CHRISTOS:

What is γ_5 in two dimensions and what is it in higher dimensions?

- COLEMAN:

In two dimensions we can use the usual Pauli matrices and define $\gamma_0 = \sigma_z$, $\gamma_1 = i\sigma_y$, and $\gamma_5 = \gamma_0\gamma_1$. Then $\gamma_5^+ = \gamma_1^+\gamma_0^+ = -\gamma_1\gamma_0 = \gamma_5$ and γ_5 is hermitian.

Also $\gamma_5^2 = 1$, $\bar{\gamma}_5 = -\gamma_5$, and $\{\gamma_5, \gamma_n\} = 0$. We see that under γ_5, $\psi \to \gamma_5 \psi$, $\bar{\psi} \to -\bar{\psi}\gamma_5$, and $\bar{\psi}\psi \to -\bar{\psi}\psi$.

A similar object exists in any even number of dimensions, but in an odd number of dimensions there is no analogue to γ_5.

- CORBO':

Suppose you have solved a theory which shows confinement. What about the asymptotic states of this theory?

- COLEMAN:

By definition confinement means the particles in question never appear in the asymptotic states although you see them at short distances. An example is two dimensional Quantum Electrodynamics with fermions (which we will call quarks). When you pull a quark and antiquark very far apart, you find a linear force between them. As you pull them farther and farther apart, eventually the energy you have fed in is greater than $2mc^2$ where m is the mass of a quark. It is then energetically favorable to materialize a quark-antiquark pair between your original quark and antiquark. Thenceforth it costs you no more energy to separate the original quark and antiquark because the new pair has shielded them from each other. You are, however, no longer dealing with a quark and antiquark anymore. What you have are two mesons. All this happens in two dimensions in good old reliable perturbation theory in the parameter e/m; the description is best for very heavy or, equivalently, very weakly interacting quarks. So you can really see the quarks at short distances (and short distances could be a million light years if you make e/m small enough), but they never appear in truly asymptotic states. We all hope something like this happens in four dimensions but of course the situation there is more complicated.

- PAFFUTI:

Can the usual renormalization program be carried out in the 1/N expansion?

- COLEMAN:

I suspect there are no problems. Unfortunately I cannot say that I have investigated the matter to all orders. If nobody has, somebody should. I would be very surprised if renormalizability were distroyed by the 1/N resummation, but I cannot provide a proof.

- PAFFUTI:

To define a composite operator like $\sigma = \bar{\psi}\psi$, isn't it necessary to control the short distance behavior of the theory?

- COLEMAN:

On a perturbative level there is no serious difficulty. In the $\bar{\psi}\psi$ theory, all graphs are finite except

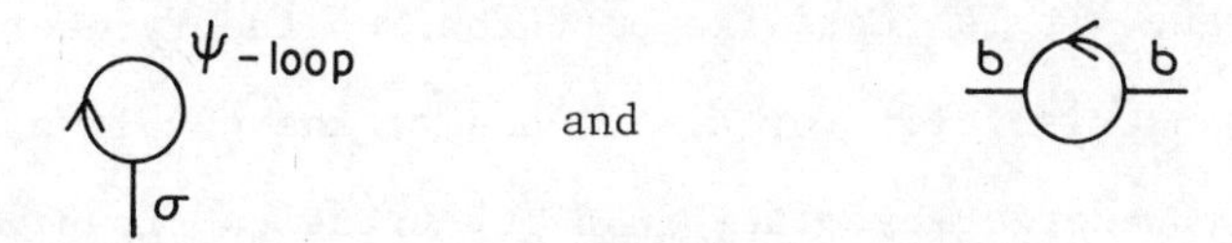

The first, a σ tadpole, vanishes by γ_5 invariance. For the second we already have a counterterm available to absorb the divergence since we have a σ^2 term in our Lagrangian ab initio. In the case of the CP^{N-1} model we have a scalar field in the problem, so the tadpole

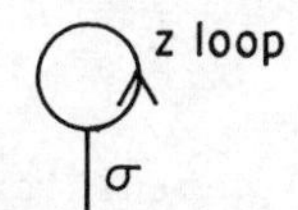

is a possible source of difficulty. However, in this model there is a term linear in σ, proportional to the bare coupling constant, available to absorb the divergence. The graph

goes as $\int \frac{d^2k}{k^4}$ since z is a scalar field. It is convergent and gives no problem. So: Yes, we have to take care of the short distance properties before we write down such a theory, and we do this with the usual perturbation counterterms.

- COQUEREAUX:

Is the planar approximation to QCD compatible with gauge invariance and if yes, how does the proof go?

- COLEMAN:

The answer is yes, trivially so. The reason is that it corresponds to leading order in an expansion in an adjustable parameter 1/N. Since the theory is gauge invariant for every possible order of N, each and every term in its asymptotic expansion should be gauge invariant. It is just like perturbation theory where, since gauge invariance is valid for all values of the coupling constant, and thus is true order by order in a perturbation expansion.

- KLEINERT:

But the Bethe-Salpeter kernel, a sum of (planar) ladder diagrams is known not to be gauge invariant.

- LOW:

I am confused about the definition of planar. If you consider the scattering of colored quarks in a given order of the gluon theory, in general both planar and non-planar diagrams (in the usual single line notation) contribute. I do not think that the planar ones by themselves would be gauge invariant. But when you decompose the gauge field into upper and lower indices, there are two ways of attaching the lines, only one of which can be associated with the truly planar diagrams (in the sense of the double line notation).

- COLEMAN:

That is correct. Planar means planar in the double line notation. A graph like

in the usual notation has associated to it the graphs

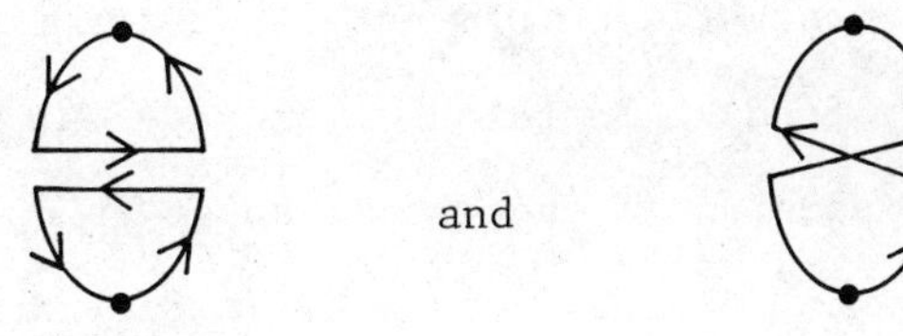

The first is planar in the double line notation and the second is down by a factor of 1/N. It is the second of these that would be grouped with a graph that is non-planar in the usual notation to preserve the Ward identity.

- OLIENSIS:

Can you show that instantons disappear in the large N limit of the CP^{N-1} model?

- COLEMAN:

The rough argument (first shown to me by Witten), is that instanton effects are proportional to the exponential of minus the instanton Euclidean action. Since the action has a factor of N in it, the instanton effects are exponentially small in N, and are seen in no finite order of the 1/N expansion. However, I am not quite sure things are this simple: the number of collective coordinates required to give the orientation of the instanton (in both CP^{N-1} and QCD) grows linearly with N for large N, and thus gives a factor of c^N, for some constant C. I would be happier had I seen a more detailed computation, but I am not aware of any such. (Note added in proof: such a computation had been performed by C. Bernard [Phys. Rev. D 19, 3013 (1979)] but I was unaware of it. It confirms the simple argument.)

- PAFFUTI:

In the CP^{N-1} model Berg and Lüscher have summed the contribution of the exact multi-instanton solutions without making a dilute gas approximation. They find no infrared divergence but mass generation is as in the 1/N expansion of the model. So it is not clear to me that all the instanton contributions will be insignificant.

- COLEMAN:

Their computation troubles me because they consider exact multi-instanton and multi-anti-instanton solutions, but they do not include mixed states of instantons and anti-instantons in their computation. It is true that these mixed states are not exact stationary points of the equations of motion but at least when the instantons and anti-instantons are relatively small and far apart they are very close to being stationary points. I do not see how a sensible computation can exclude these mixed states; indeed, the standard lore indicates that they should dominate. Perhaps there is some justification for this but it is not known to me.

DISCUSSION 3

- PAFFUTI:

Can one do an analysis of infrared divergences in four dimensional QCD at the planar level in the sense of Lee and Nauenberg, Kinoshita, etc.?

- COLEMAN:

Any analysis that is true in general is true in particular to all orders in 1/N. I do not know if one can do an intrinsic planar analysis.

- PAFFUTI:

Can you exclude the existence of a phase transition in the coupling constant in two dimensional QCD which cannot be seen in the 1/N expansion and might prevent confinement? This could perhaps be investigated using a $2-\varepsilon$ expansion.

- COLEMAN:

We have other reason to believe two dimensional QCD is confining. For example, we can investigate the Abelian theory, N = 1, which is the extreme opposite from large N. This theory has a dimensionless parameter e/m and can be examined both for $e/m \gg 1$ and $e/m \ll 1$. We find confinement in both regimes. Now since N = 1 confines and $N \to \infty$ confines, we certainly have some evidence that perhaps the theory confines for all N.

- OLIENSIS:

What would happen if there existed a single unconfined quark and what would happen to particles in the neighbourhood of this quark?

- COLEMAN:

I suppose you want me to alter the dynamics of the theory so that we can have almost confined quarks. In two dimensional QED there is no objection to giving the electromagnetic field a mass. The force would fall off exponentially and we could talk about free single quarks on distance scales short compared to the compton length of the gluon. This was worked out in detail by Parke and Steinhardt in 1+1 dimensions and also by Jaffe, Giles and De Rujula in a similar model in four space time dimensions.

- JAFFE:

I should first apologize for this model since we have in no sense made a rigorous calculation. In four dimensions these isolated quarks are large objects. You can think of them as surrounded by a spherically symmetric bag which is large compared to normal hadrons. The environment around an isolated quark is an environment where other quarks can live almost for free. The interaction between the isolated quark and baryons is both attractive and exothermic so that an isolated quark cuts nuclear matter if it passes through it until the density of matter in the region of space opened up by the quark is roughly the same as within hadronic matter.

DISCUSSION 4

- CHRISTOS:

You discussed $1/N_{color}$ for QCD. What about $1/N_{flavor}$?

- COLEMAN:

Unfortunately I cannot report on it; Veneziano worked intensively on doing a joint expansion in number of flavors and colors.

Then there is a large variety of limits you can take; for example the suppression of quark loops that formally arises as the number of colors goes to infinity with the number of flavors held fixed can be cancelled by increasing the number of flavors at the same time. Then each internal quark loop is less important for the reasons we have explained but there are more of them. There is a whole lore growing out of this which I am afraid I am not familiar with, therefore I have to refer you the work by Veneziano et al.

- CHRISTOS:

The baryon diagram you drew

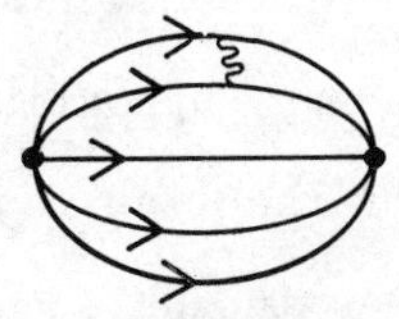

cannot be written in the double line notation as

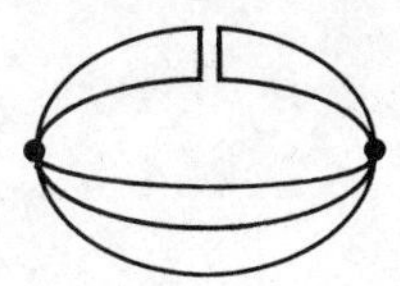

because this is zero. One must then draw the diagram as

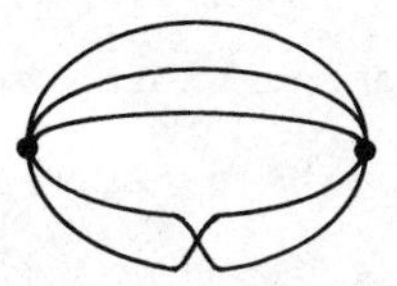

Does this mean that your result in the lecture should be suppressed by a further factor of 1/N?

- COLEMAN:

No. We started out by counting factors of 1/N by looking explicitly at the diagrams; however, we rewrote our N power counting of diagrams in terms of topological rules. You must be careful not to count powers of N from both ways, you should stick to one method in a given calculation.

- CHRISTOS:

One further thing in the 1/N expansion is that some theories that are normally unrenormalizable turn out to be renormalizable in the leading 1/N expansion (e.g. Gross-Neveu model in $2+\varepsilon$ dimensions). Does this present any problem?

- COLEMAN:

I have not studied the Gross-Neveu model in detail. I think it can happen in just the leading approximation, that does not mean anything. There are lots of theories that have terrible diseases and do not reveal themselves in the leading approximation. I would like to go at least a few terms in these theories and to see that there are no troubles. If we do the few lowest orders and the difficulties do not turn up, then we got something interesting, if this does not happen, the leading order is just meaningless. One can best see this by thinking about renormalizability in ordinary perturbation theory in g. Here one cannot argue that by taking g very small, the finitness of the Born terms imply renormalizability.

- HANSSON:

You showed us the 't Hooft equation in two dimensional QCD. A year or so ago there was another equation that was proposed by T.T. Wu for the same thing that gave completely different results. What is the present status of this?

- COLEMAN:

Shortly after 't Hooft work people attempted to derive his equation by starting in the gauge $A_1 = 0$ working in the zero momentum frame and then Lorentz transforming to the light cone frame. There were difficulties in doing this but Bars got around them by formulating the problem from the beginning in a canonical Hamiltonian form. This leads to considerable ugliness but is relatively free from ambiguity and he reproduced the 't Hooft equation.

THE BAG

R. L. Jaffe

Center for Theoretical Physics, Department of Physics and Laboratory for Nuclear Science, Massachusetts Institute of Technology, Cambridge, Mass. 02139

and CERN, Geneva

These lectures present an idiosyncratic review of the M.I.T. bag model.[1] I haven't made any attempt to be complete. Instead I've chosen a few topics which interest me or which have attracted some attention recently. I hope they illustrate the sort of things one can do with the model. Some parts of the sections on chiral symmetry are new. The rest can be found in one form or another in the literature. The very recent applications of the model discussed at this school by Prof. Low[2] apply to the spectroscopy developed in Section 3 of these notes. That section might be read in conjunction with his lectures.

The outline of these lectures is as follows:

1. Introduction
2. Formulations of bag models
    - 2.1. A new look at the traditional formulation
    - 2.2. A Lagrangian formulation with surface terms (Chodos and Thorn)
    - 2.3. A Lagrangian formulation with no explicit geometrical variables (Johnson)
3. Light quark hadrons

1. INTRODUCTION

QCD is in a remarkable state.. Most of us believe it to be the theory of strong interactions. Yet no one has succeeded in saying anything quantitative about hadron structure or interactions at ordinary mass scales ($\sim$1 GeV) on the basis of QCD alone. QCD's quantitative successes are all in the realm of very high $-Q^2$. There the effective coupling (α_c) is weak and perturbation theory may be applied. At masses of order 1 GeV α_c grows strong and one would think that large QCD renormalizations would obliterate the simple regularities observed at high momenta. In fact this is not the case. Low energy spectroscopy, weak and electromagnetic transitions, and even hadron-hadron scattering show simple regularities which are reasonably well described by the most naive quark models. As evidence of this remember that all of the ingredients of QCD: quarks, color, colored gluons and so forth were originally inferred from studies of the systematics of low energy hadron spectroscopy and interactions years before the reliability of asymptotically free perturbation theory was appreciated and exploited.

Someday, perhaps, unadorned QCD will explain low energy hadron physics. Impatient for that day, a group of us at M.I.T. and others elsewhere have attempted to develop a predictive phenomenology of hadrons by introducing confinement by hand into QCD from the beginning. The basic idea of the bag model is simple: it is assumed that wherever quark and gluon field variables are non-vanishing there is also a positive energy density B. Systems of finite energy (hadrons) perforce have finite volume. Since no color flux leaves hadrons they are all color singlets: hence confinement. Inside the bag B is passive so the dynamics is governed by the standard Yang-Mills equations. There are no new degrees of freedom. The model is Lorentz covariant (at least at the classical level). Dynamically B is a pressure; relativistically it contributes a new term to the stress tensor $-g^{\mu\nu}B$. Bagged QCD for small α_c is a quasi-free quantum gas of quarks and gluons confined by a uniform pressure exerted by the vacuum. B sets the scale for confinement phenomena. Its value, $B \cong 50\ \mathrm{MeV/fm}^3$ (or $B^{1/4} \cong 145$ MeV) determines the masses of light hadrons, the universal slope of Regge trajectories and the Hagedorn temperature. If the bag model emerges as an approximation to QCD then B itself is presumably determined from the underlying theory, which contains no dimensionful parameters, by dimensional transmutation.

In the bag model the hadronic world has two phases: First, the vacuum, which is quiescent and expels quarks and gluons; and second, the hadrons wherein quarks and gluons move more or less freely. B is the latent heat liberated when a bit of hadron is returned to vacuum. The system is like a (perfect) liquid under constant pressure at the boiling point. The liquid is the vacuum, the bag a bubble, and B the latent heat. As energy is delivered to the system, pairs are created from the vacuum. A bag is formed and expands but its temperature remains unchanged. A more fertile analogy,[3] which has been the basis of several attempts to derive the bag model from QCD,[4] is the Meissner effect in a bulk superconductor with $\vec{E}$ and $\vec{B}$ reversed. The true vacuum of QCD, like the Meissner superconductor, is thought to be dominated by complex nonperturbative field configurations. It expels color-electric flux like a superconductor expels magnetic flux. On the boundary between a normal region and a superconductor $\vec{E}$ and $\vec{B}$ obey

$$\hat{n} \times \vec{E} = 0 \qquad \text{SUPERCONDUCTOR} \qquad (1.1)$$
$$\hat{n} \cdot \vec{B} = 0 \quad .$$

According to the analogy, at the boundary of a hadron

$$\hat{n} \cdot \vec{E}_a = 0 \qquad \text{QCD} \qquad (1.2)$$
$$\hat{n} \times \vec{B}_a = 0 \quad ,$$

where a is a color index (a = 1,2...8). The covariant version of eq (1.2)

$$n_\mu F_a^{\mu\nu} = 0 \quad ,$$

is the famous bag boundary condition. So in this picture hadrons are normal regions with simple vacuum structure surrounded by a vast color-magnetic superconductor. No one has yet discovered the "Cooper pair" of QCD -- the particular field configuration which dominates the true vacuum -- though there have been many attempts.

A possible way of "rationalizing" the success of the quark model at low masses is to suppose that confinement is the only new thing which happens between asymptotic masses where α_c is small and hadron mass scales. This cannot be entirely correct because there is another obviously non-perturbative effect seen in low energy phenomena, namely approximate chiral symmetry, or rather the Nambu-Goldstone[5,6] realization of approximate chiral symmetry. The Lagrangian of QCD is invariant under the transformation $q_f \rightarrow e^{i\kappa\gamma_5/2} q_f$ in the limit of vanishing f-quark mass. Although the u, d and s-quark masses are thought to be small, hadrons made of these quarks do not form approximate chiral multiplets which transform into one another under this symmetry. Instead the (approximate) symmetry appears to be realized by the appearance of (nearly) massless pseudoscalar Goldstone bosons -- the π, K and η mesons. This is the Nambu-Goldstone realization. I will assume that the reader is at least somewhat familiar with the subject. If not, the lectures by Coleman at the 1973 Erice school give an excellent introduction.[7] Perturbation theory only admits the ordinary, multiplet realization of chiral symmetry, so the appearance of a Nambu-Goldstone realization is another non-perturbative effect in QCD.

No one has shown how QCD generates zero mass Goldstone bosons in the chiral limit, just as no one has shown how it confines. Any reasonable, complete phenomenological model for low energy hadron dynamics should incorporate chiral symmetry. As originally formulated the bag model does not. Recently there has been a revival of interest[4,8] in constructing "hybrid" models in which independent Goldstone modes are introduced into the model by hand to implement the symmetry.[9] Though this will not answer the deep question: "what are the Goldstone modes and how do they arise in QCD?", nevertheless it is an important development which one may reasonably hope will at least explain why the quark model has worked so well when it ignores the coupling of baryons to pseudoscaler mesons entirely. I will spend a later lecture describing this work.

By now there is a large literature on the bag model. It would be hopeless to attempt a comprehensive introduction in a short set of lectures. Instead I will discuss a few subjects which are of current interest. In the following section I will discuss the derivation of bag equations, first following the original pragmatic approach, and then outlining alternative Lagrangian formulations which are more useful in constructing new models. In Section 3 to show what you can do with the model I review its application to hadrons made of light quarks, especially to those made of more than three quarks. Finally in Section 4 I outline a hybrid chiral bag model incorporating both quarks and gluons and fundamental pseudoscalar Goldstone bosons.

2. FORMULATIONS OF THE BAG MODEL OLD AND NEW

2.1. A Traditional Derivation

A bagged field theory is one restricted to a finite spatial domain by means of a pressure. The most transparent way to discover the equations of motion of the bag version of a field theory is to add a pressure to the original theory and require that the modified theory conserve energy and momentum. Adding a pressure means adding a term $-g^{\mu\nu}B$[10] to the energy momentum tensor $T^{\mu\nu}$. Since this contributes a constant positive energy density to T^{00} all finite energy solutions must be of finite size. Let V be the space time region (with boundary ∂V) where the fields are non-vanishing as shown in Figure 1.

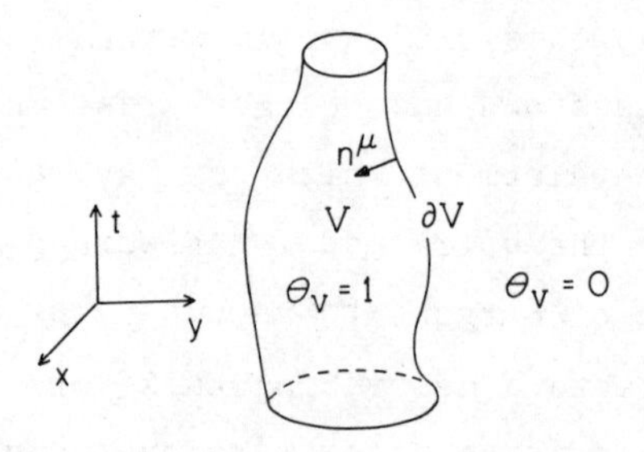

Figure 1. The space-time hypertube swept out by a 2-space dimensional bag.

Then the energy-momentum tensor for the bag model is

$$T^{\mu\nu}_{BAG} \equiv \left[T^{\mu\nu}_0 - g^{\mu\nu}B \right] \Theta_V , \tag{2.1}$$

where Θ_V is unity inside V and zero outside and $T^{\mu\nu}_0$ is the energy-momentum tensor of any field theory.

Consider, for example, a massless, non-interacting Dirac field with

$$T^{\mu\nu}_0 = \frac{i}{2}\bar{q}\gamma^\mu\partial^\nu q - \frac{i}{2}(\partial^\nu\bar{q})\gamma^\mu q + \frac{i}{2}(\bar{q}\gamma^\lambda\partial_\lambda q - (\partial_\lambda\bar{q})\gamma^\lambda q)g^{\mu\nu} . \tag{2.2}$$

The physically interesting case of QCD will be discussed below. Translation invariance requires

$$\partial_\mu T^{\mu\nu}_{BAG} = 0 \; . \tag{2.3}$$

From eq (2.3) one obtains two sorts of equations: equations of motion within V and boundary conditions on the surface ∂V. The equations of motion are the same as the underlying field theory because B is constant -- in this case merely the Dirac equation:

$$i\partial\!\!\!/ q(x) = 0 \qquad x \text{ in } V \; . \tag{2.4}$$

Boundary conditions are required when the derivative hits the shape function

$$\partial^\mu \Theta_V = n^\mu \delta_V \; , \tag{2.5}$$

where n^μ is the <u>interior</u>, covariant, unit normal to the boundary, ∂V, of the region V. In the case of the Dirac field q:

$$\frac{i}{2}\bar{q} n\!\!\!/ \partial^\nu q - \frac{i}{2}(\partial^\nu \bar{q}) n\!\!\!/ q - n^\nu B = 0 \qquad \text{on } \partial V \; . \tag{2.6}$$

Eq. (2.6) is not sufficient to uniquely define a theory. To see this suppose q obeys

$$i n\!\!\!/ q = \Gamma q \qquad \text{on } \partial V, \tag{2.7}$$

for some as yet unknown constant Dirac matrix Γ. Iterating this yields

$$\Gamma n\!\!\!/ \Gamma = n^2 n\!\!\!/ \; . \tag{2.8}$$

Since points on the bag boundary must move at less than the speed of light, there is a local rest frame where $n^\mu = (0,\hat{n})$ and therefore $n^2 = +1$. The most general solution to eq (2.8) is $\Gamma = e^{i\alpha\gamma_5/2}$ for arbitrary α. From eq (2.7) it follows immediately that

$$\bar{q} e^{i\alpha\gamma_5/2} q = 0 \qquad \text{on } \partial V, \tag{2.9a}$$

and

$$\bar{q} n\!\!\!/ q = 0 \qquad \text{on } \partial V \; . \tag{2.9b}$$

The last equation guarantees vector current conservation at the bag's surface and is a welcome result. If I now substitute eq (2.7) and (2.9a) into the original boundary condition I find

$$\left.\begin{aligned} i\not{n}q &= e^{i\alpha\gamma_5/2}q \\ n\cdot\partial\bar{q}e^{i\alpha\gamma_5/2}q &= 2B \end{aligned}\right\} \text{ on } \partial V \qquad \begin{aligned}(2.10a)\\(2.10b)\end{aligned}$$

for arbitrary α. It is easy to show that one value of α is as good as another; that is, if solutions to Dirac's equation (with mass zero) subject to eqs (2.10) exist for one choice of α we can directly construct solutions with the same energy, angular momentum, etc. and any other value of α: $q_{\alpha'} = e^{i\gamma_5(\alpha'-\alpha)/2}\, q_\alpha$. It appears that solutions with different values of α are chiral transforms of one another. The set of all solutions to eq (2.4) and (2.10) is chirally invariant; but it is also overcomplete. It is necessary to choose some value for α but by doing so one violates chiral symmetry. The situation is reminiscent of field theory models with a Nambu-Goldstone realization of chiral symmetry: there are many theories, each sufficient, all equivalent and related by chiral transformations. But where are the Goldstone bosons? For now I will choose $\alpha = 0$ and ignore the problem. Later, in Section 4, I will return to it, let α become a dynamical variable $\pi(x)$ and so-doing resurrect chiral symmetry. Until further notice the boundary conditions are

$$i\not{n}q = q \qquad \text{on } \partial V \qquad (2.11a)$$

$$\frac{1}{2}\, n\cdot\partial\bar{q}q = B \qquad \text{on } \partial V\,. \qquad (2.11b)$$

Before passing on to the more interesting case of QCD it is worth commenting on the character of solutions to equations (2.4) and (2.11). Consider a space time region generated by the translation in time of an arbitrary fixed three dimensional shape as shown schematically in Figure 2. It is well known that necessary and sufficient conditions for the solution to the Dirac equation inside such a region are the specification of non-singular initial value data, $q(\vec{x})$, at $t = t_1$, and a homogeneous boundary condition such as $i\hat{n}\cdot\vec{\gamma}q = q$,[11] on the spacelike boundaries, ∂V. This remains true for the more general case of an arbitrary prescribed spacelike surface spanning two timelike ends. Inspecting eqs (2.11) we see that the

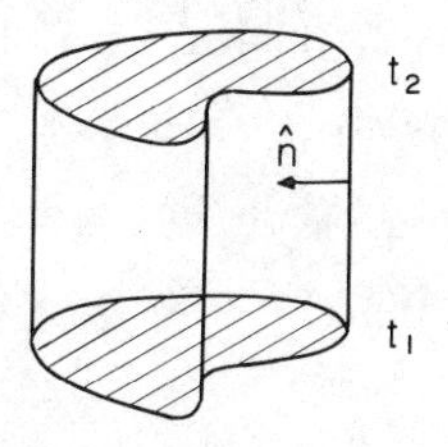

Figure 2. The space-time hypertube swept out by a 2-space dimensional rigid cavity.

bag equations appear overdetermined. The solution to this apparent problem is that eqs (2.11) together with (2.4) determine both the time evolution of the field and of the bounding surface ∂V. To see how this works in practice the reader is referred to the treatment of the one-dimensional bag in Ref. 1.

Everyone knows that rigidity is inconsistent with Lorentz covariance. The bag equations manage to specify an extended spacetime region in a covariant way by making the boundary of the region dynamical. The novel boundary condition, eq (2.11b), equates the field pressure locally on the surface of the hadron to the external pressure B. So the dynamics of the Dirac bag is indeed that of a relativistic quantum gas under constant pressure. The price of covariance is high, the boundary conditions make the system much more non-linear than ordinary quantum field theory, which is already bad enough. Furthermore a Lagrangian approach shows that the surface variables in the bag model are not independent canonical variables: they have no conjugate momenta. Instead they generate constraints. So far no one has succeeded in resolving the constraints and setting up a Hamiltonian formalism except in two dimensions. Several attempts have been made to reformulate the bag model to reduce its non-linearity. The Budapest group[12] introduce a surface tension which makes the boundary variables dynamical. This eliminates constraints but introduces new degrees of freedom associated with surface excitations. Attempts to obtain bag-like solutions from more conventional field theories[13] rely on (typically) a scalar field whose vacuum expectation value, $\langle\sigma\rangle = 0$ inside the bag and $\langle\sigma\rangle = \sigma_0$ outside, differentiates between bag and vacuum. Such models generally either end up looking like the original bag (if the transition from bag to vacuum is abrupt) or end up with additional dynamical degrees of freedom beyond quarks and gluons. Since the phenomological treatments of these alternatives are very similar to the original model I will stick to the M.I.T. formulation here.

Let me now turn to QCD. Boundary conditions and equations of motion can be derived in much the same fashion as outlined above. The results are the standard Yang-Mills equations inside the bag,

$$\begin{aligned} D_{\mu ab} F^{\mu\nu}_{b} &= g\bar{q}\gamma^{\nu}\frac{\lambda_b}{2}q \\ \left(i\partial\!\!\!/ + g\frac{\lambda_a}{2} A_{\mu a}\gamma^{\mu}\right) q &= mq \quad , \end{aligned} \tag{2.12}$$

where λ_a are the eight 3×3 matrices of Gell-Mann (normalized to $\mathrm{Tr}\lambda_a^2 = 2$), and a set of gauge invariant boundary conditions,

$$i n\!\!\!/ q = q \tag{2.13a}$$

$$n_{\mu} F^{\mu\nu}_{a} = 0 \tag{2.13b}$$

$$\frac{1}{2}\, n\cdot\partial\bar{q}q - \frac{1}{4}\sum_a F_{\mu\nu a}F^{\mu\nu}_{a} = B \quad , \tag{2.13c}$$

where $F_{\mu\nu a}$ is the usual Yang-Mills field strength tensor

$$F_{\mu\nu a} = \partial_{\mu}A_{\nu a} - \partial_{\nu}A_{\mu a} + g\varepsilon_{abc}A_{\mu b}A_{\nu c} \quad . \tag{2.14}$$

As before, the last of eqs (2.13) balances the field pressure against the external pressure, B. Notice that a bag may be "inflated" either by quark pressure, $\frac{1}{2}\, n\cdot\partial\bar{q}q$, or by gluon pressure $-\frac{1}{4}\sum_a F_{\mu\nu a}F^{\mu\nu}_{a} = \frac{1}{2}\sum_a (E_a^2 - B_a^2)$ or by both.

The bag confines color automatically. To see this calculate the color charge of any solution to QCD obeying eqs (2.12),

$$Q_a = \int_{\mathrm{BAG}} d^3x\, j^{o}_{a}(x) \quad , \tag{2.15}$$

where j^{oa} is the gauge-variant, but conserved charge density

$$j^{\mu}_{a} = \bar{q}\gamma^{\mu}\frac{\lambda_a}{2}q + f_{abc}F^{\nu\mu}_{b}A_{\nu c} \quad . \tag{2.16}$$

From eq (2.12) I have Gauss' Law, $\vec{\nabla}\cdot\vec{E}_a = gj^0_a$, which can be combined with eq (2.15) to obtain $Q_a = -\oint ds\hat{n}\cdot\vec{E}_a$. But eq (2.13b) with $\nu = 0$ implies $\vec{n}\cdot\vec{E}_a = 0$. That's it: $Q_a = 0$. Confinement is not a triumph of the bag model, it is put in by hand.

2.2. Lagrangian Formulation - 1

If one attempts to formulate a Dirac bag model by modifying the conventional Dirac Lagrangian:

$$\mathcal{L}_D \equiv \mathcal{L}_{0D} - B = \frac{i}{2}\bar{q}\gamma^{\mu}\partial_{\mu}q - \frac{i}{2}(\partial_{\mu}\bar{q})\gamma^{\mu}q - B \quad , \tag{2.17}$$

one immediately obtains $B = 0$. The root of this problem is the vanishing of $\mathcal{L}_{0D}$ by virtue of Dirac's equation. But $\mathcal{L}_{0D}$ is not unique in unbounded regions. It is determined only up to a total derivative, which does not affect the equations of motion. In bounded domains a total derivative becomes a surface term. Sometime ago Chodos and Thorn[14] developed a Lagrangian formulation of the bag employing surface terms and Lagrange multipliers. While this approach is not as physically intuitive as the traditional derivation it is less cumbersome and more easily generalizable to new systems.

Once again consider a massless Dirac field and postulate the action

$$A = \int_V d^4x\mathcal{L} = \int_V d^4x\left(\frac{i}{2}\bar{q}\gamma_{\mu}\partial^{\mu}q - \frac{i}{2}(\partial^{\mu}\bar{q})\gamma_{\mu}q - B\right) - \int_{\partial V} d^3s\bar{q}q\lambda \ , \tag{2.18}$$

where $\lambda \equiv \lambda(x)$ is a Lagrange multiplier field defined only on ∂V. V is the four-dimensional spacetime volume swept out by the bag and ∂V is its three-dimensional boundary, as before. Since no time derivatives of λ occur in A it will merely generate a constraint and eventually be eliminated in terms of the dynamical variables of the model. The surface term was chosen with malice aforethought to generate the constraint $\bar{q}q = 0$ in ∂V (see eq (2.9a) with $\alpha = 0$).

As usual to obtain equations of motion and boundary conditions one demands A to be stationary with respect to arbitrary variations in the fields, in the Lagrange multiplier and in the geometrical degrees of freedom. Variation of q (and $\bar{q}$) yields Dirac's equation in V and a boundary condition

$$\frac{i}{2}\not{n}q = \lambda q \qquad \text{on } \partial V \tag{2.19}$$

when $\overleftarrow{\partial}^{\lambda}\delta\bar{q}$ is integrated by parts (remember, n^{μ} is the interior normal). Variation of $\lambda(x)$ yields

$$\bar{q}q = 0 \qquad \text{on } \partial V \tag{2.20}$$

as expected. Finally, variation of the domain V yields

$$n\cdot\partial(\lambda\bar{q}q) + \mathcal{L}_{0D} - B = 0 \qquad \text{on } \partial V \ . \tag{2.21}$$

Eq (2.21) arises as follows. Variations in V tangent to ∂V are trivial and yield no information. Consider then a local normal variation of the surface at a point x^{μ}: $x^{\mu} \to x^{\mu}+\varepsilon n^{\mu}$ over an infinitesimal area $\Delta^3 s$. The change in the volume term of eq (2.18) is

$$\delta A_V = (\mathcal{L}_{0D} - B)\Delta^3 s\ \varepsilon \tag{2.22}$$

The change in the surface term is

$$\delta A_s = \lambda\bar{q}q\Big|_{x^{\mu}}^{x^{\mu}+\varepsilon n^{\mu}}\Delta^3 s = \varepsilon n\cdot\partial\lambda\bar{q}q\Delta^3 s \ . \tag{2.23}$$

Combining eqs (2.22) and (2.23) for arbitrary x^{μ} I obtain eq (2.21). By virtue of Dirac's equation $\mathcal{L}_{0D} = 0$. Squaring eq (2.19) one finds $\lambda^2 = 1/4$ or $\lambda = \pm 1/2$. This is part of the arbitrariness I've already discussed. In fact the full arbitrariness can be recovered by replacing $\lambda\bar{q}q$ by $\lambda\bar{q}e^{i\alpha\gamma_5}q$ for constant α in eq (2.18). Once again I propose to postpone discussing this further till Section 4. For now I choose $\lambda = 1/2$ and find that eqs (2.19) and (2.21) reduce to the familiar bag boundary conditions.

2.3. Lagrangian Formulation - 2

Last year Johnson[15] proposed another Lagrangian formulation of the bag on physical grounds. The idea is to replace the shape function Θ_V by a function of field variables, $\theta(f(\{\phi\}))$, where f is some function of quark and gluon fields (in QCD). When f is positive $\theta = 1$ and there is a contribution to the action, when f is negative θ is zero and there is no action. So $f(\{\phi\})$ plays the role of an order parameter distinguishing the "abnormal" hadronic phase from the "normal" vacuum phase of the QCD world. Apparently the bag (or bags) boundaries correspond to $f(\{\phi\}) = 0$. $f(\{\phi\})$ must be a guage invariant Lorentz scalar. Two candidates which come to mind are

$$f_1 = \bar{q}q , \tag{2.24a}$$

and

$$f_2 = -\frac{1}{4}\sum_a F_{\mu\nu a}F^{\mu\nu}_a - B . \tag{2.24b}$$

Either suffices to derive a set of confining boundary conditions. I will discuss only eq (2.24a); for a treatment of the other and of the special problem it raises see Johnson's original paper.

Consider then the following action for a massive Dirac field:

$$A = \int d^4x\,\theta(\bar{q}q)\{\ \frac{i}{2}\bar{q}\gamma^\mu\partial_\mu q - \frac{i}{2}(\partial_\mu\bar{q})\gamma^\mu q - m\bar{q}q - B\ \} . \tag{2.25}$$

Yang-Mills interactions can be incorporated simply by adding the appropriate terms in the curly brackets. Under variations in q and $\bar{q}$

$$A = \int d^4x \Big[\ \theta(\bar{q}q)\{\ \frac{i}{2}\delta\bar{q}\gamma^\mu\partial_\mu q - \frac{i}{2}(\partial_\mu\delta\bar{q})\gamma^\mu q - m\delta\bar{q}q\ \} + \delta(\bar{q}q)\delta\bar{q}q\,\mathcal{L} + \delta q\text{-variations}\ \Big] , \tag{2.26}$$

where $\mathcal{L}$ is the coefficient of $\theta(\bar{q}q)$ in eq (2.25). The second term in eq (2.26) must be integrated by parts using

$$\partial_\mu[\theta(\bar{q}q)\delta\bar{q}\gamma^\mu q] = \delta(\bar{q}q)\partial_\mu(\bar{q}q)\delta\bar{q}\gamma^\mu q + [\delta(\partial_\mu\bar{q})\gamma^\mu q + \delta\bar{q}\gamma^\mu\partial_\mu q]\theta(\bar{q}q) \tag{2.27}$$

For A to be stationary under variations in $\delta\bar{q}$ we find

$$\theta(\bar{q}q)\ (i\not{\partial} - m)q = 0 \tag{2.28}$$

and

$$\delta(\bar{q}q)\ (\mathcal{L}q + \frac{i}{2}\partial_\mu(\bar{q}q)\gamma^\mu q) = 0 \quad . \tag{2.29}$$

By virtue of eq (2.28) $\mathcal{L} = -B$, so eq (2.29) reads

$$\frac{i}{2}(\gamma_\mu \partial^\mu \bar{q}q)q = Bq \qquad \text{when } \bar{q}q = 0 \ . \tag{2.30}$$

This boundary condition can be recast in a more familiar and convenient form. One possibility, $q = 0$ on the boundary, is too restrictive a condition on Dirac's equation (see footnote 11). Instead, since $\bar{q}q = 0$ on the boundary we observe that $\partial_\mu \bar{q}q$ is normal

$$\partial_\mu \bar{q}q = an_\mu \ , \tag{2.31}$$

where a is real. Substituting in eq (2.30)

$$\frac{i}{2}a\not{n}q = Bq \ . \tag{2.32}$$

Iterating this equation we find

$$+\frac{1}{4}a^2n^2 = B^2 \ . \tag{2.33}$$

Gathering up the pieces we find the following conditions

$$\left.\begin{aligned} i\not{n}q &= q \\ n\cdot\partial\bar{q}q &= 2B \\ \text{and} \quad n^2 &= +1 \end{aligned}\right\} \quad \text{when } \bar{q}q = 0 \ . \tag{2.34}$$

These are just the Dirac boundary conditions we derived earlier -- in a more pedestrian way.

3. LIGHT QUARK HADRONS

The known quarks divide clearly into two classes: "heavy quarks" (c and b) whose Compton wavelengths are small compared to the scale of confinement ($B^{-1/4}$) and "light quarks" (u, d and s) whose Compton wavelengths are of the same order as $B^{-1/4}$ or perhaps much greater. The dynamics of heavy quarks confined to a bag may be relatively simple. They may be treated in a Born-Oppenheimer approximation as static sources of colored fields which "inflate" the bag. The energy can be estimated as a function of quark separation and used as the potential in the heavy quark Schroedinger equation. Such a scheme was suggested by a group in Budapest[12] and has been used for the $c\bar{c}$ and $b\bar{b}$ systems.[12,15] For this approach to make sense it is essential first that the quarks are localizable within hadrons and second that they move non-relativistically. In the ground states of charmonium and bottomonium these requirements seem to be adequately met.

The same approach does not make sense for light quarks. They are not localizable within hadrons and there is no sensible limit in which their relativistic kinetic energies can be ignored. [Note that the typical momentum of a Schroedinger particle confined to a sphere of radius 1 fm. is $p \sim \pi/R \sim 600$ MeV.] We at M.I.T. have abandoned any attempt to treat light quarks as non-relativistic particles and instead have gone to the opposite extreme and treated them as radiation, i.e. as the quanta of nearly massless fields obeying confining boundary conditions.

3.1. A Cavity Approximation for Light Quarks

The bag model of light quark hadrons[16] has been reviewed nicely by Johnson in his Zakopane lectures in 1975.[17] I will only outline it briefly, and then use the model to study the spin dependent forces of QCD and the spectroscopy of multiquark hadrons. The model is only intended to describe the ground state of any color singlet collection of quarks. Attempts to extend it to excitations[18] have met with limited success. The model is developed as follows:

i) The bag boundary is assumed to be static and spherical with a radius R.

ii) The QCD equations of motion are treated perturbatively in $\alpha_c \equiv g^2/4\pi$ subject to the "linear" boundary conditions eqs (2.13a and b).

iii) At zeroth order in α_c only those eigenstates which generate a spherically symmetric pressure at the bag's boundary are acceptable. All others correspond to vibrating bags and are outside the scope of the model.

iv) For states satisfying iii) the "quadratic" boundary condition, eq (2.13c), is equivalent to minimizing the energy with respect to R, thus fixing $R = R_0(E)$.

Only quark eigenstates with total angular momentum $j = 1/2$ are consistent with iii). There are four of these: particle and antiparticle with positive parity labeled $S_{1/2}$ because they reduce to $\ell = 0$ in the non-relativistic limit, and the same with negative parity labeled $p_{1/2}$ because they reduce to $\ell = 1$ in the non-relativistic limit. The latter are not orthogonal to a translation of the ground state and must be taken together with certain $j = 3/2$ modes to construct physical excitations. The same is true of radially excited $S_{1/2}$ modes. In the end only the lowest $S_{1/2}$ mode remains. Its wavefunction is

$$q_0(\vec{r},t) = \frac{N}{\sqrt{4\pi}} \begin{pmatrix} \sqrt{\frac{\omega + m}{\omega}}\, i j_0(kr) \\ -\sqrt{\frac{\omega - m}{\omega}}\, j_1(kr)\vec{\sigma}\cdot\hat{r} \end{pmatrix} U\, e^{-i\omega t} , \qquad (3.1)$$

where $\omega = \sqrt{k^2+m^2}$, m is the quark mass, j_0 and j_1 are spherical Bessel functions and U is a two component Pauli spinor. N is a normalization constant. The boundary condition $-i\vec{\gamma}\cdot\hat{r}q = q$ at $r = R$ yields the equivalue equation

$$\tan x = \frac{x}{1 - mR - (x^2 + m^2R^2)^{1/2}} , \qquad (3.2)$$

where $x = kR$. The resulting wave number is shown, versus mR in Figure 3. Notice that for a given size cavity the massless Dirac equation yields a lower momentum ($x_0 = 2.04$) than the Schroedinger equation ($x_\infty = \pi$).

In the model a hadron's energy is a function of R and is a sum of four terms:

$$E(R) = E_Q + E_V + E_0 + \Delta E . \qquad (3.3)$$

The first term is the quark kinetic energy:

$$E_Q = \frac{1}{R} \sum_{i=1}^{N} [x(m_i R)^2 + m_i^2 R^2]^{1/2} , \qquad (3.4)$$

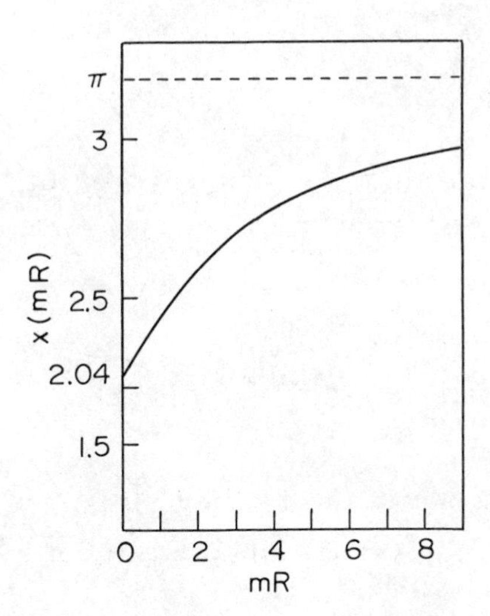

Figure 3. Eigenfrequency x(mR) of the lowest quark mode with mass m in a spherical cavity of radius R.

where the sum is over all quarks and antiquarks in the state. The next two terms, E_V and E_0, are consequences of doing field theory in a finite domain. The first,

$$E_V = \frac{4\pi}{3} BR^3 \tag{3.5}$$

is the energy associated with the confirming pressure B. The second,

$$E_0 = - Z_0/R$$

is a phenomenological estimate of quantum effects associated with fields confined to limited regions of space. To see how such a term might arise[19] consider the rudimentary model defined by E_Q and E_V above for n-massless quarks

$$E(R) = \frac{nx_0}{R} + BV \quad . \tag{3.6}$$

R_0 is fixed by setting $\left.\frac{dE}{dR}\right|_{R_0} = 0$. Since the bag is localized at

$\vec{x} = 0$ a momentum uncertainty has been introduced. Consequently $E(R_0)$ cannot directly be identified with the hadron's mass. To estimate the difference between $E(R_0)$ and M use $dE/dR|_{R_0} = 0$ to eliminate V from eq (3.6):

$$E(R_0) = \frac{4}{3}\frac{nx_0}{R_0} \quad . \tag{3.7}$$

An estimate of the mean square momentum of the bag is given by

$$\langle\vec{P}^2\rangle = \langle(\sum_i \vec{p}_i)^2\rangle = n\langle p^2\rangle \quad , \tag{3.8}$$

where $\vec{p}$, the quark momentum, is of order $\langle|\vec{p}|\rangle \sim x_0/R_0$. Then $\langle\vec{P}^2\rangle \approx nx_0^2/R_0^2$. Since $\langle P^2\rangle \ll E^2$ (at least for $n \geq 3$) we can expand

$$M \simeq E - \frac{\langle P^2\rangle}{2E} \tag{3.9}$$

and find

$$M \simeq E(R_0) - \frac{3x_0}{8R_0} \tag{3.10}$$

after substituting from eqs (3.7) and (3.8). This is only one of many quantum modifications. To allow for them we add to the energy a term of the form $-Z_0/R$. The same conclusion was reached by Rebbi[18] who studied fluctuating bags in a much more sophisticated manner. Recently Donoghue and Johnson[20] have used similar ideas to reconsider bag calculations of pseudoscalar meson masses.

The final term in eq (3.3) is the energy associated with the gluon interactions of the quarks. The long-range, strong, confining forces are provided by the bag term BV. It is not necessary for the quark-gluon coupling to become strong: Note that the bag boundaries cut off the infrared. QCD interactions in a cavity can in principle be calculated from a graphical expansion similar to ordinary Feynman perturbation theory. Though it is not difficult to write down the Feynman rules inside a cavity[21] it has not been possible so far to explicitly evaluate graphs except in lowest order. The lowest order graphs are shown in Figure 4.

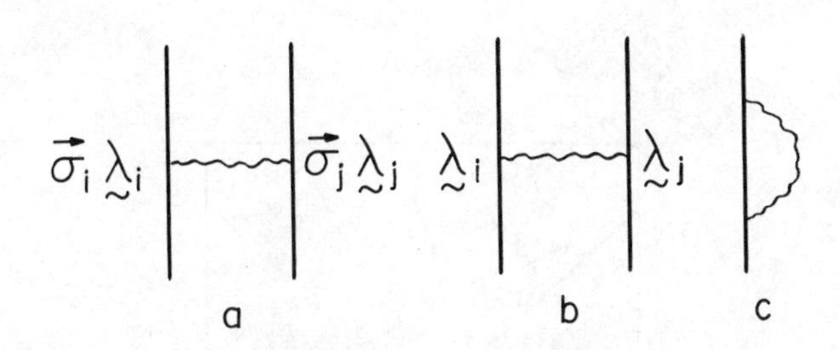

Figure 4. Lowest-order gluon interaction graphs. a) Magnetic Born graph; b) electric Born graph; c) self-interaction graph.

Only the magnetic contribution to the Born graph, Figure 4a, is important

$$\Delta E_M^{(1)} = -\frac{1}{2}\sum_a \sum_{i\neq j} \int d^3x \vec{B}_i^a(x)\cdot\vec{B}_j^a(x) \quad . \qquad (3.11)$$

Other contributions: the electric interaction energy and the self energy graphs are either small or included in the definition of the renormalized parameters of the model.[16] In eq (3.11) $\vec{B}_i^a$ is the color magnetic field generated by the current of the ith quark. Although eq (3.11) appears classical it is the true result of first order quantum-perturbation theory. The minus sign in eq (3.11) is important. Originally the color magnetic contribution to $\Delta E^{(1)}$ was

$$\Delta E_M^{(1)} = \sum_{i\neq j}\sum_a \left\{\frac{1}{2}\int d^3x \vec{B}_i^a(x)\cdot\vec{B}_j^a(x) - \int d^3x \vec{j}_i^a(x)\cdot\vec{A}_j^a(x)\right\} \quad . \qquad (3.12)$$

However, as in ordinary magnetostatics the second term can be integrated by parts using $\vec{\nabla}\times\vec{B}^a = \vec{j}^a$, $\vec{\nabla}\times\vec{A}^a = \vec{B}^a$ and $\hat{n}\times\vec{B}^a = 0$ to obtain eq (3.11).

Given the quark wavefunction of eq (3.1) $\Delta E_M^{(1)}$ can be evaluated explicitly:

$$\Delta E_M^{(1)} = -\frac{\alpha_c}{R}\sum_{i>j} \vec{\sigma}_i\cdot\vec{\sigma}_j \underset{\sim}{\lambda}_i\cdot\underset{\sim}{\lambda}_j \, h(m_iR, m_jR) \quad . \qquad (3.13)$$

λ_i^a and σ_i^k are the color and spin[22] matrices of the ith quark. [Eight component vectors in color space are denoted as $\underset{\sim}{v}$.] $h(\xi_i,\xi_j)$ results from an integral over cavity wavefunctions. It is graphed in Figure 5.

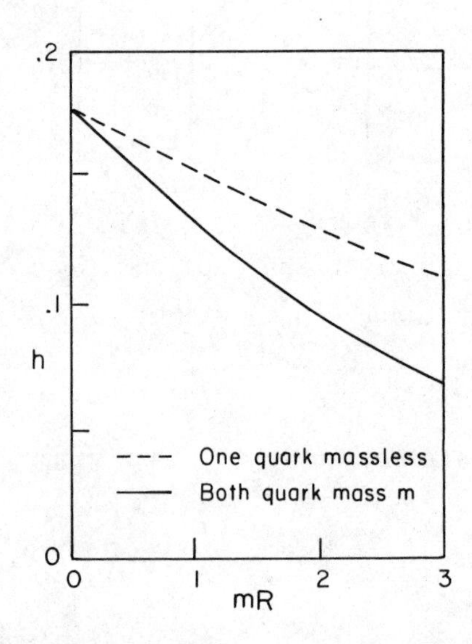

Figure 5. Magnetic gluon exchange energy of two quarks as a function of mR. The solid line gives the interaction energy between equal-mass quarks, the dashed line gives the interaction energy between a massless quark and a quark of mass m.

For large values of its arguments $h(\xi_i,\xi_j)$ reduces to a product of quark intrinsic magnetic moments and therefore falls like $1/\xi_i\xi_j$. Notice that as $m_i \to 0$ the magnetic moment of a confined quark does not diverge; instead it is scaled by R. The recipe for calculating the masses of light hadrons is then: First, construct wavefunctions properly antisymmetrised in color, flavor and spin. Second, diagonalize the energy in this basis. Third, minimize the energy eigenvalues with respect to R for each state. In practice R varies little within a given sector (fixed number of quarks and antiquarks). In the following I will treat R as a constant.

3.2. Colorspin

For the remainder of this section I will focus entirely on the color and spin dependent interaction of eq (3.13). Without this term the nucleon octet and decuplet are degenerate except for the splittings induced by giving the s-quark a larger mass than the u and d quarks: The N and Δ, the Σ, Λ and Σ* and the Ξ and Ξ* are all degenerate. Remarkably the QCD color-magnetic interaction is just right to remove these degeneracies. This is highly non-trivial and tests the non-Abelian nature of the color symmetry. To my mind it is one of the more persuasive and least appreciated pieces of evidence for color dependent quark-quark forces. After showing that these forces correctly reproduce the observed N-Δ mass splitting I will use them to outline the spectroscopy of hadrons made of more than three quarks. Since de Rujula has already discussed the color magnetic splittings of baryons in his lectures at this school [the original work may be found in Refs 23 and 16], I will concentrate on the technology necessary to discuss multiquark hadrons.

Consider the SU(3)-flavor limit. SU(3) violating effects in the baryon spectrum may be included as de Rujula has done. Multiquark spectroscopy is at too primitive a stage to worry about such subleties. In this limit the color magnetic interaction reduces to

$$\Delta E_M^{(1)} = -\frac{\alpha_c h(0,0)}{R} \sum_{i>j} \underset{\sim}{\lambda}_i \cdot \underset{\sim}{\lambda}_j \vec{\sigma}_i \cdot \vec{\sigma}_j \quad , \tag{3.14}$$

The quark operators which appear here are naturally embedded in a new SU(6) of color × spin I will call "colorspin." The 35-generators of colorspin are

$$\{\alpha\} = \left\{ \begin{array}{c} \sqrt{\frac{2}{3}}\,\sigma^k \\ \lambda^a \\ \sigma^k \lambda^a \end{array} \right\} \text{ with } \begin{array}{l} k = 1,2,3 \\ \\ a = 1,2\ldots 8 \, , \end{array} \tag{3.15}$$

normalized so that $\mathrm{Tr}\,\alpha_n^2 = 4$ for $n = 1,2\ldots35$ (since $\mathrm{Tr}(\lambda^a)^2 = \mathrm{Tr}(\sigma^k)^2 = 2$). $\Delta E_M^{(1)}$ includes the 24 generators of the form $\sigma^k\lambda^a$. It can be expressed entirely in terms of Casimir operators by adding and subtracting the missing 11 generators:

$$\sum_{i>j} \vec{\sigma}_i \cdot \vec{\sigma}_j \underset{\sim}{\lambda}_i \cdot \underset{\sim}{\lambda}_j = \sum_{i>j} \vec{\vec{\alpha}}_i \cdot \vec{\vec{\alpha}}_j - \frac{2}{3} \sum_{i>j} \vec{\sigma}_i \cdot \vec{\sigma}_j - \sum_{i>j} \underset{\sim}{\lambda}_i \cdot \underset{\sim}{\lambda}_j \quad , \tag{3.16}$$

where the double arrow denotes a (35-component) vector in colorspin space. For a hadron containing quarks or antiquarks but not both the quadratic Casimir operators of spin, color and colorspin are defined by: (If both quarks and antiquarks are present the result is a bit more complicated -- see below)

$$4S(S+1) \equiv (\sum_i \vec{\sigma}_i)^2 = 3N + 2 \sum_{i>j} \vec{\sigma}_i \cdot \vec{\sigma}_j \quad , \tag{3.17a}$$

$$C_3 \equiv (\sum_i \underset{\sim}{\lambda}_i)^2 = 8N + 2 \sum_{i>j} \underset{\sim}{\lambda}_i \cdot \underset{\sim}{\lambda}_j \quad , \tag{3.17b}$$

$$C_6 \equiv (\sum_i \vec{\vec{\alpha}}_i)^2 = \frac{70}{3} N + 2 \sum_{i>j} \vec{\vec{\alpha}}_i \cdot \vec{\vec{\alpha}}_j \quad . \tag{3.17c}$$

where N is the number of quarks or antiquarks. With this $\Delta E_M^{(1)}$ reduces to

$$\Delta E_M^{(1)} = \frac{\alpha_c h(0,0)}{R} \left(8N - \frac{1}{2} C_6 + \frac{4}{3} S(S+1)\right) \quad , \tag{3.18}$$

where I have used $C_3 = 0$ for a color singlet. In a given sector (fixed N) the lightest hadron will be the one with the largest colorspin Casimir. Spin plays little role because in general $C_6 \gg \frac{8}{3} S(S+1)$. The quantitative reason for this is that C_6 is the sum of squares of thirty-five normed generators whereas $\frac{8}{3} S(S+1)$ is the sum of only three identically normed generators.

So far all this appears rather formal. Physics enters via Fermi statistics. Since I am considering an assortment of quarks all in the same spatial state they must be antisymmetrised in color, spin, and flavor. Large Casimirs are characteristic of highly symmetric representations (think, for example, of SU(2): for any number of particles the state of highest spin is totally symmetric). Large C_6 implies highly symmetric colorspin configurations. This, via Fermi statistics, implies highly antisymmetric flavor configurations. For all cases I know of (in the limit of flavor symmetry) the lightest state in any sector (fixed N) is the one most antisymmetric in flavor, i.e. with the least peculiar flavor quantum numbers. As an example consider Q^3 -- the S-wave baryons. There are three possible SU(6) representations listed in Table 1:

TABLE 1

COLORSPIN				FLAVOR	
REPRESENTATION	TABLEAU	C_6	SU(3)×SU(2) CONTENT	TABLEAU	REPRESENTATION
[56]	□□□	90	$\{8\}^{1/2}, \{10\}^{3/2}$	□ □ □	{1}
[70]	□□ □	66	$\{8\}^{1/2}, \{8\}^{3/2}, \{10\}^{1/2}, \{1\}^{1/2}$	□□ □	{8}
[20]	□ □ □	42	$\{8\}^{1/2}, \{1\}^{3/2}$	□□□	{10}

Each is associated with a unique flavor representation by Fermi statistics. The corresponding flavor representation is shown at the right hand side of the table. The colorspin [56] contains no color singlets and consequently no physical hadrons. The lightest Q^3 baryon therefore is the spin 1/2 state found in the colorspin [70]. It is a flavor octet -- namely the nucleon. The other Q^3 baryon is the spin 3/2 state in the colorspin [20]. It is the familiar flavor decuplet. The Δ-N splitting,

$$M_\Delta - M_N = \frac{24\alpha_c}{R} h(0,0) \quad , \tag{3.19}$$

is positive. As de Rujula has emphasized this is a consequence of the non-Abelian nature of the color symmetry. Were color an additive charge like electromagnetic charge the nucleon would be heavier than the delta. For a more detailed discussion of the role of color magnetism the reader should consult the original work by the Harvard[23] or M.I.T.[16] bag collectives.

Colorspin is essential to an understanding of "multiquark

hadrons," i.e. of hadrons made of more than three quarks and antiquarks combined into total color singlets. Such states have long been expected in quark models.[24] It turns out that color magnetic interactions are such that multiquark hadrons with peculiar quantum numbers ("exotics") are made heavier, while multiquark hadrons with ordinary quantum numbers ("cryptoexotics") also obtainable from the appropriate coupling of Q^3 or $Q\bar{Q}$ are made lighter. This provides a partial explanation of our failure to find striking multiquark states in nature: the ones with clean signatures, exotics like charge-two mesons, are heavy; the light ones have conventional quantum numbers and can be misidentified as ordinary Q^3 or $Q\bar{Q}$ states. This is not the entire story. The presence of open channels for "fall apart" decay complicates the experimental signature of multiquark hadrons. This is the subject of Prof. Low's lectures at this school.[2] I won't try to summarize that work here but following him will call my multiquark cavity eigenstates "primitives." The curious reader is referred to his lectures or to our original paper.[25]

As a first example consider dibaryons,[26] i.e. Q^6 with all quarks in the S-wave of a spherical cavity. The multiplets of this sector are listed in Table 2 in a notation similar to Table 1.

TABLE 2

COLORSPIN				FLAVOR	
REPRESENTATION	TABLEAU	C_6	SPINS OF COLOR SINGLETS	TABLEAU	REPRESENTATION
[490]	□□□ □□□	144	0	□□ □□ □□	{1}
[896]	□□□ □□ □	120	1,2	□□□ □□ □	{8}
[280]	□□□ □ □ □	96	1	□□□□ □ □	{10}
[175]	□□ □□ □□	96	1	□□□ □□□	$\{\overline{10}\}$

TABLE 2 (cont'd.)

	COLORSPIN			FLAVOR	
REP.	TABLEAU	C_6	SPINS	TABLEAU	REP.
[189]		80	0,2		{27}
[35]		48	1		{35}
[1]		0	0		{28}

Masses increase as one descends the table. Note that the lighter primitives are in small, familiar SU(3)-flavor representations. The lightest of all is a flavor singlet. Since Y = S + B = 0 for a singlet this dibaryon has strangeness minus two. It is a dihyperon. Including the effects of SU(3) violation the mass of this dihyperon primitive comes out about 2150 MeV. At this mass it would be too light to decay into $\Lambda\Lambda$ and must decay weakly. The possibility of a stable dihyperon in this mass range is not excluded experimentally. The only serious search (carried out last year at Brookhaven)[27] studied the missing mass distribution in $PP \rightarrow K^+K^+X$. Data from this experiment are shown in Figure 6. There should be a threshold in M_X at $2M_\Lambda$. In the middle figure where the experimental acceptance brackets $2M_\Lambda$ one can see that there is no clear sign of a threshold. Until one sees a strong signal for continuum $\Lambda\Lambda$ production one cannot put a meaningful limit on a bound dihyperon. We must await the results of a more sensitive experiment presently being launched at Brookhaven.

The next lightest Q^6 multiplet is a flavor octet. The lightest

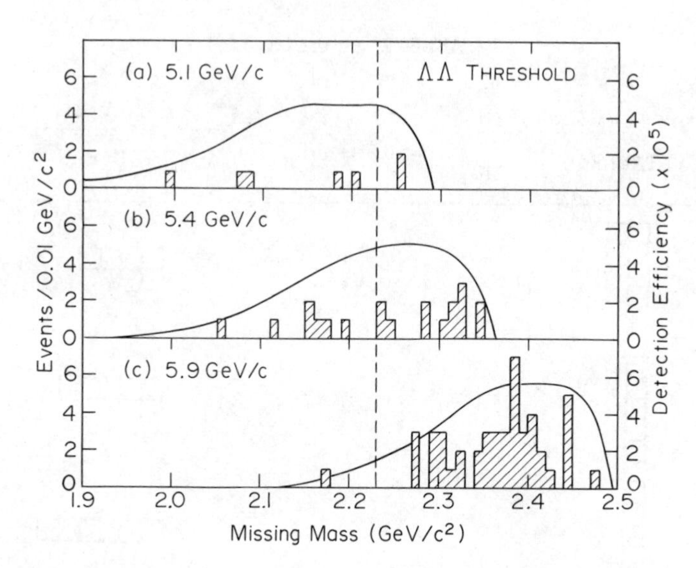

Figure 6. Missing mass recoiling against K^+K^+ system in $pp\to K^+K^+X$. The slashed boxes are the actual observed events; the solid line, the detection efficiency as a function of mass.

primitive in the octet is the Y = 1 isodoublet with S = -1. The primitive's mass is predicted to be about 2200 MeV. Although it is not widely appreciated among particle physicists, there is strong evidence for a resonance with precisely these quantum numbers at 2128 MeV in the ΛN spectrum. Typical data[28] are shown in Figure 7. Unfortunately it is not possible to unambiguously associate this state with the {8} Q^6 primitive. The problem is that it may be a strangeness -1 analog of the deuteron. In flavor SU(3) the deuteron is a member of a $\{\overline{10}\}$.[29] In the SU(3) limit the nuclear forces which just bind the deuteron would also just bind a Y = 1, I = 1/2 dibaryon in the same multiplet. This $Q^3 - Q^3$ state may also be what is seen at 2128 MeV. In any event the relation between the Q^6 primitives calculated in a spherical cavity model and physical resonances and/or bound states is far from clear. For a discussion of the subtleties once again I refer to Low's lectures at this school.[2]

As a final application of color spin dynamics I will briefly

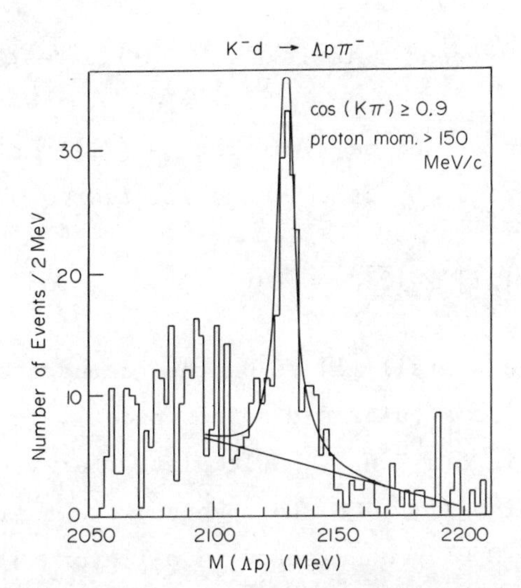

Figure 7. Results of the fit of an S-wave Breit-Wigner resonance plus polynomial background to the effective (Λp) mass distribution in the reaction $K^-d \to \Lambda p \pi^-$. Selection criteria: $\cos(K^-,\pi^-) \geq 0.9$ and proton momentum >150 MeV/c.

summarize the $Q^2\bar{Q}^2$ spectrum.[30] Here colorspin really works magic. The color magnetic interaction can again be written in terms of Casimir operators, this time of the quarks, antiquarks and total system:

$$\Delta E_M^{(1)} = \frac{\alpha_c h(0,0)}{R}\{ 8N + \tfrac{1}{2}C_6(\mathrm{TOT}) - C_6(Q) - C_6(\bar{Q}) - \tfrac{4}{3}S_{\mathrm{TOT}}(S_{\mathrm{TOT}}+1) + \tfrac{8}{3}S_Q(S_Q+1) + \tfrac{8}{3}S_{\bar{Q}}(S_{\bar{Q}}+1) + C_3(Q) + C_3(\bar{Q}) \} \quad . \tag{3.20}$$

Once again the colorspin Casimirs dominate. From this I obtain spectroscopic rules for the lightest multiplet:

1. Put quarks and antiquarks in the most symmetric colorspin SU(6) representation.
2. Couple the entire system to the smallest possible colorspin.

Fermi statistics translates the first rule into

1'. Put quarks and antiquarks into the most antisymmetric flavor SU(3) representation.

A corollary of the second rule is that the lightest multiplets have low spin.

Two quarks can either be symmetric (a [21]) or antisymmetric (a [15]) in colorspin. By rule 1, the lightest $Q^2\bar{Q}^2$ primitive is in the product:

$$[21] \times [\overline{21}] = [1] + [35] + [405] \quad . \tag{3.21}$$

Rule 2 requires the overall [1] to be the ground state. Since an SU(6) singlet is also a color and spin singlet we conclude that the $Q^2\bar{Q}^2$ ground state is a $J^{PC} = 0^{++}$ multiplet. Fermi statistics requires a colorspin [21] (□□) to combine with a flavor {$\bar{3}$} (⊟) (Rule 1´). So the $Q^2\bar{Q}^2$ ground state is a flavor $\{\bar{3}\} \times \{3\} = \{1\} + \{8\}$ -- a nonet! So the lightest $Q^2\bar{Q}^2$ multiplet is a 0^{++} nonet. Hardly exotic. The quark content of such a multiplet is peculiar

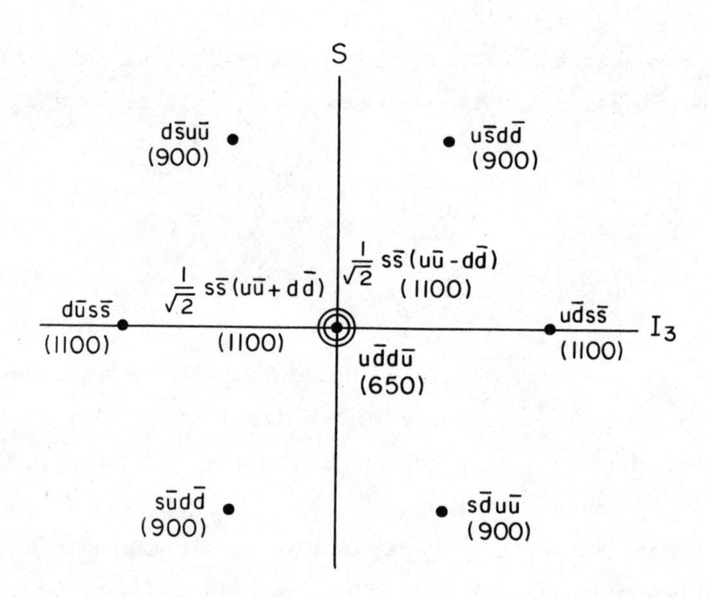

Figure 8. The lightest $Q^2\bar{Q}^2$ nonet. The quark content and predicted mass (in MeV) of the primitives are shown.

(see Fig 8) and the bag model predicts it to be very light (see Fig 8). Since they do not have exotic quantum numbers these objects could be mistakenly classified as $Q\bar{Q}$-states. Their true quark content would only be visible through anomalies in their decay couplings and masses. Low and I have shown[25] that if the bag "states" are interpreted as P-matrix poles then the predicted spectrum is in good agreement with data at least up to about 1100 MeV (above which there is insufficient data). Of course the model also predicts exotic $Q^2\bar{Q}^2$ primitives, but at higher masses. These too are in good agreement with measured phase shifts.[25] More details of the spectroscopy can be found in Refs 26 and 30 and of the P-matrix formalism necessary to interpret the spectrum in Refs 2 and 25.

4. HYBRID CHIRAL BAGS

4.1. Motivation

The Lagrangian of QCD is invariant under the chiral transformation

$$q' = e^{i\kappa\gamma_5/2}q \tag{4.1}$$

in the limit of zero quark mass. This implies the existence of a conserved axial current and time independent charge Q_5 in this limit. According to the famous Goldstone alternative,[5] either Q_5 annihilates the vacuum and hadrons come in chiral multiplets, or Q_5 does not annihilate the vacuum and there exist massless pseudoscalar bosons, one for each flavor for which $Q_5|0> \neq 0$. The latter is the Nambu-Goldstone realization of chiral symmetry.[5,6] It appears to be the one chosen by nature. Not only are the eight pseudoscalar mesons (π,κ and η) light, but also many chiral limit theorems such as the Goldberger-Treiman relation seem to be well satisfied. For a review see Ref. 31.

As I mentioned in the introduction, QCD behaves like a two phase system. In one phase, inside hadrons, the vacuum is simple and perturbative. I will call this the "free vacuum," $|\Omega\rangle$. Q_5 annihilates the free vacuum, so no Goldstone bosons are expected inside hadrons. The other phase, the "true vacuum" $|0\rangle$, is complex and dominated by as yet unknown non-perturbative field configurations which expel color electric flux. It is here in the true vacuum that $Q_5|0\rangle \neq 0$ and Goldstone bosons live. This point of view has been particularly advocated by Callan, Dashen and Gross[4] though it was first suggested in a somewhat different form by Chodos and Thorn[14] and Inoue and Maskawa.[9] [Similar ideas have been pursued in a rather different direction by Brown and his collaborators.[8]] It suggests a phenomenological model in which quarks and gluons are confined to the interior of bags and pseudoscalar Goldstone bosons live outside. I will refer to such models in general as "hybrid chiral bags" or HCB for short. It leaves open the deep question of the relation of the Goldstone bosons to the original quark-gluon degrees of freedom of QCD. For example it is not clear whether the Goldstone boson is to be identified with the lightest quark model ($Q\bar{Q}$) state with the same quantum numbers or whether one should expect two pions, two kaons, etc. Personally I believe this two phase picture makes perfectly good sense with the quark model pion identified with the Goldstone pion. In any case one is led to a hybrid model with quarks and gluons inside bags and Goldstone bosons outside.

The same model can be motivated more phenomenologically. Suppose that in the chiral limit QCD manages to produce zero mass 0^- meson bags, $Q\bar{Q}$ bound states perhaps in the fashion of the Nambu-Jona Lasinio model.[6,32] Now consider a hadronic bag of large but finite radius (R_H): a baryon would do, a quark star would do better. Inside quark and gluon degrees of freedom are complete. There are no zero energy excitations inside because the system has finite size. The zero mass Goldstone boson bags will cluster about the big bag in a way dictated by axial current conservation. Near the chiral limit the Compton wavelength (λ_c) of the Goldstone boson is much larger than its geometrical size (R_h) so it may be a good approximation to treat it as a local field. Once again we arrive at a hybrid model with quarks and gluons inside and Goldstone modes outside. From this point of view it is clear why Goldstone modes are special: other mesons are not pointlike: $\lambda_c \lesssim R_h$ for the ρ, ω, K* and heavier mesons; their couplings are not determined by an approximate symmetry and they are not long range compared to the size of hadronic sources ($\lambda_c << R_H$). This last property provides a justification for ignoring them: their effects outside of bags are damped by Yukawa factors, e^{-R_H/λ_c}, which are very small (∿.02 for the ρ-meson). It is also clear that hybrid chiral models will be useful only for very low energy phenomena. Probes with range small compared to the source (R_H) or pion (R_h) radius will be sensitive to details of the pion and source structure which the model ignores. In the end the best one can hope for from hybrid chiral models is a marrying of the effective Lagrangian approach to low energy chiral dynamics with the static quark model picture of baryons. This is, in fact, what comes out.

The original formulation of the bag model ignores the Goldstone bosons and breaks chiral symmetry in a seemingly arbitrary way. One wonders in what sense the properties of that model are approximations to a real world with approximate chiral symmetry and nearly massless pions. The hybrid model enables us to answer this question (among others). It turns out that the old bag model of baryons will emerge from a static hybrid chiral model as the first term in an expansion in a small parameter

$$\varepsilon \equiv \frac{g_A}{8\pi f_\pi^2 R_H^2}$$

where g_A is the nucleon's axial vector charge ($g_A \simeq 1.25$). The existence of such a parameter was first pointed out by Chodos and Thorn.[14] With $R_H \sim 1 - 1.25$ fm and $f_\pi = .093$ GeV, $\varepsilon \simeq \frac{1}{5} - \frac{1}{8}$. As $R_H \to \infty$ most pion effects vanish. Notable exceptions to this are quantities, like the pion-nucleon coupling constant of the Yukawa theory, which are pion pole dominated. As expected these receive contributions from the long range pion field even as $R_H \to \infty$. With these exceptions one can understand the success of the quark model in terms of the smallness of the parameter ε.

4.2. Formalism

It is important to realize at what point chiral invariance was lost in formulating the original bag model. As discussed in Section 2, the original model admits a one parameter family of fermion boundary conditions (for simplicity I consider only one flavor for the moment):

$$\left.\begin{aligned} i\not{n}q &= e^{i\alpha\gamma_5/2}q \qquad (2.10a) \\ n\cdot\partial\bar{q}e^{i\alpha\gamma_5/2}q &= 2B \qquad (2.10b) \end{aligned}\right\} \text{ on } \partial V \ .$$

Each choice of α defines a model, each with the same spectrum. Under a chiral transformation one model (choice of α) transforms into another. To calculate one must choose some α and thereby break chiral symmetry. The situation is reminiscent of a Goldstone realization with a family of equivalent vacua except, of course, for the absence of Goldstone bosons. Notice that there are real practical problems associated with the choice of α. Matrix elements of chirally odd operators such as

$$\langle\bar{q}q\rangle_B \equiv \int_{\text{BAG}} \bar{q}(x)q(x)d^3x \qquad (4.2)$$

are functions of α and therefore undetermined.

Chiral symmetry can be regained by replacing α by a dynamical variable,

$$\alpha \to 2\theta(x)/f \ , \qquad (4.3)$$

where f is a constant with dimensions of mass. If under an infinitesimal transform

$$\delta q = \frac{i\kappa\gamma_5}{2} q \qquad (4.4a)$$

and

$$\delta\theta = -\kappa f , \tag{4.4b}$$

then the boundary conditions

$$\left.\begin{aligned} i\not{n}q &= e^{i\theta\gamma_5/f}q \qquad (4.5a) \\ n\cdot\partial\bar{q}e^{i\theta\gamma_5/f}q &= 2B \qquad (4.5b) \end{aligned}\right\} \text{ on } \partial V,$$

will remain invariant.

So far θ is defined only on the surface of the bag. To obtain the hybrid chiral model I am after it is only necessary to introduce a chirally invariant Lagrangian for θ outside the bag. To derive the equations of motion and boundary conditions I will use a modification of the method of Chodos and Thorn.[14] Consider the action for a single quark flavor (the extension to SU(2) × SU(2) will be summarized later, SU(3) × SU(3) is more difficult and is left as an exercise!):

$$A = \int_V d^4x\left(\frac{i}{2}\bar{q}\gamma^\lambda\partial_\lambda q - \frac{i}{2}(\partial_\lambda\bar{q})\gamma^\lambda q - B\right) - \int_{\bar{V}} d^4x\frac{1}{2}(\partial_\mu\theta)^2 - \frac{1}{2}\oint_{\partial V} d^3s\bar{q}e^{i\theta\gamma_5/f}q . \tag{4.6}$$

Varying A with respect to q, $\bar{q}$, θ and V I obtain equations of motion and boundary conditions in a by now familiar manner:

$$i\not{\partial}q = 0 \qquad \text{in } V \tag{4.7a}$$

$$\Box\theta = 0 \qquad \text{in } \bar{V} , \tag{4.7b}$$

$$\left.\begin{aligned} i\not{n}q &= e^{i\theta\gamma_5/f}q \qquad (4.7c) \\ \frac{1}{2}n\cdot\partial\bar{q}e^{i\theta\gamma_5/f}q + \frac{1}{2}(\partial_\mu\theta)^2 &= B \qquad (4.7d) \\ n\cdot\partial\theta &= \frac{-i}{2f}\bar{q}\gamma_5 e^{i\gamma_5\theta/f}q \qquad (4.7e) \end{aligned}\right\} \text{ on } \partial V .$$

It is easy to verify that eqs (4.7) are invariant under the chiral transformation of eq (4.4). This implies the existence of a conserved axial current. To find it make a space-time dependent chiral transformation: A_μ will be the coefficient of $-\partial^\mu\kappa(x)$ in δA:[33]

$$A_\mu = \frac{1}{2}\bar{q}\gamma_\mu\gamma_5 q\Theta_V - f\partial_\mu\theta\,\Theta_{\bar{V}} \quad , \tag{4.8}$$

where $\Theta_V(\Theta_{\bar{V}})$ is 1 inside (outside) the bag and zero outside (inside). A_μ is conserved inside and outside of the bag by virtue of eqs (4.7a and b). On the surface conservation requires

$$- fn\cdot\partial\theta = \frac{1}{2}\bar{q}\not{n}\gamma_5 q \tag{4.9}$$

This follows directly from eqs (4.7c) and (4.7e). Of course the vector current:

$$J_\mu = \frac{1}{2}\bar{q}\gamma_\mu q\;\Theta_V \tag{4.10}$$

associated with the simple phase invariance $q' = e^{i\beta}q$ is also conserved.

Eqs (4.7) are highly non-linear even if the bag non-linearities are ignored. This is an expected consequence of trying to write down a chiral model of pions without the explicit appearance of scalar fields: In fact in the boson sector our model is one representation of the U(1) nonlinear sigma model. The implicit scalar field can be made more explicit by the transformations

$$\begin{aligned} \sigma &\equiv f\cos\theta/f \\ \phi &\equiv f\sin\theta/f \end{aligned} \tag{4.11}$$

applied to eqs (4.7). The resulting equations look more linear:

$$\begin{aligned} i\not{\partial}q &= \frac{1}{f}(\sigma + i\phi\gamma_5)q \\ n\cdot\partial\phi &= -\frac{i}{2f}\,\bar{q}\gamma_5 q \\ n\cdot\partial\sigma &= -\frac{1}{2f}\,\bar{q}q \end{aligned} \tag{4.12}$$

$$\frac{1}{2}(\partial\sigma^2 + \partial\phi^2) + \frac{1}{2f}\,n\cdot\partial\bar{q}(\sigma + i\phi\gamma_5)q = B$$

but because of the restriction $\sigma^2 + \phi^2 = f^2$, they are not. If this constraint is omitted the resulting σ-model now includes an unwanted elementary scalar field.

The symmetry of physical interest is either SU(2) × SU(2) corresponding to massless u and d quarks and Goldstone pions, or SU(3) × SU(3) with massless u, d and s quarks and Goldstone π's, K's and η.[34] The generalization to SU(2) follows from the replacement

$$e^{i\theta\gamma_5/f} \to e^{i\boldsymbol{\tau}\cdot\boldsymbol{\pi}\gamma_5/f} \quad , \tag{4.13}$$

where $\{\boldsymbol{\tau}\}$ are the Pauli matrices and $\boldsymbol{\pi}$ is an SU(2) triplet of pseudoscalar fields. The action

$$A = \int_V d^4x\left(\frac{i}{2}\bar{q}\gamma^\lambda\partial_\lambda q - \frac{i}{2}(\partial_\lambda\bar{q})\gamma^\lambda q - B\right) - \int_V d^4x\frac{1}{2}(D_\mu\boldsymbol{\pi})^2 - \frac{1}{2}\int_{\partial V} d^3s\,\bar{q}e^{i\boldsymbol{\tau}\cdot\boldsymbol{\pi}\gamma_5/f}q \quad , \tag{4.14}$$

is invariant under the (non-linear) chiral transformations

$$\delta q = \frac{i}{2}\boldsymbol{\kappa}\cdot\boldsymbol{\tau}\gamma_5 q \tag{4.15}$$

and

$$\delta\boldsymbol{\pi} = -f[\boldsymbol{\kappa} - (1 - x\cot x)\hat{\pi}\times(\boldsymbol{\kappa}\times\hat{\pi})] \quad , \tag{4.16}$$

where $x = |\boldsymbol{\pi}|/f$ and

$$D_{\mu ij} \equiv \left[\delta_{ij} - \left(1 - \frac{\sin x}{x}\right)(\delta_{ij} - \hat{\pi}_i\hat{\pi}_j)\right]\partial_\mu \tag{4.17}$$

Once more one obtains equations of motion and boundary conditions by varying A:

$$i\partial\!\!\!/ q = 0 \qquad \text{in } V \tag{4.18a}$$

$$\Box\boldsymbol{\pi} + \partial^\lambda\left(1 - \frac{\sin 2x}{2x}\right)\hat{\pi}\times(\hat{\pi}\times\partial_\lambda\boldsymbol{\pi}) = 0 \quad \text{in } \bar{V} \tag{4.18b}$$

$$i n\!\!\!/ q = e^{i\boldsymbol{\tau}\cdot\boldsymbol{\pi}\gamma_5/f}q \qquad \text{on } \partial V \tag{4.18c}$$

$$n\cdot\partial\pi_i = -[\delta_{ij} - (1 - \frac{2x}{\sin 2x})(\delta_{ij} - \hat{\pi}_i\hat{\pi}_j)]\frac{i}{2f}\bar{q}\tau_j\gamma_5 e^{i\tau\cdot\pi\gamma_5/f}q \tag{4.18d}$$

$$\frac{1}{2}n\cdot\partial\bar{q}e^{i\tau\cdot\pi\gamma_5/f}q + \frac{1}{2}(D_\mu\pi)^2 = B \tag{4.18e}$$

The conserved axial current is

$$A_i^\lambda = \frac{1}{2}\bar{q}\gamma^\lambda\gamma_5\tau_i q\;\Theta_V - f(\frac{\sin 2x}{2x}\delta_{ij} + (1 - \frac{\sin 2x}{2x})\hat{\pi}_i\hat{\pi}_j)\partial^\lambda\pi_j\Theta_{\bar{V}} \tag{4.18f}$$

If one considers the matrix element of A_i^λ between the true vacuum $|0\rangle$ and a one pion state $|\pi\rangle$ it is clear that f is to identified with the pion decay constant

$$f \equiv f_\pi = 93\text{ MeV}$$

I have written eqs (4.18) down in their full non-linear ugliness for two reasons. First, several attempts have been made to calculate pion corrections to bag calculations from a haphazard subset of these equations. This has resulted in inconsistent approximation schemes where some but not all of the effects at some order are included. Second, I want to emphasize that if no sensible approximation scheme exists then the situation is hopeless. If π/f is large then eqs (4.18) involve the full complexity both of the non-linear σ-model and of the bag. Fortunately, a possible perturbation expansion does exist at least for large sources and for the static cavity version of the bag model.

4.3. An Expansion for Large Source Bags

I will attempt to expand eqs (4.18) in $x = \pi/f$. To lowest order one obtains the original bag equations (eqs (2.4) and (2.11)):

$$-i\vec{\alpha}\cdot\vec{\nabla}q_0 = \omega_0 q_0 \qquad \text{in } V$$

$$i\not{n}q_0 = q_0 \qquad \text{on } \partial V$$

$$n\cdot\partial\bar{q}_0 q_0 = 2B \qquad \text{on } \partial V \ ,$$

where the subscript on q denotes that this is to be regarded as the zeroth approximation to something else. I will solve these in the static cavity approximation. The result is the model of baryons described in Section 3. This model is sensible only if the pion field generated by q_0 is in some sense small. To find it consider the first non-trivial approximation for $\underset{\sim}{\pi}$ from eqs (4.18):

$$\hat{r}\cdot\vec{\nabla}\underset{\sim}{\pi}_1 = \frac{i}{2f_\pi}\,\bar{q}_0\gamma_5\underset{\sim}{\tau}q_0 \tag{4.19}$$

$$\Box\underset{\sim}{\pi}_1 = 0$$

[Note I have used $\bar{q}_0 q_0 = 0$ on ∂V to drop a term of the form $\bar{q}_0\underset{\sim}{\tau}\cdot\underset{\sim}{\pi}_1 q_0$ on the right hand side of eq (4.19).] Using eq (3.1) for q_0 I obtain

$$\underset{\sim}{\pi}_1 = \frac{\vec{\underset{\sim}{P}}_1\cdot\hat{r}}{r^2} \qquad r \geq R_H \tag{4.20}$$

with

$$\vec{\underset{\sim}{P}}_1 = \frac{x_0}{16\pi f_\pi(x_0-1)}\,\vec{\sigma}\underset{\sim}{\tau} \ , \tag{4.21}$$

where $\vec{\sigma}$ and $\underset{\sim}{\tau}$ operate on quark variables and R_H is the radius obtained from the zeroth order approximation. To obtain, for example, the pion field about the nucleon, consider the matrix element of $\vec{P}_1$ in a nucleon state. A standard quark model calculation gives

$$\langle N|\vec{\sigma}\underset{\sim}{\tau}|N\rangle = \frac{5}{3}\,U_0^{+}\overleftrightarrow{\sigma}\underset{\sim}{\tau}U_0 \ , \tag{4.22}$$

where U_0 is a Pauli spinor-isospinor for the nucleon. Combining all this

$$\underset{\sim}{\pi}_1 = \frac{5x_0}{48\pi f_\pi(x_0-1)}\,\frac{U_0^{+}\overleftrightarrow{\sigma}\cdot\hat{r}\underset{\sim}{\tau}U_0}{r^2} \ , \qquad r \geq R_H \ . \tag{4.23}$$

Is this large? A realistic measure is the correction to q which it generates. Writing $q = q_0 + q_1$ and treating q_1 and π_1 as small in eqs (4.18) I find that q_1 obeys the boundary condition

$$(i\not{n}-1)q_1 = \frac{i\underset{\sim}{\tau}\cdot\underset{\sim}{\pi}_1}{f_\pi}\gamma_5 q_0 \qquad \text{at } r = R_H . \tag{4.24}$$

The source for q_1 is of order

$$\frac{\pi_1}{f_\pi} \simeq \frac{5x_0}{48\pi(x_0-1)} \quad \frac{1}{f_\pi^2 R_H^2} \approx \frac{1}{4\pi f_\pi^2 R_H^2} \tag{4.25}$$

since $x_0 = 2.04$. Later it will become apparent that a more accurate measure is in fact

$$\varepsilon = \frac{g_A}{8\pi f_\pi^2 R_H^2} . \tag{4.26}$$

If ε is small pion effects can be treated perturbatively. If not the model is out of control and one can say nothing. Fortunately bag estimates of R_H for a proton range from 1 to 1.4 fm. So ε is of the order 1/5 to 1/8. This is the first (and perhaps most significant) result of an HCB model: the usual quark model falls out with small corrections of order ε. Notice however that the attempt to construct a "little bag model" by forcing R_H to be small ($\sim$1/3 fm) gives $\varepsilon \sim 1$-2.

All of the parameters of the model have systematic expansions in ε:

$$q = q_0 + \varepsilon q_1 + \varepsilon^2 q_2 + \ldots$$

$$\frac{1}{f_\pi}\underset{\sim}{\pi} = \varepsilon(\underset{\sim}{\pi}_1/f) + \varepsilon^2(\underset{\sim}{\pi}_2/f) + \ldots$$

$$E = E_0 + \varepsilon E_1 + \varepsilon^2 E_2 + \ldots$$

$$g_{\pi N} = g_{\pi N}^0 + \varepsilon g_{\pi N}^1 + \ldots$$

and so on. Notice that the Yukawa coupling, $g_{\pi N}$, has a contribution at zeroth order in ε despite the fact that it vanishes (trivially) if pions are ignored. This is as it should be: $g_{\pi N}$ is sensitive to the pion pole, which is not suppressed as $R_H \to \infty$. Having set up this expansion let us calculate some effects at $O(\varepsilon^0)$ and $O(\varepsilon^1)$ and see how much the original bag results are modified.

4.4. Applications

a. Goldberger-Treiman relation, axial charge and $g_{\pi N}$:

As a start I will calculate the nucleon matrix element of $\underset{\sim}{A}^{\mu}$ to lowest order in ε. The object of this exercise is threefold: first, to show $\underset{\sim}{A}^{\mu}$ is indeed conserved as required in a chiral model; second, to calculate the axial charge, g_A; and third, to calculate the pion nucleon coupling constant by virtue of the expected Goldberger-Treiman relation:

$$g_{\pi N} = g_A M_N / f_\pi \tag{4.27}$$

Consider the nucleon matrix element of the spatial component of the axial current:

$$\langle N(P_2)|\vec{\underset{\sim}{A}}|N(P_1)\rangle \equiv \sqrt{\frac{M^2}{E_1 E_2}}\, \bar{U}(P_2)\frac{\underset{\sim}{\tau}}{2}[g_A(q^2)\vec{\gamma}\gamma_5 + h_A(q^2)\vec{q}\gamma_5]\, U(p_1) \tag{4.28}$$

where $q^\mu \equiv P_2^\mu - P_1^\mu$. In the chiral limit $h_A(q^2)$ should have a pion pole whose residue is $-2f_\pi g_{\pi N}$. To allow for this write

$$h_A(q^2) \equiv \frac{d_A(q^2)}{q^2} \tag{4.29}$$

Now take the nonrelativistic reduction ($q^\mu \to 0$) of eq (4.28) in the nucleon rest frame:

$$\lim_{q^\mu \to 0} \langle N(P_2)|\vec{\underset{\sim}{A}}|N(P_1)\rangle \equiv \vec{\underset{\sim}{A}}(0) = U_0^+(g_A(0)\vec{\sigma} + \frac{d_A(0)}{2M}\vec{\sigma}\cdot\hat{q}\hat{q})\frac{\underset{\sim}{\tau}}{2}\, U_0 \tag{4.30}$$

In the ordinary bag model $d_A(0) = 0$[35] since there is no pion pole to be found. To calculate $g_A(0)$ and $d_A(0)$ in the HCB model one explicitly Fourier transforms the static matrix element of $\vec{\underset{\sim}{A}}(\vec{r})$ (see Ref 35 for a review of this method)

$$\vec{\underset{\sim}{A}}(0)_{BAG} = \lim_{\vec{q}\to 0} \int d^3r e^{i\vec{q}\cdot\vec{r}} \vec{\underset{\sim}{A}}(r) \tag{4.31}$$

$$= \lim_{\vec{q}\to 0} \int d^3r \bar{q}\vec{\gamma}\gamma_5 \frac{\underset{\sim}{\tau}}{2} q - \int_V d^3r f_\pi \vec{\nabla}\underset{\sim}{\pi} \tag{4.32}$$

$$\equiv \underset{\sim}{\vec{A}}(0)_{QUARK} + \underset{\sim}{\vec{A}}(0)_{PION} \quad . \tag{4.33}$$

To order ε^0

$$\underset{\sim}{\vec{A}}(0)_{QUARK} = \int_V d^3r \bar{q}_0 \vec{\gamma}\gamma_5 \frac{\underset{\sim}{\tau}}{2} q_0 \tag{4.34a}$$

and

$$\underset{\sim}{\vec{A}}(0)_{PION} = -f_\pi \int_V d^3r \, \vec{\nabla}\underset{\sim}{\pi}_1 \quad . \tag{4.34b}$$

The contribution of q_0 to $\vec{A}(0)_{QUARK}$ was calculated long ago[36]

$$\underset{\sim}{\vec{A}}_{QUARK}(0) = \frac{5}{9} \frac{x_0}{(x_0-1)} U_0^+ \vec{\sigma} \frac{\underset{\sim}{\tau}}{2} U_0 \quad . \tag{4.35}$$

The pion contribution is new:

$$\underset{\sim}{\vec{A}}_{PION}(0) = -\lim_{\vec{q}\to 0} \int_V d^3r e^{i\vec{q}\cdot\vec{r}} f_\pi \vec{\nabla} \frac{\langle N|\underset{\sim}{\vec{P}}_1 \cdot \vec{r}|N\rangle}{r^3} \quad . \tag{4.36}$$

Integrating by parts I find

$$\underset{\sim}{\vec{A}}_{PION}(0) = \lim_{\vec{q}\to 0} \left\{ f_\pi \int d^3r (\vec{\nabla} e^{i\vec{q}\cdot\vec{r}}) \frac{\langle N|\underset{\sim}{\vec{P}}_1 \cdot \vec{r}|N\rangle}{r^3} \right.$$

$$\left. -f_\pi \oint_{S_\infty} d^2s \hat{r} e^{i\vec{q}\cdot\vec{r}} \frac{\langle N|\underset{\sim}{\vec{P}}_1 \cdot \vec{r}|N\rangle}{r^3} + f_\pi \oint_{S_{R_H}} d^2s \hat{r} e^{i\vec{q}\cdot\vec{r}} \frac{\langle N|\underset{\sim}{\vec{P}}_1 \cdot \vec{r}|N\rangle}{r^3} \right\} . \tag{4.37}$$

S_{R_H} and S_∞ are spheres at the bag's surface and at ∞ respectively. The surface at infinity does not contribute. The other two contributions yield

$$\underset{\sim}{\vec{A}}_{PION}(0) = 4\pi f_\pi \langle N|\underset{\sim}{\vec{P}}_1 \cdot \hat{q}\hat{q} - \underset{\sim}{\vec{P}}_1/3|N\rangle \ , \tag{4.38}$$

where the second term comes from the surface integral. Using eq (4.22) to evaluate the nucleon matrix element of the quark operator

$\vec{P}_1$ I find

$$\vec{A}_{PION}(0) = -\frac{5}{6}\frac{x_0}{x_0-1} U_0^+ \frac{\tau}{2} (\vec{\sigma}\cdot\hat{q}\hat{q} - \frac{\vec{\sigma}}{3}) U_0 \quad . \tag{4.39}$$

Combining $\vec{A}_{QUARK}(0)$ and $\vec{A}_{PION}(0)$ finally

$$\vec{A}_{HCB}(0) = \frac{x_0}{x_0-1} U_0^+ \frac{\tau}{2} (\frac{5}{9}\vec{\sigma} - \frac{5}{6}(\vec{\sigma}\cdot\hat{q}\hat{q} - \frac{\vec{\sigma}}{3})) U_0 \quad . \tag{4.40}$$

Note that $\vec{A}_{HCB}$ is conserved: $\lim_{\vec{q}\to 0} q\, \vec{A}_{HCB}(q) = 0$ (a good check on the algebra). The nucleon's axial vector charge may be read off eq (4.40):

$$g_A^{HCB} = \frac{5}{6}\frac{x_0}{x_0-1} = \frac{3}{2} g_A^0 \ , \tag{4.41}$$

where g_A^0 is the quark contribution to g_A in the old bag model. Comparing eq (4.40) with eq (4.30) I find

$$\frac{d_A^{HCB}(0)}{2M} = -g_A^{HCB} \quad . \tag{4.42}$$

Since $d_A(0)$ is precisely $-2g_{\pi N}f_\pi$ eq (4.42) is the expected Goldberger-Treiman relation.

The Yukawa coupling constant could equally well have been calculated directly rather than via Goldberger-Treiman. From eq (4.23) the pion field far from a nucleon source is

$$\lim_{r\to\infty} \pi_1(r) = \frac{5x_0}{48\pi(x_0-1)f_\pi} \frac{U_0^+\vec{\sigma}\cdot\hat{r}\tau U_0}{r^2} \quad . \tag{4.43}$$

In pseudoscalar meson theory the pion field is easily shown to be

$$\lim_{r\to\infty} \pi(r) = \frac{g_{\pi N}}{8\pi M} \frac{U_0^+\vec{\sigma}\cdot\hat{r}\tau U_0}{r^2} \quad . \tag{4.44}$$

Comparing the two I find

$$g_{N\pi}^{HCB} = \frac{5x_0}{6(x_0-1)} \frac{M}{f_\pi} = g_A^{HCB} \frac{M}{f_\pi} \quad . \tag{4.45}$$

From known values of x_0, M and f_π I find $(g_{\pi N}^{HCB})^2/4\pi \simeq 20$ as compared to the experimental value of about 14. Not bad for an admittedly crude model. The problem is that g_A^{HCB} is too large. g_A^0 (= 1.09) was a bit below the experimental value $g_A \simeq$ 1.25 but the pion correction has overshot.

The lessons I would like to draw from this calculation are: first, that the model is capable of generating a large Yukawa coupling $g_{\pi N}$ even though the coupling of pions to quarks, ε, is small; second, g_A is not a reliable prediction of quark models. It is modified to zeroth order in ε by the pions required by chiral symmetry. The modification is, however, a surface effect and the calculation I have given cannot be trusted numerically. Now it should be clear why I defined ε as I did in eq (4.26). From pseudoscalar meson theory (eq (4.43)) and the Goldberger-Treiman relation one obtains a measure of the pion field strength at a hadron's surface independent of details of the hybrid chiral bag model:

$$\underset{\sim}{\pi}(r)\Big|_{r=R_H} = \frac{g_A}{8\pi f_\pi R_H^2} U_0^\dagger \vec{\sigma}\cdot\hat{r}\underset{\sim}{\tau} U_0 \quad . \tag{4.46}$$

[Eq (4.23) is identical to this with g_A^{HCB} replacing g_A.] It is now obvious that quark models of baryons which ignore pions are successful because baryons are big.

As a second and final (for these notes) application of hybrid chiral bags I will calculate the lowest order (in ε) energy shift of the nucleon and Δ due to the external pion field in the chiral limit. This calculation is made much easier by the following theorem:

$$E - E_0 \equiv \frac{1}{2}\int_V d^3x(-q_0^\dagger i\vec{\alpha}\cdot\vec{\nabla}q_1 - q_1^\dagger i\vec{\alpha}\cdot\vec{\nabla}q_0 + \text{h.c.})$$

$$+ \frac{1}{2} \int_V d^3x |\vec{\nabla}\underset{\sim}{\pi}_1|^2 + O(\varepsilon^2) \tag{4.47}$$

$$= - \frac{1}{2} \oint_{\partial V} \underset{\sim}{\pi}_1 \, (\hat{n}\cdot\vec{\nabla})\underset{\sim}{\pi}_1 + O(\varepsilon^2) \quad , \tag{4.48}$$

where $\underset{\sim}{\pi}_1$ is given by eq (4.20) and (4.21) in terms of quark variables. Eq (4.48) is obtained by careful manipulation of the equations of motion and boundary conditions obeyed by q_0, q_1 and $\underset{\sim}{\pi}_1$ all of which have been given above. The surface average is trivial leaving

$$E = E_0 - \varepsilon \frac{3g_A}{50R_H} \sum_{i,j} \vec{\sigma}_i\cdot\vec{\sigma}_j \underset{\sim}{\tau}_i\cdot\underset{\sim}{\tau}_j + O(\varepsilon^2) \ . \tag{4.49}$$

Diagrammatically the energy shift comes from quark diagrams as shown in Figure 9a. These correspond to both N and Δ intermediate states (Figure 9b). If ε were zero the N and Δ would be degenerate. The interaction with external pions lifts the degeneracy. Using Fermi statistics for the quarks it is not hard to show

$$\sum_{i,j} \vec{\sigma}_i\cdot\vec{\sigma}_j \underset{\sim}{\tau}_i\cdot\underset{\sim}{\tau}_j = 3N^2 + 12N - 4S(S+1) - 4I(I+1) \ , \tag{4.50}$$

where N = 3 for the N or Δ and S and I are the total spin and isospin respectively. Combining eqs (4.49) and (4.50) for $R_H \simeq 1.2$ fm I find an energy shift of -110 MeV and -65 MeV for the nucleon and Δ respectively. This is an approximately 10% effect. It depends sensitively on the choice of R_H. One should not take the actual numbers very seriously because of this. Nevertheless it is amusing that the Δ-N splitting generated by pion effects is in the same direction as the one generated by QCD. To attribute the <u>whole</u> Δ-N splitting to pions would require huge shifts in the N and Δ masses individually. This can be done by arbitrarily fixing R_H small enough

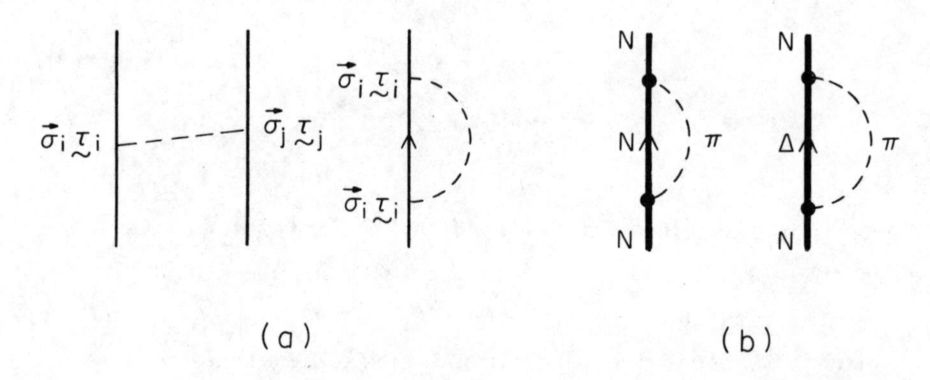

Figure 9. a) "Feynman" diagrams for the lowest order pion contribution to baryon masses; b) corresponding physical intermediate states.

but then ε is large and the whole model is out of control. As it is, about 15% of the Δ-N splitting is attributable to pions. The rest presumably comes from QCD. As yet no one has systematically redone baryon spectroscopy in a hybrid chiral model.

It is very interesting now to introduce chiral summetry violation into the model by giving both the quarks and the pseudoscalar mesons mass. Time doesn't permit me to do this here. In closing, though, I will quote one tantalizing result: the pseudoscalar bosons outside a nucleon contribute to the famous sigma term (Σ) of πN scattering. Their contribution is of just the right size and flavor -SU(3) structure to account for the deviation of Σ from naive quark model estimates![37]

It should be clear to the reader that hybrid chiral models are of limited theoretical interest. They are entirely ad hoc (since chiral symmetry is put in by hand), and restricted to the low energy regime. Nevertheless they reassure us that standard quark models were not ignoring huge effects and may yet shed some light on a stubborn problem such as the πN-Σ term.

ACKNOWLEDGEMENTS

I would like to thank S. Coleman, N. Cottingham and C.B. Thorn for useful discussions on the material in Section 4. I am also grateful to Professor R.H. Dalitz and the Department of Theoretical Physics at Oxford University and to CERN for support during this work.

REFERENCES

1. A. Chodos, R. L. Jaffe, K. Johnson, C. B. Thorn and V. F. Weisskopf, Phys. Rev. D9, 3471 (1974).
2. F. E. Low, M.I.T. Preprint MIT-CTP-805.
3. S. Mandelstam, Phys. Rev. D19, 2391 (1979).
4. C. Callen, R. Dashen and D. Gross, Phys. Rev. D19, 1826 (1979).
5. J. Goldstone, Nuovo Cimento 19, 154 (1960).
6. Y. Nambu, Phys. Rev. Letters 4, 380 (1960).
7. S. Coleman, "Secret Symmetry" in Laws of Hadronic Matter, A. Zichichi, ed. (Academic Press, New York, 1975).
8. G. E. Brown and M. Rho, SUNY Stony Brook Preprint.
   G. E. Brown, M. Rho and V. Vento, SUNY Stony Brook Preprint.
9. The original idea is due to A. Chodos and C. B. Thorn, Phys. Rev. D12, 2733 (1975) and T. Inoue and T. Maskawa, Prog. Theor. Phys. 54, 1833 (1975).

10. I use the metric $-g^{00} = g^{ii} = 1$, but otherwise the conventions of J. D. Bjorken and S. D. Drell, Relativistic Quantum Field Theory (McGraw-Hill, New York, 1965).

11. As is well known a Dirac spinor has only two independent components, the other two being determined in terms of the first by Dirac's equation. Thus it is impossible to impose a boundary condition like $q = 0$ on Dirac's equation. The boundary condition I have quoted -- $(i\hat{n}\cdot\gamma - 1)q = 0$ -- constrains only two components. To see this define the projection operators $P_{\pm}(x) = \frac{1}{2}(1 \pm i\hat{n}\cdot\vec{\gamma})$ at each point on the surface. If $P_{\pm}q \equiv q_{\pm}$ then the boundary condition is $q_{-} = 0$ with q_{+} unconstrained as it should be.

12. P. Grädig, P. Hasenfratz, J. Kuti and A. S. Szalay, Phys. Letters 64B, 62 (1975); P. Hasenfratz, J. Kuti and A. S. Szalay, Proc. Xth Rencontre de Moriond, Meribel, March 1975, Vol. II, p. 209.

13. M. Creutz, Phys. Rev. D10, 1749 (1974); M. Creutz and K. S. Soh, Phys. Rev. D12, 443 (1975). Friedberg and T. D. Lee, D18, 2623 (1978).

14. A. Chodos and C. B. Thorn, Phys. Rev. D12, 2733 (1975).

15. K. Johnson, Proceedings of the 1977 Coral Gables Conference and Phys. Lett. 78B, 259 (1978).

16. T. A. DeGrand, R. L. Jaffe, K. Johnson and J. Kiskis, Phys. Rev. D12, 2060 (1975).

17. K. Johnson, Acta Physica Polonica B6, 865 (1975).

18. T. A. DeGrand and R. L. Jaffe, Ann. Physics (N. Y.) 100, 425 (1976). T. A. DeGrand, Ann. Physics (N. Y.) 101, 496 (1976). C. Rebbi, Phys. Rev. D12, 2407 (1975) and Phys. Rev. 14, 2362 (1976). T. A. DeGrand and C. Rebbi, SLAC Preprint, 1978.

19. R. L. Jaffe and F. E. Low, unpublished.

20. J. Donoghue and K. Johnson, M.I.T. Preprint MIT-CTP-802.

21. T. D. Lee, Phys. Rev. D19, 1802 (1979).

22. Actually these σ-matrices are the generators of rotations of the relativistic $j = \frac{1}{2}$ quark mode of eq (3.1). They are not generators of spin (internal) angular momentum which, of course, does not commute with the Dirac Hamiltonian. For convenience I refer to this as "spin" anyway.

23. A. de Rujula, S. L. Glashow and H. Georgi, Phys. Rev. D12, 147 (1975).

24. One of the most persuasive arguments is Rosner's, J. Rosner, Phys. Rev. Letters 21, 950 (1968), based on duality, or quark line diagrams.

25. R. L. Jaffe and F. E. Low, Phys. Rev. D19, 2105 (1979).

26. R. L. Jaffe, Phys. Rev. Letters 38, 195 (1977).

27. A. S. Carroll, I-H. Chiang, R. A. Johnson, T. F. Kycia, K. K. Ki, L. S. Littenberg and M. D. Marx, Phys. Rev. Letters 41, 777 (1978).

28. O. Braun, F. Gandini, H. J. Grimm, V. Hepp, C. Kiesling, D. E. Plane, H. Stroebele, C. Thoel, T. J. Thouw and W. Wittek, Nucl. Phys. B124, 45 (1977).

29. R. J. Oakes, Phys. Rev. 131, 2239 (1963).

30. R. L. Jaffe, Phys. Rev. D15, 267, 281 (1977).

31. H. Pagels, Phys. Reports 16C, 219 (1975).

32. Y. Nambu, J. Jona-Lasinio, Phys. Rev. 122, 345 (1961).

33. See, for example, R. Jackiw "Field Theoretic Investigations in Current Algebra" in Lectures on Current Algebra and Its Applications (Princeton University, Princeton, 1972).

34. I will ignore the "U(1)-problem," i.e. I assume that though n-quarks are massless only an SU(n) not a U(n) chiral symmetry results. See G. t'Hooft, Phys. Rev. Letters 37, 8 (1976), for an explanation of this in the framework of QCD.

35. A discussion of $h_A(q^2)$ in the cavity approximation to the bag model as well as a general review of the techniques used in this calculation can be found in S. L. Adler, E. W. Colglazier, Jr., J. B. Healy, I. Karliner, J. Lieberman,

Y. J. Ng and H-S. Tsao, Phys. Rev. D11, 3309 (1975).

36. P. N. Bogoliubov, Ann. Inst. H. Poincaré 8, 163 (1967).

37. R. L. Jaffe, to be published.

DISCUSSIONS

CHAIRMAN: Prof. R.L. Jaffe

Scientific Secretaries: G. Adkins, A. Kreymer and J. Szwed

DISCUSSION 1

- MAHARANA:

What are the corrections to the zero point energy in a bag due to quantum fluctuations?

- JAFFE:

Since I have not even shown you how to calculate the lowest order energy in a bag due to quantum fluctuations, that is what I will do now. To begin with, we do not really understand the quantum mechanics of both geometrical and field degrees of freedom - except in one dimension where the problem can be solved completely using tricks. What we can do in three dimensions is to fix a cavity and calculate the energy of quantum fluctuations inside (this calculation for parallel plates was first performed by Casimir and Polder). For a spherical cavity we have shown that the zero point energy has a power series expansion in a cutoff Ω :

$$E_0 = \sum_{\substack{\text{quark} \\ \text{and gluon modes } n}} \frac{1}{2} \hbar \omega_n = c_4 \Omega^4 V + c_3 \Omega^3 S + c_2 \Omega^2 R + \cdots - \frac{Z_0}{R}$$

The first term can be absorbed into a renormalization

of the bag constant, and the second is absent for massless fermion and vector particles. We assumed that the other divergencies were likewise absent (although this is still an open question), and found a zero point energy that is proportional to 1/R.

- GINSPARG:

I have first a comment. My interpretation of this is that we have a situation with quantum contributions from all momentum scales. What you have done is to lump all the effects of strong coupling into the bag itself and to treat the weak coupling effects by ordinary perturbation theory, e.g. via the Feynman diagrams described in lecture. The problem from this standpoint is to include the intermediate coupling regime.

Now for my question: what is the dual argument you said Rosner used to show the existence of exotics at some level and how is it consistent with the $\frac{1}{N}$ expansion which Coleman has asserted both looks like a dual theory and excludes exotics to leading order in $\frac{1}{N}$?

- JAFFE:

To reply to the second part of your question as to whether there is an inconsistency between the exotics which appear in our model and the absence of exotics up to order $\frac{1}{N}$ in the $\frac{1}{N}$ expansion, I think that Sidney and I are in agreement that there is no discrepancy between the models. A description of the correspondence would require more time than is available here. To reply to the first part of your question, consider baryon-antibaryon scattering in terms of quark line diagrams. If you think that this scattering is dominated by t-channel exchange of conventional meson trajectories, then this is dual to the s-channel $q^2\bar{q}^2$ resonances.

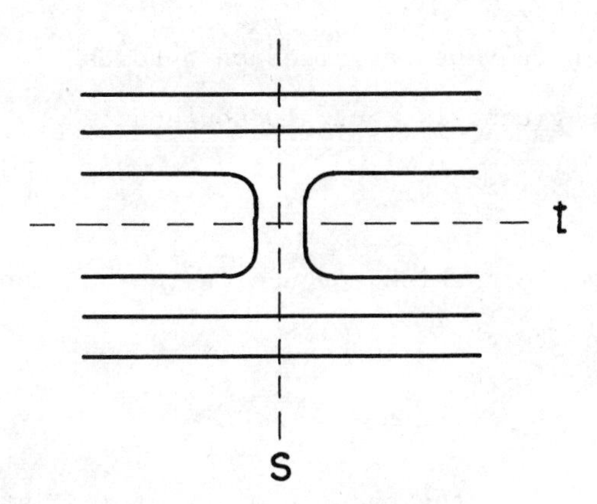

- HANSSON:

What can you say about glueballs in this model?

- JAFFE:

You can attempt to build the same kind of model for gluon states as we have built for quark states. In fact Johnson and I have calculated the spectrum of gluon states confined in a spherical cavity. In coulomb gauge the field equations are

$$\partial^2 \vec{A} = 0 \quad , \quad \vec{\partial} \cdot \vec{A} = 0$$

with the boundary conditions

$$\vec{n} \cdot \vec{E} = 0 \quad , \quad \vec{n} \times \vec{B} = 0$$

and with the quadratic boundary condition

$$\frac{1}{2} (\vec{E}^2 - \vec{B}^2) = B$$

imposed only in an average sense. However, this spherical calculation cannot be taken very seriously, for, as Giles has pointed out, the quadratic boundary condition requires a non-vanishing tangential electric field on the surface (unless the variuos color components of $\vec{E}$ vanish at different points). This is impossible on a sphere, and can only be satisfied on a torus. So in fact glueballs may not be glueballs, but gluedoughnuts.

- SZWED:

Is the volume term which you use in the bag model better than a surface tension term?

- JAFFE:

I have not studied the phenomenology of models with surface tension in detail. I do not know how you would deal with the degrees of freedom introduced by the surface. Oscillations of the surface transverse to itself are real dynamical degrees of freedom, and there is no room for them in the spectrum.

- SZWED:

What about surface oscillations in the volume model?

- JAFFE:

In the volume model the surface variables are not independent degrees of freedom but can be expressed in terms of the quark fields. This can be seen explicitly in the one dimensional case where the model is exactly soluble.

- COLEMAN:

Do not the people who try to derive the bag as a limit of field theory have a surface tension term in their effective Lagrangian? This would come from the fact that it costs energy to change the field from one configuration to another. Does this give them trouble?

- JAFFE:

The Princeton group is the only group which claims to do something in QCD. They claim to have a very sharp surface, and therefore, I suspect, a high surface tension, which elevates the surface excitations to high energy.

- COLEMAN:

What about T.D. Lee and collaborators who attempt to derive it from a theory with real scalar fields?

- SZWED:

In these derivations of the bag, the appearance of volume and surface terms depends on the parameters of the Lagrangian.

- COLEMAN:

In that case the elimination of unobserved surface modes serves as a strong constraint on such models.

- OLIENSIS:

What happens if you calculate the first correction to the ground state energy considering only the first few modes?

- JAFFE:

No one has actually done such a calculation. The validity of this as an approximation to the actual answer is unclear because it is not a sum of positive terms. I would be tempted to say that this contribution would be much smaller than the entire sum.

- KOH:

What can be said about the pion mass in this model?

- JAFFE:

It is my opinion that this formulation of the bag model can say nothing about the pion. In order to say anything meaningful about the pion you have to have a model which has explicit chiral symmetry I broke chiral symmetry from the start by choosing a definite value of α .

- GINSPARG:

Could you explain the relationship between constituent quark masses and current algebra masses?

- JAFFE:

There are actually three different things which are called quark masses.

(i) The mass parameters in the Lagrangian, either bare masses which are infinite or Lagrangian masses which are renormalized with a specific renormalization prescription. They depend on the scale with which you measure the momentum. With a momentum scale μ I will call these masses $m^{\circ}(\mu)$. These are the current algebra masses.

(ii) The effective quark masses $m(\mu)$ are masses defined in a non-perturbative way, for example as coefficients in a suitably normalized inverse fermion propagators, à la Georgi and Politzer. These are the effective quark masses.

(iii) The quark energies E which occur in the relativistic bag model, which are often mistakedly called quark masses.

The approximate values of these masses are as follows:

QUARK TYPE	$m^{\circ}(\mu)$ (MeV)	$m(\mu)$ (MeV)	E (MeV)
u	10	10 + c	10 + 300
d	20	20 + c	20 + 300
s	300	300 + c	300 + 300

where μ is a typical mass in hadron mass splittings ($\mu \sim 1$ GeV), and where c is an approximately flavor dependent constant - proportional to $\langle o | \bar{q}q | o \rangle$ - which may or may not be zero - it is zero inside the bag in Princeton version - but it is not as large as 300 MeV.

DISCUSSION 2

- HANSSON:

How do you treat the old bag model pion now that you have introduced field pions surrounding the bag?

- JAFFE:

The model I have developed is a model of pions surrounding a large, static, semiclassical source. By definition, in this model a quark model pion is not a large, static, semiclassical source of pions. It is certain that this model does not apply to the dynamics of a quark model pion.

- PAFFUTI:

The stress-energy tensor of the bag model has a term $Bg_{\mu\nu}$ which resembles a cosmological term. What is the meaning of this term when you couple your theory to gravity?

- JAFFE:

You are correct that this term has the structure of a cosmological term. However, it is just a part of the stress-energy tensor of matter, and probably has no particular ramifications for general relativity. However, there is an interesting analogy between Einstein's theory, where a cosmological term is introduced to close the space and the bag model, where a cosmological-like term is introduced to confine the quarks.

QUARK MODEL STATES AND LOW ENERGY SCATTERING

F. E. Low

Center for Theoretical Physics, Laboratory for Nuclear Science and Department of Physics
Massachusetts Institute of Technology
Cambridge, Massachusetts 02139

I. INTRODUCTION

In these lectures, based on work with R. Jaffe, I wish to discuss the connection between the discrete mass eigenstates which are calculated in an approximate quark-gluon model and the observed low energy hadron-hadron scattering in the same mass region The problem we claim to have solved can be stated crudely: where are the exotics? The quark model predicts exotic states, for example, in the I=2 $\pi\pi$ system at 1.15 GeV; the I=2 $\pi\pi$ S-wave phase shift, on the other hand, shows no sign of interesting behavior up to an energy of about 1.5 GeV, by which time it is approximately -.5 radians. Similar comments apply to other exotic channels. Our solution of this problem rests on a new method of analyzing low energy hadron-hadron scattering, applicable to multi-two-body channels but not to genuine three body channels. It is simplest in its application to single channel systems, and therefore most reliably applicable to low energies and exotic channels.

The problem is the following: If a calculated mass eigenvalue is below the threshold for two particle emission, we have a true discrete state, a strongly stable object which can be directly observed, or indirectly observed through its possible weak or electromagnetic decay. If the mass eigenvalue is above the two particle threshold, the situation is more complicated. In this case, we call the discrete eigenstate a "primitive". Its very existence depends on having made an approximation, that is, on having in some way decoupled it from its decay channels. Now, if the primitive is in fact sufficiently weakly coupled to its decay channels, we know that a narrow resonance will appear in the scattering and reaction amplitudes, and that the width of the resonance will go to zero with the coupling. The resonant energy will lie close to the primitive eigenvalue, as will the real part of the pole in the complex energy plane. An example of this well understood situation is provided by the lowest vector meson nonet -- the ρ, ω, k* and ϕ. For these states, the approximation that creates the primitives might be the rstriction to the $q\bar{q}$ sector of the Hilbert space for the vector mesons as well as for each of the pseudoscalar mesons into which they decay. The decays can proceed only via the two quark-four quark coupling, and are in addition inhibited by the $\ell=1$ angular momentum barrier of the final state. Similar remarks of course apply to the baryon decuplet.

On the other hand, if the primitive is strongly coupled to its decay channels, then understanding its reflection in the observed reaction and scattering amplitudes requires an analysis that may depend in detail on the approximation that created the primitive in the first place. Of course, if we had a full understanding of the system, we could simply calculate the observed amplitudes themselves. Lacking that we exploit our (hoped for) understanding of the short range interaction of quarks and gluons to determine the properties of the primitives, and then analyze the low energy amplitudes to measure the masses and coupling of these primitives. Our technique is a modification of the Wigner-Eisenbud[1] formalism suited to the case of primitives created by confining boundary conditions.

The kind of primitive we have in mind here would consist of four, five, six or more quarks and antiquarks in SU(3) singlet states, i.e. $q^2\bar{q}^2$, $q^4\bar{q}$, q^6, etc. The approximation that creates the discrete primitives consists in confining the quarks and gluons to a spherical volume of radius R with bag boundary conditions at that radius.

In the following we shall consider in particular the primitives created by four quarks in even parity, j=1/2 states of the bag model with total angular momentum zero. These are then coupled to the S states of the pseudoscalar meson system.

We digress now to give a simple example of how a confining boundary condition creates primitives, and how these primitives can be studied in the scattering. Consider the S-wave scattering of a non-relativistic particle by a weak attractive square well of radius b and depth $V_o = \frac{U_o\hbar^2}{2m}$. It is clear that no resonant or bound states are created by this potential. However, if one imposes a boundary condition requiring the wave function to vanish at b, one creates an infinite set of primitive internal states at momenta k_n where $b\sqrt{k_n^2+U}=n\pi$,

---

1. E.P. Wigner and L. Eisenbud, Phys. Rev. <u>72</u>, 19 (1947)

$n=1,2,\ldots$. Now, although the phase shift does nothing spectacular as k varies from zero to infinity, we can precisely identify the primitives by looking for the zeros of the quantity

$$\sin[kb+\delta(k)]$$

which occur precisely at the k_n's of the primitives. This is because the external wave function $\sin(kr+\delta)$ clearly vanishes (by continuity) at $r=b$ when the energy is such that the internal wave function satisfies the vanishing boundary condition at $r=b$. Therefore the measured phase shift can give us information about an artificial problem, to wit the problem of the discrete states of the original system supplemented by the vanishing boundary condition at $r=b$.

We shall see later that in a many channel problem it is useful to look for poles of a matrix quantity P which is analogous to $k\cot(kb+\delta)$. We note here that $P=1/R$, where R is the derivative matrix introduced by Wigner and Eisenbud. It was shown by these authors that for a non-relativistic theory, R could be expanded in a sum of poles, provided the interaction vanishes outside of b. A similar result holds for $P=1/R$, except that a single subtraction constant is needed. The distinction is that in matching R to the internal wave-function one must expand the scattering wave-function in a set of eigenfunctions whose derivatives vanish at b, whereas in the case of P it is the eigenfunctions themselves which vanish at b, causing a more serious singularity at $r=b$ in the expansion. We give the argument for P, restricting ourselves to a non-relativistic one channel $\ell=0$ problem. The generalization to several discrete channels and arbitrary ℓ is trivial. However, relativistic kinematics, the finiteness of b and continuous inelastic channels will necessarily imply more distant left hand and inelastic cuts.

We consider the radial scattering wave function normalized for r>b, to

$$\psi_o = \cos k(r-b) + \frac{P}{k}\sin k(r-b) \tag{1.1}$$

which, with

$$P = k\cot(kb+\delta) \tag{1.2}$$

is equal to

$$\psi_o = \frac{\sin(kr+\delta)}{\sin(kb+\delta)} \quad . \tag{1.3}$$

Evidently, when k is such that P has a pole, $\frac{\sin k(r-b)}{k}$ dominates ψ_o, and therefore, with appropriate renormalization, ψ vanishes at r=b; hence the internal wave function also vanishes there by continuity. We attempt to exploit this fact by expanding ψ, for r<b, in a set of orthonormal functions ϕ_n, with energy E_n, on which we impose the boundary condition of vanishing at r=b. Thus, for r<b,

$$\psi = \sum_n C_n \phi_n (r) \tag{1.4}$$

and

$$C_n = \int_0^b \phi_n^* (r)\psi \, dr . \tag{1.5}$$

We evaluate c_n by noting that $H\phi_n = E_n\phi_n$, and $H\psi = E\psi$, so that, with 2m=1,

$$\begin{aligned}(E_n - E)C_n &= \int_0^b [(H\phi_n)^*\psi - \phi^* H\psi]dr \\ &= -\int_0^b dr\frac{d}{dr}\left(\phi_n^{*\prime}\psi - \phi_n^*\psi'\right) = -\phi_n^{*\prime}(b)\end{aligned} \tag{1.6}$$

and

$$\psi(r) = -\sum_n \frac{\phi_n^*(b)\phi_n(r)}{E_n - E} \quad . \tag{1.7}$$

Now as remarked earlier, since $\psi(r) \neq 0$ at $r=b$ and each $\phi_n(b)=0$, the expansion (1.7) will have trouble at $r=b$. However, the energy derivative of ψ vanishes at b, since $\psi(r)=\frac{\sin(kr+\delta)}{\sin(kb+\delta)}$ so that $\psi(b)=1$ and $\frac{d\psi(b)}{dE}=0$.

Thus, if we differentiate both sides of (1.7) with respect to E, the expansion can be expected to converge at $r=b$. The values of the inside and outside functions are now automatically equal at $r=b$ (they are both zero), so we have only to match radial derivatives:

$$\frac{\partial}{\partial r}\frac{\partial \psi_o}{\partial E}\bigg|_{r=b} = -\sum_n \phi_n'^*(b)\phi_n'(b)\frac{d}{dE}\frac{1}{E_n - E} \quad . \tag{1.8}$$

But, from (1.1), $\frac{\partial\psi}{\partial r}\Big|_{r=b} = P$, so that (1.8) becomes

$$\frac{dP}{dE} = -\sum_n \left|\phi_n'(b)\right|^2 \frac{d}{dE}\frac{1}{E_n - E} \tag{1.9}$$

and

$$P - P(E_o) = -\sum_n \left|\phi_n'(b)\right|^2 \left|\frac{1}{E_n - E} - \frac{1}{E_n - E_o}\right| \tag{1.10}$$

with $P(E_o)$ undetermined. Thus P can be expanded in a sum of poles with one subtraction. Note that $\frac{dP}{dE} < 0$; this latter result can be made plausible in an arbitrary theory, as we shall see in section III.

The expansion (1.10) becomes particularly simple in the case of the square well discussed above; in fact

$$P = k\cot(kb+\delta) = q \cot qb \tag{1.11}$$

where $q^2 = k^2 + U_o$

and the expansion (1.10) is simply the well known expansion of the cotangent:

$$x \cot x = 1+2x^2 \sum_{n=1}^{\infty} \frac{1}{x^2-n^2\pi^2} \tag{1.12}$$

Notice that for a weak potential the primitives at $x=n\pi$ are artifacts of the boundary condition; in particular, for $U_o=0$, there is an infinite set of poles at $k_n = \frac{n\pi}{b}$ and no interaction!

On the other hand, if the internal dynamics is such that P has a pole with a small residue, so that nearby

$$\frac{P}{k} \sim A + \frac{\lambda}{E-E_n} \tag{1.13}$$

$$|E_n - E_{n+1}| >> \lambda > o$$

then the S matrix will have a pole very nearby in the complex plane, and we will be dealing with a conventional narrow resonance. To see this, we note that

$$S = e^{2i\delta} = -e^{2ikb} \frac{1-iP/k}{1+iP/k} \tag{1.14}$$

as is easily verified by substituting $P = k\cot(kb+\delta)$. Therefore, S has a pole at

$1+iP/k = 0$, or, from (1.13),

$$E = E_n - i\frac{\lambda}{1+iA} \quad . \tag{1.15}$$

We make contact with the quark-gluon bag model calculations by noting that in the four quark state the two pion relative probability density

$$\rho_Q(\vec{r}) = \int \delta\left(\frac{\vec{r}_1+\vec{r}_2}{2} - \frac{\vec{r}_3+\vec{r}_4}{2}\right)\rho(\vec{r}_1)\rho(\vec{r}_2)\rho(\vec{r}_3)\rho(\vec{r}_4)\, d\vec{r}_{1234}$$

vanishes very strongly at r=2R, where R is the bag radius. In fact, ρ_Q is almost indistinguishable from the density

$$\rho_M = |\phi(r)|^2$$

where

$$\phi = \frac{1}{\sqrt{2\pi b}} \quad \frac{\sin\pi r/b}{r}$$

with b adjusted to give the same mean-square radius as ρ_Q. The resulting relationship is $b \sim 1.4R$. ρ_Q and ρ_M are shown in Fig. 1. We therefore interpret bag model calculations of primitives as corresponding to a vanishing wave-function at r=b: hence we study, for elastic scattering,

$$P = k\cot(bk+\delta)$$

and for multi-channel reactions the matrix analogue which we shall discuss later. We take R (and hence b) from the calculations of Jaffe[2]: R is related to the expected mass of the primitive by the virial theorem

$$R = 5M^{1/3} \text{ GeV}^{-1}$$

and hence

$$b = 7M^{1/3} \text{ GeV}^{-1}, \text{ with M in GeV.}$$

Before we look at data, we observe that a primitive which occurs at a wave-number k_c such that $k_c b=\pi$ requires a zero phase-shift at $k=k_c$, since if $P = \frac{k\cos(kb+\delta)}{\sin(kb+\delta)}$ has a pole at $k_c b=\pi$, $\delta(k_c)$ must be zero. We call this phenomenon compensation and refer to $M(k_c)=M_c$ as the compensation mass. Now, if $M_n<M_c$, then $\delta(k_n)>0$; if $M_n>M_c$, then $\delta(k_n)<0$. Thus a positive phase shift signals a primitive below the compensation energy, a negative phase shift signals a primitive above the compensation energy. Using $k=\pi/b(M)$ and $b(m)=7M^{1/3}$ we find compen-

---

2. R. Jaffe, Phys. Rev. D15, 267 and 281 (1977).

sation masses of .95 for the $\pi\pi$ and 1.08 for the πk systems.

The general rule then is the following: a single pole in P with a residue of order unity and no other pole within a characteristic energy variation interval will result in a moderately varying phase shift, attractive or repulsive according to whether the pole location is below or above the compensation energy. An isolated pole with small residue produces a narrow resonance near the primitive. Finally, it is evident that with two nearby primitives -- nearby meaning much closer than the characteristic energy variation interval -- $\cot(\delta+kb)$ will vary from $+\infty$ to $-\infty$ between them, so that δ must vary rapidly by π. This is sensible, since one primitive can be an artifact of a boundary condition, but two nearby ones cannot.

We shall see in Section III that there is an analogous statement that can be made in the many channel case.

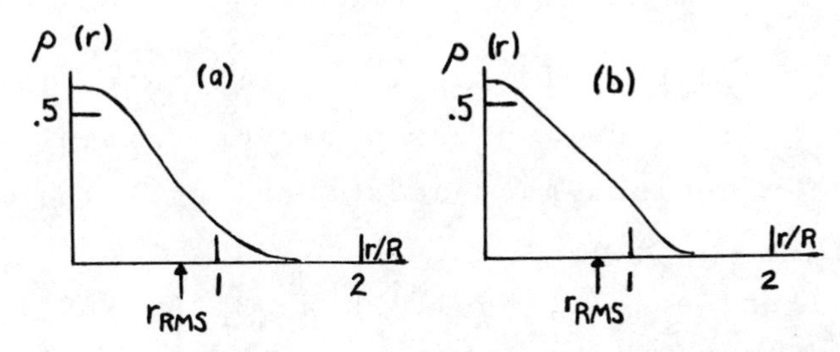

Figure 1. (a) The density for tne meson-meson separation in the ground-state spherical $Q^2\bar{Q}^2$ bag. (b) The density for a free meson-meson wavefunction with its first zero at r=b. b is chosen so that the rms radius of the wavefunction matches the rms radius in Fig. 1(a).

II. Single Channel Analysis of Meson-Meson Scattering

We consider first the exotic channels. Fig. 2(a)[3] shows the I=2 S-wave $\pi\pi$ phase shift and Fig. 2(b) the corresponding P matrix with a pole at 1.04. The Jaffe prediction is 1.15.

Fig. 3(a)[4] shows the I=3/2 S-wave $k\pi$ phase shift and Fig. 3(b) the corresponding P matrix with a pole at 1.19. The Jaffe prediction is 1.35. Note the repulsive phase shifts corresponding to primitives above the compensation energy.

We consider next the non-exotic states below their inelastic thresholds. Fig. 4(a)[5] shows the $\pi\pi$ I=0 S-wave phase shift below $\bar{k}k$ threshold and Fig. 4(b) the P matrix element for the same region. There is a pole at .69; the Jaffe four quark prediction is .65. Going up in energy, there is a second pole at .98; the Jaffe prediction is 1.10 -- it is presumed to be a state with hidden strangeness -- an $s\bar{s}$ pair -- and the remaining $u\bar{u}$ and $d\bar{d}$ in an I=0 combination.

We finally have the I=1/2 $k\pi$ system. The S-wave phase shift is shown in Fig. 5(a)[3], the P matrix element in Fig. 5(b). The pole is at .96, the Jaffe prediction at .90.

We may add to this section preliminary results obtained by M. Roiesnel on meson-baryon scattering and M. Schatz on nucleon-nucleon scattering. The single

---

3. W. Hoogland et al., Nucl. Phys. B126, 109 (1977).
4. P. Estabrooks et al., Nulc. Phys. B133, 490 (1978).
5. S.D. Protopopescu et al., Phys. Rev. D7, 1279 (1973); G. Gruyer et al., Nucl. Phys. B75, 189 (1974). Particle Data Group, Phys. Lett. 75B, 1 (1978).

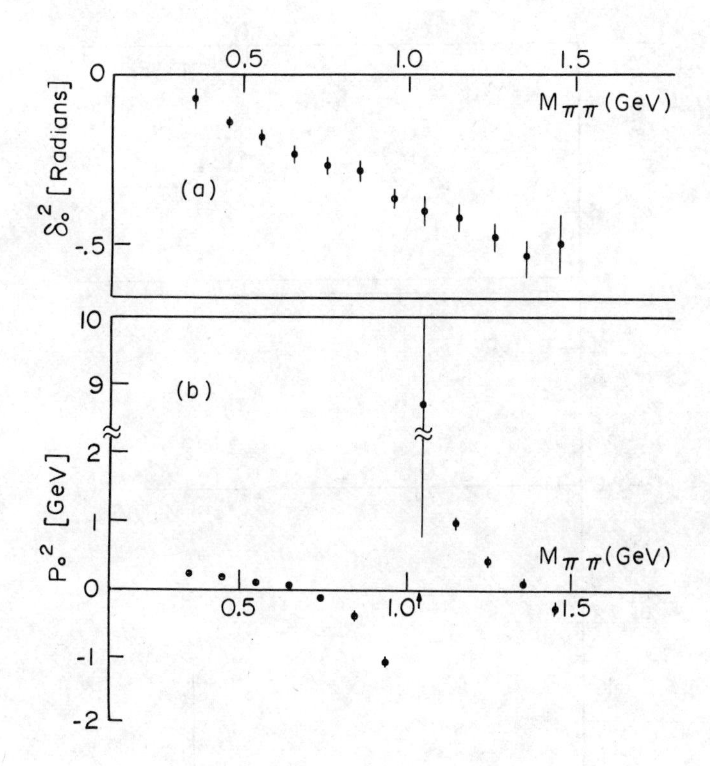

Figure 2. (a) $\pi\pi I=2$ (exotic) S-wave phase shift. (b) $\pi\pi I=2$ S-wave P-matrix element. The arrow marks the P-matrix pole.

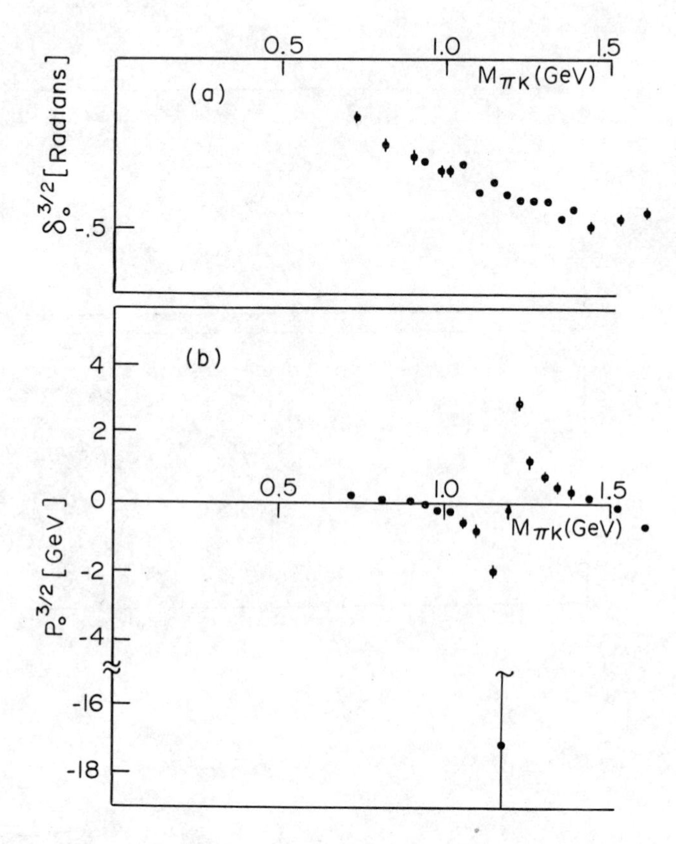

Figure 3. (a) $\pi K I=\frac{3}{2}$ (exotic) S-wave phase shift. (b) $\pi K I=\frac{3}{2}$ S-wave P-matrix element. The arrow marks the P-matrix pole.

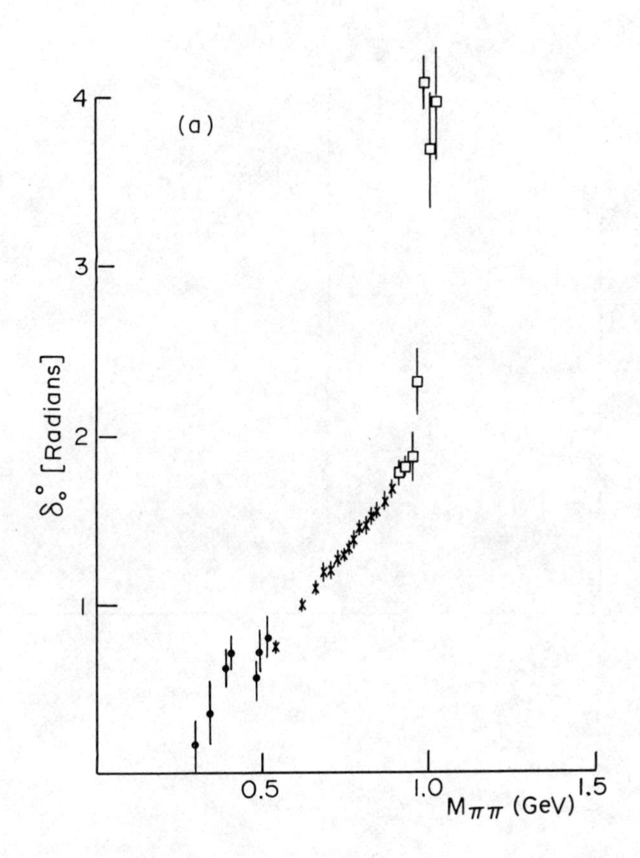

Figure 4(a). $\pi\pi I=0$ (nonexotic) S-wave phase shift for $\sqrt{s} \lesssim 2M_K$.

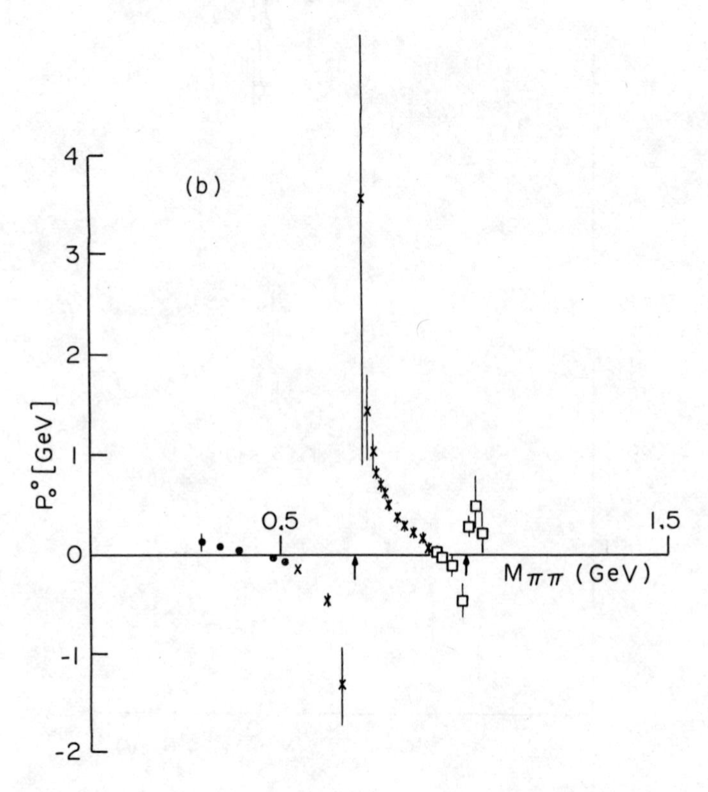

Figure 4(b). $\pi\pi I=0$ (nonexotic) S-wave P-matrix element for $\sqrt{s} \lesssim 2M_K$. The arrow marks the P-matrix poles.

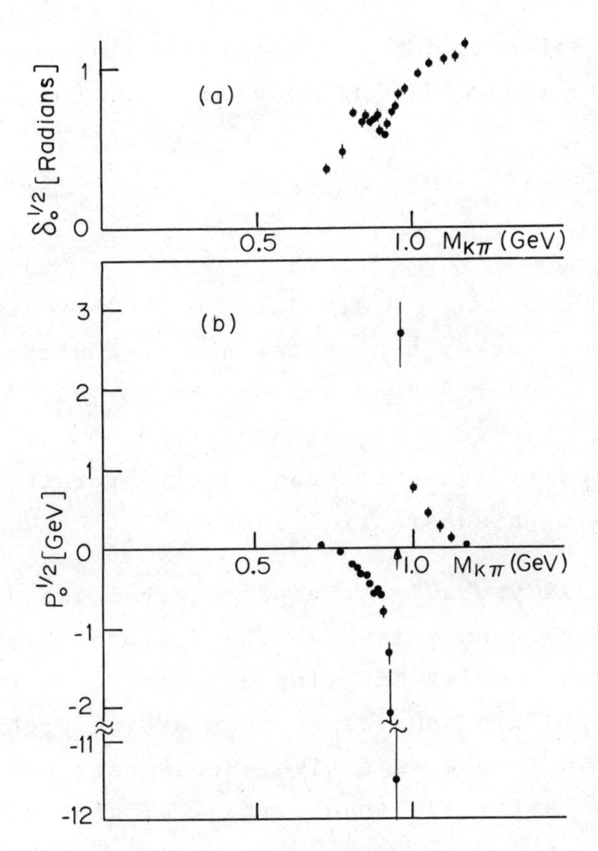

Figure 5. (a) $\pi K I=\frac{1}{2}$ (nonexotic) S-wave phase shift. (b) $\pi K I=\frac{1}{2}$ (nonexotic) S-wave P-matrix element. The arrow marks the P-matrix poles.

channel analysis is complicated in the meson-baryon case by the rapid onset of inelasticity -- nucleons shake off pions very easily. The best channels from this point of view turn out to be the exotic kn I=0 and I=1 states, and the non-exotic πn I=1/2 state. In each of these states M. Roiesnel finds a primitive corresponding reasonably with an MIT bag model 5 quark calculation:

state	primitive	theory
I=0,kn	1.7 GeV	1.7
I=1,kn	1.78	1.9
I=1/2,πn	1.43	1.5

The low energy n-n system has an additional problem, which is the near degeneracy of the $^1S_0$ and $^3S_1$ phases for all but the very lowest energies. On the other hand, the spherically symmetric bag model separates the two primitives by a substantial energy: $E_{^1S_0} = 2.23$, $E_{^3S_1} = 2.15$. It is probable, therefore, that the correct P matrix analysis must involve the feature which differentiates the triplet and singlet states, that is the S-D mixing. This remains an interesting problem for the reader. In the meanwhile, Mr. Schatz has worked out the $^1S_0$ primitive (the virtual singlet state); it lies at $E_{^1S_0} = 2.04$.

III. Multi-channels

We assume as before that outside a relative separation b in the center of mass the n-channel two-hadron system is free and that continuum channels are unimportant. We may then choose n independent ℓ=0 wave functions:

$$\psi_{i,j}(r_j) = \delta_{ij}\cos k_j(r_j-b) + \frac{P_{ji}}{k_j}\sin k_j(r_j-b) \tag{3.1}$$

(where the first index on ψ labels the state, the second the channel) so that a pole of P corresponds to a state whose wave function vanishes at r=b, and hence to the kind of boundary condition on the internal wave functions which we wish to impose. We make contact with the S matrix by noting that

$$\psi_{i,j}(r_j) = \sum_{\ell} \frac{A_{\ell i}}{\sqrt{K_j}} \left[\delta_{j\ell} e^{-iK_j r_j} - S_{j\ell} e^{iK_j r_j}\right] \tag{3.2}$$

for some matrix A. Solving for S,

$$S = -e^{-ikb} \frac{1 - \frac{i}{\sqrt{k}} P \frac{1}{\sqrt{k}}}{1 + \frac{i}{\sqrt{k}} P \frac{1}{\sqrt{k}}} e^{-ikb} . \tag{3.3}$$

In the one-channel case, of course, P reduces to $k\cot[kb+\delta(k)]$.

Note again that P=1/R, where R is the matrix introduced by Wigner and Eisnebud. The extension of the Wigner-Eisenbud type formalism to particle reactions was proposed earlier by Breit and Bouricius and by Feshbach and Lomon[6]. The Feshbach-Lomon F. matrix is essentially our P matrix, although the physics motivation is quite different.

The factor $\frac{1}{\sqrt{k_j}}$ in Eq. (3.2) is arbitrary, and depends on an arbitrary normalization of channel wavefunctions. It, and the corresponding $\frac{1}{k_j}$ in Eq. (3.1), are chosen to make P hermitian, symmetric and free of two body threshold singularities, as we shall now see. We solve Eq. (3.3) for P:

---

6. G. Breit and W.G. Bouricius, P.R. 75, 1029 (1949) H. Feshbach and E.L. Lomon, Ann. Phys. (N.Y.) 29, 19 (1964).

$$P = -i\sqrt{k}\,\frac{1+\tilde{S}}{1-\tilde{S}}\,\sqrt{k}\ , \tag{3.4}$$

with

$$\tilde{S} = e^{ikb}\,S e^{ikb}\ . \tag{3.5}$$

a) Since $\tilde{S}$ is unitary and symmetric, P is hermitian and symmetric, and hence real.

b) The residue of a pole in P factorizes. We see this trivially from Eq. (3.4) for P. Since $\tilde{S}$ is unitary and symmetric we may write $\tilde{S} = e^{2i\tilde{\delta}}$, where $\tilde{\delta}$ is hermitian and symmetric, so that

$$P = \sqrt{k}\,\cot\tilde{\delta}\,\sqrt{k}\ .$$

Expanding in eigenfunctions of $\tilde{\delta}$, we have

$$P_{ij} = \sqrt{k_i}\sqrt{k_j}\,\sum_n \langle i|n\rangle\langle n|j\rangle \cot\delta_n\ .$$

The poles of P_{ij} arise from the poles of $\cot\delta_n$; barring accidental degeneracy, only one of these will occur at a given energy, and its residue will therefore be proportional to

$$\sqrt{k_i}\sqrt{k_j}\langle i|n_o\rangle\langle n_o|j\rangle\ .$$

c) P has no two body threshold singularities. To see this, we define the scattering amplitude matrix f by the equation

$$S = 1 + 2i\sqrt{k}\,f\sqrt{k} \tag{3.6}$$

so that the unitarity of S implies

$$(1-2i\sqrt{k}\,f^{+}\sqrt{k})(1+2i\sqrt{k}\,f\sqrt{k}) = 1 \tag{3.7}$$

or

$$\frac{f - f^{+}}{2i} = f^{+}kf \tag{3.8}$$

and

$$\frac{1}{2}\left(\frac{1}{f} - \frac{1}{f^{+}}\right) = -i \ . \tag{3.9}$$

The scattering amplitude f is closely related to the invariant Møller amplitude $\mathcal{M}$ defined via

$$S = 1 + \frac{(2\pi)^4 \, i \, \delta^4(P_f - P_i)}{\sqrt{\pi_f 2E_f \, \pi_i 2E_i}} \mathcal{M} \tag{3.10}$$

for which the unitarity equation (in the two body space) is

$$\frac{(\mathcal{M} - \mathcal{M}^{+})_{ba}}{2i} = \frac{(2\pi)^4}{2} \int \frac{d\vec{p}_1 \, d\vec{p}_2}{(2\pi)^6 \, 2E_1 \, 2E_2} \, \delta^4(P_1 + P_2 - P)$$

$$\langle b|\mathcal{M}^{+}|1,2\rangle\langle 1,2|\mathcal{M}|a\rangle \tag{3.11}$$

Integration over the intermediate states for ℓ=0 gives

$$\frac{\mathcal{M} - \mathcal{M}^{+}}{2i} = \frac{1}{8\pi W} \mathcal{M}^{+} \, k \, \mathcal{M} \tag{3.12}$$

where k is the intermediate state momentum. Comparison with (3.8) shows that

$$\mathcal{M} = 8\pi W f \ , \tag{3.13}$$

where W is the total c. of m. energy, $W = \sqrt{s}$.

The ℓ=0 Møller matrix has very simple analyticity properties. Each element is a real analytic function of s except for threshold, inelastic and left-hand cuts, and so therefore is $f = \frac{\mathcal{M}}{8\pi W}$ (the singularity at W=0 is a left hand cut in f). The inverse matrix $\frac{1}{f}$ can in addition have poles. If we now write

$$\frac{1}{f} = \frac{s}{\pi}\int \frac{ds'}{s'-s} \; \frac{\rho(s')}{s'} + \text{poles} \tag{3.14}$$

with $\rho(s')$ a symmetric, hermitian and real matrix, Eq. (3.9) tells us that

$$\rho(s') = -\,k(s') \tag{3.15}$$

in the physical region and below the three-body threshold. Therefore, since

$$-\frac{s}{\pi}\int \frac{ds'}{s'-s} \; \frac{k(s')}{s'} + ik(s)$$

has no discontinuity in the same region, the two body branch points in $\frac{1}{f}$ can be removed by writing

$$\frac{1}{f} = -ik + C(s) = D \tag{3.16}$$

where C(s) is a real, hermitian, symmetric matrix without singularities except for left-hand and three or more body cuts and poles. The function $-ik(s)$ is of course the real analytic diagonal matrix.

$$-ik = \sqrt{-k^2\,(s)} \tag{3.17}$$

for which each element is defined to be positive and real just below the threshold of that channel. The matrix k^2 is given by

$$k^2 = [s^2 - 2(m_1^2+m_2^2)s + (m_1^2-m_2^2)^2]/4s \tag{3.18}$$

with m_1 and m_2 the channel masses.

We can now substitute Eq. (3.16) into Eq. (3.6) for the S matrix, and then substitute S so calculated into Eq. (3.4) for P, and thus express P in terms of k, b, and $\mathcal{C}$. We find after some algebra

$$P = (\cos kb\,\mathcal{C} - k\sin kb)\frac{1}{\frac{\sin kb}{k}C + \cos kb} \tag{3.19}$$

which obviously has the desired property, since coskb, ksinkb and $\frac{\sin kb}{k}$ are all functions of k^2, and hence without threshold singularities. Eq. (3.19) is derived in Appendix A.

We note from (3.8) that at b=0,

$$P = C$$

and from Eq. (3.15) for C and (3.6) for f, that

$$C = \sqrt{k}\,\cot\delta\sqrt{k}$$

where the phase matrix δ is defined by $S = e^{2i\delta}$.

Also, although P will have left hand cuts, potential models give us some confidence that, provided b is sufficiently large, their importance should be drastically cut down. Unfortunately we cannot make as strong a statement about the inelastic cuts.

d) As we move below a threshold, the S matrix continues to be given by Eq. (3.3), with P continued analytically. Since P has no threshold singularities, it will vary slowly, and can be used to connect the S matrix slightly above and slightly below a threshold. In particular, one can derive a general formula for the effective open channel P matrix due to higher closed channels.

The mathematical problem is to calculate

$$S_{oo} = -e^{-ikb}\left[\left(1 - \frac{i}{\sqrt{k}}P\frac{1}{\sqrt{k}}\right)\frac{1}{\left(1 + \frac{i}{\sqrt{k}}P\frac{1}{\sqrt{k}}\right.}\right]_{oo} e^{-ikb} \tag{3.20}$$

where of course k is diagonal in the channel space. The

solution to this problem is given in Appendix B. It is

$$S_{oo} = -e^{-ikb} \frac{\left(1 + \frac{i}{\sqrt{k}} P_{eff} \frac{1}{\sqrt{k}}\right) e^{-ikb}}{\left(1 + \frac{i}{\sqrt{k}} P_{eff} \frac{1}{\sqrt{k}}\right)} \tag{3.21}$$

where k and P_{eff} are matrices in the open channels, and

$$P_{eff} = P_{oo} - P_{oc} \left(\frac{1}{\sqrt{-k^2} + P}\right)_{cc} P_{co} \tag{3.22}$$

with o,c representing open and closed channels, respectively. We shall see in the appendix that the true pole in P_{oo} is not present in P_{eff}; instead a displaced pole appears due to the denominator $(\sqrt{-k^2}+P)^{-1}_{cc}$.

e) The residue of a pole is a positive projection. To see this, we turn again to Eq. (3.4) for P and differentiate with respect to b:

$$\frac{dP}{db} = -i\sqrt{k}\left[\frac{d\tilde{S}}{db} \frac{1}{1 - \tilde{S}} + (1 + \tilde{S}) \frac{1}{1 - \tilde{S}} \frac{d\tilde{S}}{db} \frac{1}{1 - \tilde{S}}\right]\sqrt{k} \tag{3.23}$$

Now $\frac{d\tilde{S}}{db} = ik\tilde{S} + \tilde{S}ik$ so that

$$\frac{dP}{db} = +\sqrt{k}\left[\{k,\tilde{S}\} \frac{1}{1 - \tilde{S}} + (1 + \tilde{S}) \frac{1}{1 - \tilde{S}} \{k,\tilde{S}\} \frac{1}{1 - \tilde{S}}\right]\sqrt{k} \tag{3.24}$$

Now substitute $S = -\left(1 - \frac{i}{\sqrt{k}} P \frac{1}{\sqrt{k}}\right) \Big/ 1 + \frac{i}{\sqrt{k}} P \frac{1}{\sqrt{k}}$

and there results

$$\frac{dP}{db} = -k^2 - P^2 \tag{3.25}$$

A pole in P will have a residue p, so that $P = \frac{p}{E-E_o}$ + finite and, to leading order, as $E \to E_o$

$$P/(E - E_o)^2 \, dE_o/db = - P^2/(E - E_o)^2 + \sim \frac{1}{E - E_o} + .. \tag{3.26}$$

Thus, with $p = - \frac{dE_o}{db}\alpha$, Eq. (3.26) shows that

$$\alpha = \alpha^2, \tag{3.27}$$

and the residue p is proportional to a projection operator. The coefficient is $- \frac{dE_o}{db}$, which we expect to be positive, since as b increases, we constrain the system less and hence, for a reasonable system, lower the energy. Note also that for a genuine narrow resonance E_o will be insensitive to k, and the residue will be small, as stated earlier.

f) Our last general observation relates to a remark in Sect. I, to wit, that a single primitive (in a one channel problem) can be an artifact of a boundary condition, but two nearby primitives cannot. Correspondingly, we saw that two nearby primitives required a rapid variation of the scattering phase between them.

The corresponding statement for an N channel problem is obvious: n nearby primitives, with $n \leq N$, can be artifacts of a boundary condition, and hence, barring either small residues or near degeneracy -- that is approximately linearly dependent residues -- no rapid variation of scattering or reaction amplitudes should be expected However, if $n > N$, then we must effectively have more than one primitive per channel, and some anomalous amplitude behavior must be expected.

Let us recall first the basis of anomalous behavior in the one channel problem with $P \sim \text{const} + \frac{\lambda_1}{E-E_1} + \frac{\lambda_2}{E-E_2}$, $E_1 - E_2$ small. Since k and kb are assumed to vary slowly between E_1 and E_2 we must have

$$\cot(\delta + kb) = P/k$$

vary rapidly, from $-\infty$ to $+\infty$, between E_1 and E_2 and hence δ vary over π between those limits.

More formally,

$$\tilde{S} = e^{2ikb} S = -\frac{1 - iP/k}{1 + iP/k}$$

is $\sim +1$ near E_1 or E_2, but somewhere in between, when $P=0$, $\tilde{S} \sim -1$. Hence a rapid variation in $\tilde{S}$ is produced.

The N channel case goes as follows. Suppose

$Q_{ij} = \text{const} + \sum_{k=1}^{n} \frac{\lambda_i^k \lambda_j^k}{E-E_k}$ where $Q = \frac{i}{\sqrt{k}} P \frac{1}{\sqrt{k}}$ with all E_k's near each other and E in that neighborhood. Then

$$\tilde{S} = -\frac{1-Q}{1+Q} = +1 - \frac{2}{1+Q} \tag{3.28}$$

and

$$\tilde{S} = +1 - \frac{2T}{\det(1+Q)} \tag{3.29}$$

where T is the minor of 1+Q.

We consider separately three cases: $n<N$, $n=N$, and $n>N$. For the first two cases, for $E-E_k$ small, $k=1,\ldots n$, the leading term in the derminant is $\sim a/(E-E_1)(E-E_2)\cdots(E-E_n)$, since the residue factorization forbids any terms of the form $\frac{1}{(E-E_1)^2}\cdots$.

For $n<N$, the leading term in T will also be $\sim b/(E-E_1)(E-E_2)\ldots(E-E_n)$, and $\tilde{S} \sim +1 - 2\,b/a$ in the region between E_1 and E_2.

For $n=N$, since the minor T has rank $N-1$, the leading terms in T will be $\frac{1}{(E-E_1)}\cdots \cdot \frac{1}{(E-E_n)}$ with one denominator left out, and we will therefore have

$$\tilde{S} \sim 1$$

for E near $E_1, \ldots E_n$.

These conclusions rest on the linear independence of the residues. If the residues are linearly dependent, or almost so, the conclusions fail to hold. For example, with two states, in two dimensions, the $\frac{1}{\Delta E_1 \Delta E_2}$ term has the coefficient $(\lambda^2_1\lambda^1_2 - \lambda^2_2\lambda^1_1)$. If this coefficient is anomalously small, then the leading term will be $\sim \frac{1}{\Delta E_1} + \frac{1}{\Delta E_2}$ which can again generate rapidly varying S matrix elements.

Finally, for n>N, the λ^k's must be linearly dependent and the leading terms in the determinant will be a sum of terms with successive denominators left out, and hence will produce rapidly varying S matrix elements.

IV. A Two Channel Analysis

We wish to illustrate the considerations of Sect.III, and choose the I=0 S-wave meson-meson system above the $\bar{k}k$ threshold and below the $\eta\eta$ threshold -- that is between 1.0 and 1.1 GeV.

The 2x2 S matrix can be parameterized as

$$S = \begin{pmatrix} \eta e^{2i\delta_1} & i\sqrt{1-\eta^2}e^{i(\delta_1+\delta_2)} \\ i\sqrt{1-\eta^2}e^{i(\delta_1+\delta_2)} & \eta e^{2i\delta_2} \end{pmatrix}$$

leading to an expression for

$$P_{ij} = N_{ij}/D \tag{4.1}$$

with

$$N_{11} = k_1[\eta\sin(\tilde{\delta}_1-\tilde{\delta}_2) - \sin(\tilde{\delta}_1+\tilde{\delta}_2)] \tag{4.2}$$

$$N_{12} = N_{21} = (k_1 k_2)^{1/2}\,(1-\eta^2)^{1/2} \tag{4.3}$$

$$N_{22} = k_2 [\eta \sin(\tilde{\delta}_2 - \tilde{\delta}_1) - \sin(\tilde{\delta}_1 + \tilde{\delta}_2)] \quad (4.4)$$

and

$$D = \cos(\tilde{\delta}_1 + \tilde{\delta}_2) - \eta \cos(\tilde{\delta}_1 - \tilde{\delta}_2) \quad (4.5)$$

with $\tilde{\delta}_i = \delta_i + k_i b$.

In principal, one can obtain δ_1, δ_2 and η from the data, and hence calculate N_{ij} and D. Fig. 6[7] shows D and N_{ij} between 1.0 and 1.1 GeV. The pole of P is now at 1.04, 60 MeV higher than the displaced pole found earlier in the elastic data. The two numbers appear consistent, but the inelastic data are not sufficiently accurate to permit an extrapolation.

Appendix A

From Eq. (3.16), (3.6) and (3.4) we have

$$P = -i\sqrt{k}\,\frac{\left[1 + e^{ikb}\{1 + 2i\sqrt{k}\frac{1}{D}\sqrt{k}\}e^{ikb}\right]}{1 - e^{ikb}\{1 + 2i\sqrt{k}\frac{1}{D}\sqrt{k}\}e^{ikb}}\sqrt{k} \quad (A.1)$$

We write the denominator second, and factor out $e^{ikb}/\sqrt{k}$ on the left of numerator and denominator. Then

$$P = -i\left[e^{-ikb} + e^{ikb} + 2ike^{ikb}\frac{1}{D}\right]\sqrt{k}e^{ikb}$$

$$\times \frac{1}{\left[\frac{e^{-ikb} - e^{ikb}}{\sqrt{k}} - 2i\sqrt{k}\, e^{ikb}\frac{1}{D}\right]\sqrt{k}\, e^{ikb}} . \quad (A.2)$$

7. G. Gruyer et al., ibid: M.J. Corden et al., Rutherford Laboratory Report No. RL-78-067, 1978 (unpublished); P. Estabrooks, Carleton University Report, 1978 (unpublished).

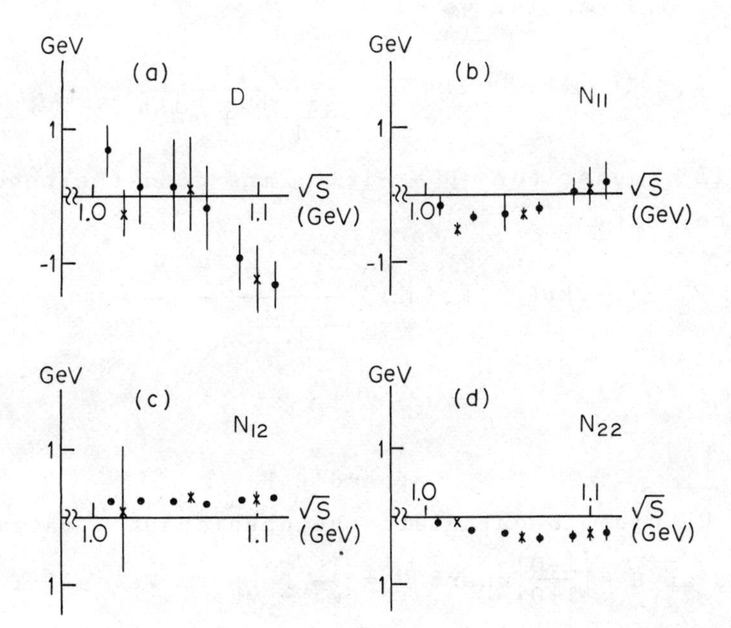

Figure 6. P-matrix elements for $\pi\pi$ and $K\bar{K}$ $I=0$ S wave for $1.01<\sqrt{s}<1.11$ GeV. We assumed (somewhat unrealistically) no error in η.

Now

$$AB\,\frac{1}{CB} = A\cdot\frac{1}{C} \tag{A.3}$$

so that

$$P = (\cos kb + ike^{ikb}/D)\,\frac{1}{\sin kb/k + e^{ikb}/D}\,. \tag{A.4}$$

Using (A.3) again, we find

$$P = (\cos kbD + ike^{ikb})\,\frac{1}{\frac{\sin kb}{k} + e^{ikb}}\,. \tag{A.5}$$

Finally, substitute $D = -ik+C$, and find the result given in the text:

$$P = (\cos kbC - k\sin kb)\,\frac{1}{\frac{\sin kb}{k}C + \cos kb}\,.$$

Appendix B

We wish to calculate the open channel matrix-elements of $-\tilde{S} = \left(\frac{1-Q}{1+Q}\right)$ where $Q = \frac{i}{\sqrt{k}}\,P\,\frac{1}{\sqrt{k}}$. We call $I = \frac{1}{1+Q}$, and so

$$-\tilde{S}_{oo} = (1-Q)_{oo}\,I_{oo} - Q_{oc}\,I_{co} \tag{B.1}$$

while

$$(1+Q)_{oo}\,I_{oo} + Q_{oc}\,I_{co} = 1 \tag{B.2}$$

and

$$(1+Q)_{cc}\,I_{co} + Q_{co}\,I_{oo} = o\,. \tag{B.3}$$

Thus,

$$I_{co} = -\left(\frac{1}{1+Q}\right)_{cc} Q_{co}\, I_{oo} \tag{B.4}$$

and

$$\left[(1+Q)_{oo} - Q_{oc}\left(\frac{1}{1+Q}\right)_{cc} Q_{co}\right] I_{oo} = 1 \quad . \tag{B.5}$$

We find directly from B.1), (B.4) and (B.5)

$$-\hat{S}_{oo} = (1-Q_{eff})\,(1+Q_{eff}) \tag{B.6}$$

with

$$Q_{eff} = Q_{co} - Q_{oc}\left(\frac{1}{1+Q}\right)_{cc} Q_{co} \quad . \tag{B.7}$$

Now $Q = \frac{i}{\sqrt{k}}\, P\, \frac{1}{\sqrt{k}}$

so that

$$Q_{eff} = \frac{i}{\sqrt{k}}\left\{ P_{oo} - iP_{oc}\frac{1}{\sqrt{k_c}}\left(\frac{1}{1+\frac{i}{\sqrt{k}}P\frac{1}{\sqrt{k}}}\right)_{cc}\frac{1}{\sqrt{k_c}}\, P_{co}\right\}\frac{1}{\sqrt{k}} \tag{B.8}$$

$$P_{eff} = P_{oo} - P_{oc}\left(\frac{1}{k_{c/i} + P}\right)_{cc} P_{co} \tag{B.9}$$

we recall here that $-ik_c = +\sqrt{-k_c^2}$ with $-k_c^2>0$

so that

$$P_{eff} = P_{oo} - P_{oc}\left(\frac{1}{\sqrt{-k^2} + P}\right)_{cc} P_{co} \quad . \tag{B.10}$$

We also wish to show the cancellation of the original pole in P_{eff}, together with the appearance of a new, displaced one. This cancellation rests on the factorization of the residue of the original pole in D. Thus, the residue of the original pole in P_{eff} is given by

$$r_{oo'} = \lim_{E\to E_o} (E-E_o)\left\{\frac{\lambda_o\lambda'_o}{E-E_o} - \frac{\lambda_o\lambda_c}{E-E_o}\left(\frac{1}{\sqrt{-k^2}+A+\frac{p}{E-E_o}}\right)_{cc'}\frac{\lambda_{c'}\lambda'_o}{E-E_o}\right\} \tag{B.11}$$

where the projection operator $p_{cc'} \equiv \lambda_c\lambda_{c'}$ and A is that part of P_{cc} which is not singular at $E=E_o$. Taking the limit, we find

$$r_{oo'} = \lambda_o\lambda_{o'}\left\{1 - \lim_{E\to E_o} \mathrm{Tr}\,\frac{1}{(\sqrt{-k^2}+A)E-E_o+p}\cdot p\right\} \tag{B.12}$$

There is no difficulty taking the limit, since the projection operator projects onto the subspace for which $p \neq 0$. Then the limit exists and vanishes. The new pole is given by the zero of the closed channel determinant

$$\det\{(\sqrt{-k^2}+A)(E-E_o)+p\}$$

which for small p will be close to E_o.

DISCUSSIONS

CHAIRMAN: Prof. F. Low

Scientific Secretaries: L. Anderson, I. Caprini and F. Pauss

DISCUSSION 1

- CHRISTOS:

What is the difference in the role of the boundary conditions in the $q\bar{q}$ and $q^2\bar{q}^2$ systems?

- LOW:

In the $q\bar{q}$ case the color confining forces make the natural wave function fall rapidly with the relative separation r between the two quarks, and therefore imposing the bag boundary condition is irrelevant. On the contrary, the $q^2\bar{q}^2$ system contains color singlet subunits which are not bound to each other by color forces and can be separated as real hadrons. In this case the bag boundary conditions become real conditions that are imposed on the system and modify it. However, it can be shown using our formalism that the four-quark component of the ρ -meson wave function is not large, so although the boundary condition is important for the four-quark component, the location of the state is not particularly determined by this boundary condition.

- GINSPARG:

Do you consider multiparticle intermediate states, or just two-meson states?

- LOW:

The method is applicable to multi-two-body channels, but not to genuine multi-body channels. This approximation is however very good for low-energy meson-meson scattering, which is known phenomenologically to be dominated by two-body channels. Moreover, the $\pi\pi$ exotic channels turn out to be single-channel problems up to high energies.

- GINSPARG:

How can you justify your procedure of considering the primitive discrete states as dominating the amplitude?

- LOW:

I did not say that they dominate the scattering as one might for poles dominating the scattering amplitude. They are only useful, as in the Wigner-Eisenbud model, for making the connection between the internal description given by the bag model and the external two-pion wave function. The main point is that, when the true wave function vanishes at the position where the wall is introduced, then the wall does not make any difference; then the energy of the primitive state inside the wall must be the same as the energy of the external wave function vanishing at the wall position, and rigorously speaking, only at that energy can the phase shift be calculated. However, the boundary matching procedure is soft, in the sense that b is not exactly determined, and therefore this part of the analysis contains an unavoidable uncertainty.

- PAFFUTI:

Is there a spectrum of physical states corresponding to the surface excitations of the bag itself?

- LOW:

In the present phenomenological approach these excitations are not considered and it is difficult to say how seriously one should take the quantum mechanics of the surface.

- JAFFE:

If the question is whether there are surface excitations independent of the quark and gluon degrees of freedom, there is a naïve argument against this at the lagrangian level. The time derivates of the surface variables do not appear in the Lagrangian, so it looks as if the surface variables should only be variables of constraint and that the quark degrees of freedom are complete. Also, Rebbi's calculations, at the level of small fluctuations about the semiclassical solutions show that all of the surface oscillations can be identified with the quark degrees of freedom.

- PATRASCIOIU:

In your phenomenology is the bag taken as a rigid sphere with free quark fields inside, or do you consider also quark interactions?

- LOW:

The bag is taken as a rigid sphere with magnetic interactions treated in the first order.

- HANSSON:

Could a potential model be used in this analysis instead of the bag?

- LOW:

The technique of matching the inside to the outside regions of the configuration space could be used for a large variety of internal systems, as long as a vanishing condition is imposed at the boundary.

- HANSSON:

Is the fact that the bag has a distinct boundary important?

- LOW:

No, because the effective boundary used ($b=1.4R$) is well inside the boundary $2R$ imposed by the bag.

- HANSSON:

Would it be more convenient to use a potential model for analyzing orbitally excited four-quark states?

- LOW:

This is a dynamical question about the internal state, which is not directly related to the analysis I presented. I do not think that it would be correct, since we believe, at least for highly excited angular momenta, that the angular momentum associated with rotating color flux lines is an important aspect of the system.

DISCUSSION 2

- CAPRINI:

Did you find primitives of the bag model, with energies in the domain investigated in your analysis, which do not correspond to phenomenological poles in the P-matrix?

- LOW:

There certainly are higher states that we did not identify, but, as I said, when you go to higher energies more channels open up. Even if they are two-body channels, in order to carry out the analysis we must know the entire S-matrix, which we do not. In addition, genuine three-body channels are opening, and therefore the formalism can not be applied. As far as I know, this is a successful method of analysis for all the expected states in the energy range where it can be applied.

- CAPRINI:

Are you justified in neglecting the inelastic and left hand cuts of the P-matrix?

- LOW:

The left hand cuts correspond in some way to potentials, so taking b outside the finite range of the force, we believe, decreases the importance of the left hand cuts in the finite complex plane. In the non-relativistic limit for a finite range potential, the P-matrix can be expanded uniquely as a sum of poles plus a subtraction constant - no other singularities. The P-matrix has also an inelastic cut.

- SZWED:

Does the six-quark bound state with strangeness -2, predicted by Jaffe, survive your analysis?

- LOW:

Yes, it should be there, but with a mass below that of the primitive. It would be extremely interesting to find it, but the matrix element for its production is probably very small.

- GINSPARG:

If the P-matrix has a pole at the compensation energy and the residue is approximately equal to the free value, will the primitive have a negligible effect on the scattering near the compensation energy?

- LOW:

Indeed, if there is a pole in the P-matrix at the compensation energy and the residue is essentially the residue of kcot (kb), then in fact the scattering phase shift should be zero over some energy range. Of course, there are also subtraction constants and the effect depends on how large they are.

- PAFFUTI:

How rapidly does the number of states grow with mass in the bag model? Is it similar to that predicted by dual models?

- LOW:

The number of primitives grows very rapidly, probably exponentially. Presumably the highly excited angular momentum states have string-like characteristics, the tension being given by the

α' of the particle trajectories.

- HANSSON:

How do the results of the analysis depend on b, if it is taken as a free parameter?

- LOW:

The bag calculations make sense only in the neighbourhood of a minimum of the bag energy as a function of the bag radius R. That is, for larger b, the spherical cavity approximation becomes very bad, since the system will tend to separate into color singlets; for smaller b, the interaction will certainly not vanish for $r > b$. Therefore in our analysis R was fixed by this requirement, so that although in principle b is a free parameter, in practice it is confined to a narrow range. Therefore a potential model of excitation is not likely to be successful.

- COLEMAN:

Given that you have to make a P-matrix analysis, how well does the bag model agree with the actual experimental data?

- JAFFE:

One might argue that the agreement with experiment is trivial. The argument is as follows: in the absence of the gluon corrections a four-quark bag looks roughly like twice the two-quark bag, i.e. $M_{4q} \simeq 2M_{2q}$, this being just the kinetic energy of the quarks. But also the compensation mass $M_{comp} \simeq 2M_{2q}$, since it is given by $2\,k_{comp} \simeq 2\,\frac{\pi}{b}$, where $\frac{\pi}{b}$ is a typical momentum of a particle confined in a configuration region scaled by b, which corresponds actually to the size of the meson. So perhaps the agreement between experiment and theory will be more or less guaranteed. To see this is not so, I have tabulated the results of two theoretical calculations and of two "experimental" obervations for $\pi\pi$ channels:

Channel	Exp		Bag	
	$\delta = 0$	δ_{exp}	magnetic corrections	no magnetic corrections
$(\pi\pi)^0$	.95	.69	.65	1.46
$(\pi\pi)^2$	.95	1.04	1.15	1.46

The first and the second columns contain the compensation masses and the experimental primitives, i.e. the P-matrix poles, respectively . In the third and fourth columns the theoretical calculations of the bag model with and without color magnetism are given. To compare theory and experiment I have calculated the RMS deviations for the four possible cases (in MeV):

	$\delta = 0$	δ_{exp}
no color magnetism	500	550
with color magnetism	250	80

Notice that the real bag calculation (with color magnetism) agrees much better with the real experiment than the naïve bag estimate (no color magnetism) agrees with a world with no meson-meson interaction. The analysis really is predictive. The same conclusion is obtained for the two πK channels.

EXACT RESULTS IN THE THEORY OF NEW PARTICLES

A. Martin

CERN

Geneva, Switzerland

In this written version of my lectures, which are in a way a continuation of my lectures at Erice in 1977, I shall not repeat the general description of the situation of "quarkonia", the J/ψ family, or the Υ family. Figure 1 summarizes the present knowledge of the J/ψ spectrum. The most noticeable improvement is the disappearance of the state at 2.83 GeV, decaying mostly into 2γ, and its replacement by a state X at 2.97 GeV, observed in $\psi' \to \gamma + X$ and $J/\psi \to \gamma + X$, decaying mostly into hadrons. For theorists this is a relief, because it was difficult to understand the large energy gap between 2.83 and the J/ψ, the too low transition rate observed, and the dominance of the $\gamma\gamma$ decay mode. On the other hand, as I said during my lectures, it would have been a catastrophe if a candidate for the 0^{-+} brother of the J/ψ had not been observed.

The situation (at the time of writing, which is March 1980!) of the upsilon spectrum is summarized in Fig. 2. It is extremely impressive.

Two points I would like to discuss are

1) the validity of the potential-model, non-relativistic, description of these systems as heavy quark-antiquark bound states;

2) the flavour independence of the forces between quarks, and its consequences.

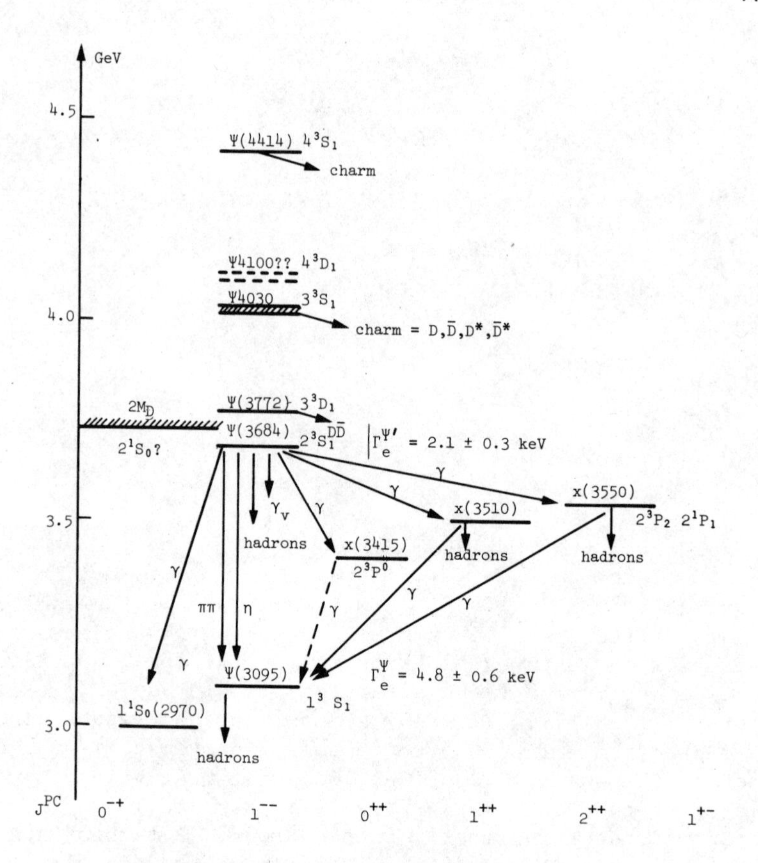

Fig. 1 The J/ψ spectrum

		$\Gamma_{ee}/\Gamma_{ee}(\Upsilon)$	Γ_{ee}	Γ_{tot}
Υ'''	10.58*	0.2 ± ?*		∿ 20 MeV
Υ''	10.35*○	0.32 ± 0.04* 0.35 ± 0.04○		
Υ'	10.02*○□	0.39 ± 0.06* 0.44 ± 0.06○		
Υ	9.46*○□	1	1.2 keV○	$40 {+50 \atop -20}$□ keV

MASS SCALE: * CUSB
DESY ○ CLEO
□ DESY

Fig. 2 The ϒ spectrum

Concerning point (1), there is some new material to present. Until recently the justification for non-relativistic treatment was *a posteriori:* you try to fit the J/ψ and/or the ϒ spectra and leptonic widths by a potential and a choice of the quark mass, and you check that the mean value of v^2/c^2 is not too large. I present here an inequality and an estimate which make it possible to see directly, from the knowledge of the energy levels, whether this approximation is reasonable or not. The inequality is

$$\langle T\rangle_{1S} \geq \frac{3}{4}\,(E_{1P} - E_{1S}) \,, \tag{1}$$

where T designates the kinetic energy; it goes in the wrong direction, and all we can say is that if the r.h.s. of (1) is too large, the non-relativistic approximation is unacceptable. However, (1) becomes an equality for a potential which is a harmonic oscillator well, and an approximate equality within a few per cent for potentials of the form r^α, $1 < \alpha < 4$. An estimate, inspired by inequality (1) and adjusted to be exact for a pure Coulomb potential, is

$$\langle T\rangle_{1S} \simeq \frac{3}{4}\,(E_{1P} - E_{1S})\left[1 + \frac{7}{9}\left(\frac{E_{1S} + E_{2S} - 2E_{1P}}{E_{2S} - E_{1S}}\right)^2\right] . \tag{2}$$

This estimate has been checked for usual potentials, such as the mixture $-(A/r) + Br$, and found to be good to a few per cent. When applied to the J/ψ system, where we take for E_{1P} the mean energy of the P states, with weight 2J + 1, we find $\langle T\rangle_{1S} \simeq 0.375$ GeV, which for a charmed quark mass of 1.5 gives for the velocities of each quark,

$$\left\langle \frac{v^2}{c^2} \right\rangle = 0.25 \,. \tag{3}$$

This is small enough to justify a non-relativistic treatment in first approximation.

We expect that this non-relativistic treatment will be much better justified for the ϒ, in which the quarks are much heavier.

We now turn to the question of flavour independence. In the framework of QCD, the coupling between quarks and gluons is flavour independent. Indeed gauge invariance requires that this coupling be equal to the three-gluon coupling, characteristic of the non-Abelian character of the gauge group. It follows that the quark-antiquark potential will be, at least approximately, flavour-independent. Therefore the $b\bar{b}$ and the $c\bar{c}$ potentials will be approximately the same. Let us remark that in the initial versions of the potential, the short range part was taken to be

$$-\frac{4}{3}\frac{\alpha_s}{r} \tag{4}$$

and, in this case, α_s should depend on the relevant mass in the problem. It seems better to take into account asymptotic freedom in the potential itself by softening the short distance part:

$$V = -\frac{8\pi}{27\, r \log\left(\frac{r_0}{r}\right)} \; . \tag{5}$$

If one takes this kind of precaution, one finds (and this has been done by several groups) that one can discover a unique potential fitting the upsilon and charmonium spectra. From this kind of analysis, Quigg and Rosner find, for instance,

$$M_b - M_c = 3.455 \text{ GeV} \; . \tag{6}$$

In fact one can find rigorous inequalities on this mass difference, such as $M_b - M_c > 3.29$ GeV (Grosse and Martin).

If one accepts the fact that the potential is flavour-independent, one can discuss the question of the number of bound states of quarkonia that are stable with respect to the Zweig rule as a function of the mass of the quarks.

The first remark is that the mesons with visible "heavy flavour" are expected to have a binding energy which is independent, in first approximation, of the mass of the heavy quark. Indeed, a system $(Q\bar{q})$ (where $Q = c, b$, and hopefully t, and $\bar{q} = \bar{u}, \bar{d}$) is a system with an almost fixed centre which will have a mass $M_Q + \Delta$, where Δ is independent of the heavy flavour even if the light quark u or d is relativistic.

Then a $Q\bar{Q}$ system will become unstable if

$$M_{Q\bar{Q}} \geq 2(M_{Q\bar{q}}) = 2\Delta + 2M_Q \; , \tag{7}$$

and since $M_{Q\bar{Q}} = 2M_Q + E$, where E is the binding energy, (7) becomes

$$E \geq 2\Delta \; . \tag{8}$$

If we accept the validity of the WKB approximation, the nth $\ell = 0$ state will be such that

$$n - \frac{1}{4} = \frac{1}{\pi} \int \sqrt{2M_Q (E - V)}\, dr \ , \tag{9}$$

where the integral extends to the region where E - V is positive. Therefore the number of bound states that are stable with respect to the Zweig rule will be given by

$$n = \frac{1}{4} + \sqrt{2M_Q}\, \frac{1}{\pi} \int \sqrt{2\Delta - V}\, dr \ . \tag{10}$$

Since we know that n = 2 for charmonium, we conclude that

$$n \simeq \frac{1}{4} + \frac{7}{4} \sqrt{\frac{M_Q}{M_c}} \tag{11}$$

for the upsilon system; with $M_Q \simeq 4.5\text{-}5$ and $M_c \sim 1.2\text{-}1.5$ we get

$$n \simeq 3\text{-}3.5 \ . \tag{12}$$

This agrees with experiment. We do not yet have measurements of the total widths of the Υ, Υ', and Υ'', but if we assume that their production cross-sections in p + nucleus collisions are of the same order of magnitude and note that their leptonic widths are all of the same order of magnitude theoretically -- as shown by an argument of Bertlmann, Grosse and myself -- and experimentally confirmed, we conclude by combining these results with the observations of the Lederman group that the total widths are all of the same order of magnitude, i.e. not more than a few hundred keV, as in the case of the Υ for which a measurement has been made at DORIS.

The theoretical argument concerning the leptonic widths is the following. The most simple-minded estimate of the leptonic widths is given by the Weisskopf-Van Royen formula:

$$\Gamma_{ee} = 16\pi\alpha^2 e_Q^2 \frac{|\psi(0)|^2}{(M_{res})^2} \ . \tag{13}$$

Assume now that the potential has a power behaviour r^α (this hypothesis is not as wild as it looks, because only the medium range of the potential counts for not too high levels and not too heavy quark masses), and assume furthermore the validity of the WKB approximation. Then we get

$$|\psi_n(0)|^2 \simeq C\left(n - \frac{1}{4}\right)^\lambda , \tag{14}$$

where λ depends in a known way on α. By using three values of n, one can eliminate C and λ. So

$$\left|\frac{\psi_n(0)}{\psi_1(0)}\right|^2 \simeq \left(\left|\frac{\psi_2(0)}{\psi_1(0)}\right|^2\right)^{\log[(4n-1)/3]/\log(7/3)} . \tag{15}$$

When this is applied to the case n = 3 it gives the leptonic width of the Υ'' as a function of the leptonic widths of the Υ and of the Υ'. Experiments with various potentials show that (15) is good to within 15%.

Another consequence of (7) is that it gives another test of flavour independence. In (7) what counts is that the forces between light and heavy quarks are flavour independent, and this leads to a prediction of the quark mass difference. For instance,

$$M_{(b\bar{u})} - M_{(c\bar{u})} = M_b - M_c . \tag{16}$$

This mass difference between the quarks should coincide with the one determined from comparison of the upsilon and charmonium spectra, given by (6), and which is obtained by assuming equality of the $b\bar{b}$ and $c\bar{c}$ forces. The coincidence of the two results would constitute a non-trivial test of flavour independence. If we believe the indications of an experiment done at CERN, according to which there is a state of mass 5.3 GeV decaying into $J/\psi + \bar{K}$ [a likely decay for a $(b\bar{u})$ or $(b\bar{d})$ object, according to Fritzsch], we get

$$M_b - M_c \simeq 5.3 - 1.87 = 3.43 \text{ GeV} , \tag{17}$$

in very good agreement with (6). Even if the observation is not confirmed, we believe that a state $(b\bar{u})$ or $(b\bar{d})$ should have a mass very close to 5.3 GeV. Its absence would be a major catastrophe for flavour independence.

REFERENCES

H. Grosse and A. Martin, Preprint CERN TH. 2674 (1979), to appear in Phys. Reports.

M. Krammer and H. Kraseman, Acta Physica Austriaca Suppl. XX (1979).

A. Martin, "Ettore Majorana" International School of Subnuclear Physics, 15th Course: The Whys of Subnuclear Physics (ed. A. Zichichi) (Plenum Press, New York and London, 1979), p. 395.

C. Quigg and J. Rosner, Phys. Reports 56, 167 (1979).

C. Quigg and J. Rosner, communication to the EPS Int. Conf. on High-Energy Physics, Geneva, 1979.

H. Schopper, "Ettore Majorana" International School of Subnuclear Physics, 15th Course: The Whys of Subnuclear Physics (ed. A. Zichichi) (Plenum Press, New York and London, 1979), p. 203.

DISCUSSIONS

CHAIRMAN: Prof. A. Martin

Scientific Secretaries: J. Maharana and A.K. Raina

DISCUSSION 1

- VANNUCCI:

Experimentally it is attractive to look for the decay $\eta_c \to \varphi\varphi$ giving 4K in the final state. Do you have an idea about $\Gamma(\eta_c \to \varphi\varphi)/\Gamma(\eta_c \to \gamma\gamma)$?

- MARTIN:

I have no comments.

- OLIENSIS:

Can you explain again the two approaches to the charmonium and upsilon potentials in the short distance region?

- MARTIN:

There are two approaches. One approach takes the potential coming from one gluon exchange with a coupling constant dependent on quark mass and hence different for charmonium and upsilon system. In the other one takes the Fourier transform of the qq scattering amplitude in the one gluon exchange approximation.

This gives const./rlog(r) behaviour for the potential which is then less singular at short distances. The potential is now the same for both systems. Effectively the two approaches are similar since the upsilon, being heavier and more concentrated near the origin, is more sensitive to the logarithm.

- PREPARATA:

I am worried about the non-relativistic approximation when distances become too small. Am I right to worry?

- LOW:

The heavy quark system is non-relativistic only because the coupling constant gets weak at high mass via asymptotic freedom.

- GINSPARG:

I am interested in Maiani's prediction of the charmonium mass from the K° mass splitting which you mentioned this morning. We expect to find a similar CP violating system in the B meson sector; would it be possible to estimate the t-quark mass from the analogous mass splitting?

- PREPARATA:

Yes. The relations in the Kobayashi-Maskawa model have been calculated by Ellis, Gaillard and Nanopolous.

- VANNUCCI:

In the comparison of the $c\bar{c}$ and $b\bar{b}$ systems there is a large difference in hadronic productions, namely:

$B\sigma(\Upsilon')/B\sigma(\Upsilon) \simeq 40\%$ while $B\sigma(\psi')/B\sigma(\psi/J) \simeq 2\%$.

Do you have a clue to this puzzle?

- MARTIN:

A large fraction of J/ψ's are produced by the decay $\chi \rightarrow J/\psi + \gamma$.

- VANNUCCI:

But the same argument could hold for P states between Υ and Υ'.

- STERLING:

There should be P states lying above the Υ' which will decay into $\Upsilon' + \gamma$ and $\Upsilon + \gamma$ for which there is no analogy in the ψ system. This increases the production of Υ'.

- MARTIN:

Lederman has measured the production cross section for Υ, Υ' times their leptonic branching ratios. We also know the leptonic widths to reasonable accuracy. Theoretical arguments could then conclude from this data that Υ, Υ', Υ'' all have the same approximate total width, which is strange.

- PATON:

What is the advantage of this analytic method of estimating the bound-state energies over a simple numerical evaluation for a given potential?

- MARTIN:

It is a matter of taste.

- RAINA:

Can you explain how the lower bound on the kinetic energy in the ground state could enable you to conclude that the non-relativistic approximation was justified?

- MARTIN:

Although we have a lower bound, numerical tests show that it gives a good estimate. We see that relativistic velocities are not involved.

DISCUSSION 2

- VANNUCCI:

Can you apply the model to the ss system?

- MARTIN:

The strange quark system is too light for the non-relativistic approximation to be reasonable.

- OLIENSIS:

In the model fittings to the charmonium and bottonium systems, how important is the long-range potential compared to the short-range potential?

- MARTIN:

The linear potential is very important. In fact the charmonium system can almost be fitted by a linear potential. Potentials increasing faster than linear for large r are hard to reconcile with the assumption that the bumps at 4.03 and 4.4 GeV are higher radial excitations.

- GINSPARG:

I have a question related to the following observation: if one interpolates between the Coulomb and linear parts of the potential with a logarithmic potential, one can ensure equal $\psi - \psi'$ and $\Upsilon - \Upsilon'$ splittings since the states are found to be characterized by a length-scale lying in the logarithmic region. My question is whether it is possible to relax your conditions that the potential satisfies for all r to only a limited range of r in proving, e.g. $E_{1D} > E_{2S}$.

- MARTIN:

Yes. My conditions are merely sufficient. I should add that the potentials I discussed may give almost equal splitting for $J/\psi - \psi'$ and $\Upsilon - \Upsilon'$ systems but would predict a deviation from equal splitting for the toponium system. Here the prediction is different from the equal spacing predicted by a logarithmic potential.

- HANSSON:

What are the effects of the long range force on spin-dependent effects?

- MARTIN:

I have disregarded spin-dependent forces.

- LOW:

I have a comment. In states like $c\bar{u}$, the heavy quark mass dependence will depend on the light quark energy rather than its mass, which means 300 MeV.

- MARTIN:

Thank you. The considerations nevertheless remain valid.

- GINSPARG:

It seems that your theorem regarding the increasing number of stable states with increasing quark mass depends crucially on the flavor independence of the heavy quark - light quark binding energy. How do you prove this flavor independence?

- MARTIN:

We can suppose that the heavy quark acts like a fixed center, even if we take the energy of the light quark rather than its mass for comparison as suggested by Prof. Low. Since the force is flavor independent then Δ should be independent of quark mass.

- GINSPARG:

From the equality of the $c\bar{c}$ and $b\bar{b}$ forces you derived an inequality for the c and b quark masses. Under what conditions does this become an equality?

- MARTIN:

For a Coulomb potential.

- SCHMIDT:

Knowing that the logarithmic potential is a convenient solution, or near solution, to the "dilemma" of the $J/\psi - \psi'$ and $\Upsilon - \Upsilon'$ splittings, if when $t\bar{t}$ states are found, they are found to have the same spacing, would this be a problem?

- MARTIN:

Yes, because we know that the short distance behaviour of the potential is like 1/r, we expect flavoronium to probe this behaviour more deeply and that the levels should not be equally spaced.

- MAHARANA:

In the decay $\psi' \to \eta_c \gamma$ the transition is M1 and the matrix element vanishes due to orthogonality of 1s-2s wave functions. Do you think 1s-2s mixing due to the spin-dependent contact term will give rise to appreciable effects in $\psi' \to \eta_c \gamma$?

- MARTIN:

$\psi' \to \eta_c \gamma$ matrix element is nonzero when recoil effects are taken into account. Spin-dependent interactions can mix ψ' and J/ψ states.

- LOW:

The charmonium P states give some estimates of the tensor force strength. How important is that in your considerations? I understand that your theorems do not include tensor forces.

- MARTIN:

The P-wave splittings due to tensor forces and spin-orbit forces are well accounted for by the contributions coming from one gluon exchange. It is not necessary to appeal to a particular spin dependence of the confining force. This is very well explained in the DESY report 79/20 by M. Krammer and A. Krasemann.

NEUTRINO INTERACTIONS: NEUTRAL AND INCLUSIVE CHARGED CURRENTS

F. Dydak

Institut für Hochenergiephysik der Universität

Heidelberg, Germany

1. NEUTRAL CURRENTS

1.1 Introduction

Weak neutral currents were discovered in 1973 by the Gargamelle (GGM) Collaboration[1]. Four years elapsed until 1977, when data were presented by the GGM[2] and the CERN-Dortmund-Heidelberg-Saclay (CDHS) Collaboration[3] which convincingly unravelled the isospin and the space-time structure of this new interaction. Since then, important new results and improvements of old results have been reported, but our picture of the neutral-current interaction did not change. Today, this picture is well supported by many experiments which explore various aspects of neutral currents.

The processes which can occur via the weak neutral-current interaction are depicted in the "Sakurai tetragon"[4] which is shown in Fig. 1. It is an analogue to the Puppi triangle for charged-current interactions. The coupling constants governing the various neutral-current interactions are based on the assumption that the neutral-current interaction is effectively of the current-current form where the current is made up of a linear combination of vector and axial vector covariants, and the hadronic weak current comprises isoscalar

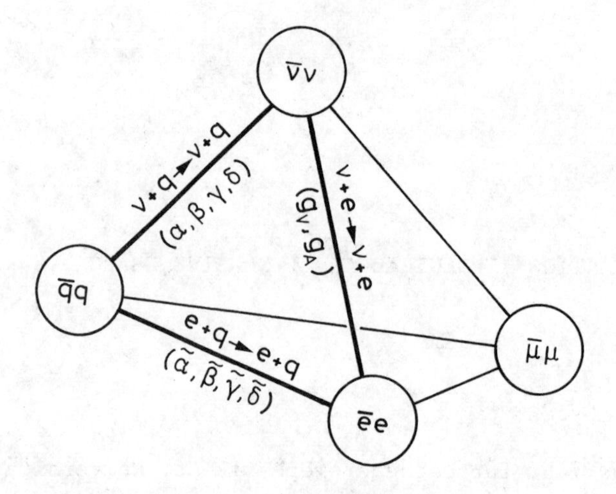

Fig. 1 The Sakurai tetragon of neutral-current interactions.

and isovector pieces only. The effective Lagrangians of the neutral-current interactions which have been explored experimentally are:

for the process $\nu + e \rightarrow \nu + e$:

$$L = -\frac{G}{\sqrt{2}} \left[\bar{\nu}\gamma_\lambda\left(1 + \gamma_5\right)\nu\right]\left[\bar{e}\gamma^\lambda\left(g_V + g_A\gamma_5\right)e\right] ;$$

for the process $e + q \rightarrow e + q$ (parity-violating parts only):

$$L = -\frac{G}{\sqrt{2}} \left\{\bar{e}\gamma_\lambda\gamma_5 e\left[\frac{\tilde{\alpha}}{2}\left(\bar{u}\gamma^\lambda u - \bar{d}\gamma^\lambda d\right) + \frac{\tilde{\gamma}}{2}\left(\bar{u}\gamma^\lambda u + \bar{d}\gamma^\lambda d\right)\right]\right.$$

$$\left. + \bar{e}\gamma_\lambda e\left[\frac{\tilde{\beta}}{2}\left(\bar{u}\gamma^\lambda\gamma_5 u - \bar{d}\gamma^\lambda\gamma_5 d\right) + \frac{\tilde{\delta}}{2}\left(\bar{u}\gamma^\lambda\gamma_5 u + \bar{d}\gamma^\lambda\gamma_5 d\right)\right]\right\} ;$$

and for the process $\nu + q \rightarrow \nu + q$:

$$L = -\frac{G}{\sqrt{2}} \bar{\nu}\gamma_\lambda(1 + \gamma_5)\nu \left[\frac{1}{2}\bar{u}\gamma^\lambda(\alpha + \beta\gamma_5)u - \frac{1}{2}\bar{d}\gamma^\lambda(\alpha + \beta\gamma_5)d\right.$$

$$\left. + \frac{1}{2}\bar{u}\gamma^\lambda(\gamma + \delta\gamma_5)u + \frac{1}{2}\bar{d}\gamma^\lambda(\gamma + \delta\gamma_5)d\right] .$$

This definition of the coupling constants is due to Hung and Sakurai[5]. The relation between these coupling constants and the Lorentz covariants and the isospin components of the weak hadronic current is given in Table 1.

Table 1

The relation of the coupling constants (Hung and Sakurai notation) to the Lorentz covariants and the isospin components of the weak hadronic current

	Lorentz covariant	Isospin component
$\alpha(\tilde{\alpha})$	Vector	Isovector
$\beta(\tilde{\beta})$	Axial vector	Isovector
$\gamma(\tilde{\gamma})$	Vector	Isoscalar
$\delta(\tilde{\delta})$	Axial vector	Isoscalar

An alternative very useful notation has been made popular by Sehgal[6]. The "chiral coupling constants" are defined via the effective Lagrangian for the process $\nu + q \to \nu + q$:

$$L = - \frac{G}{\sqrt{2}} \bar{\nu}\gamma_{\lambda}(1 + \gamma_5)\nu \left\{ \bar{u}\gamma^{\lambda}\left[u_L(1 + \gamma_5) + u_R(1 - \gamma_5) \right] u \right.$$

$$\left. + \bar{d}\gamma^{\lambda}\left[d_L(1 + \gamma_5) + d_R(1 - \gamma_5) \right] d \right\} .$$

The meaning of the chiral coupling constants in terms of the chirality of the weak hadronic current and their relation to the coupling constants defined above is given in Table 2.

The experimental challenge is to determine the various neutral-current coupling constants as precisely as possible, and to compare them with theoretical predictions. As far as theory is concerned, there is only one model that has survived[6,7] a series of high-precision experiments: the model due to Salam[8] and Weinberg[9],

Table 2

Meaning of the chiral coupling constants

	Chirality	Quark
$u_L = \frac{1}{4}(\alpha+\beta+\gamma+\delta)$	Left-handed (V-A)	u
$d_L = \frac{1}{4}(-\alpha-\beta+\gamma+\delta)$	Left-handed (V-A)	d
$u_R = \frac{1}{4}(\alpha-\beta+\gamma-\delta)$	Right-handed (V+A)	u
$d_R = \frac{1}{4}(-\alpha+\beta+\gamma-\delta)$	Right-handed (V+A)	d

based on the gauge group $SU(2) \otimes U(1)$, extended by the Glashow-Iliopoulos-Maiani (GIM)[10] scheme to include the weak interactions between the u-, d-, s-, and c-quarks. Hereafter it will be referred to simply as the "Salam-Weinberg" or "standard" model.

The standard model has achieved such a degree of respectability that experimental results are normally quoted in terms of $\sin^2\theta_w$, which is essentially the only free parameter of the model at presently available energies. This is a way of making a comparison with theory. However, it is more objective to determine the set of coupling constants experimentally and to compare them individually with theory. This is possible because of the high precision of today's neutral-current experiments.

We note in passing that we have included only the contributions from u- and d-quarks in the effective Lagrangians. Today, high-precision experiments ought to apply small corrections for the s-quark content of the sea (assuming that the neutral-current coupling of the s- and d-quark is the same, as predicted by the standard model). Future experiments, however, should provide experimental information on the neutral-current coupling of s-, c, ... quarks.

In Section 1.2 we shall describe the main features of the Salam-Weinberg model. In Sections 1.3 to 1.5 we shall discuss in some detail the experimental results on neutral current neutrino-electron, electron-quark, and neutrino-quark scatterings. The results will be interpreted in terms of the neutral-current coupling constants.

1.2 Salam-Weinberg Model

The Salam-Weinberg model is an example of a renormalizable theory of weak and electromagnetic interactions. It was proposed before neutral currents had been seen, and strongly motivated their discovery.

The theory unifies the weak and the electromagnetic interaction. It needs four vector bosons: two charged ones for ordinary charged weak interactions, one neutral for the neutral weak interaction, and one for the photon. The smallest gauge group which accommodates these fields is $SU(2) \otimes U(1)$.

The fermion fields are placed in weak isospin doublets of left-handed fermions and in weak isospin singlets of right-handed fermions, as shown in Table 3. The table includes the τ-neutrino and the "top"

Table 3

The weak isospin assignments of the fermion fields

Leptons	Quarks	Weak isospin
$\begin{pmatrix} \nu_e \\ e^- \end{pmatrix}_L$ $\begin{pmatrix} \nu_\mu \\ \mu^- \end{pmatrix}_L$ $\begin{pmatrix} \nu_\tau \\ \tau^- \end{pmatrix}_L$	$\begin{pmatrix} u \\ d_C \end{pmatrix}_L$ $\begin{pmatrix} c \\ s_C \end{pmatrix}_L$ $\begin{pmatrix} t \\ b_C \end{pmatrix}_L$	$I_{3L} = \frac{1}{2}$ $I_{3L} = -\frac{1}{2}$
— — — e^-_R μ^-_R τ^-_R	u_R c_R t_R $(d_C)_R$ $(s_C)_R$ $(b_C)_R$	$\left.\right\} I_R = 0$

quark, two examples of well established but hitherto unobserved particles. The subscript "C" refers to Cabibbo-rotated states. Right-handed neutrinos do not exist because all neutrino masses are taken to be zero. The leptons are colour singlets whereas the quarks exist in three colour states.

The intermediate bosons and the fermions acquire mass through the agency of a scalar Higgs particle of isospin ½, the existence of which is essential for the theory.

The Lagrangian of the theory is invariant under transformations of the group SU(2) ⊗ U(1). It contains two coupling constants g and g′ which are related to the factors SU(2) and U(1) of the gauge group. It has become customary to define an angle by the relation

$$\mathrm{tg}\,\theta_W = \frac{g'}{g} ,$$

where θ_W is usually called the Weinberg angle. The coupling constants g and g′ are related to the electric charge e by

$$\frac{1}{e^2} = \frac{1}{g^2} + \frac{1}{g'^2} ,$$

with

$$e = g \sin\theta_W .$$

The ordinary charged-current interaction is mediated by $W^{\pm}$ exchange with a coupling strength which is given by

$$\frac{G}{\sqrt{2}} = \frac{g^2}{8M_W^2} = \frac{e^2}{8M_W^2 \sin^2\theta_W} .$$

It follows for the mass of the charged vector boson $W^{\pm}$,

$$M_W = \left[\frac{e^2}{4\sqrt{2}\,G}\right]^{\frac{1}{2}} \frac{1}{\sin\theta_W} = \frac{37.3\ \mathrm{GeV}}{\sin\theta_W} .$$

The weak and electromagnetic coupling constants are of the same order of magnitude. The apparent huge difference in the strength of these interactions is due to the difference of the masses of the intermediate bosons ($M_\gamma = 0$, $M_W \geq 37.3$ GeV).

The $W^\pm$, Z^0, and the fermions acquire mass through the same Higgs spontaneous symmetry-breaking mechanism. As a consequence, the parameter

$$\rho = \frac{M_W^2}{M_Z^2 \cos^2\theta_w}$$

equals unity. This parameter can be measured since it determines the effective strength of the charged-current versus the neutral-current process at low energies.

In the Salam-Weinberg model, $\rho = 1$ yields the relation

$$M_W = M_Z \cos\theta_w ,$$

thereby predicting the Z^0 mass to be

$$M_Z = \frac{37.3 \text{ GeV}}{\sin\theta_w \cos\theta_w} .$$

The resulting weak neutral current is a mixture of the third (neutral) isospin component of the charged weak current, and of the electromagnetic current:

$$J_\lambda^{NC} = J_\lambda^3 - \sin^2\theta_w J_\lambda^{em} .$$

The left- and right-handed coupling constants of the fermions are given by

$$f_L = I_{3L} - Q \sin^2\theta_w$$

$$f_R = - Q \sin^2\theta_w .$$

Here Q is the electric charge of the fermion and I_{3L} is the third component of its weak isospin ($I_{3L} = +\frac{1}{2}$ for the "up" member of a doublet, and $I_{3R} = -\frac{1}{2}$ for the "down" member; see Table 3). As a consequence, only the neutrinos (Q = 0) couple with pure chirality (V-A).

With the relation $g_V = f_L + f_R$ and $g_A = f_L - f_R$ arising from the equality (see Section 1.1),

$$\bar{\psi}\gamma_\lambda(g_V + g_A\gamma_5)\psi = \bar{\psi}\gamma_\lambda[f_L(1+\gamma_5) + f_R(1-\gamma_5)]\psi ,$$

the vector- and axial-vector coupling constants of the fermions are given by

$$g_V = I_{3L} - 2Q\sin^2\theta_w$$

$$g_A = I_{3L} .$$

The resulting coupling constants for the most important fermions (from the experimental point of view) are summarized in Table 4.

By construction, the GIM-mechanism[10] causes the neutral currents to conserve flavour: strangeness-changing or charm-changing neutral currents are absent. The same mechanism causes the physical quark states to appear in the neutral-current Lagrangian, whereas the Cabibbo-rotated quark states appear in the charged-current Lagrangian.

All sequential neutrinos (negative leptons) have the same couplings as $\nu_\mu(\mu^-)$. Similarly, in the quark sector, all quarks with charge $^2/_3$ have the same coupling as the u-quark, and all quarks with charge $-^1/_3$ have the same coupling as the d-quark.

The Salam-Weinberg model contains as an additional aspect the Higgs meson with the peculiar property that its coupling strength to fermions is proportional to the fermion mass. The mass of the Higgs meson is not predicted.

Table 4

Neutral-current coupling constants in the Salam-Weinberg model

	ν_μ	μ^-	u	d
I_{3L}	$\frac{1}{2}$	$-\frac{1}{2}$	$\frac{1}{2}$	$-\frac{1}{2}$
Q	0	-1	$\frac{2}{3}$	$-\frac{1}{3}$
f_L	$\frac{1}{2}$	$-\frac{1}{2} + \sin^2\theta_W$	$\frac{1}{2} - \frac{2}{3}\sin^2\theta_W$	$-\frac{1}{2} + \frac{1}{3}\sin^2\theta_W$
f_R	0	$\sin^2\theta_W$	$-\frac{2}{3}\sin^2\theta_W$	$\frac{1}{3}\sin^2\theta_W$
g_V	$\frac{1}{2}$	$-\frac{1}{2} + 2\sin^2\theta_W$	$\frac{1}{2} - \frac{4}{3}\sin^2\theta_W$	$-\frac{1}{2} + \frac{2}{3}\sin^2\theta_W$
g_A	$\frac{1}{2}$	$-\frac{1}{2}$	$\frac{1}{2}$	$-\frac{1}{2}$

1.3 Neutrino Scattering on Electrons

The particularly attractive feature of νe scattering is that there is no hadronic structure involved. Theoretical predictions are straightforward and unambiguous. A major drawback is the very small cross-section. All experiments performed so far have obtained only small data samples.

Four reactions are possible in neutral-current νe scattering:

$$\nu_e + e^- \rightarrow \nu_e + e^-$$

$$\bar{\nu}_e + e^- \rightarrow \bar{\nu}_e + e^-$$

$$\nu_\mu + e^- \rightarrow \nu_\mu + e^-$$

$$\bar{\nu}_\mu + e^- \rightarrow \bar{\nu}_\mu + e^- .$$

Of these, ν_e and $\bar{\nu}_e$ can scatter via both neutral and charged currents, whereas ν_μ and $\bar{\nu}_\mu$ scatter only via the neutral current.

The $\overset{(-)}{\nu}_\mu e$ scattering process is studied both in bubble chambers and counter experiments. Bubble chambers have the advantage of a good electron signature and low background, but suffer from small event numbers and scanning biases. Counter experiments are expected to accumulate several hundred events in the near future, because of their higher target mass. They have no problem with isolated γ background, as bubble chambers do. But other backgrounds are large, and the biases due to tight selection criteria have to be carefully examined.

1.3.1 The process $\nu_\mu + e^- \rightarrow \nu_\mu + e^-$. The experimental results[11-15] on the process $\nu_\mu + e^- \rightarrow \nu_\mu + e^-$ are summarized in Table 5. This process aroused a lot of interest in 1978 when a group working with Gargamelle reported[16] a cross-section, based on a subsample of their statistics, which was too large to be

Table 5

Summary of experiments on $\nu_\mu e$ scattering

Experiment	Sample of $\nu_\mu + N \rightarrow \mu^- + X$	$\nu_\mu e$ candidates	Background	σ/E [a)]
GGM[11)] CERN-PS		1	0.3 ± 0.1	< 3 (90% c.l.)
AP[12)] Counter exp.		32	20.5 ± 2.0	1.1 ± 0.6
GGM[13)] CERN-SPS	64,000	9	0.5 ± 0.2	$2.4^{+1.2}_{-0.9}$
CB[14)] FNAL 15'	83,700	8	0.5 ± 0.5	1.8 ± 0.8
CHARM[15)] Counter exp.	56,000	11	4.5 ± 1.4	2.6 ± 1.6
Average of the experiments				1.6 ± 0.4
Prediction of the standard model ($\sin^2\theta_W = 0.23$)				1.5

a) In units of 10^{-42} cm^2/GeV.

compatible with the standard model prediction with accepted values of $\sin^2\theta_W$. Shortly afterwards, Cnops et al.[14] reported a result based on an exposure to a neutrino flux bigger by a factor of 4. Their cross-section was in disagreement with the result of the Gargamelle group, but in agreement with the standard model with $\sin^2\theta_W = 0.2$.

This discrepancy turned out to be largely due to a statistical fluctuation in the data sample of the Gargamelle group. The first result was based on 10 observed events in a sample of 24,000 charged-current interactions. Later on, two events were removed because they were possibly due to bremsstrahlung from muons passing the chamber. This was accomplished by a cut in the fiducial volume around such muons, with a loss of a few percent in the fiducial volume. The increase in the statistics of charged-current events from 24,000 to 64,000 resulted in only one more observed event, with a final sample of 9 events. This result is in good agreement with all other experiments as well as with the prediction of the standard model. The average of the slope of the cross-section from all experiments is

$$\sigma/E = (1.6 \pm 0.4) \times 10^{-42}\ \text{cm}^2/\text{GeV} ,$$

yielding a Weinberg angle

$$\sin^2\theta_W = 0.22\ {}^{+\,0.08}_{-\,0.05} .$$

A possible second solution at large values of $\sin^2\theta_W$ is excluded by experiments on the process $\bar{\nu}_\mu + e^- \rightarrow \bar{\nu}_\mu + e^-$ (see Section 1.3.2).

Very recently, the fine-grain calorimeter detector of the CERN-Hamburg-Amsterdam-Rome-Moscow (CHARM) Collaboration installed at the CERN-SPS, came up with a preliminary result on the cross-section of $\nu_\mu e$ scattering. As is evident from Table 5, this first result is in good agreement with other results. The authors hope soon to accumulate much more statistics and to increase the precision of their result substantially.

1.3.2 The process $\bar{\nu}_\mu + e^- \rightarrow \bar{\nu}_\mu + e^-$. The great interest in the process $\nu_\mu + e^- \rightarrow \nu_\mu + e^-$ initiated an intense search for the process $\bar{\nu}_\mu + e^- \rightarrow \bar{\nu}_\mu + e^-$. The results of these experiments[17-19] are summarized in Table 6, together with the results of two earlier experiments[11-12].

Table 6

Summary of experiments on $\bar{\nu}_\mu e$ scattering

Experiment	Sample of $\bar{\nu}_\mu + N \rightarrow \mu^+ + X$	$\bar{\nu}_\mu e$ candidates	Background	σ/E [a)]
GGM[11)] CERN-PS		3	0.4 ± 0.1	$1.0 {+2.1 \atop -0.9}$
AP[12)] Counter exp.		17	7.4 ± 1.0	2.2 ± 1.0
GGM[17)] CERN-SPS	7400	0	< 0.03	<2.7 (90% c.l.)
FMMS[18)] FNAL 15'	8400	0	0.2 ± 0.2	<2.1 (90% c.l.)
BEBC TST[19)] CERN-SPS	7500	1	0.5 ± 0.2	<3.4 (90% c.l.)
Average of the experiments[b)]				1.3 ± 0.6
Prediction of the standard model ($\sin^2\theta_W = 0.23$)				1.3

a) In units of 10^{-42} cm^2/GeV.

b) This average is obtained by adding the number of events observed in the experiments and dividing by the sum of the effective antineutrino fluxes.

All three recent experiments performed in the high-energy domain reported only upper limits for σ/E. The results are in agreement with the earlier low-energy experiments[11-12], and with the prediction of the standard model with $\sin^2\theta_W = 0.23$. The average of the slope of the cross-section from all experiments is

$$\sigma/E = (1.3 \pm 0.6) \times 10^{-42} \text{ cm}^2/\text{GeV} ,$$

yielding a Weinberg angle

$$\sin^2\theta_W = 0.23 \begin{matrix} +\ 0.09 \\ -\ 0.23 \end{matrix} .$$

The cross-sections of all four possible νe scattering processes in the framework of the standard model have been calculated by 't Hooft[20] and are shown in Fig. 2. Since nature has chosen a value of $\sin^2\theta_W$ close to 0.25, a precise determination of $\sin^2\theta_W$ from the $\overset{(-)}{\nu}_\mu e$ cross-section is very hard because of its weak dependence on $\sin^2\theta_W$.

1.3.3 <u>The coupling constants for neutrino-electron scattering.</u> In terms of the coupling constants g_V and g_A defined above, the differential cross-section for νe scattering is given for high energies by

$$\frac{d\sigma}{dy} = \frac{G^2 m_e E_\nu}{2\pi} \left[(C_V + C_A)^2 + (C_V - C_A)^2 (1 - y)^2 \right] ,$$

where $y = E_e/E_\nu$ is the fraction of the neutrino energy transferred to the target electron. The relation of C_V and C_A to the coupling constants g_V and g_A is given in Table 7 for all four possible νe scattering processes. The total cross-section is obtained by integration over y between y = 0 and y = 1.

Table 7

Relation between C_V and C_A and the coupling constants g_V and g_A

Process	C_V	C_A
$\nu_e + e^- \to \nu_e + e^-$	$1 + g_V$	$1 + g_A$
$\bar{\nu}_e + e^- \to \bar{\nu}_e + e^-$	$1 + g_V$	$-1 - g_A$
$\nu_\mu + e^- \to \nu_\mu + e^-$	g_V	g_A
$\bar{\nu}_\mu + e^- \to \bar{\nu}_\mu + e^-$	g_V	$-g_A$

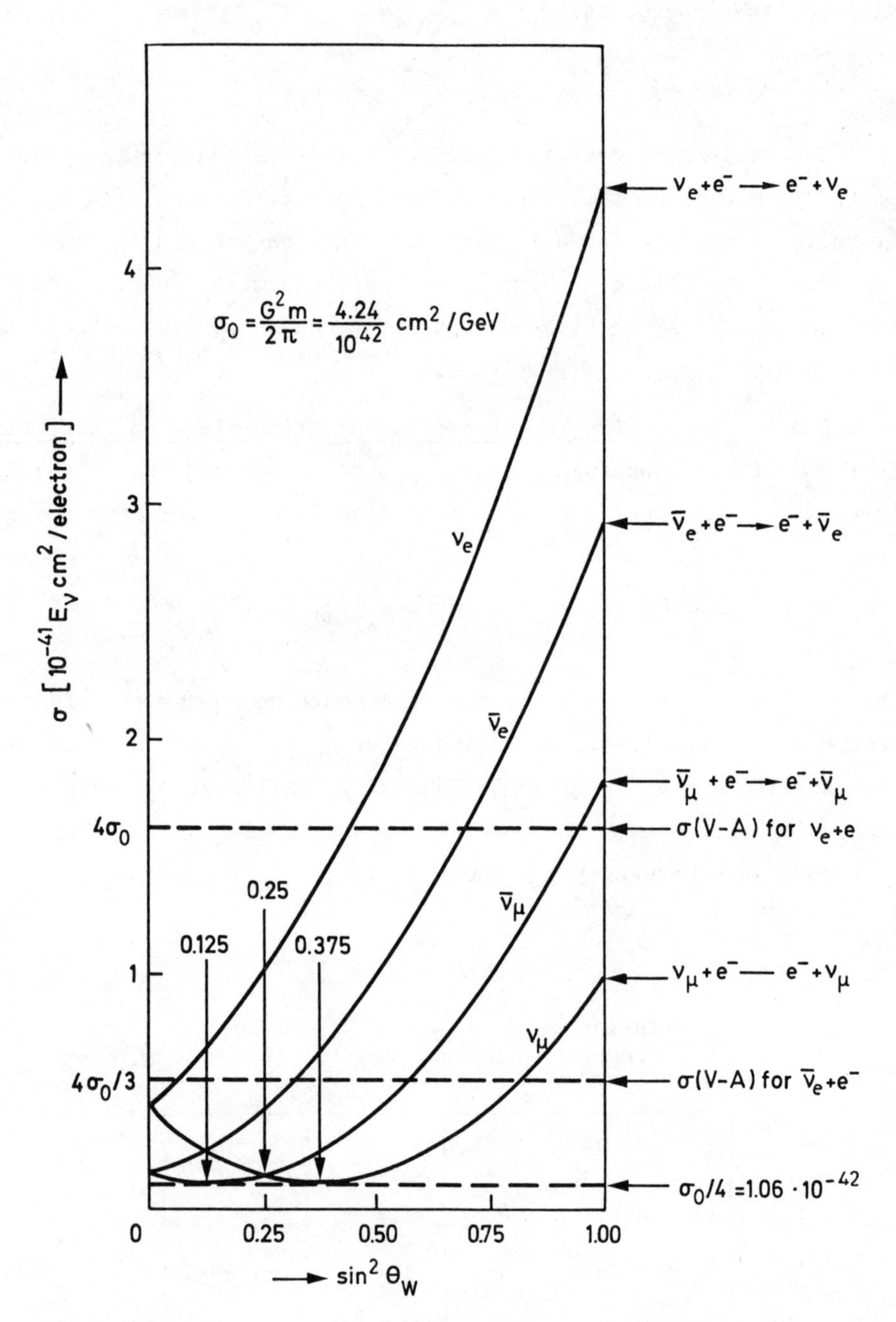

Fig. 2 Neutrino-electron scattering cross-sections in the Salam-Weinberg model.

The cross-section defines an ellipse in the g_V, g_A plane. As can be seen in Fig. 3, the elliptic domains allowed by the measured $\nu_\mu e$ and $\bar{\nu}_\mu e$ cross-sections define four regions of overlap in part due to a sign ambiguity. The signs of the coupling constants g_V and g_A can be determined experimentally because of the interference between the charged- and neutral-current amplitudes in the $\nu_e e$ and $\bar{\nu}_e e$ scattering processes. The latter process has been observed by Reines et al.[21)] at the Savannah River fission reactor. Although their measured cross-section is not significantly different from the V-A cross-section of the charged-current process, in the framework of the standard model the mixing angle is constrained to $\sin^2\theta_W =$ $= 0.29 \pm 0.05$ thanks to the strong dependence of the cross-section on $\sin^2\theta_W$ (see Fig. 2). The elliptic domain allowed by the $\bar{\nu}_e e$ cross-section restricts the allowed domains of g_V, g_A to two. This remaining ambiguity cannot be resolved with νe scattering experiments alone. The two possible solutions correspond to a dominant vector or a dominant axial-vector current. A discrimination between the two solutions can be obtained with electron-quark and neutrino-quark scattering experiments, but not in a model-independent way (see Section 1.4.3).

The solution with axial vector dominance is in good agreement with the prediction of the standard model with $\sin^2\theta_W = 0.23$, as can be seen from Table 8.

Table 8

Summary of the results on the coupling constants g_V, g_A

	Best fit value a)	Standard model	$\sin^2\theta_W = 0.23$
g_V	0.06 ± 0.08	$-\frac{1}{2} + 2\sin^2\theta_W$	-0.040
g_A	-0.52 ± 0.06	$-\frac{1}{2}$	-0.500

a) Solution with axial vector dominance only.

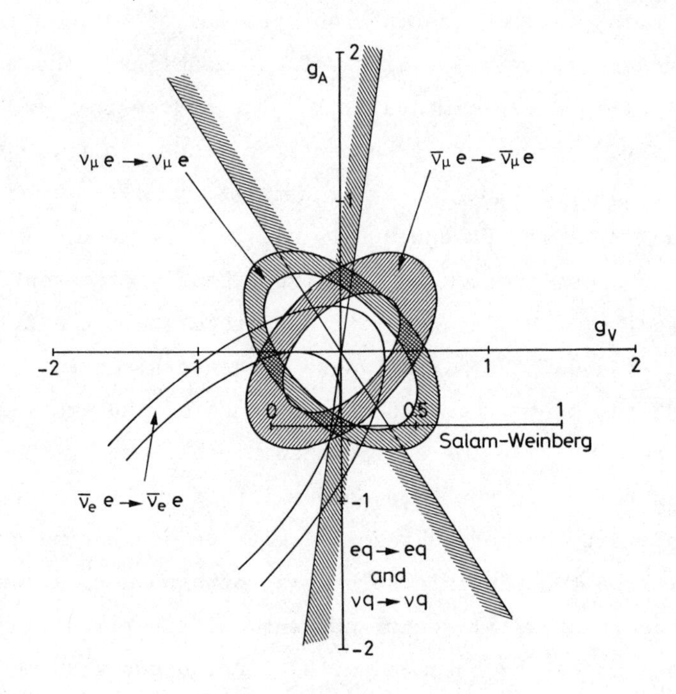

Fig. 3 Domains in the g_V, g_A plane allowed by νe scattering experiments.

In summary, all known results on νe scattering are consistent and in agreement with the predictions of the standard model. The precision of the experiments ought to be improved although there is little hope of getting a precise determination of $\sin^2\theta_W$ from $\nu_\mu e$ scattering experiments. Good precision on $\sin^2\theta_W$ is in principle expected from $\nu_e e$ scattering, which is the only channel not yet explored experimentally.

1.4 Electron Scattering on Quarks

Very recently, substantial progress has been made in the understanding of eq scattering via weak neutral current. This progress stems mainly from new results on parity-violating effects in the inelastic scattering of polarized electrons at SLAC, but also from new results on parity-violating effects in optical transitions of heavy atoms.

1.4.1 Polarized electron scattering on deuterium. The SLAC-Yale group reports on an extension of the previous measurement[22] of a parity-violating asymmetry in the inelastic scattering of longitudinally polarized electrons of about 20 GeV beam energy from unpolarized deuterium nuclei. The first measurement was done essentially for one value of the inelasticity y, namely $y \sim 0.2$. The new measurements[23] give the asymmetry as a function of $y = (E_e - E_e')/E_e$ covering the range $0.15 \leq y \leq 0.36$.

The new measurements open up new possibilities of testing the predictions of specific gauge models. The asymmetry for the inelastic scattering of right- or left-handed electrons on deuterium,

$$A = \frac{\sigma_R - \sigma_L}{\sigma_R + \sigma_L} \quad ,$$

can be written as[24]

$$\frac{A}{Q^2} = a_1 + a_2 \frac{1 - (1-y)^2}{1 + (1-y)^2} \quad ,$$

where Bjorken scaling and R = 0 are assumed (R is the ratio of the absorption cross-sections for longitudinal and transverse photons). The coefficients a_1 and a_2 depend in general on kinematic parameters; but with an isoscalar target such as deuterium they are expected to be constants, and can be expressed in terms of the relevant coupling constants $\tilde{\alpha}$, $\tilde{\beta}$, $\tilde{\gamma}$, and $\tilde{\delta}$, as follows[25]:

$$a_1 = (G/\sqrt{2}e^2)(9\tilde{\alpha} + 3\tilde{\gamma})/5 \ , \qquad a_2 = (G/\sqrt{2}e^2)(9\tilde{\beta} + 3\tilde{\delta})/5 \ .$$

An order of magnitude estimate for a_1 and a_2 is given by the constant $G/\sqrt{2}e^2 \sim 10^{-4}$ GeV^{-2}, which is the ratio of the weak to the electromagnetic amplitude giving rise to the observed interference effect.

The measurement of the y-dependence of the asymmetry permits a separate determination of the coefficients a_1 and a_2. The fit yielded[23]

$$a_1 = (-9.7 \pm 2.6) \times 10^{-5} \text{ GeV}^{-2}$$

and

$$a_2 = (4.9 \pm 8.1) \times 10^{-5} \text{ GeV}^{-2} ,$$

which gives in terms of the coupling constants the linear relations[25]

$$\tilde{\alpha} + \frac{\tilde{\gamma}}{3} = -0.60 \pm 0.16 ,$$

$$\tilde{\beta} + \frac{\tilde{\delta}}{3} = 0.31 \pm 0.51 .$$

Further experimental information is needed to determine the coupling constants individually.

Figure 4 shows the measured asymmetry as a function of y. The authors compare their results with the predictions of two gauge models, which differ in the assignment of the right-handed electron. In the standard model, right-handed quarks and leptons are placed in singlets. In the hybrid model, invented to explain the absence of parity violation in heavy atoms[26-27], the right-handed electron is placed in a doublet with a hypothesized neutral heavy lepton E^0, making the electronic current pure vector. As can be seen from Fig. 4, the data support the standard model predicting a small value of the slope a_2 with currently accepted values of $\sin^2\theta_W$, whereas the hybrid model predicting a large slope and an intercept $a_1 = 0$ appears to be ruled out.

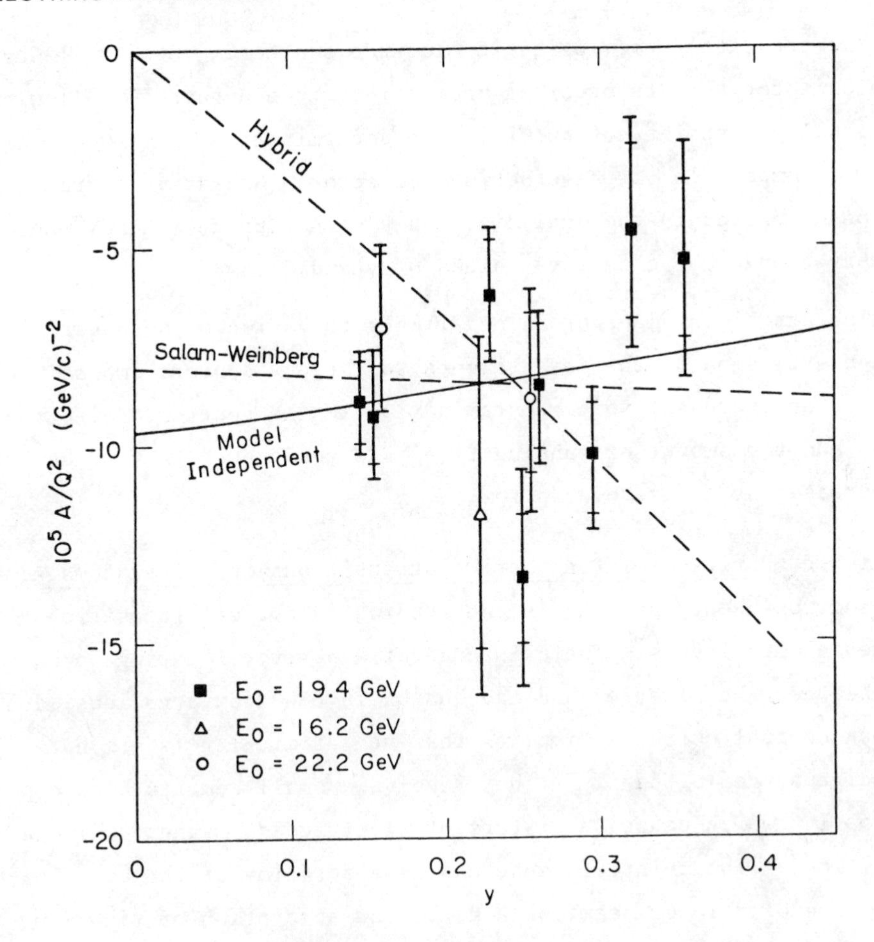

Fig. 4 Asymmetry of polarized electron scattering on deuterium as a function of the inelasticity $y = (E_e - E_e')/E_e$ (from Ref. 23).

The value of $\sin^2\theta_W$ as determined from a_1 and a_2 is $\sin^2\theta_W = 0.224 \pm 0.020$, where the error includes about equal contributions from statistical and systematic sources. The excellent agreement of this value of $\sin^2\theta_W$ with values determined in neutrino experiments is a great success for the standard model.

The interpretation of this experiment in terms of $\sin^2\theta_W$ depends on the validity of the quark-parton model, which is not likely to describe accurately the inelastic electron scattering at $Q^2 \sim 1\ \text{GeV}^2$. To account for this theoretical uncertainty, the authors[23] allow an additional uncertainty of ±0.01 in the determination of $\sin^2\theta_W$, which is about comparable to the experimental error. Radiative corrections have been applied in the analysis. They do not produce an asymmetry but they change the effective values of y and Q^2.

The experimental error is not likely to be improved substantially. An extension of the measurements to large y appears impossible owing to an increasing pion contamination in the scattered electron yield, and the amount of running time is already such that it cannot be increased by a sizeable factor.

1.4.2 Parity-violating optical transitions in heavy atoms. The electronic current can also be studied in optical transitions between atomic levels. The existence of a parity-violating potential between the electron and the quarks in the atomic nucleus, due to weak neutral currents, implies that the atomic levels are not pure eigenstates of parity. They receive a small admixture of opposite parity which causes a mixture of electric and magnetic dipole transitions. Their interference causes a rotation of the polarization plane of a laser beam, or a different absorption of right- or left-circularly polarized laser light.

Parity violation in heavy atoms is primarily sensitive to the weak neutral charge[28] which is given in the notation of Hung and Sakurai[25] by

$$Q = -\tilde{\alpha}(Z - N) - 3\tilde{\gamma}(Z + N) ,$$

which depends only on the product of the axial electron current and the hadronic vector current.

The experimental situation on parity violation in atoms is not satisfactory. The experiments carried out at Seattle[26] and Oxford[27] measuring the rotation of the polarization plane of laser light going through bismuth vapour, showed essentially null results, with small statistical errors. A Novosibirsk team[29], however, reported evidence for a non-zero result in Bi, being consistent with the standard model prediction. The ratio of the experimental to the theoretical rotation (standard model with $\sin^2\theta_W = 0.25$) is 1.07 ± 0.14, which is clearly incompatible with parity conservation. Note that the Novosibirsk experiment is carried out on the same optical transition as the Oxford experiment, with conflicting results.

Recently, another experiment carried out at Berkeley[30] also reports parity violation observed in atomic thallium, although the effect has only a 2σ significance. The authors hope to improve the accuracy substantially in the near future.

The results of the four existing experiments are summarized in Table 9. The agreement is poor. All experiments are continuing to take data.

Table 9

Summary of experiments on parity violation in atoms

Experiment	Atom	Transition (nm)	R [a]
Seattle[26]	Bi	876	0.0 to 0.2
Oxford[27]	Bi	648	0.0 to 0.1
Novosibirsk[29]	Bi	648	1.07 ± 0.14
Berkeley[30]	Tl	293	$2.3^{+3.1}_{-1.4}$

a) Ratio of the experimental result to the prediction of the standard model, with reasonable values of $\sin^2\theta_W$.

1.4.3 The coupling constants for electron-quark scattering. It is difficult to choose between the conflicting experimental results in order to determine the eq coupling constants. Tentatively, we go along with the positive results from the Novosibirsk and Berkeley groups and hope that the future development will justify this step (it cannot be justified at present on clear-cut experimental grounds).

Following Hung and Sakurai[25], we compare the weak neutral charges for the Bi and Tl nuclei

$$Q(Bi) = 43\tilde{\alpha} - 627\tilde{\gamma}$$

and

$$Q(Tl) = 42\tilde{\alpha} - 612\tilde{\gamma}$$

to the measured values[29-30]

$$Q(Bi) = -140 \pm 40$$

and

$$Q(Tl) = -280 \pm 140 \ .$$

It should be noted that the experimental error of the Novosibirsk experiment (see Table 9) allows a smaller error on Q(Bi), but the theoretical uncertainty of 15 to 20% in the atomic physics calculation of Bi [31] requires the quoted error for Q(Bi).

The SLAC polarized electron scattering experiment and the atomic physics experiments define nearly orthogonal linear relations between $\tilde{\alpha}$ and $\tilde{\gamma}$, as shown in Fig. 5. The domain in the $\tilde{\alpha}$, $\tilde{\gamma}$ plane which satisfies all experiments is given by

$$\tilde{\alpha} = -0.72 \pm 0.25$$
$$\tilde{\gamma} = \ 0.36 \pm 0.28 \ .$$

Both coupling constants agree with the prediction of the standard model for $\sin^2\theta_W = 0.23$.

A separation of the coupling constants $\tilde{\beta}$ and $\tilde{\delta}$ is not yet possible in a model-independent way. In a single Z-boson model, only seven independent parameters denoting the coupling strengths of $u_{L,R}$, $d_{L,R}$, $e_{L,R}$, and ν_L are needed, compared to the 10 coupling parameters which appear in the Sakurai tetragon (Fig. 1). As pointed out by Hung and Sakurai[32], there must be three independent "factorization relations" among the 10 constants. They may be taken to be[25]

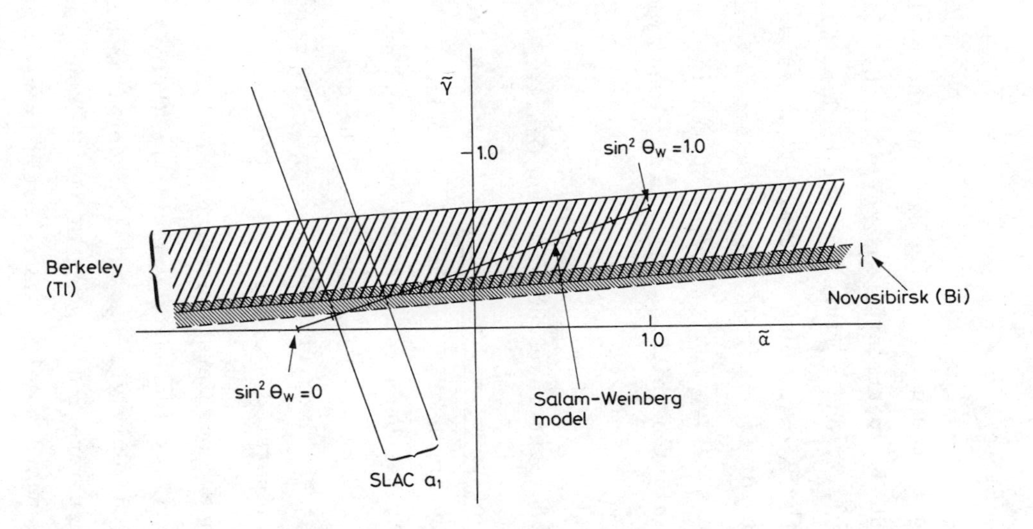

Fig. 5 Domains in the $\tilde{\alpha}$, $\tilde{\gamma}$ plane allowed by polarized electron scattering and atomic physics experiments (from Ref. 25).

$$\frac{\tilde{\gamma}}{\tilde{\alpha}} = \frac{\gamma}{\alpha}\ , \quad \frac{\tilde{\delta}}{\tilde{\beta}} = \frac{\delta}{\beta}\ , \quad \frac{g_V}{g_A} = \frac{\alpha\tilde{\beta}}{\tilde{\alpha}\beta}\ .$$

A combination of these three relations gives

$$\frac{g_V}{g_A} = \frac{(\alpha + \gamma/3)(\tilde{\beta} + \tilde{\delta}/3)}{(\tilde{\alpha} + \tilde{\gamma}/3)(\beta + \delta/3)}\ ,$$

which can be tested in the g_V, g_A plane. Using the ratio a_2/a_1 as determined in the SLAC experiment[23], $a_2/a_1 = -0.50 \pm 0.74$, and using the results for α, β, γ, and δ given in Section 1.5, we can draw the allowed region in the g_V, g_A plane (see Fig. 3). As a consequence, the vector dominant solution for g_V, g_A appears to be ruled out, yielding the unique solution for g_V and g_A quoted in Table 8. Once more, this is the solution which is in good agreement with the standard model with $\sin^2\theta_W = 0.23$.

1.5 Neutrino Scattering on Quarks

1.5.1 <u>The reaction $\bar{\nu}_e + D \rightarrow \bar{\nu}_e + P + N$.</u> Recently, a 4σ signal of the weak disintegration of the deuteron via a neutral current, $\bar{\nu}_e + D \rightarrow \bar{\nu}_e + P + N$, has been reported by an Irvine group[33]. A well-shielded target of 268 kg D_2O was exposed to the high $\bar{\nu}_e$ flux originating from the Savannah River fission reactor. The reaction was identified via the liberated neutrons.

The process under consideration is, because of the low energy involved, essentially forbidden for vector-type (Fermi) interactions[34]. It proceeds only via the axial vector (Gamow-Teller) interaction and is therefore independent of the Weinberg angle, in the framework of the standard model. The reported cross-section is

$$(3.8 \pm 0.9) \times 10^{-45}\ \text{cm}^2\ ,$$

in agreement with the prediction of the standard model, 5.0×10^{-45} cm^2. The experiment is being continued to reduce the statistical error, which is claimed to dominate the uncertainties.

1.5.2 <u>Inclusive neutral-current reactions on isoscalar targets.</u> Inclusive neutral-current reactions on (nearly) isoscalar targets allow the most precise measurement of neutral-current couplings. The measured quantities are the ratios of the inclusive neutral-to-charged-current cross-sections R_ν and $R_{\bar{\nu}}$, respectively. A study of the y-distribution ($y = E_{had}/E_\nu$) has given information on the space-time structure of hadronic neutral currents. A comparison of the results of different experiments can be found elsewhere[35]. The hadronic neutral current on isoscalar targets is known to be dominated by V-A, with a small ($\sim$ 10%) admixture of V+A. Thus the current is neither pure in parity nor in chirality. It should be noted, however, that the V+A admixture has been seen so far only at a 4σ level by the CDHS experiment[3]. It is an experimental challenge to improve the significance of the V+A contribution and to rule out a pure V-A structure of the hadronic current. The most sensitive place to look for this is the $\bar{\nu}$ y-distribution, which at large y receives about equal contributions from V-A scattering from the antiquark content and from V+A scattering from the quark content of the nucleon.

The simplest and most precisely measurable quantities are R_ν and $R_{\bar{\nu}}$. The precision of $\sin^2\theta_w$, as determined in neutrino experiments, is essentially given by R_ν. It therefore seems appropriate to discuss in more detail the problems, both experimental and theoretical, of the measurement of R_ν and $R_{\bar{\nu}}$.

Experimental problems:

The precision measurement of R_ν and $R_{\bar{\nu}}$ is a domain of counter experiments because large event numbers are important. The dominant sources of systematic errors are for R_ν the K_{e3} correction (ν_e from K_{e3} decays fake neutral-current events in a calorimeter detector) and the high-y charged-current background for neutral-current events. The latter correction is due to charged-current events where the muon is hidden in the hadronic shower. This is visualized in Fig. 6,

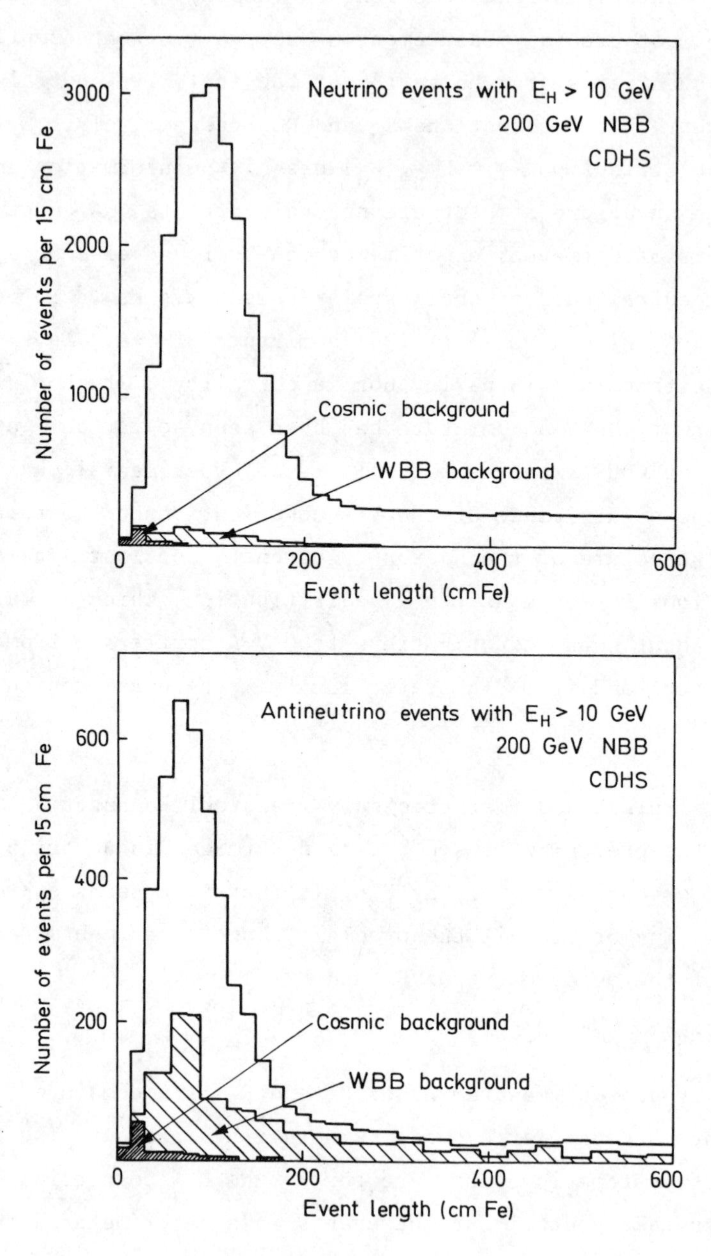

Fig. 6 Penetration length in Fe for ν and $\bar{\nu}$ events (CDHS data).

where the maximum penetration length in Fe is plotted for the new CDHS data[36]. Here a fine-grain calorimeter such as the CHARM detector offers an advantage since the minimum muon track length can be much reduced. At present, a limit of ±0.005 in the accuracy of R_ν, mostly due to uncertainties in the parent π/K ratio showing up in the K_{e3} correction, seems appropriate. This corresponds to a limit in accuracy for $\sin^2\theta_W$ of ±0.008.

The main source of uncertainty for $R_{\bar\nu}$ is the "wide-band" background of narrow-band neutrino beams arising from the decay of π's or K's before sign and momentum selection. This background is determined experimentally in separate "closed collimator" runs and statistically subtracted (see Fig. 6), which requires a precise knowledge of the relevant fluxes. The "new source" which has been discovered in the CERN beam dump experiments[37], and which is presumably due to a prompt neutrino flux from semileptonic charm decay, is automatically subtracted with this method.

At present, a precision of ±0.015 for $R_{\bar\nu}$ seems within reach. This number is important as a check for the validity of the standard model, which predicts $R_{\bar\nu}$ once $\sin^2\theta_W$ is known from R_ν. However, $R_{\bar\nu}$ is not useful for a determination of $\sin^2\theta_W$, since $R_{\bar\nu}$ is very much independent of $\sin^2\theta_W$ around $\sin^2\theta_W = 0.23$.

Theoretical problems:

The experimental information is of such a high precision that theoretical uncertainties in the analysis start to be of the same magnitude as the experimental errors. In this section we discuss the problems of radiative corrections, scaling violations, uncertainties in the sea, and elastic and quasi-elastic events. We restrict the discussion to y-distributions and R_ν and $R_{\bar\nu}$, since structure functions (x-distributions) from neutral-current scattering are measured so far with low precision.

As long as the measurement of the hadronic energy is "electromagnetically" inclusive, as is the case in calorimeter targets (the final hadronic state comprises both hadrons and photons), only the radiative correction due to the lepton leg is to be considered in the "leading-log" approximation[38]. This correction is obviously different for neutral- and charged-current scattering and is expected to be in specific kinematical domains such as large y at the level of $\alpha/\pi \cdot \ln(Q^2/m_\mu^2) \sim$ several percent. This starts to be relevant for the experimental subtraction of the high-y charged-current background in the neutral-current sample (see Fig. 6) for the determination of R_ν and $R_{\bar{\nu}}$, and for a study of the space-time structure from the difference of the neutral- and charged-current y-distributions. Fortunately, the only relevant $\bar{\nu}$ y-distribution is less affected by radiative corrections at large y than is the ν y-distribution[38] (this holds for charged currents, whereas neutral currents need not be corrected).

Experimentalists have ignored the radiative correction problem up to now. Future precision experiments ought to worry about it.

In the analysis of hadronic neutral currents, the quark-parton model is employed. Gentle deviations from scaling, consistent with expectations from QCD, are established[39]. As a consequence of the slight changes in the amount of valence- and sea-quarks, and of the difference in the charged and neutral coupling to various quark flavours, scaling violation effects are to be considered for high-precision neutral-current work.

Fortunately, the effects of scaling violation are not very important at the present level of precision. Buras and Gaemers[40] and the Aachen-Bonn-CERN-London-Oxford-Saclay (ABCLOS) (BEBC) Collaboration[41] have shown, employing the Buras-Gaemers parametrization[40] of scaling violations with $\Lambda \sim 0.5$ GeV, that the effects at SPS and FNAL energies are small: R_ν changes by < 0.002 compared to a quark-parton model analysis, at $\sin^2\theta_W = 0.23$. Hence the determination of $\sin^2\theta_W$ is virtually unaffected by scaling violations.

On the contrary, $R_{\bar{\nu}}$ is more affected by scaling violations. They reduce $R_{\bar{\nu}}$ by typically 0.015, which is not negligible compared to the experimental error. This means that for future interpretations of $R_{\bar{\nu}}$ in terms of theoretical models, scaling violations ought to be taken into account.

The uncertainty in the amount of sea-quarks in the nucleon and in its flavour composition is in first approximation also irrelevant for the interpretation of R_{ν} (hence also for the determination of $\sin^2\theta_W$). However, $R_{\bar{\nu}}$ changes by about ±0.01 if the relative amounts of all antiquarks and the strange antiquarks are changed within reasonable limits around the central values[39]:

$$\frac{\int x(\bar{u} + \bar{d} + 2\bar{s})dx}{\int x(u + d)dx} = 0.18$$

and

$$\frac{\int x2\bar{s}dx}{\int x(u + d)dx} = 0.03 \ .$$

In these estimates the coupling of the different quark flavours to the neutral current has been taken as predicted by the standard model, since there is no experimental information on the s-quark coupling.

We may conclude from this discussion that there is no point in substantially improving the measurement of $R_{\bar{\nu}}$ for a high-precision check of the standard model, without at the same time improving the precision of our knowledge of the nucleon structure.

Elastic and quasi-elastic events are not accounted for in the quark-parton model. They may confuse the comparison with the standard model at few GeV energies, but this should not matter very much at high energies. All neutral-current experiments require a minimum amount of hadronic energy. This cut removes essentially the elastic and quasi-elastic events from the sample, and the comparison with the standard model should be done with the same cut applied. This method, employing the quark-parton model with scaling violations, radiative corrections, and the correct quark structure of the nucleon, works well and yields a precise value for $\sin^2\theta_W$ derived from $R_\nu(E_{had} > E_{had}^{min})$. As an alternative, a model-independent method has been proposed by Paschos and Wolfenstein[42] long ago:

$$\frac{\sigma_\nu^{NC} - \sigma_{\bar{\nu}}^{NC}}{\sigma_\nu^{CC} - \sigma_{\bar{\nu}}^{CC}} = \frac{1}{2} - \sin^2\theta_W ,$$

where the total cross-sections may be replaced by a partial cross-section representing any kinematical domain. Hence experimental data need not be corrected for losses due to selection criteria. In practice, however, there are some problems to be overcome: since neutrino experiments use a continuous energy spectrum of incident particles, equal flux shapes for ν and $\bar{\nu}$ and the same energy-dependence of the cross-sections are necessary. This forbids the application at the threshold of a new quark flavour.

New experimental results

Preliminary new values of R_ν and $R_{\bar{\nu}}$ have recently been reported from the CDHS[36] and CHARM[15] Collaborations. The CHARM Collaboration uses a fine-grain calorimeter consisting of marble plates alternating with planes of drift-tubes and scintillators, which recently came into operation. The CDHS detector consists of iron plates sandwiched with scintillator sheets and drift chambers. Its fiducial mass is larger by a factor of 5, but the granularity of the CHARM detector and its construction allows some corrections to the neutral-current signal to be significantly reduced. The new results are consistent with the previous world average, as shown in Table 10.

Table 10

Experimental results on R_ν and $R_{\bar{\nu}}$

Experiment	R_ν a)	E_{had} cut	$R_{\bar{\nu}}$ a)	E_{had} cut
CDHS[36) preliminary	0.307 ± 0.008 (0.003)	10 GeV	0.373 ± 0.025 (0.014)	10 GeV
CHARM[15) preliminary	0.30 ± 0.02 (0.006)	∿8 GeV	0.39 ± 0.02 (0.014)	∿5 GeV
Previous average[35)	0.29 ± 0.01	None	0.35 ± 0.025	None

a) The statistical error is given in brackets.

Both new results are in good agreement with the standard model, as shown in Fig. 7. The CDHS result for $\sin^2\theta_W$, determined from the Paschos-Wolfenstein formula[42), is $\sin^2\theta_W = 0.228 \pm 0.018$, which is in good agreement with the previous world average[35) $\sin^2\theta_W =$ $= 0.23 \pm 0.02$.

In terms of the chiral coupling constants, R_ν and $R_{\bar{\nu}}$ can be used to extract[6) the combinations $u_L^2 + d_L^2$ and $u_R^2 + d_R^2$, respectively. This extraction from R_ν and $R_{\bar{\nu}}$ is independent of a possible violation of the Callan-Gross relation, provided it is the same for neutral and charged currents.

A new dimension in the study of the hadronic neutral current is opened up by the coming into operation of large devices capable of measuring the hadronic energy flow and thus determining the neutral-current structure function.

A first attempt to explore the structure function of the nucleon with neutral currents has been reported by the CHARM Collaboration[15). Figure 8 shows their measured ratio of neutral-to-charged-current events as a function of the Bjorken variable x. For both neutral

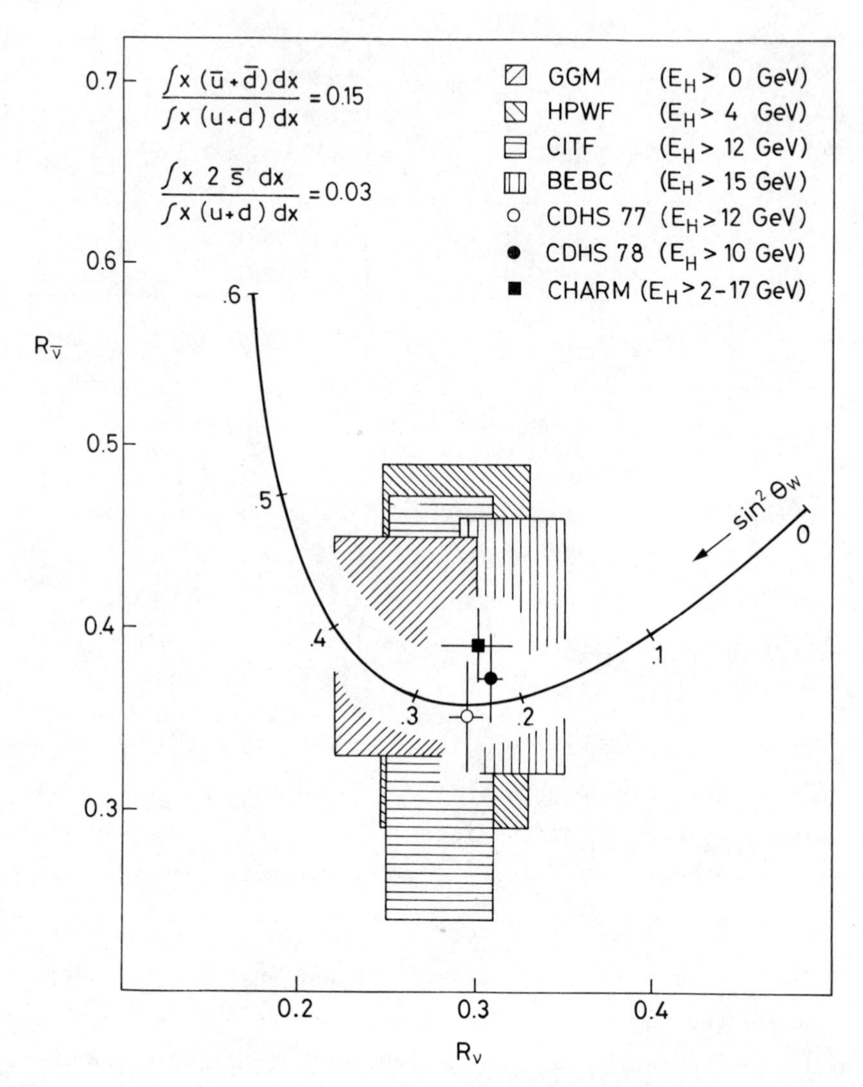

Fig. 7 Comparison of the results of various experiments on R_{ν} and $R_{\bar{\nu}}$ with the Salam-Weinberg model. The dimension of the rectangles indicate the error bars for R_{ν} and $R_{\bar{\nu}}$. The theoretical curve is drawn for the experimental conditions of the CDHS experiment. Corrections for scaling violation or radiative effects are not applied.

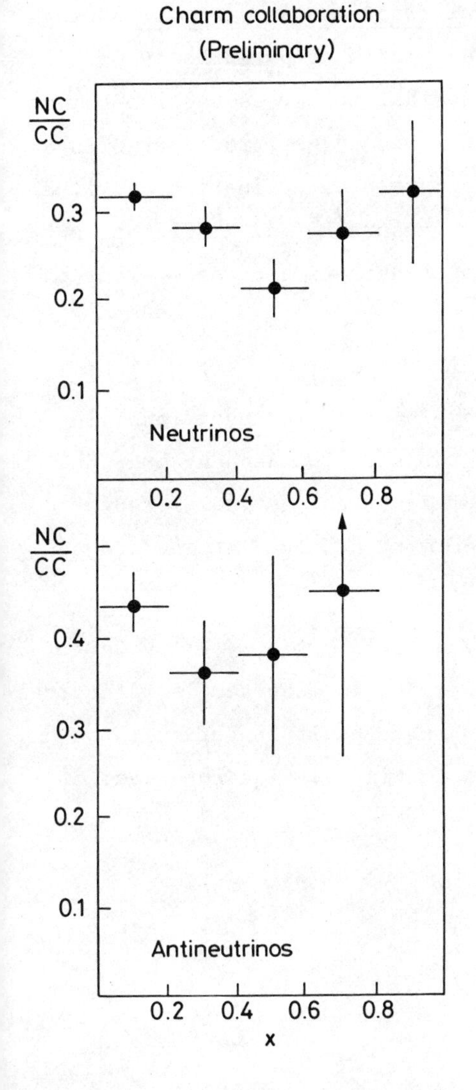

Fig. 8 Ratio of neutral- to charged - current events as a function of x (from Ref. 15).

and charged currents, x has been determined from the hadronic energy flow. Although this result is preliminary and only a first step, it may be concluded that there is no large anomaly in the nucleon structure as determined by neutral currents. A qualitatively similar result, based on much smaller statistics, has recently been reported by a Columbia-Rutgers-Stevens group[43].

1.5.3 Inclusive neutral-current reactions on protons. Owing to the different quark content of the proton compared to an isoscalar target, a measurement of R_ν^p and $R_{\bar{\nu}}^p$, the cross-section ratios of neutral-to-charged currents on a proton target, together with a measurement of R_ν and $R_{\bar{\nu}}$, allows a separate determination of u_L^2, d_L^2, u_R^2, and d_R^2. Previous attempts[44] to measure R_ν^p and $R_{\bar{\nu}}^p$ did not put stringent constraints on the coupling constants because of lack of precision.

Neglecting the sea of $q\bar{q}$ pairs in the nucleon, the neutral-current coupling strength is given by $2u_L^2 + d_L^2$ for neutrinos, in contrast to $u_L^2 + d_L^2$ on an isoscalar target. The Aachen-Bonn-CERN-Munich-Oxford (ABCMO) (BEBC) Collaboration has just completed a measurement[45] of R_ν^p with BEBC filled with hydrogen exposed to a ν wide-band beam.

The result is $R_\nu^p = 0.52 \pm 0.04$, where a cut $p_T^{had} > 1.9$ GeV/c is applied. This cut was found to provide a clean sample of neutral-current events. The interpretation of the result in terms of the coupling constants u_L^2 and d_L^2 is seen in Fig. 9. The intersect of the allowed domains for u_L^2 and d_L^2 yields

$$u_L^2 = 0.15 \pm 0.05$$

and

$$d_L^2 = 0.16 \pm 0.07 \ .$$

The result is consistent with the standard model, but requires a value of $\sin^2\theta_W = 0.18 \pm 0.03$, being somewhat lower than the currently accepted value.

1.5.4 Elastic neutrino scattering on protons. Elastic neutrino scattering on protons is analogous in its importance to neutron β-decay, for the study of the isospin and space-time properties of the hadronic neutral current.

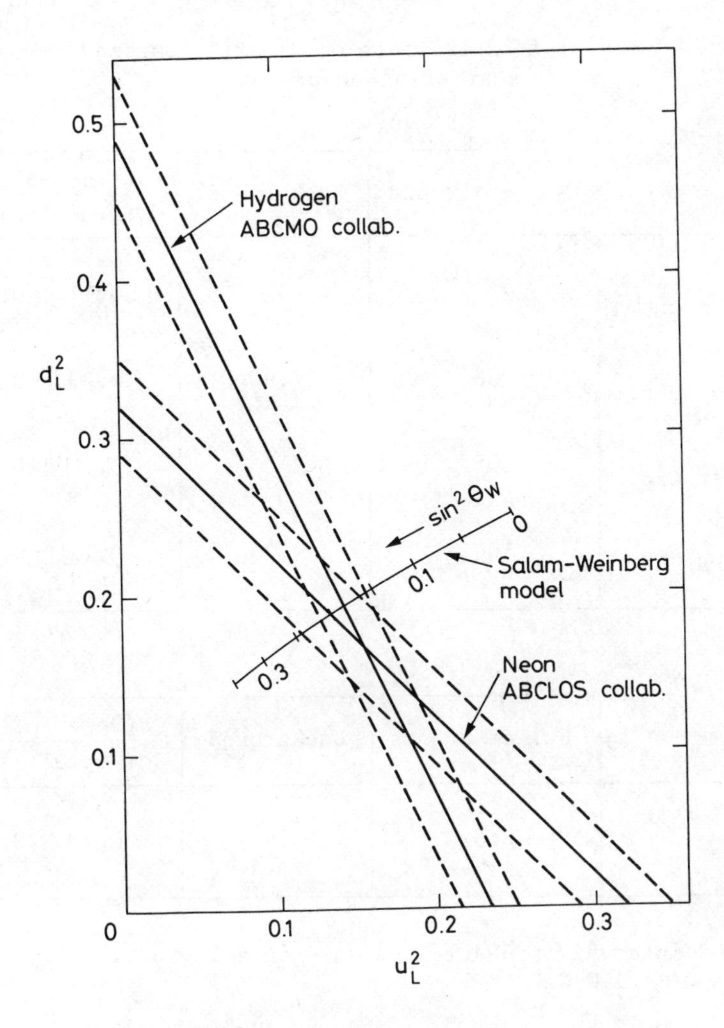

Fig. 9 Domains in the u_L^2, d_L^2 plane allowed by measurements of R_ν and R_ν^p (from Ref. 45).

The process $\nu + p \to \nu + p$ has been observed in four experiments, and $\bar{\nu} + p \to \bar{\nu} + p$ in one experiment. The results are summarized in Table 11. To indicate the experimental difficulties, the total number of observed events is given as well as the background estimate. The agreement between different experiments is good.

Table 11

Summary of experiments on elastic neutrino scattering on protons

	No. of events	Background	$\frac{\sigma(\nu+p\to\nu+p)}{\sigma(\nu+n\to\mu^-+p)}$ a)
HPB[46] Counter exp.	217	82	0.11 ± 0.02
CIR[47] Counter exp.	38	19	0.23 ± 0.09
AP[48] Counter exp.	155	110	0.10 ± 0.04
GGM[49] CERN-PS	100	62	0.12 ± 0.06

	No. of events	Background	$\frac{\sigma(\bar{\nu}+p\to\bar{\nu}+p)}{\sigma(\bar{\nu}+p\to\mu^++n)}$ a)
HPB[46] Counter exp.	66	28	0.19 ± 0.05

a) The Q^2 interval is nearly the same in all experiments: $0.3 < Q^2 < 1.0$ GeV^2.

The results of the Harvard-Pennsylvania-Brookhaven (HPB) Collaboration[46] include an analysis of the Q^2-dependence of the cross-section. The precision is such that stringent limits can be put for the neutral-current coupling constants[50].

We write the matrix element of the hadronic neutral current in the form[51] (with $Q^2 = -q^2$)

$$\left\langle p' \middle| J_\lambda^{NC} \middle| p \right\rangle \propto \bar{u}(p')\left[\gamma_\lambda F_1(Q^2) + \frac{i}{2M_p}\sigma_{\lambda\nu}q^\nu F_2(Q^2) + \gamma_\lambda\gamma_5 G_1(Q^2)\right]u(p)$$

and replace the vector form factors $F_1(Q^2)$ and $F_2(Q^2)$ by the "Sachs form factors"

$$G_E(Q^2) = F_1(Q^2) - \frac{Q^2}{2M_p^2} F_2(Q^2)$$

$$G_M(Q^2) = F_1(Q^2) + F_2(Q^2) \ .$$

The reason for this is to avoid the strong error correlation of the "Dirac form factors" $F_1(Q^2)$ and $F_2(Q^2)$. We assume further that G_E, G_M, and G_1 have the same Q^2-dependence as their charged-current counterparts, i.e. a dipole form $1/(1 + Q^2/M^2_{V,A})^2$, with $M_V = 0.84$ GeV and $M_A = 0.90$ GeV.

The form factors $G_E(0)$, $G_M(0)$, and $G_1(0)$ are given in Table 12 in terms of the neutral-current coupling constants, together with the results of the HPB fit to the measured differential cross-section $d\sigma/dQ^2$. They are in good agreement with the predictions of the Salam-Weinberg model, with $\sin^2\theta_W = 0.23$. Note, however, that the quoted errors are only statistical; systematic errors are estimated to be comparable to the statistical ones.

Table 12

Results for the form factors in elastic neutrino scattering on protons

Form factor	Exp. result[46]	Salam-Weinberg model[7,52]	For $\sin^2\theta_W = 0.23$
$G_E(0)$	$0.5 \ {}^{+\ 0.25}_{-\ 0.5}$	$\frac{1}{2}(\alpha + 3\gamma)$	0.04
$G_M(0)$	$1.0 \ {}^{+\ 0.35}_{-\ 0.40}$	$\frac{4.7}{2}(\alpha + 0.56\gamma)$	1.07
$G_1(0)$	$0.5 \ {}^{+\ 0.2}_{-\ 0.15}$	$\frac{1.25}{2}(\beta + 0.6\delta)$	0.63

1.5.5 Charge ratios of final-state hadrons. The analysis of charge ratios of final-state hadrons provides another way to disentangle u_L^2, d_L^2, u_R^2, and d_R^2. This method rests on the assumption that the composition of the hadronic system reflects in the "current fragmentation region" the flavour of the struck quark. As is visualized in Fig. 10 for neutrinos, the charged and neutral currents couple with different strengths to the quarks of an isoscalar target:

charged current: transforms $d \rightarrow u$ with coupling strength 1;

neutral current: couples with strength $u_L^2 + \frac{1}{3} u_R^2$ to u quarks and with $d_L^2 + \frac{1}{3} d_R^2$ to d quarks.

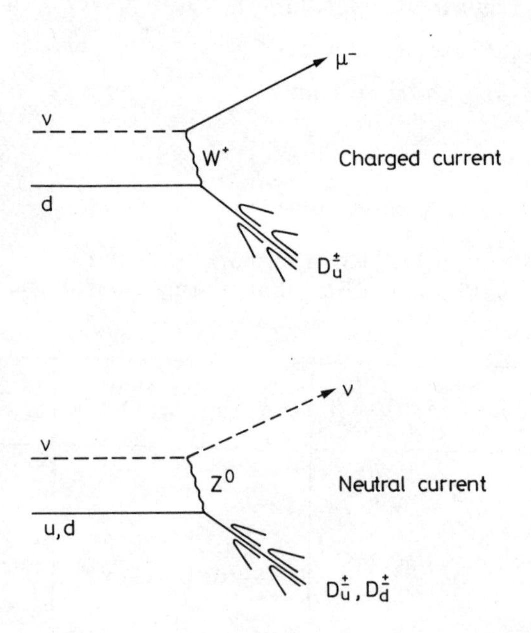

Fig. 10 Fragmentation of quarks struck by charged and neutral currents in ν and $\bar{\nu}$ reactions.

Further necessary ingredients are the fragmentation functions $D_q^{\pm}(z)$, which denote the probability for the quark q to give a positive or a negative hadron which carries a fraction z of the quark energy. Neglecting again for simplicity the sea of $q\bar{q}$ pairs in the nucleon, we get for the charge ratios of final-state hadrons in ν-induced reactions on isoscalar targets:

$$\frac{h^+}{h^-} = \frac{\left(u_L^2 + \frac{1}{3}\,u_R^2\right)D_u^+ + \left(d_L^2 + \frac{1}{3}\,d_R^2\right)D_d^+}{\left(u_L^2 + \frac{1}{3}\,u_R^2\right)D_u^- + \left(d_L^2 + \frac{1}{3}\,d_R^2\right)D_d^-}$$

($L \leftrightarrow R$ for $\bar{\nu}$-induced reactions). The fragmentation functions can be determined from charged-current ν and $\bar{\nu}$ reactions. The positive and negative hadrons are predominantly pions, but kaons and protons are included in the data sample since in a bubble chamber the particles cannot be distinguished at momenta > 1 GeV/c.

The Aachen-Bonn-CERN-Demokritos-London-Oxford-Saclay (ABCDLOS) (BEBC) group[53] has reported a recent determination of h^+/h^- ratios at large z. They used BEBC filled with a heavy-neon mixture and exposed to ν and $\bar{\nu}$ narrow-band beams. From 300 ν and 140 $\bar{\nu}$ neutral-current events they deduce, together with the knowledge of $u_L^2 + d_L^2$ and $u_R^2 + d_R^2$ as determined in an earlier experiment[41],

$$u_L^2 = 0.11 \pm 0.05\ , \qquad d_L^2 = 0.21 \pm 0.05$$

$$u_R^2 = 0.01 \pm 0.03\ , \qquad d_R^2 = 0.03 \pm 0.03\ .$$

The coupling constants are in agreement with the predictions of the standard model for $\sin^2\theta_W = 0.23$.

Sehgal[6] was the first to determine the squares of the individual coupling constants, employing π^+/π^- ratios measured in the fragmentation region of low-energy GGM data[54]. Since the validity of the quark fragmentation model may be questioned there, the confirmation of his earlier conclusions with high-energy data is gratifying.

Still, one should not overestimate the systematic precision of the measurement of final-state charge ratios: the event selection is difficult, and corrections for various backgrounds are large.

The final-state charge ratios as determined in the ABCDLOS experiment are given in Table 13. The result for $\bar{\nu}$-induced hadrons agrees well with a recent measurement from the Fermilab-Michigan-Moscow-Serpukhov (FMMS) Collaboration[55] performed in the FNAL 15′ bubble chamber filled with a heavy-neon mixture and exposed to a $\bar{\nu}$ wide-band beam.

Table 13

Recent measurements of positive-to-negative final-state charge ratios

Experiment	$(h^+/h^-)_\nu$	$(h^+/h^-)_{\bar{\nu}}$
ABCDLOS[53]	1.07 ± 0.17 $(z > 0.3)$	1.54 ± 0.45 $(z > 0.3)$
FMMS[55]	-	1.60 ± 0.27 $(0.3 < z < 0.9)$

1.5.6 Limits on charm-changing neutral currents. In the standard model, the hadronic neutral current conserves flavour by construction. A search for charm-changing reactions of the type

$$\bar{\nu} + N \to \bar{\nu} + C \quad \text{with} \quad C \to e^+ + \nu_e + \text{anything}$$

has recently been carried out by the FMMS Collaboration[55]. The detector was the FNAL 15′ bubble chamber filled with a heavy-neon mixture and exposed to a $\bar{\nu}$ wide-band beam. Only one candidate for a $\Delta C = 1$ neutral current was found, yielding an upper limit

$$\frac{\sigma(\bar{\nu} + N \to \bar{\nu} + C)}{\sigma(\text{all neutral currents})} < \frac{0.87 \times 10^{-3}}{0.38 \times 0.1} = 2.3\%$$

at the 90% confidence level (where a semileptonic branching ratio of the charmed particle of 0.1 is assumed). This upper limit is similar in accuracy to a corresponding earlier limit for charm-changing reactions of the type

$$\nu + N \rightarrow \nu + C \quad \text{with} \quad C \rightarrow \mu^+ + \nu_\mu + \text{anything} ,$$

which has been given by the CDHS Collaboration[56]:

$$\frac{\sigma(\nu + N \rightarrow \nu + C)}{\sigma(\text{all neutral currents})} < 2.6\%$$

at the 90% confidence level.

1.5.7 The coupling constants for neutrino-quark scattering. The concept of fitting the coupling constants of the most general Lagrangian to experimental data was pioneered by Hung and Sakurai, Sehgal, and Abbott and Barnett. In 1978, this work resulted in the first complete set of neutral-current coupling constants[6,7,35], employing inclusive and semi-inclusive data from isoscalar targets, elastic scattering on protons, inclusive scattering on protons, and single-pion production.

Since then, many other authors have contributed to this field. Recently, Liede and Roos[57] and Langacker et al.[58] presented the results of global fits to all available neutral-current data. The fit gives a reasonable χ^2, which is rather surprising since the experimental errors quoted by the various experiments are presumably not Gaussian. All (but two) experimental results of the last two years are in qualitative agreement with each other and with the standard model. Thus one single parameter, $\sin^2\theta_W$, is able to describe a host of experimental results. Properly adjusted small modifications of the standard model (extension of the Higgs doublet, right-handed doublets of weak isospin, more than one Z) cannot be ruled out, but the standard model is at present sufficient.

Table 14 gives a summary of the best-fit values[25,58] for the neutral-current coupling constants. The fit did not yet include the very recent measurements on R_ν, $R_{\bar{\nu}}$, and R^p_ν.

Table 14

Neutral-current coupling constants

Coupling constant	Best-fit value	Ref.	Salam-Weinberg model	For $\sin^2\theta_W = 0.23$
u_L	0.32 ± 0.03	58	$\frac{1}{2} - \frac{2}{3}\sin^2\theta_W$	0.347
d_L	-0.43 ± 0.03		$-\frac{1}{2} + \frac{1}{3}\sin^2\theta_W$	-0.423
u_R	-0.17 ± 0.02		$-\frac{2}{3}\sin^2\theta_W$	-0.153
d_R	-0.01 ± 0.05		$\frac{1}{3}\sin^2\theta_W$	0.077
α	0.58 ± 0.14	25	$1-2\sin^2\theta_W$	0.540
β	0.92 ± 0.14		1	1
γ	-0.28 ± 0.14		$-\frac{2}{3}\sin^2\theta_W$	-0.153
δ	0.06 ± 0.14		0	0
$\tilde{\alpha}$	-0.72 ± 0.25	25	$-1 + 2\sin^2\theta_W$	-0.540
$\tilde{\beta}$			$-1 + 4\sin^2\theta_W$	-0.080
$\tilde{\gamma}$	0.36 ± 0.28		$\frac{2}{3}\sin^2\theta_W$	0.153
$\tilde{\delta}$			0	0
g_V	0.06 ± 0.08		$-\frac{1}{2} + 2\sin^2\theta_W$	-0.040
g_A	-0.52 ± 0.06		$-\frac{1}{2}$	-0.500

The central value of $\sin^2\theta_W$ which best describes all data has not changed since one year[35], but the error has decreased:

$$\sin^2\theta_W = 0.230 \pm 0.015 \ .$$

The central value is rounded in the last digit and represents a compromise between 0.229 ± 0.014[57], 0.232 ± 0.009[58] (this fit value includes the new results), and 0.228 ± 0.018 from the new CDHS measurement of R_ν (see Section 1.5.2). The quoted error on $\sin^2\theta_W$ contains an estimate of the systematic error from theoretical uncertainties as discussed in Section 1.5.2.

In the standard model, the $W^\pm$ and Z^0 masses are related by $\rho = M_W^2/M_Z^2 \cos^2\theta_W = 1$. If we extend the standard model so as to include two Higgs doublets, ρ becomes a parameter to be determined by experiment. The simultaneous fit of ρ and $\sin^2\theta_W$ to all data yields[59]

$$\sin^2\theta_W = 0.22 \pm 0.03$$

and

$$\rho = 0.99 \pm 0.03$$

where again theoretical uncertainties have been included in the quoted errors. Hence the extension of the Higgs sector does not seem to be necessary.

1.6 Summary on Neutral-Current Reactions

i) All (but two) more recent experimental results are consistent with each other.

ii) The Salam-Weinberg model gives a good description of all (but two) experimental results, with only one free parameter: $\sin^2\theta_W = 0.230 \pm 0.015$.

iii) Eight out of the ten phenomenological parameters governing νe, eq, and νq scattering are determined from data. All of them are consistent with the predictions of the Salam-Weinberg model with $\sin^2\theta_W = 0.23$.

iv) The experimental precision has reached such a level that theoretical uncertainties in the data analysis can no longer be ignored.

v) For the immediate future it is still worth while to improve the experimental accuracy. The experimental challenges are: a better determination of the V+A admixture on isoscalar targets; a better determination of the isoscalar-isovector interference (inclusive cross-section ratios on proton and neutron targets, charge ratios in final states); a study of the structure function for neutral currents; and a better determination of $\bar{\nu}_\mu e$ scattering.

vi) A new dimension of neutral-current physics is opened up with the advent of high-energy colliding machines. The first step is the detection of the forward-backward asymmetry in $e^+e^- \to \mu^+\mu^-$ due to weak and electromagnetic interference. The big challenge is to find $W^\pm$ and Z^0 at the predicted masses, and to find the Higgs boson with its peculiar coupling properties.

2. INCLUSIVE CHARGED CURRENTS

2.1 Introduction

The study of the inclusive charged-current scattering of neutrinos on nucleons is interesting for two reasons: firstly, the structure of the weak interaction is tested at large momentum transfers; secondly, once the weak interaction is understood, the neutrino serves as an excellent probe for the internal structure of the nucleon. The charged-current reaction selects well-defined quark flavours, its chiral nature enables the distinction of quarks and antiquarks among the nucleon constituents, and the cross-section is undamped at presently available momentum transfers by virtue of the large W-mass. These features guarantee the important role of the neutrino in the exploration of the internal structure of the nucleon, and complement analogous studies using electrons and muons as projectiles.

The deep inelastic scattering process (Fig. 11) is described by three independent variables,

$$s = (k+p)^2 = 2ME + M^2$$
$$Q^2 = -q^2 = -(k-k')^2 = 4EE_\mu \sin^2 \frac{\theta}{2}$$
$$\nu = \frac{pq}{M} = E - E_\mu = E_H - M ,$$

where E denotes the energy of the incoming neutrino, M the nucleon mass, and θ the scattering angle of the outgoing muon.

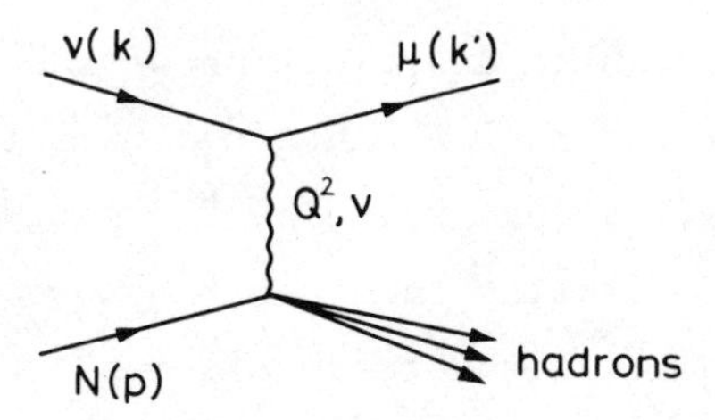

Fig. 11 The deep inelastic scattering process.

Experimentally, one measures the momentum and the angle of the muon, and the energy of the hadronic final state (the measured quantity is ν but is usually referred to as hadron energy E_H by experimentalists). Alternatively, the neutrino energy E and the Bjorken scaling variables

$$x = \frac{Q^2}{2M\nu} \quad \text{and} \quad y = \frac{\nu}{E}$$

are often used: y is the fractional neutrino energy transferred to the hadrons, and x is the fraction of the maximum Q^2 possible for the particular y.

Assuming a local V-A coupling, averaging over initial polarizations, summing over final polarizations, and neglecting the muon mass, the most general form of the double differential cross-section for the inclusive neutrino scattering on nucleons is given by

The double differential cross-section then becomes

$$\frac{d^2\sigma^\nu}{dxdy} = \frac{G^2ME}{\pi}\left[q(x,Q^2) + (1-y)q_L(x,Q^2) + (1-y)^2\bar{q}(x,Q^2)\right]$$

$$\frac{d^2\sigma^{\bar{\nu}}}{dxdy} = \frac{G^2ME}{\pi}\left[(1-y)^2q(x,Q^2) + (1-y)q_L(x,Q^2) + \bar{q}(x,Q^2)\right].$$

From the y-dependence of the cross-section of the scattering of point-like particles, one identifies (in the limit of the QPM) q(x) with the momentum distribution of the quark content of the nucleon, and $\bar{q}(x)$ with the momentum distribution of the antiquark content of the nucleon (the fraction of the nucleon momentum which is carried by the struck quark, is just x). The "longitudinal" structure function q_L vanishes, reflecting the spin ½ nature of the quarks. This is equivalent to the "Callan-Gross relation"

$$F_2 = 2xF_1 ,$$

which leaves, in the QPM limit, only terms with flat y and with a $(1-y)^2$ dependence in the double differential cross-section.

Expressing the quark momentum distributions in terms of the different quark flavours up to the charmed quark,

$$q(x) = u(x) + d(x) + s(x) + c(x)$$

and

$$\bar{q}(x) = \bar{u}(x) + \bar{d}(x) + \bar{s}(x) + \bar{c}(x) ,$$

we then get the following differential cross-sections, in the QPM limit and for an isoscalar target,

$$\frac{d^2\sigma^\nu}{dxdy} = \frac{G^2ME}{\pi}\left[q+s-c + (1-y)^2(\bar{q}-\bar{s}+\bar{c})\right]$$

$$\frac{d^2\sigma^{\bar{\nu}}}{dxdy} = \frac{G^2ME}{\pi}\left[(1-y)^2(q-s+c) + \bar{q}+\bar{s}-\bar{c}\right].$$

$$\frac{d^2\sigma}{dxdy} = \frac{G^2ME}{\pi}\left[(1 - y - \frac{Mxy}{2E})F_2(x,Q^2) + \frac{y^2}{2}\,2xF_1(x,Q^2) \pm \right.$$
$$\left. \pm\,(y - \frac{y^2}{2})xF_3(x,Q^2)\right] .$$

The upper sign refers to neutrinos, the lower to antineutrinos. The three structure functions $F_i(x,Q^2)$ describe the structure of the hadron vertex, and are different for neutrinos and antineutrinos, and for neutrons and protons. Assuming charge symmetry, the following relations hold between the structure functions: $F_i^{\nu n} = F_i^{\bar{\nu}p}$ and $F_i^{\bar{\nu}n} = F_i^{\nu p}$. Using in addition an isoscalar target, the structure functions are averaged over protons and neutrons. The number of structure functions is thus reduced from twelve to three.

In the Bjorken scaling limit, i.e. for $Q^2,\nu \to \infty$ while $x = Q^2/2M\nu$ remains fixed, the structure functions become functions of the dimensionless scaling variable x only: $F_i(x,Q^2) \to F_i(x)$. The approximate validity of the scaling hypothesis, as observed in electron, muon, and neutrino scattering on nucleons, has inspired the quark-parton model (QPM) for the structure of the nucleon. It is assumed that the nucleon is made up of point-like, spin ½ constituents called partons, which are identified with the quarks. It is further assumed that the partons interact freely and independently, without interference. After the scattering, the partons rearrange themselves to form the final-state hadrons. The total cross-section is the incoherent sum of the elementary parton cross-sections.

The nucleon consists of three valence-quarks and a sea of quark-antiquark pairs. In this context, it is convenient to define three new structure functions as follows:

$$q = \frac{1}{2}\,(2xF_1 + xF_3)$$

$$\bar{q} = \frac{1}{2}\,(2xF_1 - xF_3)$$

$$q_L = F_2 - 2xF_1 \ .$$

Here u(x) denotes the fractional momentum carried by the up-quarks in the proton (which is assumed to be equal to the fractional momentum carried by the down-quarks in the neutron) etc. Expressing the structure functions $F_2(x)$ and $xF_3(x)$ in terms of the quark momentum densities, we get

$$F_2^{\nu}(x) = F_2^{\bar{\nu}}(x) = q(x) + \bar{q}(x)$$

$$xF_3^{\nu}(x) = q(x) - \bar{q}(x) + 2\bar{s}(x) - 2\bar{c}(x)$$

$$xF_3^{\bar{\nu}}(x) = q(x) - \bar{q}(x) - 2\bar{s}(d) + 2\bar{c}(x) .$$

Hence $F_2(x)$ represents the fractional momentum carried by all quarks in the nucleon, and $xF_3(x)$ is essentially the fractional momentum carried by the valence quarks only. Since the strange quark content of the nucleon is presumably larger than the charmed quark content, owing to the larger mass of the charmed quark, small corrections are applied by experimentalists to account for this difference, as well as for small deviations from isoscalar targets. The measured structure function xF_3 is the average of the neutrino and antineutrino structure functions: $xF_3 = \frac{1}{2}(xF_3^{\nu} + xF_3^{\bar{\nu}})$.

The QPM picture of neutrino scattering on nucleons has been shown by experiments to be approximately valid. It serves as a reference frame also for the more general case where the (small) Q^2-dependence of the structure functions is explored.

In Section 2.2 we shall discuss the measurement of the muon polarization which supports the V-A structure of the charged weak current also at large Q^2. In Sections 2.3 to 2.6 we shall present the experimental evidence for the approximate validity of the QPM, and the results for the structure functions. The scaling violation observed in neutrino scattering and its interpretation in terms of quantum chromodynamics (QCD) will be discussed in Sections 2.7 and 2.8.

2.2 Muon Polarization

From weak decay studies at low energies it is known that the weak charged current has a V-A structure. At high energies, the observed y-distributions of the form $d\sigma/dy \propto a + b(1-y)^2$ are consistent with a V-A structure, too (see Section 2.4). However, such y-distributions can also be described by an appropriate mixture of S, P, and T covariants. Measurements of the helicity of the outgoing muon in the neutrino scattering process can resolve this ambiguity since V and A terms preserve the helicity of the incoming neutrino, whereas S, P, and T terms flip the neutrino helicity. Hence, a μ^+ emerging from an $\bar{\nu}$ interaction will have a polarization P = +1 in the V, A case, or P = -1 in the S, P, T case.

A measurement of the polarization of μ^+ produced by the interaction of $\bar{\nu}$ in iron has been reported by the CHARM Collaboration[60] (μ^- cannot be measured in the same way because of their quick capture). In 3400 $\bar{\nu}$ interactions in iron, a μ^+ was stopped and its helicity was determined from the decay asymmetry. The forward-backward decay asymmetry has been measured by the spin precession method. The amplitude and the phase of the asymmetry (Fig. 12) of

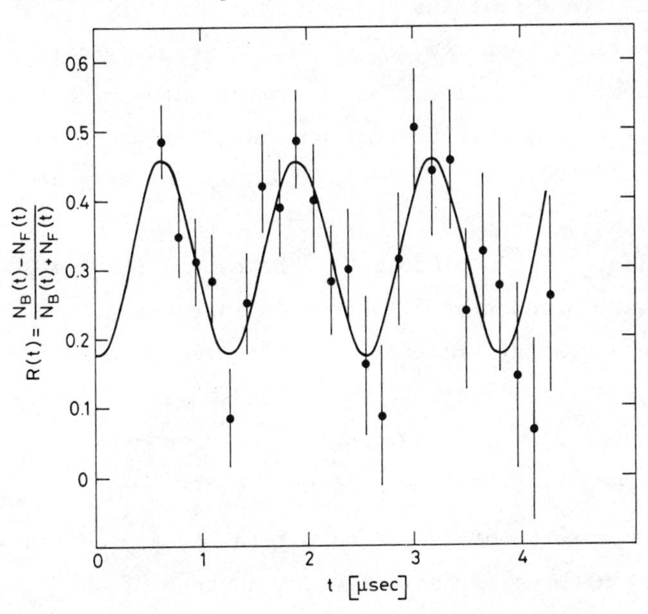

Fig. 12 Time-dependence of the forward-backward decay asymmetry of μ^+ produced in $\bar{\nu}$ interactions.

the e^+ emission provides information on the polarization and on its sign. The polarization was found to be $P = +(1.09 \pm 0.22)$ at an average momentum transfer $\langle Q^2 \rangle = 3.2\ \text{GeV}^2$, consistent with the expectation $P = +1$ from the V-A structure. An upper limit $\sigma_{S,P,T} <$ $< 0.18\ \sigma_{tot}$ (95% c.l.) has been quoted on the contribution of other than V, A couplings.

2.3 Total Cross-Sections

The standard local current-current weak interaction theory predicts that the cross-section of neutrino scattering off point-like spin ½ objects grows linearly with energy. At very high energy, a departure from linearity is expected owing to the propagator term from the W-boson exchange. On the other hand, the QPM predicts the deep inelastic structure functions of the nucleon to scale in the dimensionless variable x. Hence the neutrino-nucleon cross-section is expected to grow linearly with energy as well. The study of ν and $\bar{\nu}$ total cross-sections on nucleons therefore provides a fundamental test both of the weak interaction theory and of the structure of nucleons.

The observed total cross-sections show an approximately linear rise with energy. The results on the slopes σ/E measured by different experiments are shown in Fig. 13. The agreement between the experiments is good. There is some indication that σ^{ν}/E is slightly larger at low energies ($\lesssim 10$ GeV). The data are compared with a QCD calculation[61] incorporating the onset of charm production and the scaling violation predicted for $\Lambda = 0.5$ GeV. The curve with $\Lambda = 0$ shows the effect of charm production alone, which increases the cross-section with higher energy. Altogether, the predicted deviation of the cross-section from linearity is rather small, in agreement with the experimental data. No damping of the cross-section up to 260 GeV is seen, pointing to a very large W-mass.

The deviation of the ratio of the $\bar{\nu}$ and ν total cross-sections from ⅓ is interpreted in terms of the antiquark content of the nucleon. The experimental results are consistent with a 10-15%

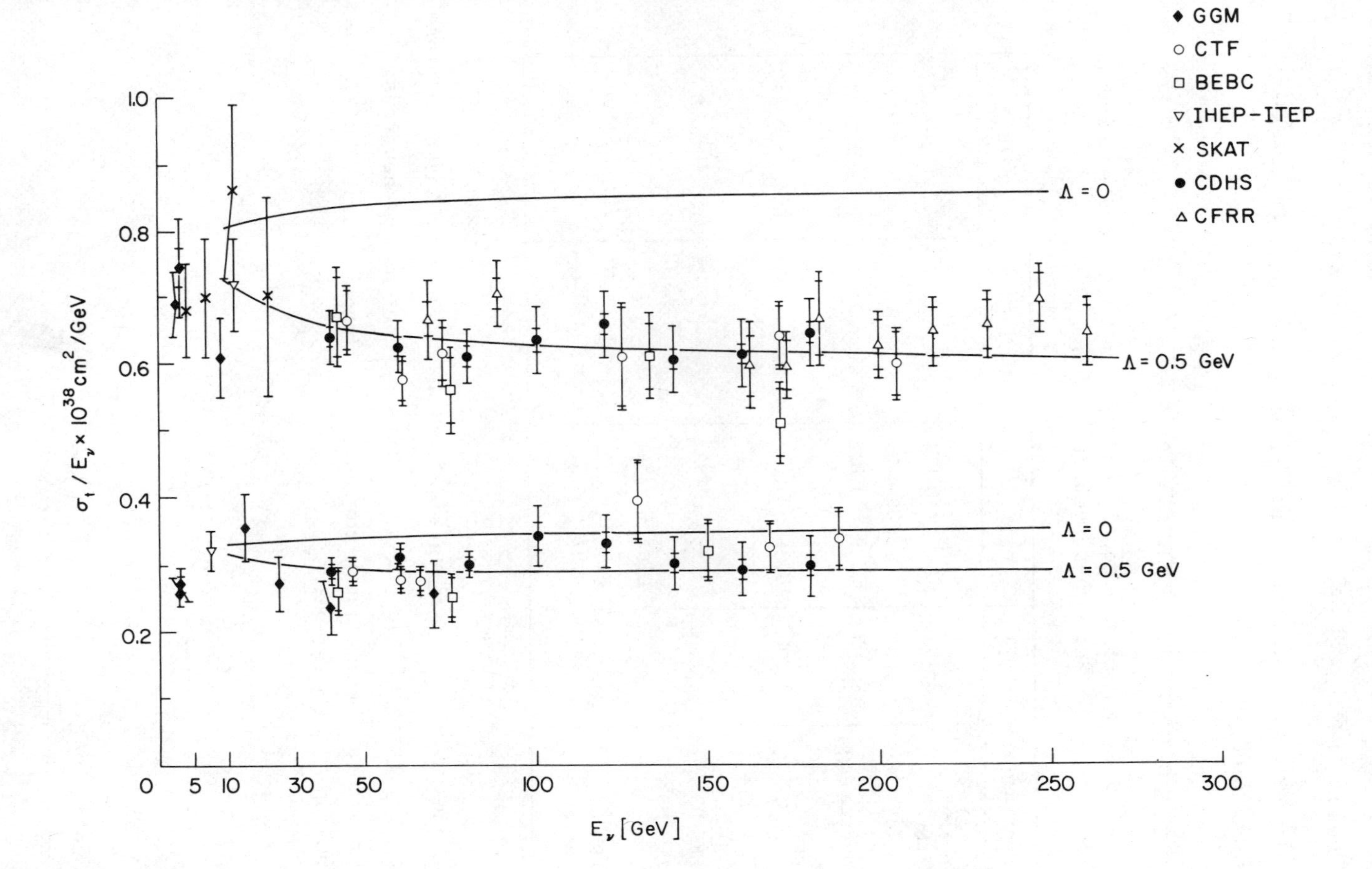

Fig. 13 The slopes σ/E for the ν and $\bar{\nu}$ total cross-section on nucleons. The curves are explained in the text.

antiquark contribution in the nucleon, with a rising tendency in the energy range 2-200 GeV.

The most precise experimental results on total cross-sections are summarized in Table 15.

Table 15

Slopes σ/E of the total neutrino-nucleon cross-sections, and cross-section ratios $\sigma^{\bar{\nu}}/\sigma^{\nu}$

Experiment	E(GeV)	$\sigma^{\bar{\nu}}/E$ a)	σ^{ν}/E a)	$\sigma^{\bar{\nu}}/\sigma^{\nu}$
GGM[62] CERN-PS	2-10	0.27 ± 0.03	0.74 ± 0.07	0.38 ± 0.025
CFR[63] Counter exp.	45-205	0.29 ± 0.015	0.61 ± 0.03	0.48 ± 0.02
CDHS[64] Counter exp.	30-200	0.30 ± 0.02	0.62 ± 0.03	0.48 ± 0.02
GGM[65] CERN PS	3	0.69 ± 0.05		
"	9	0.61 ± 0.06		
CFRR[66] Counter exp.	70-260	0.67 ± 0.04		

a) In units of 10^{-38} cm^2/GeV.

A considerable experimental effort has gone into the determination of the ratio of the cross-sections of protons and neutrons. Owing to the different quark content, one expects roughly $\sigma^{\nu n}/\sigma^{\nu p} = 2$ and $\sigma^{\bar{\nu}n}/\sigma^{\bar{\nu}p} = 0.5$. A more detailed analysis, taking into account the sea quarks, yields

$$\frac{\sigma^{\nu n}}{\sigma^{\nu p}} = \frac{\int[u(x) + s(x) + \frac{1}{3}\bar{d}(x)]dx}{\int[d(x) + s(x) + \frac{1}{3}\bar{u}(x)]dx} \sim 1.8 ,$$

and

$$\frac{\sigma^{\bar{\nu}n}}{\sigma^{\bar{\nu}p}} = \frac{\int[\frac{1}{3}d(x) + \bar{u}(x) + \bar{s}(x)]dx}{\int[\frac{1}{3}u(x) + \bar{d}(x) + \bar{s}(x)]dx} \sim 0.6 ,$$

where u(x) and d(x) represent the fractional momentum of all up- and down-quarks in the proton [$u(x) \simeq 2d(x)$]. These predictions are further modified by 10-20% if the shapes of u(x) and d(x) are taken somewhat different, as observed experimentally in ep and en scattering. The experimental results, summarized in Table 16, are consistent with the expectation.

Table 16

Neutron-to-proton cross-section ratios

Experiment	Target	E(GeV)	$\sigma^{\nu n}/\sigma^{\nu p}$	$\sigma^{\bar{\nu} n}/\sigma^{\bar{\nu} p}$
ANL 12' [67]	D_2	1.5-6	1.95 ± 0.21	
BNL[68]	H_2/D_2	< 10	1.48 ± 0.17	
GGM[69]	CF_3Br	1-10	2.08 ± 0.15	
BEBC-TST[70]	H_2-Ne	> 10	1.97 ± 0.38	
FNAL 15' [71]	D_2	> 10	1.74 ± 0.25	
GGM[72]	CF_3Br	< 5		0.46 ± 0.10
FNAL 15' [73]	Ne/H_2	> 10		0.51 ± 0.10

2.4 y-Distributions

The study of y-distributions is meaningful only in the framework of the QPM, i.e. if one neglects the small Q^2-dependence (which is equivalent to a y-dependence) of the structure functions which is observed experimentally (see Section 2.7).

In Fig. 14, the distributions in the inelasticity y is shown which has been measured by the CDHS Collaboration[64] in the energy range from 30 to 200 GeV. A special cut in Q^2 and in the invariant mass of the hadron system ($Q^2 > 1$ GeV^2, $W^2 = M^2 + 2M\nu - Q^2 > 4$ GeV^2) is applied to the data in order to remove quasi-elastic events from

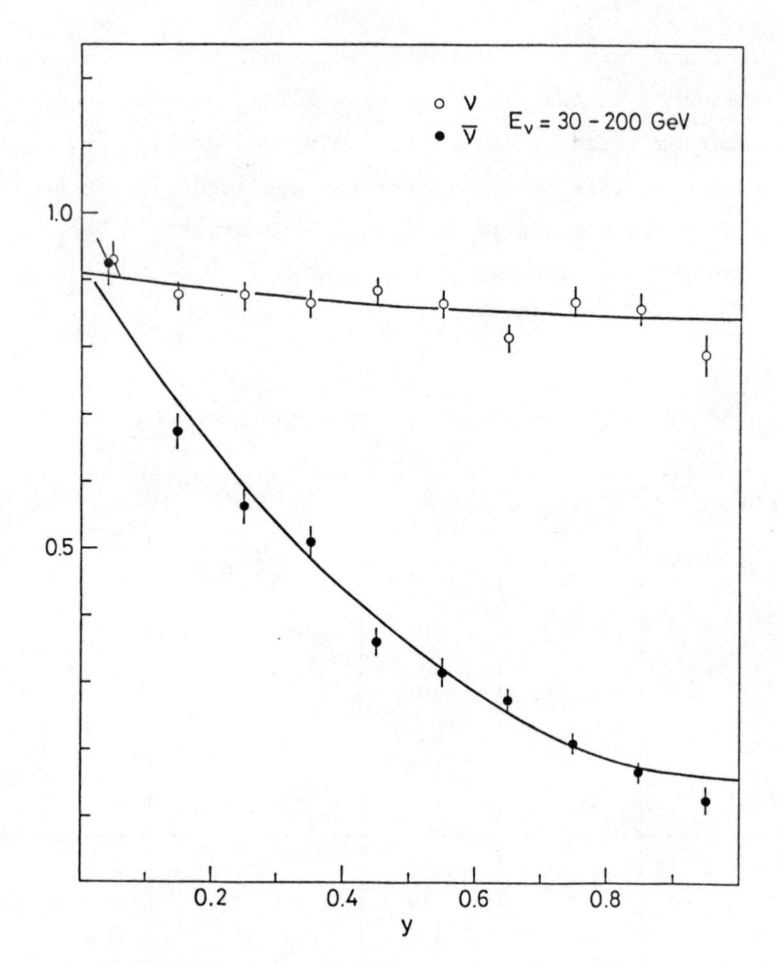

Fig. 14 Distributions in y for ν and $\bar{\nu}$.

the sample. The shapes of the y-distributions do not change significantly in the range from 30 to 200 GeV, and are consistent with the form $d\sigma/dy \propto a + b(1-y)^2$, supporting the QPM as a good approximation for the structure of the nucleon.

By virtue of the chiral nature of the weak interaction, the fits of the form $a + b(1-y)^2$ to the y-distributions determine the fraction of the antiquarks in the nucleon. The experimental results for the antiquark content of the nucleon are summarized in Table 17, together with the method of its determination and the relevant energy range. The antiquark content seems to increase with energy, and to settle around 15% at high energies. It should also be pointed out that radiative corrections cannot be ignored here, but their application is not straightforward. For example, the recipe of Barlow and Wolfram[74)] changes the fractional antiquark content, as determined from the CDHS $\bar{\nu}$ y-distribution,

$$\frac{\int[\bar{q}(x) + \bar{s}(x)]dx}{\int[q(x) + \bar{q}(x)]dx},$$

from 0.16 ± 0.02 (raw value) to 0.15 ± 0.02. The radiative correction is therefore of the same order as the experimental uncertainty.

Table 17

Results on the fractional antiquark momentum of the nucleon

Experiment	E(GeV)	Method	$\frac{\bar{Q} + \bar{S}}{Q + \bar{Q}}$ a)	Radiative correction
GGM[62)] CERN-PS	2-10	Total cross-section	0.07 ± 0.04	No
ABCLOS (BEBC)[75)]	30-200	From F_2 and xF_3	0.11 ± 0.03	No
CDHS[64)] Counter exp.	30-200	$\bar{\nu}$ y-distribution	0.15 ± 0.02	Yes
		$\bar{\nu}$ high-y cross-section	0.16 ± 0.01	Yes
HPWFOR[76)] Counter exp.	10-45	$\bar{\nu}$ y-distribution	0.11 ± 0.02	No
	80-220	"	0.17 ± 0.02	No
FMMS[77)] FNAL 15′	∿ 30	$\bar{\nu}$ y-distribution	0.13 ± 0.02	No

a) $\frac{\bar{Q} + \bar{S}}{Q + \bar{Q}} = \frac{\int[\bar{q}(x) + \bar{s}(x)]dx}{\int[q(x) + \bar{q}(x)]dx}$.

In the QPM, the antiquark contributions to neutrino and antineutrino scattering differ by twice the strange antiquark distribution (neglecting the charmed quark contribution). However, for reasons of statistical accuracy, but also for reasons of systematic errors due to scaling violation and radiative corrections, it is better to determine the strange antiquark contribution from dimuon data (see Section 2.5). For the same reason, also a determination of a linear term (1-y) in $d\sigma/dy$ is not the best method for measuring a violation of the Callan-Gross relation, although the study of the y-distribution certainly excludes a large violation.

2.5 Structure Functions

The structure functions $F_2(x)$ and $xF_3(x)$ are obtained from the sum and the difference of the ν and $\bar{\nu}$ cross-sections:

$$F_2(x) = \left[\frac{d^2\sigma^{\nu}}{dxdy} + \frac{d^2\sigma^{\bar{\nu}}}{dxdy}\right] \frac{\pi}{G^2ME} \frac{1}{1 + (1-y)^2 - y^2R'}$$

$$xF_3(x) = \left[\frac{d^2\sigma^{\nu}}{dxdy} - \frac{d^2\sigma^{\bar{\nu}}}{dxdy}\right] \frac{\pi}{G^2ME} \frac{1}{1 - (1-y)^2} \cdot$$

In these expressions, small corrections for the contribution of the strange sea and for a deviation from an isoscalar target are dropped for reasons of simplicity. High-precision measurements of the structure functions so far exist only for heavy-target materials with almost isoscalar nuclei.

The term y^2R' with $R' = (F_2 - 2xF_1)/F_2$ accounts for a violation of the Callan-Gross relation. Assuming $R' = 0$ and neglecting the Q^2-dependence of the structure functions, the CDHS[64] and the Harvard-Pennsylvania-Wisconsin-Fermilab-Ohio-Rutgers (HPWFOR)[78] Collaborations have obtained the structure functions $F_2(x)$ and $xF_3(x)$ averaged over the energy range of the experiments (30-200 GeV and 20-200 GeV, respectively). The results are shown in Figs. 15 and 16. The measurements agree with each other reasonably well.

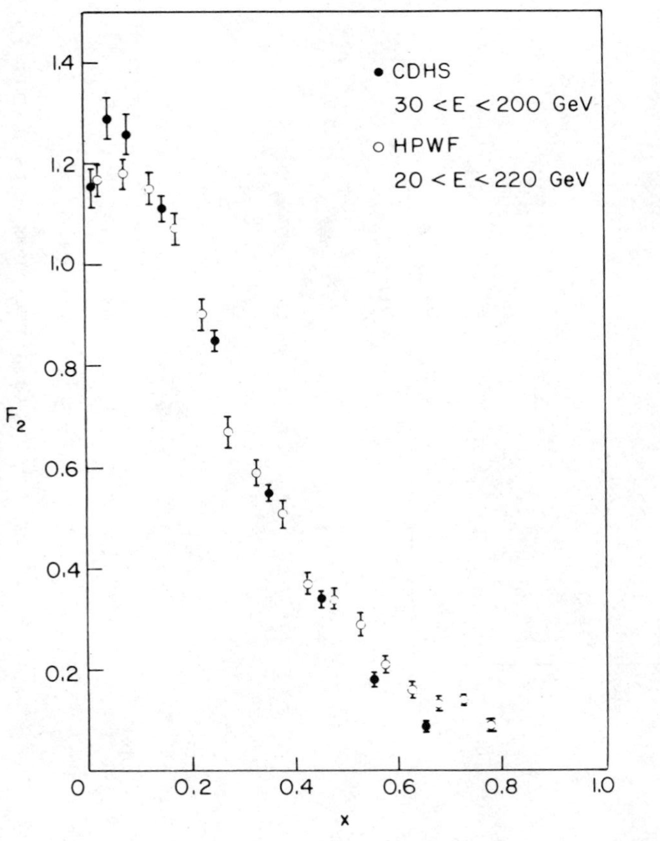

Fig. 15 The structure function $F_2(x)$ for an isoscalar target.

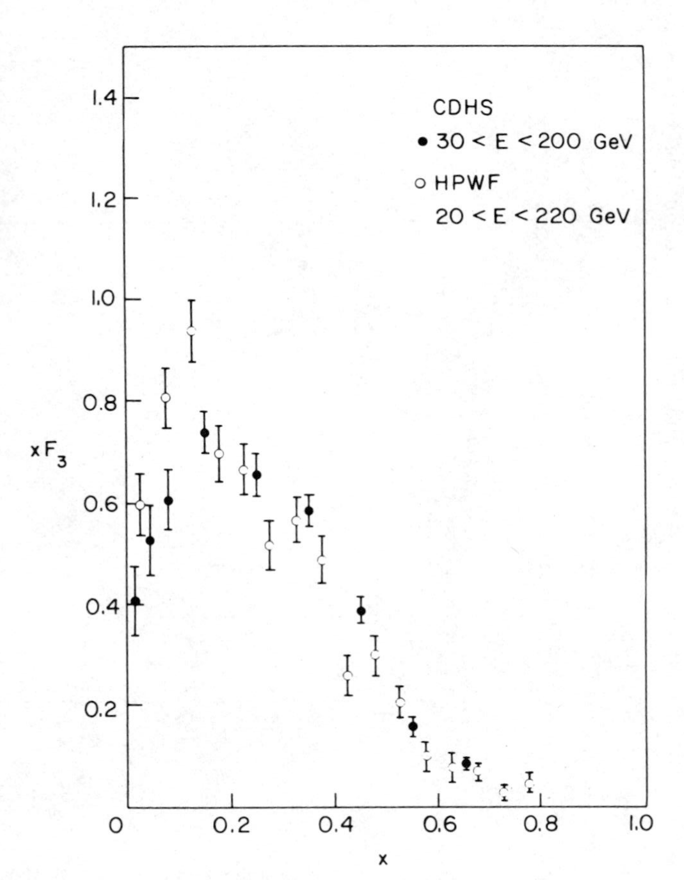

Fig. 16 The structure function $xF_3(x)$ for an isoscalar target.

By fitting the x-dependence of the y-distribution the Purdue-Argonne-Carnegie Mellon (PACM) Collaboration[79)] has obtained the structure function $F_2^{\bar{\nu}p}(x)$ from $\bar{\nu}$ data on hydrogen, which is shown in Fig. 17. In terms of quark momentum distributions, $F_2^{\bar{\nu}p}(x)$ is

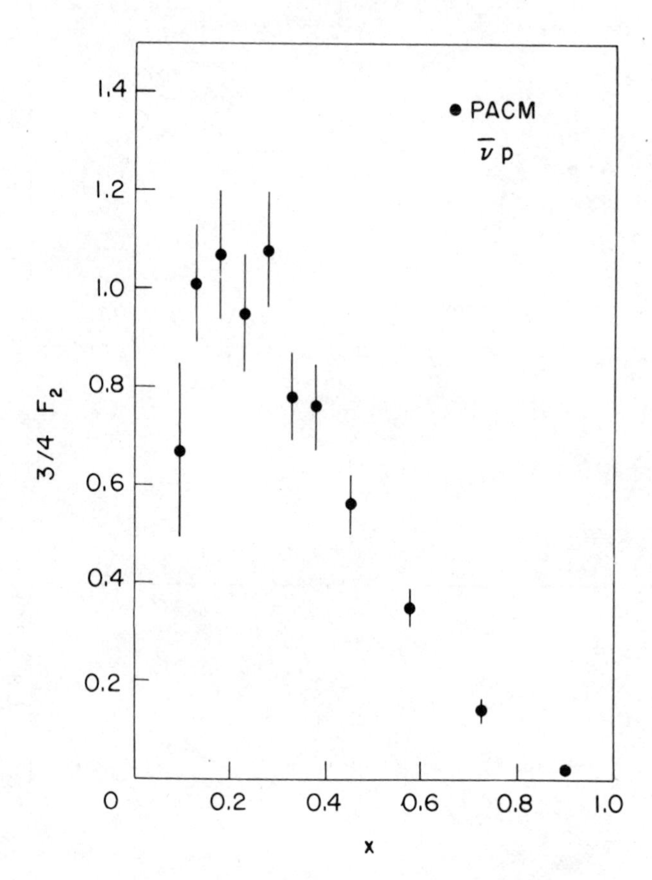

Fig. 17 The structure function $F_2^{\bar{\nu}p}(x)$.

given by $F_2^{\bar{\nu}p} = 2[u(x) + \bar{d}(x) + \bar{s}(x)]$, whereas $F_2(x) = u(x) + d(x) + \bar{u}(x) + \bar{d}(x) + 2\bar{s}(x)$. With $u(x) = 2d(x)$, a scale factor of $^3/_4$ applied to $F_2^{\bar{\nu}p}$ allows, in principle, a direct comparison with $F_2(x)$. However, since the data of the PACM Collaboration are normalized with analogous arguments to electron scattering data, no conclusion can be drawn on the absolute value of $F_2^{\bar{\nu}p}(x)$. As far as the shape is concerned, $F_2^{\bar{\nu}p}(x)$ appears to be somewhat wider than $F_2(x)$, supporting an up-quark distribution in the proton which is wider than the down-quark distribution.

The structure function $xF_3(x)$ can be parametrized in the form $xF_3(x) \propto \sqrt{x}(1-x)^n$, with an exponent $n = 3.5 \pm 0.5$. The integrals of the structure functions are for neutrino energies between 30 and 200 GeV:

$$\int F_2(x)dx = 0.44 \pm 0.02$$

and

$$\int xF_3(x)dx = 0.31 \pm 0.03 ,$$

giving the fractional nucleon momentum which is carried by all quarks and antiquarks, and by the valence quarks only. The missing 56% of the nucleon momentum is believed to be carried by gluons.

The integral $\int F_3(x)\,dx$ represents the total number of valence quarks in the nucleon and is therefore expected to be close to three [Gross-Llewellyn Smith sum rule[80]]. The valence quark density $F_3(x)$ is obtained by dividing the structure function $xF_3(x)$ by x. The results are consistent with the QPM expectation:

$$\int F_3(x)\,dx = 3.2 \pm 0.5 \qquad \text{CDHS}^{64)}$$
$$= 2.8 \pm 0.45 \qquad \text{HPWFOR}^{78)}$$
$$> 2.7 \pm 0.4 \qquad \text{ABCLOS (BEBC)}^{75)} .$$

The structure function of the sea can be determined from the $\bar{\nu}$ cross-section at large y:

$$\bar{q}(x) + \bar{s}(x) = \frac{\pi}{G^2ME}\left[\frac{d^2\sigma^{\bar{\nu}}}{dxdy} - (1-y)^2\,\frac{d^2\sigma^{\nu}}{dxdy}\right]_{y\to 1} .$$

The antiquark structure function which has been determined with this method by the CDHS Collaboration[64], is shown in Fig. 18. In contrast to the structure functions F_2 and xF_3, the antiquark structure function is much more concentrated at small x. The shape of the structure function can be parametrized in the form $(1-x)^m$, with an exponent m = 6.5 ± 0.5. In the same Fig. 18, also the data of the HPWFRO Collaboration[78] are shown. The agreement between the HPWFOR data and the earlier CDHS data is poor.

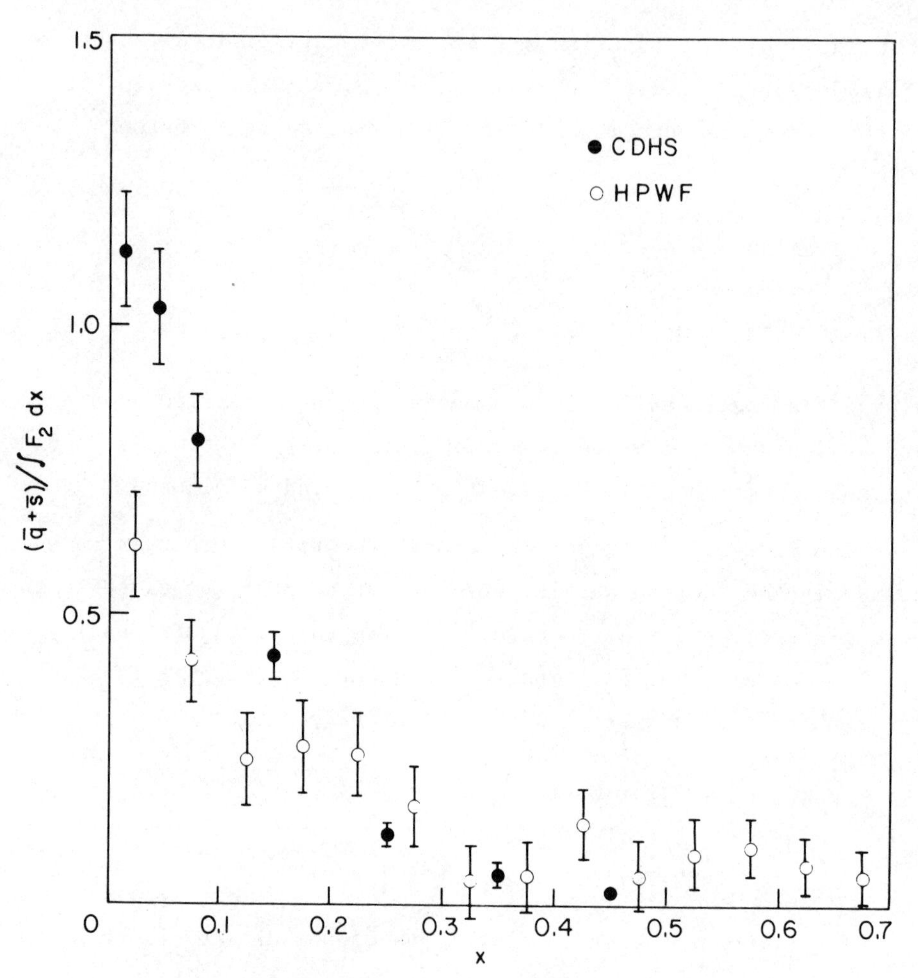

Fig. 18 The antiquark structure function $\bar{q}(x) + \bar{s}(x)$.

The structure function of the strange antiquarks $\bar{s}(x)$ can be determined via the charmed quark production by $\bar{\nu}$ which in the Glashow Iliopoulos-Maiani (GIM) model[10] is given by

$$\frac{d^2\sigma^{\bar{\nu}}}{dxdy} = \frac{G^2ME}{\pi}\left\{\left[\bar{u}(x) + \bar{d}(x)\right] \sin^2\theta_C + 2\bar{s}(x) \cos^2\theta_C\right\},$$

where θ_C is the Cabibbo angle. Since $\sin^2\theta_C << \cos^2\theta_C$, charm production by $\bar{\nu}$ probes the strange antiquark content of the nucleon. The signature for charm production is an e^- or μ^- from the semi-leptonic decay $\bar{c} \rightarrow \bar{s}\ell^-\bar{\nu}$ (ℓ = e or μ) in addition to the leading μ^+. From a sample of opposite-sign $\bar{\nu}$-induced dimuon events, the CDHS Collaboration[81)] has determined a preliminary x-distribution of the leading muons. The result is shown in Fig. 19 and compares well with the average antiquark x-distribution determined earlier by the same group (see Fig. 18).

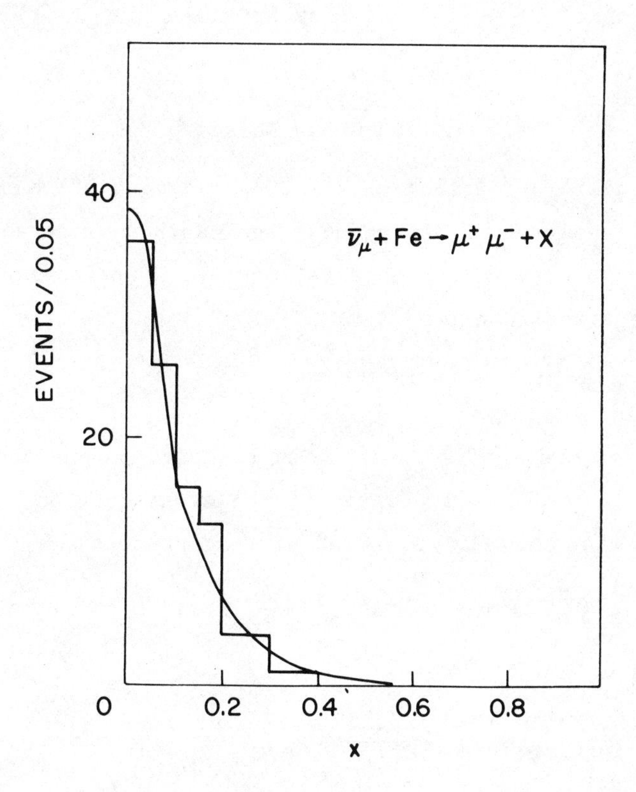

Fig. 19 The x-distribution of the leading μ^+ from $\bar{\nu}$ induced dimuon events.

The amount of the strange antiquark content can be obtained from the ratio of the ν- and $\bar{\nu}$-induced opposite-sign dimuon cross-sections:

$$\frac{\int 2\bar{s}(x)dx}{\int[u(x) + d(x)]dx} = \frac{\int 2\bar{s}(x)dx}{\int[q(x) - \bar{q}(x)]dx} \cong \frac{\tan^2\theta_C}{\left[\frac{\sigma(\nu N \to \mu^-\mu^+X)}{\sigma(\bar{\nu}N \to \mu^+\mu^-X)} - 1\right]} .$$

By this method, the CDHS Collaboration[82] obtained a result for the momentum carried by the strange sea, normalized to the total quark and antiquark momentum, of $\int 2\bar{s}(x)dx / \int[q(x) + \bar{q}(x)]dx = 0.05 \pm 0.02$. A new preliminary result of the same group[81], based on a sample of higher statistics, is 0.03 ± 0.01. These results, compared to a total antiquark momentum of $\sim 0.15 \times \int[q(x) + \bar{q}(x)]dx$ at high energies, suggest that the sea is not flavour-symmetric $[\int\bar{s}(x)dx/\int\bar{u}(x)dx \sim 0.3]$.

2.6 Callan-Gross Relation

The Callan-Gross relation[83] $F_2(x,Q^2) = 2xF_1(x,Q^2)$ has been assumed in all analyses of charged-current neutrino interactions, but it has been poorly tested in experiments. Neutrino scattering experiments test the quantity

$$R' = \frac{F_2 - 2xF_1}{F_2} ,$$

whereas electron and muon scattering experiments on nucleons test the ratio $R = \sigma_L/\sigma_T$, i.e. the ratio of the cross-sections for longitudinally and transversely polarized virtual photons. With

$$R(x,Q^2) = \frac{F_2(x,Q^2)[1 + (4M^2x^2/Q^2)] - 2xF_1(x,Q^2)}{2xF_1(x,Q^2)}$$

we get the relation

$$R' = \frac{R - 4M^2x^2/Q^2}{1 + R} = \frac{R - Q^2/\nu^2}{1 + R} ,$$

i.e. $R' < R$. The ratio R is expected to be small, but non-zero owing to the transverse momentum of the quarks: $R = 4\langle p_\perp^2\rangle/Q^2$ [84].

Although R' measured in neutrino interactions is smaller than R and may be even somewhat negative at small values of ν, R' will in general be non-zero and indicate a violation of the Callan-Gross relation. A Q^2-dependent $R' \neq 0$ is also predicted by QCD, and provides a very sensitive test of the theory. This test is particularly interesting because electron[85] and muon scattering[86] experiments have reported large values of R which can hardly be accommodated by first-order QCD calculations.

R' can be obtained from the y-dependence of the sum of ν and $\bar{\nu}$ cross-sections:

$$\frac{d^2\sigma^{\nu}}{dxdy} + \frac{d^2\sigma^{\bar{\nu}}}{dxdy} = \frac{G^2ME}{\pi} F_2(x,Q^2) \left[1 + (1-y)^2 - y^2R'\right] .$$

Since the structure function $F_2(x,Q^2)$ is known to be somewhat Q^2-dependent and therefore y-dependent (see Section 2.7), the fit of R' ought to be done for fixed Q^2. Alternatively, a fit of the y-dependence for fixed ν is independent of corrections due to scaling violations as well ($Q^2 = x.2M\nu$). However, the result on R' is in any case strongly dependent on radiative corrections[38].

Figure 20 shows the result for R' from a fit of the CDHS data[87] in fixed ν bins (the average Q^2 in each ν bin is about half of the value of ν). The errors are large because of the limited range in y in each ν bin, and because of uncertainties in the flux normalization. The average value of R' obtained with this method is $R' =$ $= 0.02 \pm 0.12$. The result is consistent with a QCD calculation[88] which is also shown in Fig. 20. The results on R' and R from different experiments are summarized in Table 18. The results are not inconsistent with each other, but more precise results are necessary in order to arrive at a definite conclusion on the size and Q^2-dependence of the Callan-Gross violation. Furthermore, the x- and Q^2-ranges explored in the experiments are different. For example, the GGM value[75] is taken from data with $Q^2 < 1$ GeV^2, and the CHIO value[86] is taken in the kinematical domain $1 < Q^2 < 12$ GeV^2 and x

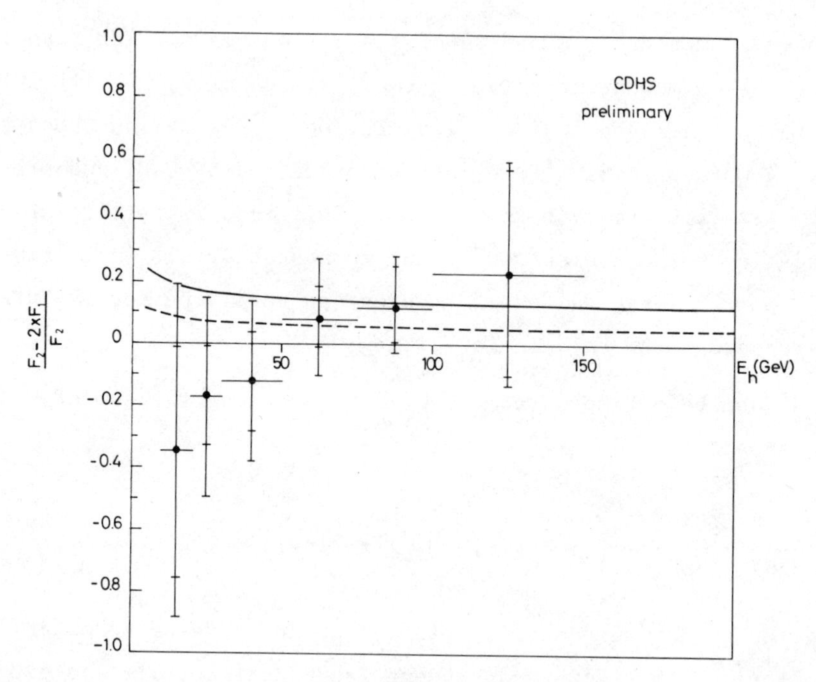

Fig. 20 R′ as a function of $\nu \cong E_H$, compared with a QCD calculation with (—) and without (--) the contribution from the gluon distribution.

Table 18

Summary of measurements of R′ and R

Experiment	R′	R
ABCLOS (GGM) [75]	0.32 ± 0.15	
ABCLOS (BEBC) [75]	0.11 ± 0.14	
CFR [63]	0.17 ± 0.09	
HPWFOR [78]	0.18 ± 0.07 [a]	
CDHS [87]	0.02 ± 0.12 [a]	
FMMS (FNAL 15′) [77]	-0.12 ± 0.16	
SLAC (ep) [85]		0.21 ± 0.10 [a]
CHIO (μp) [86]		0.52 ± 0.35 [a]

a) Radiative corrections applied.

below 0.1. This fact, together with the absence of radiative corrections in part of the experiments, makes the quantitative comparison of the results difficult.

2.7 Scaling Violation

In the scaling limit, the structure functions no longer depend on Q^2 and ν, where either variable may be replaced by the dimensionless ratio $x = Q^2/2M\nu$, but only on x. This hypothesis can be tested experimentally by comparing the x-dependence of the structure functions at different values of Q^2 or, alternatively, ν.

Scaling violation has first been seen in inelastic muon scattering data at FNAL, and was confirmed in electron scattering data at SLAC. Early neutrino scattering experiments failed to see similar scaling violation effects, partly because of inadequate instruments, partly because it was customary to use the neutrino energy as the variable in the study of differential cross-sections. Since Q^2 is believed to be the best variable to exhibit scaling violations, the neutrino energy is less suited for this purpose since in each energy bin all Q^2 from zero up to 2ME contribute. As a consequence, Q^2-dependent effects are washed out.

The ABCLOS Collaboration[75)] working with BEBC at the CERN-SPS was the first to observe scaling violation in neutrino scattering. This scaling violation had the same pattern as the one observed before in electron and muon scattering. In a thoughtful analysis, the observed scaling violation was shown to be compatible with the predictions of first-order QCD. However, the scaling violation was only convincing when the high-energy data points of BEBC were taken in conjunction with the (reanalysed) low-energy data points of GGM.

On the basis of high-energy data alone ($30 < E < 200$ GeV), the high-statistics CDHS counter experiment[64)] demonstrated a significant scale-breaking effect in the structure functions, and showed the compatibility of the observed scaling violation with first-order QCD

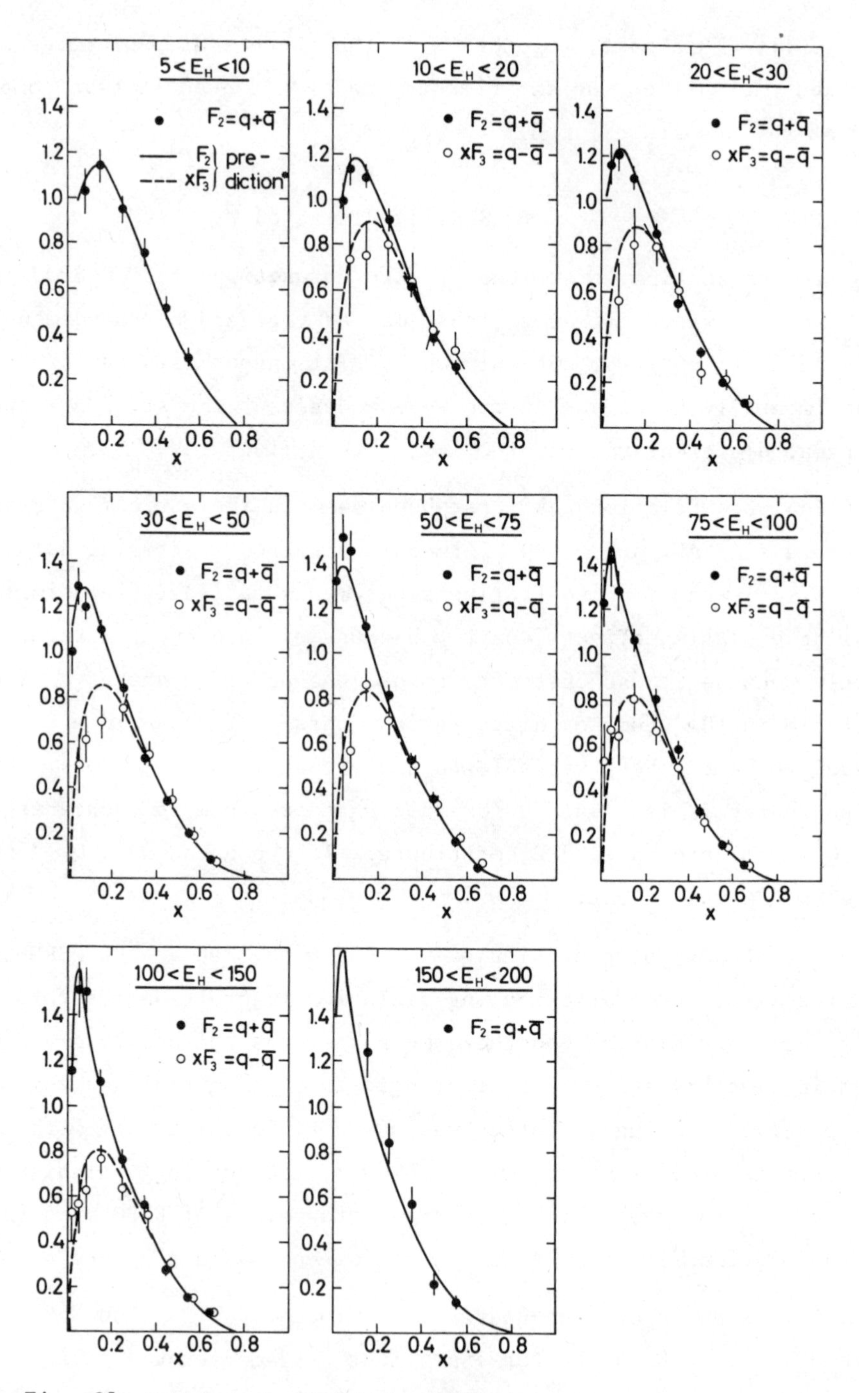

Fig. 21 $F_2(x)$ and $xF_3(x)$ for different $\nu \cong E_H$. The lines represent test functions (inspired by QCD) to reproduce the data.

calculations. Since then, scaling violation effects have been observed in other experiments too, but detailed analyses so far exist only from the ABCLOS (BEBC)[75] and CDHS[64] Collaborations. In the following, we will concentrate more on the CDHS results because they are based on much larger statistics, whereas the systematic errors are comparable in the two experiments. The results of both analyses are consistent with each other.

The scaling violation of the structure functions F_2 and xF_3 determined by the CDHS experiment is shown in Fig. 21. The structure functions are shown for different ranges of ν, equivalent to different ranges of $\langle Q^2 \rangle$. The reason for the choice of this variable is that i) for each ν-bin the whole range of x is available, and ii) in the comparison of the shape of the structure functions errors due to neutrino flux normalization do not matter. It is clearly visible that the structure functions shrink towards low x as ν increases. From the theoretical point of view, however, the Q^2-dependence of the structure functions is more immediately calculable than is the ν-dependence. From the experimental point of view, bins of fixed Q^2 have the disadvantage of the limited range in x owing to an experimental cut-off at very low hadronic energy (missing low Q^2, large x) and the maximum available energy of the neutrino beam (missing large Q^2, low x). The accessible kinematical domain in x and Q^2 is shown in Fig. 22 for the conditions of the CDHS experiment.

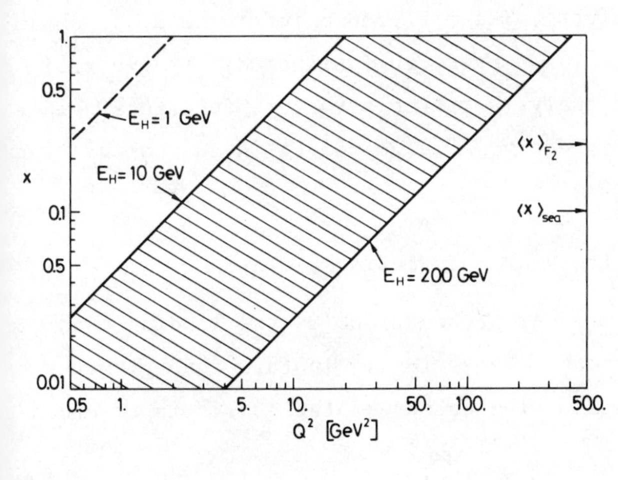

Fig. 22 Kinematical domain of x and Q^2 accessible for the CDHS experiment.

Converting ν to Q^2 using the relation $Q^2 = x.2M\nu$, one gets the Q^2-dependence of the structure functions $F_2(x,Q^2)$ and $xF_3(x,Q^2)$ which is shown in Figs. 23 and 24. The only significant scaling violation is observed for $F_2(x,Q^2)$, whereas scaling violation in $xF_3(x,Q^2)$ is almost hidden by the experimental errors.

In order to compare F_2 with F_2^{ed}, the structure function obtained in ed scattering experiments at SLAC, the following normalization relation has been used:

$$F_2^{ed} = \frac{5}{9}\left\{1 - \frac{3}{5}\,\frac{(s + \bar{s})}{(q + \bar{q})}\right\} F_2 \ .$$

This relation corrects for the difference of the electromagnetic and weak charges of the quarks inside the nucleon. The structure function F_2 agrees well with F_2^{ed} in the region of overlap, and the ν data appear as good continuation of the ed data towards higher Q^2. This observation provides a strong support for the QPM. Moreover, it suggests that the same scale-breaking mechanism is at work in electron and neutrino scattering.

The structure functions determined by the ABCLOS (BEBC)[75] Collaboration agree in the common range of Q^2 with the CDHS data, except perhaps for large values of x where the ABCLOS data have a tendency to fall below the CDHS points. The difference in the data, as shown in Fig. 25, is not dramatic. However, this slight difference will propagate into the moment analysis of the structure functions, since the data points at large x receive a large weight in the calculation of higher moments.

2.8 QCD Analysis of the Scaling Violation

2.8.1 QCD predictions. At present, QCD is the leading candidate for the theory of strong interactions. In this theory quarks exist in three colour states. The colour of the quarks is regarded

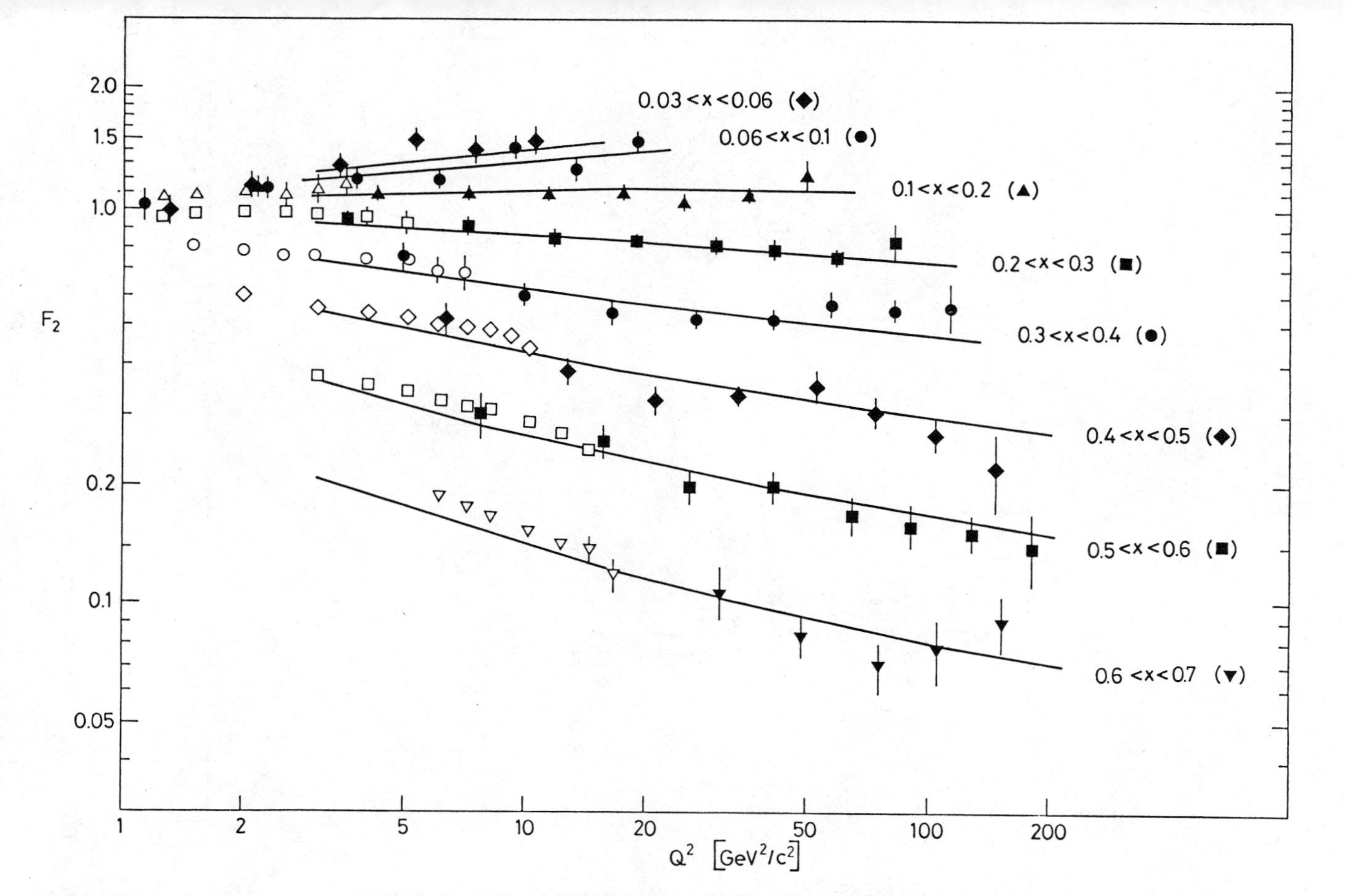

Fig. 23 Comparison of the structure function $F_2(x,Q^2)$ as obtained by the CDHS experiment (black points) with data from ed scattering (open points), and a QCD fit.

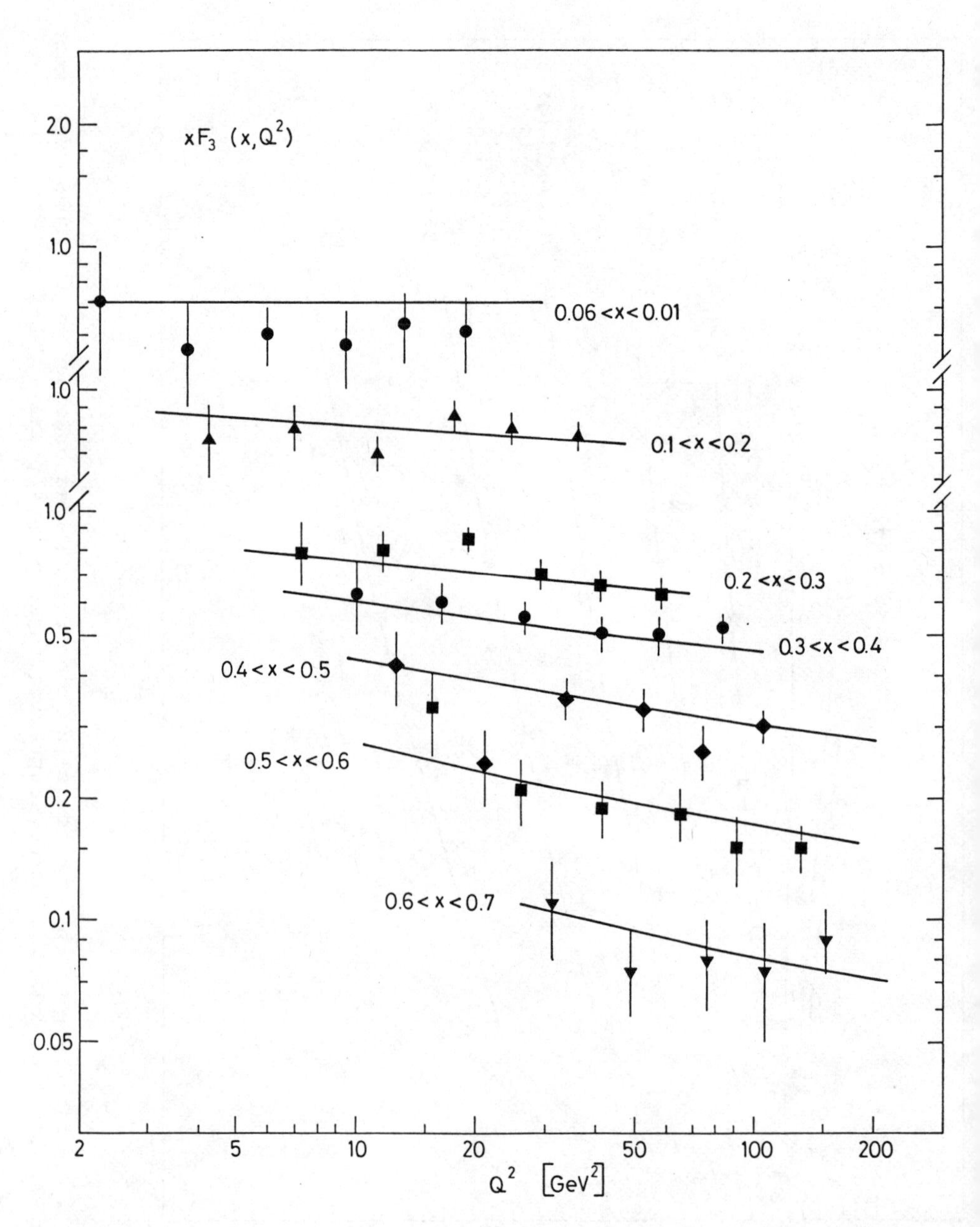

Fig. 24 Comparison of the structure function $xF_3(x,Q^2)$ as obtained by the CDHS experiment, with a QCD fit.

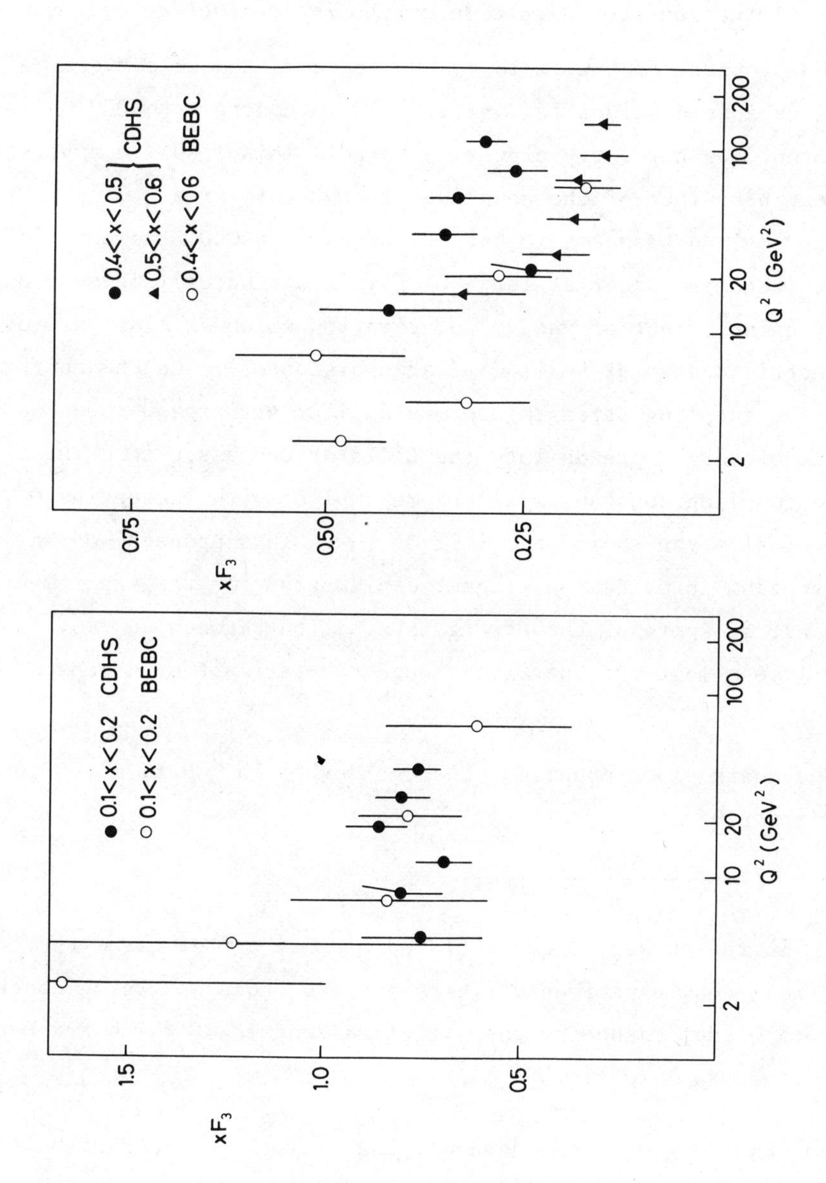

Fig. 25 Comparison of $xF_3(x,Q^2)$ as obtained by the CDHS experiment (black points), and the ABCLOS (BEBC) experiment (open points).

as the source of the field which mediates the force between the quarks. The field quanta are eight massless gluons which carry the quantum numbers of a $(q\bar{q})$ pair. The gluons are singlets with respect to flavour but constitute an octet with respect to colour.

The property of the field quanta to carry the colour charge themselves is a distinct feature of QCD, in contrast to QED where the photon does not carry electric charge. This leads to a very different behaviour of the coupling strength at large Q^2, or equivalently small distances, in both theories. In QED, the coupling strength becomes larger at large Q^2 (small distances), because of the screening effect of vacuum polarization which shields part of the electric charge at small Q^2 (large distances). On the contrary, in QCD the coupling strength becomes smaller at large Q^2 because the colour charge spreads into the field around a source. The effective coupling constant will become smaller with increasing Q^2, and eventually vanish in the limit $Q^2 \to \infty$. This property of "asymptotic freedom" provides an elegant explanation for the experimental fact that, in spite of the strong interaction between quarks, they behave like almost free particles when they are hit by a lepton at large Q^2.

The quark-gluon coupling constant $\alpha_s(Q^2)$ in QCD is given to lowest order by

$$\alpha_s(Q^2) = \frac{12\,\pi}{(33 - 2m)\,\ln\,(Q^2/\Lambda^2)} ,$$

where m is the number of quark flavours and Λ a free parameter to be determined by experiment. There are two processes which contribute to a Q^2-dependence of the structure functions, which are both proportional to $\alpha_s(Q^2)$:

i) quarks and gluons emit gluons, thus shifting the fractional momentum distributions towards small x; and

ii) gluons create $(q\bar{q})$ pairs which feed the fractional momentum distributions of q and $\bar{q}$ at small x. QCD does not predict the shape of the structure functions, but it predicts its evolution with Q^2. It is this evolution with Q^2 which can be tested by experiments.

The Altarelli-Parisi equation[89] connects the Q^2-evolution of the structure function with its shape at larger x:

$$Q^2 \frac{\partial}{\partial Q^2} \frac{F(x,Q^2)}{x} = \frac{\alpha_s(Q^2)}{2\pi} \int_x^1 \frac{dz}{z} \left[\frac{F(z,Q^2)}{z} P_{qq}\left(\frac{x}{z}\right) + \frac{G(z,Q^2)}{z} P_{qg}\left(\frac{x}{z}\right)\right];$$

$[\alpha_s(Q^2)/2\pi][P_{qq}(x/z)]$ is the probability of seeing a quark with fractional momentum x arising from a quark with fractional momentum z, when probing with Q^2; $[\alpha_s(Q^2)/2\pi]\,[P_{qg}(x/z)]$ is the analogous probability for the transition gluon to quark; $G(z,Q^2)$ is the fractional momentum distribution of the gluons. The above equation is coupled with a second equation for the Q^2-evolution of the gluon x-distribution:

$$Q^2 \frac{\partial}{\partial Q^2} \frac{G(x,Q^2)}{x} = \frac{\alpha_s(Q^2)}{2\pi} \int_x^1 \frac{dz}{z} \left[\frac{F(z,Q^2)}{z} P_{gq}\left(\frac{x}{z}\right) + \frac{G(z,Q^2)}{z} P_{gg}\left(\frac{x}{z}\right)\right].$$

The functions P_{qq}, P_{qg}, P_{gq} and P_{gg} are given by QCD. If F denotes the flavour-non-singlet structure function xF_3, the gluon term is absent. The predictions then do not depend on the (unknown) gluon x-distribution, and are therefore more reliable.

So far, three methods have been used to compare the predicted Q^2-dependence of the structure functions with the one exhibited by the data:

i) For the case of the non-singlet structure function xF_3, Abbot and Barnett[90] integrated the Altarelli-Parisi equation numerically and made a comparison with the Q^2-dependence of the CDHS[64] and ABCLOS (BEBC)[75] data. They found agreement between the predicted and the observed Q^2-dependence of xF_3, for $Q^2 >$ > 5 GeV^2, with results similar to those described in Section 2.8.2.

ii) One could determine the shape of the structure functions F_2 and xF_3 at a particular value Q_0^2, and then compare the predicted and the actual Q^2-evolution. The disadvantage of this method is that only a small fraction of the data at $Q^2 = Q_0^2$ is used as a basis for the comparison at other Q^2. To overcome this problem and to allow a global fit of the whole data, the structure functions are parametrized in terms of a few Q^2-dependent parameters. This method of Buras and Gaemers[91] is more transparent but less rigorous than the previous method.

iii) The third approach is the moment method to solve the Altarelli-Parisi equation. Taking the Nth moment of xF_3,

$$M_3(N,Q^2) = \int_0^1 x^{N-2} \, xF_3(x,Q^2)dx \; ,$$

one gets the following result for its Q^2-evolution,

$$M_3(N,Q^2) = M_3(N,Q_0^2) \left[\frac{\ln Q^2/\Lambda^2}{\ln Q_0^2/\Lambda^2}\right]^{-d_N} \; ,$$

with the "anomalous dimension" d_N given by QCD as

$$d_N = \frac{4}{33 - 2m}\left[1 - \frac{2}{N(N+1)} + 4 \sum_{j=2}^{N} \frac{1}{j}\right] .$$

The predicted Q^2-dependence of the moments does not depend on a specific form of xF_3 as function of x, and allows for quantitative checks both on the logarithmic Q^2-dependence and on the numerical value of the anomalous dimension.

2.8.2 QCD fits to xF_3 and F_2. A QCD fit employing the parametrization of the structure functions proposed by Buras and Gaemers[91] has been done by the CDHS Collaboration[92]. The structure function of the valence quarks has been assumed to satisfy, within the accessible Q^2 range, the ansatz

$$xF_3(x,Q^2) \propto x^{\eta_1(s)}(1-x)^{\eta_2(s)} ,$$

with $s = \ln\left[\ln(Q^2/\Lambda^2)/\ln(Q_0^2/\Lambda^2)\right]$. The normalization is such that the number of valence quarks is three. Similarly, the structure function of the antiquarks was parametrized as

$$\bar{q}(x) = A(s)(1-x)^{P(s)} .$$

Buras and Gaemers have demonstrated that these parametrizations are good enough to satisfy the relevant QCD moment equations with sufficient accuracy.

The fits to the CDHS data are shown in Figs. 23 and 24. They reproduce satisfactorily the Q^2-dependence of the data with a scale parameter $\Lambda = 0.5 \pm 0.2$ GeV.

Since the Q^2-evolution of F_2 is dependent upon the momentum distribution of the gluons $G(x,Q^2)$, a common fit to xF_3 and F_2 gives also a result on the third gluon moment, or equivalently, on the average fractional momentum carried by the gluons. The fit yields the following results[92], at $Q^2 = 5$ GeV^2:

$$xF_3(x) \propto x^{0.56\pm0.02}(1-x)^{2.7\pm0.1}$$

$$\bar{q}(x) \propto (1-x)^{8.1\pm0.7}$$

$$\langle x\rangle_{\text{gluons}} = 0.21 \pm 0.04 .$$

The Q^2-dependence of the area and the average values of the momentum distributions of the nucleon constituents are shown in Fig. 26. The data suggest a gluon x-distribution which is less steep than the sea distribution, but more confined to small x compared to the valence quark distribution.

2.8.3 Moment analysis of xF_3. The principal difficulty of a moment analysis is the need to integrate the structure functions between $x = 0$ and $x = 1$, at fixed Q^2. In order to minimize the gap of missing data which are outside the kinematical limits, the ABCLOS (BEBC)[75] Collaboration has chosen to combine the high-energy BEBC

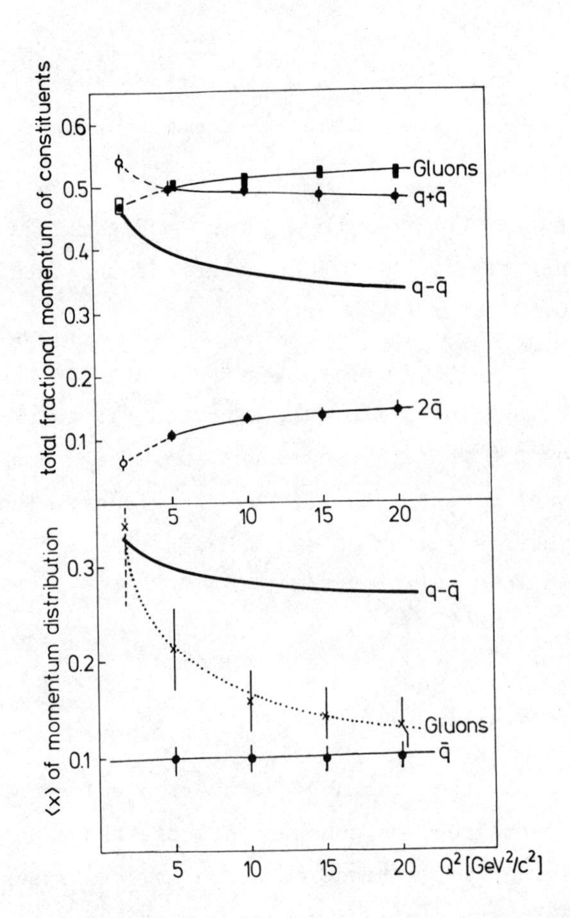

Fig. 26 Total functional momentum and average values of the fractional momentum of the nucleon constituents.

data with low-energy GGM data. The CDHS Collaboration[93] has chosen to combine the high-energy data with ed scattering data from SLAC[94]. Even then, the x-range is not complete and one has to rely on extrapolations of the structure functions to x = 0 and to x = 1.

A moment analysis of xF_3 has been presented both by the ABCLOS (BEBC)[75] and the CDHS[93] Collaborations. In these analyses, a functional form $x^{\alpha}(1-x)^{\beta}$ for xF_3 has been used for the extrapolation into experimentally inaccessible regions. All moments for which the extrapolation contributes more than 25% of the integral have been dropped in the CDHS analysis.

The higher moments of xF_3 below $Q^2 \sim 10\ \mathrm{GeV}^2$ depend upon whether or not quasi-elastic events (x = 1) are included or not[90]. Also, target mass effects are important at presently used Q^2 values, since M^2 is in general not negligible compared to Q^2. In order to absorb the target mass, the Bjorken scaling variable x is replaced by the variable

$$\xi = \frac{2x}{1 + \sqrt{1 + 4M^2x^2/Q^2}} ,$$

and the ordinary moment by the "Nachtmann moment"[95]

$$\tilde{M}_3(N,Q^2) = \int_0^1 \frac{\xi^{N+1}}{x^3}\, xF_3(x,Q^2)\, \frac{1 + (N+1)\sqrt{1 + 4M^2x^2/Q^2}}{N + 2}\, dx .$$

The use of the Nachtmann moment takes into account not only the target mass but also, at least in part, the contribution of quasi-elastic and resonance events near x = 1[90]. Therefore, quasi-elastic events should rather be included when calculating Nachtmann moments of xF_3.

A comparison of the Nachtmann moments of xF_3 from the ABCLOS (BEBC) and CDHS Collaborations is given in Fig. 27. The data are consistent, although the ABCLOS (BEBC) moments have a tendency to

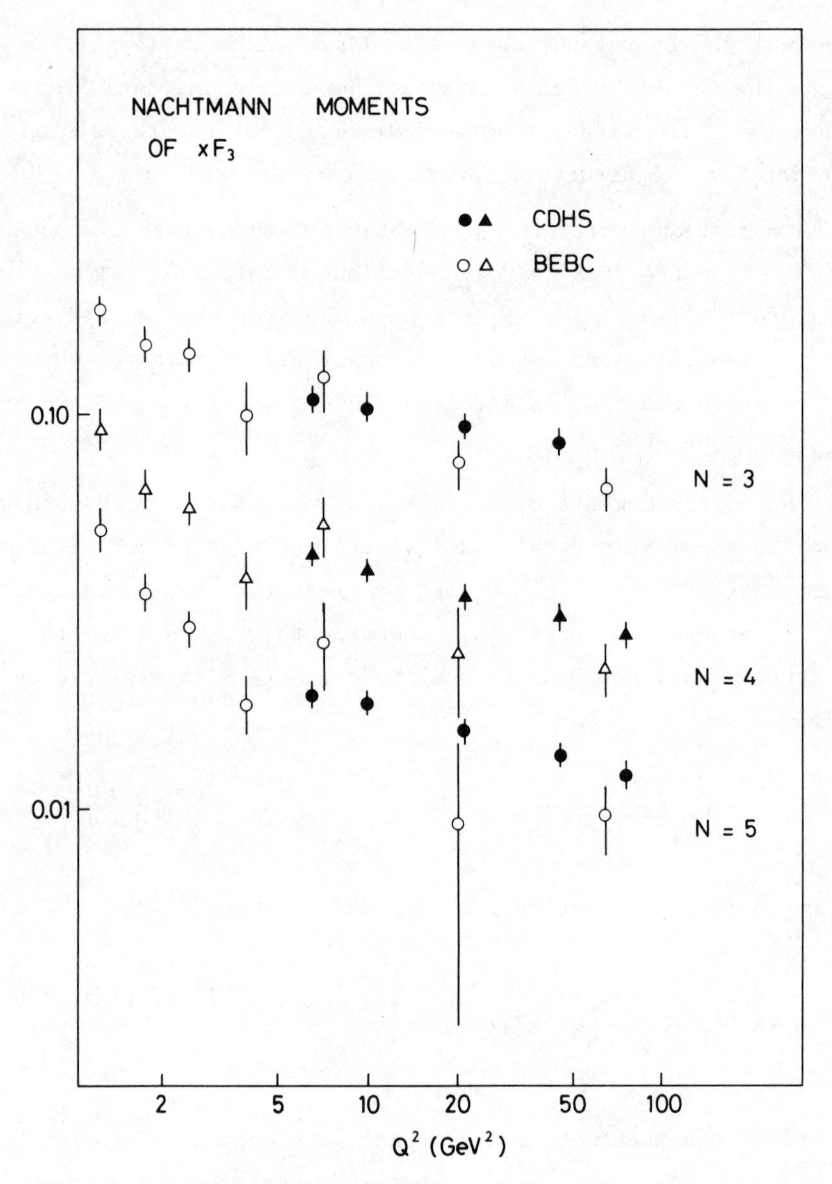

Fig. 27 Comparison of Nachtmann moments of xF_3 as obtained by the CDHS experiment (black points), and the ABCLOS (BEBC) experiment (open points).

fall below the CDHS moments in the common Q^2 range, reflecting the difference of xF_3 at large x in the two data samples (see Fig. 25).

The numerical value of the anomalous dimensions can be tested as follows. If one plots the logarithm of a moment $\tilde{M}_3(K,Q^2)$ against the logarithm of another moment $\tilde{M}_3(L,Q^2)$, the points are expected to fall on a straight line with a slope d_K/d_L independent of the value of Λ and the number of quark flavours:

$$\ln M_3(K,Q^2) = \frac{d_K}{d_L} \ln M_3(L,Q^2) + \text{const.}$$

The data points shown in Fig. 28 are consistent with this expectation. The results for the slopes d_K/d_L are summarized in Table 19, together

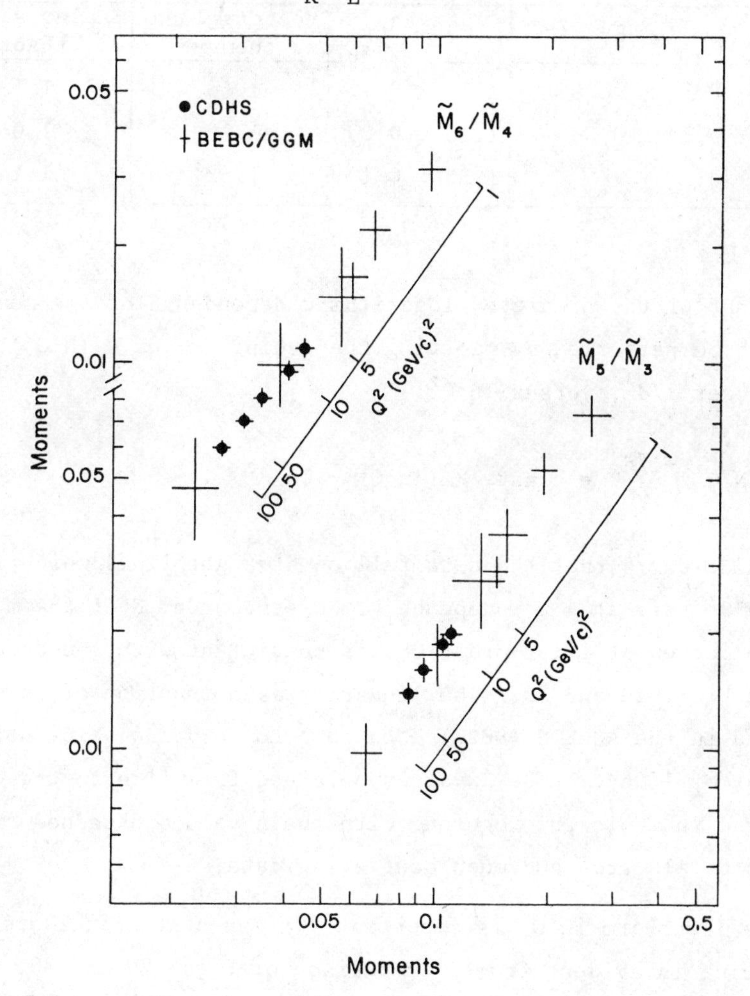

Fig. 28 A log-log plot of various Nachtmann moments of xF_3 and the ratios of anomalous dimensions as predicted for vector gluons. The Q^2 scale refers to the CDHS experiment.

with the predictions of QCD for vector gluons, and with the predictions of a scalar gluon theory. The data favour vector gluons over scalar gluons but are not precise enough to be conclusive.

Table 19

Ratios of anomalous dimensions

d_K/d_L	ABCLOS (BEBC)[75]	CDHS[83]	Vector gluon theory	Scalar gluon theory
d_5/d_3	1.50 ± 0.08	1.34 ± 0.12	1.46	1.12
d_6/d_4	1.29 ± 0.06	1.18 ± 0.09	1.29	1.06
d_6/d_3	-	1.38 ± 0.15	1.62	1.14

A test of the predicted logarithmic dependence of the moments on Q^2 can be performed by plotting the reciprocal of $M_3(N,Q^2)$, raised to the power $1/d_N$, versus $\ln Q^2$:

$$M_3(N,Q^2)^{-1/d_N} = \text{const.} \times (\ln Q^2 - \ln \Lambda^2) \ .$$

The prediction is that the data fall on straight lines which intersect the Q^2 axis at Λ^2, independently of the order N of the moment. This prediction of first-order QCD is consistent with the data as shown in Figs. 29 and 30. The linearity as a function of $\ln Q^2$ is satisfied in the ABCLOS (BEBC) data down to very low values of Q^2 of the order of 1 GeV^2. The Λ values obtained from fits to the data are summarized in Table 20, together with the Λ values obtained from QCD analyses of electron and muon scattering data.

The lever-arm in Q^2 is important for the statistical precision of Λ, which is evident from a comparison of Figs. 29 and 30. However, at low Q^2, higher-order QCD corrections may be important. The results with second-order QCD corrections for Λ are included in Table 20 for comparison.

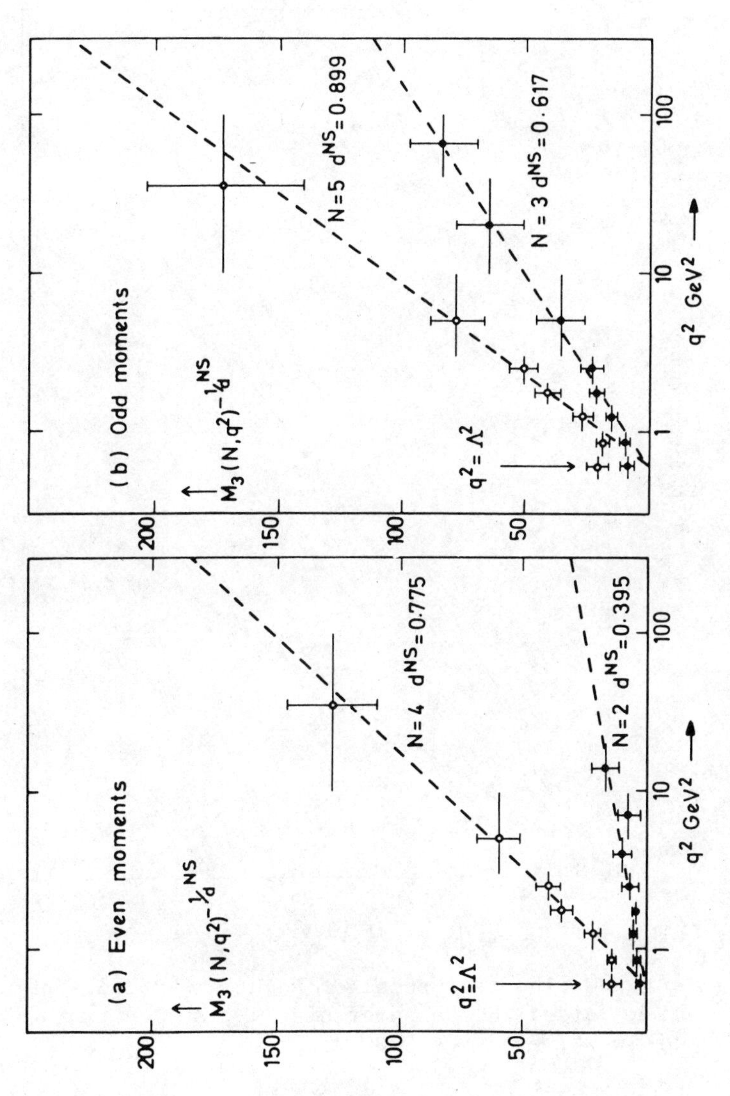

Fig. 29 The reciprocal of Nachtmann moments of xF_3, raised to the power of $1/d_N$, as function of $\ln Q^2$, from the ABCLOS (BEBC) data.

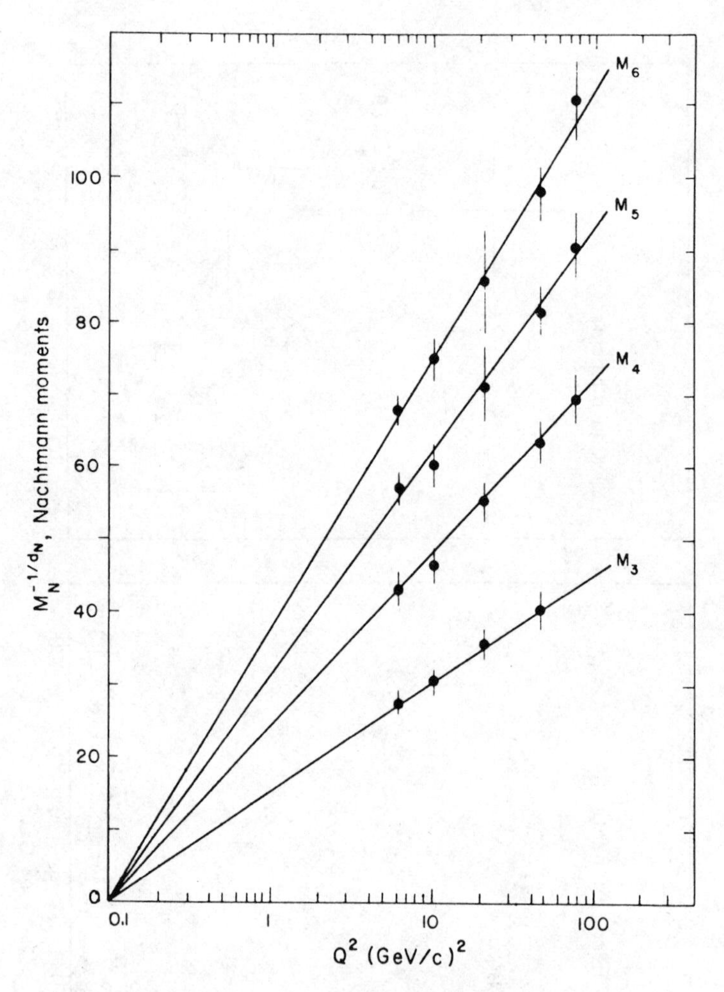

Fig. 30 The reciprocal of Nachtmann moments of xF_3, raised to the power of $1/d_N$, as function of ln Q^2, from the CDHS data.

Table 20

Results on Λ in first order and with second-order corrections

Experiment	Λ (GeV)		
	1st order result	With 2nd order correction	
ABCLOS (BEBC)[75]	0.74 ± 0.05		$Q^2 > 1$
CDHS[93]	0.33 ± 0.20	0.20 ± 0.05	$Q^2 > 6$
CHIO[86]	0.64 ± 0.15	0.46 ± 0.11	$Q^2 > 3$

The value for Λ quoted by the ABCLOS (BEBC) Collaboration is larger than the one quoted by the CDHS Collaboration. Roughly half of the discrepancy is explained by differences in the two analyses: inclusion or exclusion of quasi-elastic events, number of excited quark flavours, application of radiative corrections, and correction for the Fermi motion of the nucleons inside heavy target nuclei. The other half of the discrepancy (which is not significant) seems to have its origin in the data.

Taking higher-order QCD corrections into account, the value of Λ determined by the first-order formula given above, should vary with the order of the moment, N. This is because at least part of the higher-order corrections can be absorbed into a redefinition of Λ[96]. To illustrate this point, Fig. 31 shows Λ as determined from the xF_3 moments of the ABCLOS (BEBC) and CDHS data, as a function of the order of the moment, N. The agreement of the data with the expected N-dependence is poor (note that the data points are highly correlated because for each N the same data are used).

Throughout this discussion "higher twist" effects have been neglected, i.e. terms with a Q^2-dependence like a power of $1/Q^2$. These effects are departures from the simple QPM because they account for effects such as the gluon exchange between struck and spectator quarks, quark transverse momentum, resonance production, and quasi-elastic scattering. One cannot even exclude, on the basis

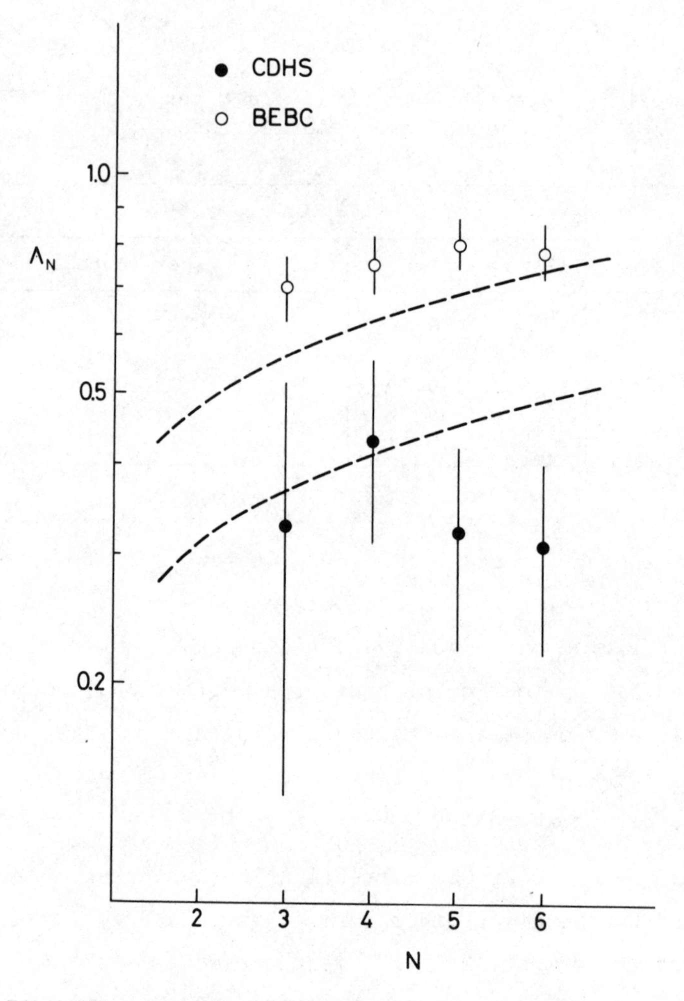

Fig. 31 Values of Λ extracted from the xF_3 moments using the first-order algorithm only, and the QCD prediction (Ref. 96). The data are from the CDHS (black points) and ABCLOS (BEBC) experiments (open points).

of the available data, that these effects account for most of the observed scaling violations[90], so that the real value of Λ is much smaller than $\sim$ 0.5 GeV. Presently available data do not unambiguously demonstrate the typical logarithmic Q^2-dependence predicted by QCD, in contrast to a power law dependence on Q^2. More precise data at higher Q^2 are needed to enable decisive tests on QCD.

2.9 Summary on Inclusive Charged-current Reactions

i) All data are consistent with a V-A structure of the charged weak current.

ii) At large Q^2, the "naïve" quark-parton model is still a good approximation for the internal structure of the nucleon.

iii) Small scaling violations of the nucleon structure functions have definitively been observed.

iv) The scaling violation of the structure functions has been explained successfully by first-order QCD calculations, with a scale parameter $\Lambda = 0.5 \pm 0.2$ GeV.

v) Although there is no serious contradiction between data and QCD, it is too early to state that QCD is the ultimate theory of strong interactions. The observed scaling violation might largely be due to trivial effects, and the true QCD scale parameter Λ might be smaller than assumed at present.

Acknowledgements

I wish to thank Professor A. Zichichi for his kind invitation and for the nice stay at Erice. I would like to express my appreciation to my colleagues from the CDHS Collaboration for their help and contributions to these lectures. Finally, I thank Mrs. M. Guarisco for her careful and patient typing of the manuscript.

REFERENCES

1) F. Hasert et al., Phys. Lett. 46B, 138 (1973).

2) W. Krenz et al., Nucl. Phys. B135, 45 (1978).

3) M. Holder et al., Phys. Lett. 71B, 222 (1977); 72B, 254 (1977).

4) J.J. Sakurai, Proc. Int. Conf. on Neutrino Physics and Neutrino Astrophysics, Baksan Valley, USSR, 1977 (Nauka, Moscow, 1978), Vol. 2, p. 242.

5) P.Q. Hung and J.J. Sakurai, Phys. Lett. 63B, 295 (1976).

6) L.M. Sehgal, Proc. "Neutrino-78", Int. Conf. on Neutrino Physics, Purdue Univ., 1978 (Purdue University, West Lafayette, Indiana, 1978), p. 253.

7) J.J. Sakurai, Proc. Topical Conf. on Neutrino Physics at Accelerators, Oxford, 1978 (Rutherford Laboratory, Chilton, Didcot, 1978), p. 338.

8) A. Salam, Proc. 8th Nobel Symposium, Aspenäsgården, 1968 (Almqvist and Wiksell, Stockholm, 1968), p. 367.

9) S. Weinberg, Phys. Rev. Lett. 19, 1264 (1967); Phys. Rev. D 5, 1412 (1972).

10) S.L. Glashow, J. Iliopoulos and L. Maiani, Phys. Rev. D 2, 1285 (1970).

11) J. Blietschau et al., Nucl. Phys. B114, 189 (1976); Phys. Lett. 73B, 232 (1978).

12) H. Faissner et al., Phys. Rev. Lett. 41, 213 (1978).

13) P. Alibran et al., High-energy elastic ν_μ scattering off electrons in Gargamelle, preprint CERN-EP/79-38 (1979).

14) A.M. Cnops et al., Phys. Rev. Lett. 41, 357 (1978).

15) A. Rosanov, talk given at the EPS Int. Conf. on High-Energy Physics, Geneva, 1979 (CERN, Geneva, in press), p. 177.

16) P. Alibran et al., Phys. Lett. 74B, 422 (1978).

17) D. Bertrand et al., An upper limit to the cross-section for the reaction $\bar{\nu}_\mu e^- \rightarrow \bar{\nu}_\mu e^-$ at SPS energies, preprint PITHA 79/07 (1979).

18) J.P. Berge et al., A search at high energies for antineutrino-electron elastic scattering, preprint Fermilab Pub 79/27-EXP (1979).

19) N. Armenise et al., Phys. Lett. 81B, 385 (1979).

20) G. 't Hooft, Phys. Lett. 37B, 195 (1971).

21) F. Reines, H.S. Gurr and H.W. Sobel, Phys. Rev. Lett. 37, 315 (1976).

22) C.Y. Prescott et al., Phys. Lett. 77B, 347 (1978).

23) C.Y. Prescott et al., Further measurements of parity non-conservation in inelastic electron scattering, preprint SLAC Pub. 2319.

24) J.D. Bjorken, Phys. Rev. D 18, 3239 (1978).

25) P.Q. Hung and J.J. Sakurai, Evidence for factorization and coupling constant determination in neutral-current processes, preprint UCLA/79/TEP/9.
J.J. Sakurai, talk given at "Neutrino-79", Int. Conf. on Neutrinos, Weak Interactions and Cosmology, Bergen, 1979.

26) L.L. Lewis et al., Phys. Rev. Lett. 39, 795 (1977).

27) P.E.G. Baird et al., Phys. Rev. Lett. 39, 798 (1977).

28) M.A. Bouchiat and C.C. Bouchiat, Phys. Lett. 48B, 111 (1974).
J. Bernabéu and C. Jarlskog, Nuovo Cimento 38A, 295 (1977).

29) L.M. Barkov and M.S. Zolotorev, JETP Lett. 26, 379 (1978); and Proc. "Neutrino-78". Int. Conf. on Neutrino Physics, Purdue Univ., 1978 (Purdue University, West Lafayette, Indiana, 1978), p. 423.
L.M. Barkov and M.S. Zolotorev, Phys. Lett. 85B, 308 (1979).

30) R. Conti et al., Phys. Rev. Lett. 42, 343 (1979).

31) V.N. Novikov, O.P. Sushkov and I.B. Khriplovich, Sov. Phys. JETP 44, 872 (1976).

32) P.Q. Hung and J.J. Sakurai, Phys. Lett. 69B, 323 (1977).

33) H.W. Sobel, talk given at the "Neutrino-79", Int. Conf. on Neutrinos, Weak Interactions and Cosmology, Bergen, 1979.

34) A. Ali and C.A. Dominguez, Phys. Rev. D 12, 3673 (1975); and references cited therein.

35) C. Baltay, Proc. Int. Conf. on High-Energy Physics, Tokyo, 1978 (Physical Society of Japan, Tokyo, 1979), p. 882.

36) C. Geweniger, talk given at the "Neutrino-79", Int. Conf. on Neutrinos, Weak Interactions and Cosmology, Bergen, 1979.

37) T. Hansl et al., Phys. Lett. 74B, 139 (1978).
P.C. Bosetti et al., Phys. Lett. 74B, 143 (1978).
P. Alibran et al., Phys. Lett. 74B, 134 (1978).

38) A. de Rújula, R. Petronzio and A. Savoy-Navarro, Nucl. Phys. B154, 394 (1979).

39) R. Turlay, Charged weak currents, Rapporteur's talk given at the EPS Int. Conf. on High-Energy Physics, Geneva, 1979 (CERN, Geneva, in press), p. 50.

40) A.J. Buras and K.J.F. Gaemers, Nucl. Phys. B132, 249 (1978); Phys. Lett. 71B, 106 (1977).

41) P.C. Bosetti et al., Phys. Lett. 76B, 505 (1978).
H. Deden et al., Nucl. Phys. B149, 1 (1979).

42) E.A. Paschos and L. Wolfenstein, Phys. Rev. D 7, 91 (1973).

43) H. French, talk given at the "Neutrino-79", Int. Conf. on Neutrinos, Weak Interactions and Cosmology, Bergen, 1979.

44) F. Harris et al., Phys. Rev. Lett. 39, 437 (1977).
M. Derrick et al., Phys. Rev. D 18, 7 (1978).

45) L. Pape, talk given at the EPS Int. Conf. on High-Energy Physics, Geneva, 1979 (CERN, Geneva, in press), p. 155.

46) A. Entenberg et al., Phys. Rev. Lett. 42, 1198 (1979).

47) W. Lee et al., Phys. Rev. Lett. 37, 186 (1976).

48) H. Faissner et al., Proc. "Neutrino-78", Int. Conf. on Neutrino Physics, Purdue Univ., 1978 (Purdue University, West Lafayette, Indiana, 1978), p. C89; H. Faissner et al., to be published.

49) M. Pohl et al., Phys. Lett. 72B, 489 (1978).

50) H.H. Williams, Proc. Int. Conf. on High-Energy Physics, Tokyo, 1978 (Physical Society of Japan, Tokyo, 1979), p. 325.

51) G. Ecker, Nucl. Phys. B123, 293 (1977).

52) C.H. Albright et al., Phys. Rev. D14, 1780 (1976).
E.A. Paschos, Phys. Rev. D 19, 83 (1979).

53) H. Deden, talk given at the "Neutrino-79", Int. Conf. on Neutrinos, Weak Interactions and Cosmology, Bergen, 1979.

54) H. Kluttig, J.G. Morfin and W. Van Doninck, Phys. Lett. 71B, 446 (1977).

55) B.P. Roe, talk given at the "Neutrino-79", Int. Conf. on Neutrinos, Weak Interactions and Cosmology, Bergen, 1979.

56) M. Holder et al., Phys. Lett. 74B, 277 (1978).

57) I. Liede and M. Roos, A neutral-current evaluation of the standard model, contributed paper to the "Neutrino-79", Int. Conf. on Neutrinos, Weak Interactions and Cosmology, Bergen, 1979; and references cited therein.
M. Roos, talk given at the "Neutrino-79", Int. Conf. on Neutrinos, Weak Interactions and Cosmology, Bergen, 1979.

58) P. Langacker, talk given at the "Neutrino-79", Int. Conf. on Neutrinos, Weak Interactions and Cosmology, Bergen, 1979.
H.H. Williams, private communication.

59) J.E. Kim, P. Langacker, M. Levine, H.H. Williams and D.P. Sidhu, preprint COO-3071-243.

60) M. Jonker et al., Phys. Lett. 86B, 229 (1979).

61) K.J.F. Gaemers, private communication.

62) T. Eichten et al., Phys. Lett. 46B, 274 (1973).

63) B.C. Barish et al., Phys. Rev. Lett. 39, 1595 (1977).

64) J.G.H. de Groot et al., Z. Phys. C1, 143 (1979).

65) S. Ciampollilo et al., Phys. Lett. 84B, 281 (1979).

66) J. Ludwig, talk given at the "Neutrino-79", Int. Conf. on Neutrinos, Weak Interactions and Cosmology, Bergen, 1979.

67) S.J. Barish et al., preprint ANL-HEP-PR-78-30 (1978).

68) N.P. Samios, Proc. Int. Symposium on Lepton and Photon Interactions at High Energies, Stanford, 1975 (SLAC, Stanford, 1975), p. 527.

69) W. Lerche et al., Nucl. Phys. B142, 65 (1978).

70) J. Guy, talk given at the "Neutrino-79", Int. Conf. on Neutrinos, Weak Interactions and Cosmology, Bergen, 1979.

71) J. Hanlon et al., paper submitted to the "Neutrino-79", Int. Conf. on Neutrinos, Weak Interactions and Cosmology, Bergen, 1979.

72) O. Erriquez et al., Phys. Lett. 80B, 309 (1979).

73) V.I. Efremenko et al., Phys. Lett. 84B, 511 (1979).

74) R. Barlow and S. Wolfram, Oxford preprint OUNP 24/78 (1978).

75) P.C. Bosetti et al., Nucl. Phys. B142, 1 (1978).

76) A. Benvenuti et al., Phys. Rev. Lett. 42, 149 (1979).

77) D. Sinclair, talk given at the "Neutrino-79", Int. Conf. on Neutrinos, Weak Interactions and Cosmology, Bergen, 1979.

78) A. Benvenuti et al., Phys. Rev. Lett. 42, 1320 (1979).

79) E. Fernandez et al., talk given at the "Neutrino-79", Int. Conf. on Neutrinos, Weak Interactions and Cosmology, Bergen, 1979.

80) D.J. Gross and C.H. Llewellyn Smith, Nucl. Phys. B14, 337 (1969).

81) H.J. Willutzki, talk given at the "Neutrino-79", Int. Conf. on Neutrinos, Weak Interactions and Cosmology, Bergen, 1979.

82) M. Holder et al., Phys. Lett. 69B, 377 (1977).

83) C.G. Callan and D.J. Gross, Phys. Rev. Lett. 22, 156 (1969).

84) R.P. Feynman, Photon-hadron interactions (Benjamin, New York, 1972).

85) M.D. Mestayer, Thesis, Report SLAC-214 (1978).

86) H.L. Anderson et al., Fermilab-Pub-79/30-Exp. (1979).

87) A. Savoy-Navarro, talk given at the "Neutrino-79", Int. Conf. on Neutrinos, Weak Interactions and Cosmology, Bergen, 1979.

88) G. Altarelli and G. Martinelli, Phys. Lett. 76B, 89 (1978).
A. de Rújula et al., Ann. Phys. 103, 315 (1977).

89) G. Altarelli and G. Parisi, Nucl. Phys. B126, 298 (1977).

90) L.F. Abbott and R.M. Barnett, SLAC-Pub-2325 (1979).

91) A.J. Buras and K.J.F. Gaemers, Nucl. Phys. B132, 249 (1978).

92) J.G.H. de Groot et al., Phys. Lett. 82B, 456 (1979).

93) J.G.H. de Groot et al., Phys. Lett. 82B, 292 (1979).

94) E.M. Riordan et al., SLAC-Pub-1634 (1975).

95) O. Nachtmann, Nucl. Phys. B63, 237 (1973).

96) A. Para and C.T. Sachrajda, preprint TH. 2702-CERN.

D I S C U S S I O N S

CHAIRMAN: Prof. F. Dydak

Scientific Secretary: Michael P. Schmidt

DISCUSSION 1

- NICOLAIDIS:

You have quoted the value $\sin^2\theta_W = 0.23$, for the Weinberg angle. How does this compare with the values given by the grand unified theories?

- DYDAK:

Our experimental value is $\sin^2\theta_W = 0.23 \pm 0.015$, while the prediction given by the SU(5) scheme of grand unification is ~ 0.20. I would like to emphasize, that we believe the experimental determination is valid, and that the difference is significant. It is not possible for us to reconcile the two values.

- WEISSKOPF:

Why is the lifetime of the proton dependent on the Weinberg angle?

- GINSPARG:

In grand unified theories the $SU(2) \otimes U(1)$ Salam Weinberg model is imbedded in a larger gauge group, and hence the Weinberg angle is predictable at the unification mass essentially as a Clebsch coefficient. The fermion spectrum implied by the usual sequential scheme generates a value $\sin^2\theta_W = 3/8 = 0.375$. This value can be renormalized down to experimentally accessible energy scales by using the separate renormalization group equations for the U(1) and SU(2) coupling constants. The answer is as low as $\sin^2\theta_W \sim 0.20$ if the unification scale is as high as 10^{15} GeV.

In grand unified theories proton decay is given typically by graphs like:

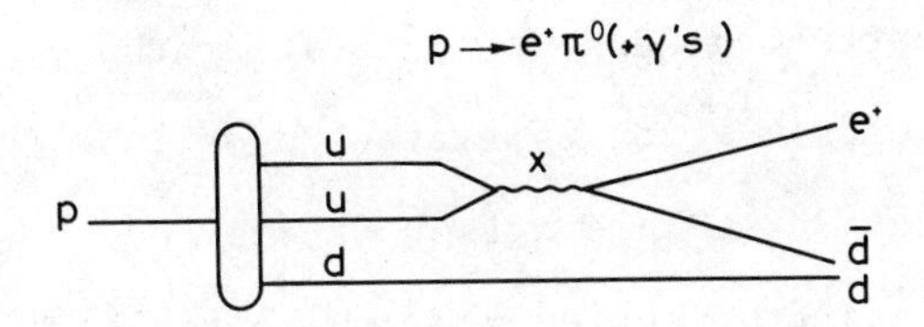

where the superheavy X - boson has a mass M_X of order 10^{15} GeV, the unification scale. The amplitude for such a graph is suppressed by a factor $1/M_X^2$ from the X - boson propagator and thus using the value of the Weinberg angle to determine the unification scale gives a dependence of the proton lifetime on the Weinberg angle.

- SZWED:

You observed a necessity to include a 10% V + A interaction for neutral currents on an isoscalar target. Do you know of any other process where you need V + A?

- DYDAK:

In neutrino-nucleon scattering, this is the only case known to me where we can so clearly isolate the V + A contribution. This we observe at the level of 3 to 4 standard deviations.

DISCUSSION 2

- DYDAK:

I would like to make a comment in answer to de Rújula. He said that the antiquark distribution in the nucleon is not known because of radiative corrections. I would like to answer him with his own paper which he wrote about one month ago.

What we are doing essentially in neutrino physics is looking at antineutrino scattering, and we are testing the y distribution. What we see is an offset which is flat, and on top of it a $(1 - y)^2$ distribution. We interpret this offset as just due to the scattering of antineutrinos from antiquarks, which yields a flat y distribution, assuming that the interaction is V - A. The antiquark fraction, $(\bar{Q} + \bar{S})/(Q + \bar{Q})$, is then given by the ratio of the heights of the y distribution at y = 1 and at y = 0. The value obtained, $(\bar{Q} + \bar{S})/(Q + \bar{Q}) = 0.16$, is brought to about 0.15 by radiative corrections applied to the outgoing muon.

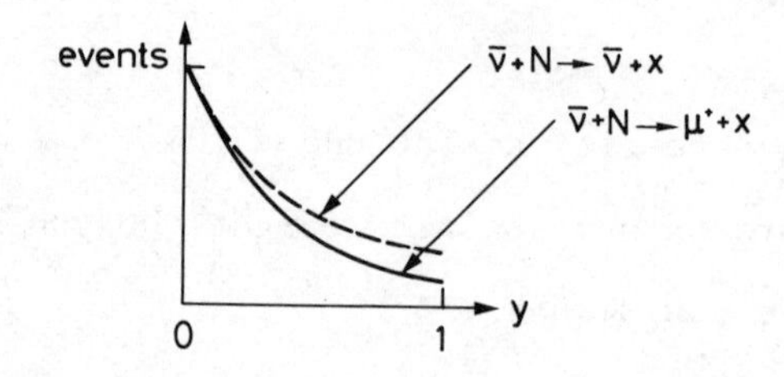

- WEISSKOPF:

You told us that the sea contribution is 30%. This is difficult to understand since you measure the momentum weighted sea. We know that the sea has low x. This means that actually there is very large sea density, not compatible with the valence quark model.

- JAFFE:

The sea content of the proton is a sensitive function of Q^2 in QCD. It evolves particularly rapidly at low Q^2. Llewellyn Smith has reported to me model calculations in which the proton is primarily composed of valence quarks at a mass scale of 1 GeV. If first order renormalization group equations are used to run these distributions up to higher Q^2, it is found that substantial gluon and quark sea components are generated. Thus it is qualitatively possible to have a valence quark model at low mass scales and still have a large amount of glue and sea at higher Q^2.

- PREPARATA:

Does your determination of antiquark densities from structure functions and the y dependence agree?

- DYDAK:

Yes.

PHYSICS OF e^+e^- REACTIONS

E. Lohrmann

University of Hamburg
and Deutsches Elektronen-Synchrotron DESY
Hamburg

1. The Fundamental Fermions

There is now a consistent body of evidence, which points towards the following fundamental building blocks of matter:

i) Three doublets of leptons:

charge			
0	ν_e	ν_μ	ν_τ
-1	e^-	μ^-	τ^-

The τ neutrino ν_τ has not been directly observed, but information on the τ decays makes its existence very likely.[1,2]

ii) Five (possibly six)[3] flavors of quarks, which can again be arranged in three doublets according to their charge:

charge			
2/3	u	c	t*
-1/3	d	s	b

Each quark flavor comes as a color triplet. With flavor and color there are altogether 5 × 3 = 15, possibly 6 × 3 = 18 quark states — so far. Ordinary hadrons are color neutral ("white").

---

*Not yet seen.

One believs that forces between quarks are mediated by 8 massless neutral vector particles called gluons. Gluons have color, they are octets in color SU3. They have no flavor, couple equally to all flavors.[4]

The interaction of gluons with quarks is in analogy with the interaction of photons with leptons (quantum chromodynamics QCD).

There are, however, also essential differences:

i) There is a gluon-gluon interaction (three gluon vertex):

ii) The quark-gluon coupling g is not a constant,[5] but depends on the squared momentum transfer Q:[2]

$$\frac{g^2 (Q^2)}{4\pi} = \alpha_s (Q^2) = \frac{12\pi}{(33-2N_f) \ln Q^2/\Lambda^2} \tag{1}$$

Λ = the only adjustable constant;
N_f = number of quark flavors whose threshold is below the energy in question.

From scale-breaking effects in inelastic neutrino scattering[6] one has $\Lambda \sim 0.4$ GeV at moderate values of Q^2. Observation on charmonium point towards a somewhat smaller value ($\Lambda \sim 0.2$ GeV). However, these values can only be considered as very rough guesses, due to the many uncertainties in the computations. In the limit of very large values of $Q^2 \to \infty$ (short distances) the coupling "constant" $\alpha_s (Q^2)$ becomes very small. Then the quarks inside a nucleon behave more and more like free particles. This may explain the success of the parton model for deep inelastic scattering of muons, electrons, and neutrinos (asymptotic freedom). For small values of Q^2 (large distances) the coupling becomes strong.

For small values of α_s the methods of QED become applicable to QCD, and many things can be computed, although with considerable complications due to higher order corrections and to the gluon-gluon interaction.

The leptons and quarks have spin 1/2 and, being fundamental, are pointlike. Therefore they are coupled in a well defined and calculable way (by QED) to the photon (with the exception of the neutrinos), and therefore evidence for them must appear in e^+e^--collisions in a straightforward way.

2. Cross Section for Annihilation

A simple example is the annihilation of e^+e^- into a muon pair: $e^+e^- \to \mu^+\mu^-$ (Fig. 1).

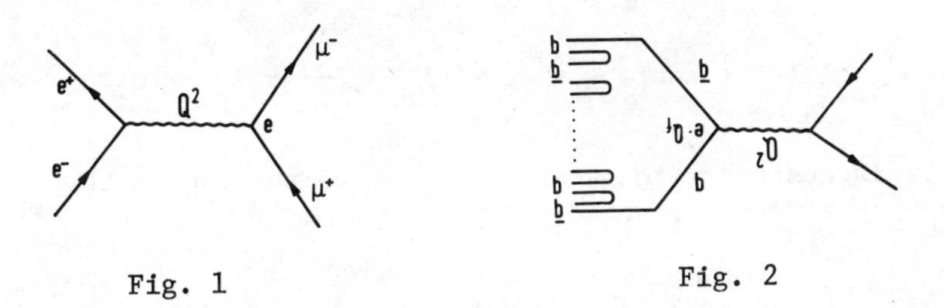

Fig. 1 Fig. 2

The differential cross section in the CM system is (in the lowest order of perturbation theory):

$$\frac{d\sigma}{d\Omega}(\mu\mu) = \frac{r_e^2 m_e^2}{4s} \beta_\mu \left[(1 + \cos^2\theta) + (1 - \beta_\mu^2)\sin^2\theta\right] \tag{2}$$

where r_e = classical electron radius ($2.82 \cdot 10^{-13}$ cm); m_e = electron mass; $s = E^2_{CM}$, E_{CM} = CM energy; β_μ = muon velocity in the CM system; $\beta_\mu = P_\mu/E_\mu$; θ = angle between e^- and μ^-.

In the high energy limit $E_{CM} >> m_\mu$, $\beta_\mu \rightarrow 1$ we have

$$\frac{d\sigma}{d\Omega}(\mu\mu) = \frac{r_e^2 m_e^2}{4s} (1 + \cos^2\theta) \tag{3}$$

and the total $\mu^+\mu^-$ cross section is

$$\sigma_{\mu\mu} = \int \frac{d\sigma}{d\Omega}(\mu\mu)\, d\Omega = \frac{4\pi}{3} r_e^2 \frac{m_e^2}{s} = \frac{4\pi\alpha^2}{3s} = \frac{86.8}{s} \text{ nb GeV}^2 \tag{4}$$

For the production of a quark-antiquark pair ($q\bar{q}$) analogous formulae apply. The total production cross section is:

$$\sigma(q\bar{q}) = \frac{4\pi}{3} \cdot r_e^2 \frac{m_e^2}{s} \cdot Q_q^2 \tag{5}$$

where Q_q is the charge of the quark, measured in units of e. It is presently believed that the very strong forces between quarks prevent them from emerging directly in the annihilation process. Rather they will fragment into normal hadrons (Fig. 2).

In order to get the total cross section for annihilation into hadrons one has to sum over all kinds of quarks:

$$\sigma(e^+e^- \rightarrow \text{hadrons}) = \sigma_h = \sum_{\text{quark } i} \frac{4\pi}{3} \cdot r_e^2 \frac{m_e^2}{s} \cdot Q_i^2 \tag{6}$$

It is customary to present this in comparison to the $\mu^+\mu^-$ cross section by giving the ratio

$$R = \frac{\sigma_h}{\sigma_{\mu\mu}} = \sum_{\text{colour, flavour}} Q_i^2 = 3 \cdot \sum_{\text{flavour}} Q_i^2 \tag{7}$$

The sum goes over all quark charges Q_i. If the CM energy is below threshold for producing charmed particles, the sum will just go over the u, d, s-quarks and we have

$$R(<c) = 3[(\tfrac{2}{3})^2 + (\tfrac{1}{3})^2 + (\tfrac{1}{3})^2] = 2 \tag{8}$$

The factor 3 is there because every quark flavour comes in three colours. Above charm threshold the value of R is

$$R(>c) = 3[(\tfrac{2}{3})^2 + (\tfrac{1}{3})^2 + (\tfrac{1}{3})^2 + (\tfrac{2}{3})^2] = \frac{10}{3} \tag{9}$$

Equation (7) is an approximation which does not take into account the strong interaction between quarks, which is probably mediated by gluons. Effects due to gluons may come in a way similar to radiative corrections in QED, e.g., the first order annihilation diagram of Fig. 2 may be modified as follows in Fig. 3.

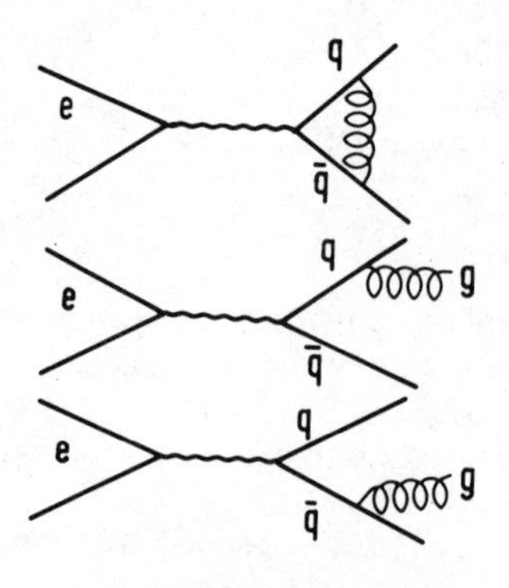

Fig. 3

Table 1. Storage Rings

Installation	Beam energy, GeV	Remarks
ACO, Orsay	0.2-0.55	Out of service
VEPP, Novosibirsk	0.2-0.65	
ADONE, Frascati	0.6-1.55	
DCI, Orsay	0.7-1.1	Energy expected to go to >1.7 GeV
SPEAR, Stanford	1.2-4	
DORIS, Hamburg	1.4-5	Energy could go to <1.4 GeV
PETRA, Hamburg	5-18	Working up to 16 GeV beam energy
PEP, Stanford	5-18	Under construction
CESR, Cornell	4-8	Starting up, tests

g denotes a gluon. In QED the radiative corrections are of the order

$$\frac{\alpha}{\pi} \ln\left(\frac{E}{m_e}\right), \quad \frac{\alpha}{\pi} \ln\left(\frac{E}{\Delta E}\right)$$

One expects the same order of magnitude for the gluonic corrections, with α replaced by the α_s of Eq. (1). Under rather general assumptions one can show that in first approximation the correction is α_s/π, so that Eq. (7) is modified to [7]

$$R = 3\cdot\sum_{\text{flavour}} Q^2_i \cdot \left(1 + \frac{\alpha_s}{\pi} + O(\alpha^2_s)\right) \tag{10}$$

For $\sqrt{s}$ = 10 GeV we have from Eq. (10), with N_f = 4, Λ = 0.4 GeV, $\alpha_s \approx 0.23$ leading to a 7% correction.

What do the experiments have to say about this? Data have come from the first seven storage rings listed in Table 1.

Figure 4 shows a typical measurement of the total hadronic cross section.[8]

One sees that below charm threshold (E_{CM} < 4 GeV) $R \cong 2$ as expected. Around charm threshold the cross section exhibits a complicated structure, probably a series of resonances. Above that region it settles down to a value of $R \approx 4.5$. This has to be compared with the value $R = 3^1/_3$ expected from Eq. (9). There is, however, a contribution to the measurements from the process

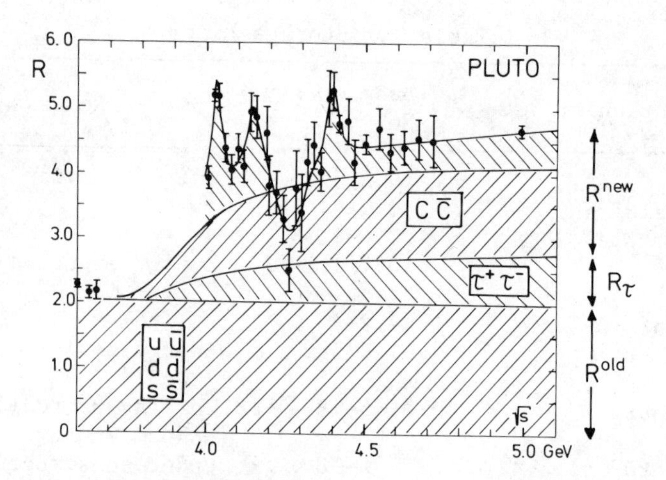

Fig. 4. The total cross section for e^+e^--annihilation in units of $R = \sigma_h/\sigma_{\mu\mu}$, as function of CM energy $\sqrt{s}$. Contribution of $e^+e^- \rightarrow \tau^+\tau^-$ is shown explicitly and should be subtracted (from Ref. 8).

$e^+e^- \rightarrow \tau^+\tau^-$, which has $R \approx 1$ and which must be subtracted to get a number for the truly hadronic cross section. This reduces the experimental value to $R_{had} \approx 3.5$, in fair agreement with the theoretical expectation.

Figures 5 and 6 show the experimental situation in detail. There appear slight discrepancies between the three different high-energy experiments, but they are probably within the systematic errors of about 15% quoted by the groups. Figure 5 shows new data at the low energy end which are in good agreement with theoretical expectations ($R \approx 2$).

Figure 7 shows the first measurements of R made with PETRA. The cross section is in good accord with theory using a contribution of u, d, s, c and possibly b quarks.

3. Jets

Can we see the quarks more directly? Figure 2 gives a clue. At high energies the hadrons emerging from the quark fragmentation should form two jets of particles which can be recognized, if the angular spread of the jet is sufficiently small. The jet axis should preserve the direction of the original quark and therefore has an angular distribution like Eq. (3), i.e., $1 + \cos^2\theta$. Another clue comes from the distribution of the azimuth angle ϕ of the jet around the beam axis. If the electrons are polarized trans-

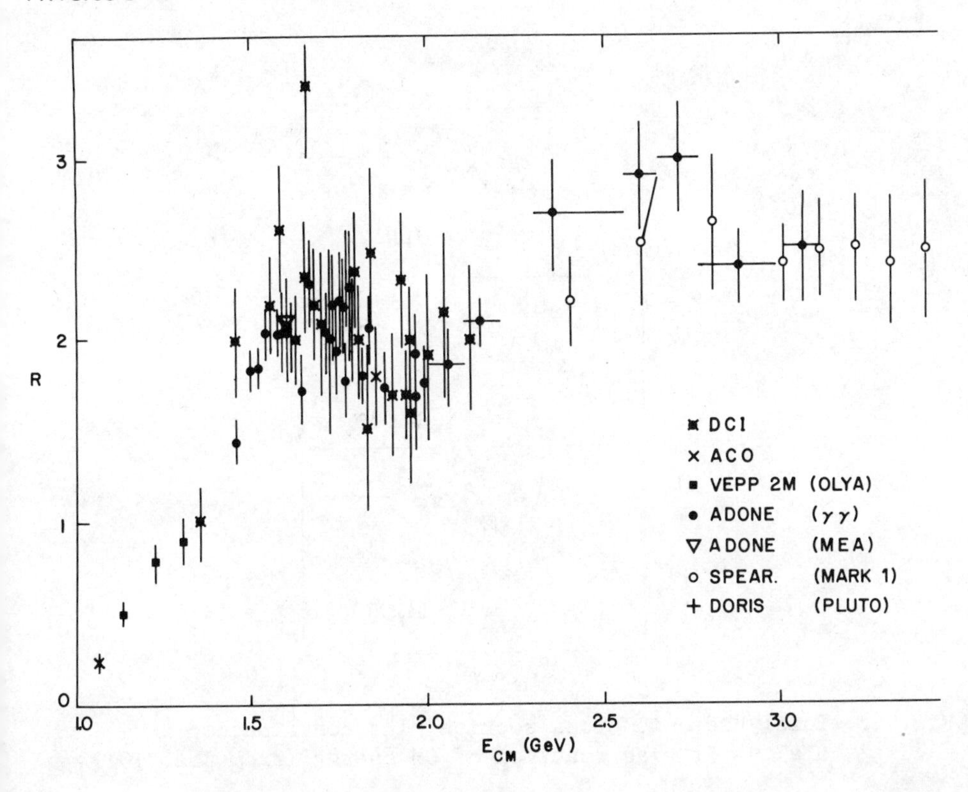

Fig. 5. Total hadronic annihilation cross section for low energies in units of $R = \sigma_h/\sigma_{\mu\mu}$, as function of CM energy (from Ref. 9).

verse to the beam direction and if P_1 and P_2 are the degrees of polarization of the two beams, then the angular distribution is given by (electron polarization perpendicular to the storage ring plane):

$$\frac{d\sigma}{d\Omega} = \sigma_0 \; (1 + \alpha \cos^2\theta) + P_1P_2 \; \alpha \sin^2 \theta \cos 2\phi] \qquad (11)$$

with $|\alpha| \leqslant 1$, $\phi = 0$ in the horizontal plane (= plane of the storage ring). For spin 1/2 particles such as quarks $\alpha = 1$ and

$$\sigma_0 = \frac{r_e^2 \; m_e^2}{4s} \cdot 3 \cdot \sum_{\text{flavour}} Q^2 f \qquad (12)$$

Experimentally the problem is first to establish the existence of a jet-like structure and to determine the jet-axis. To this end

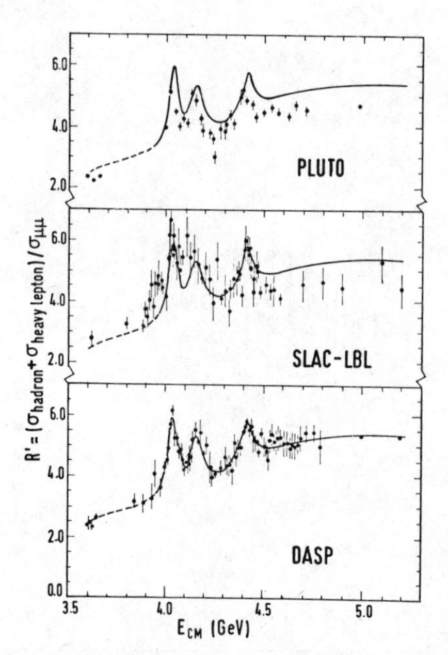

Fig. 6. Total hadronic cross section plus contribution from $e^+e^- \rightarrow \tau^+\tau^-$, as function of CM energy (from Ref. 10).

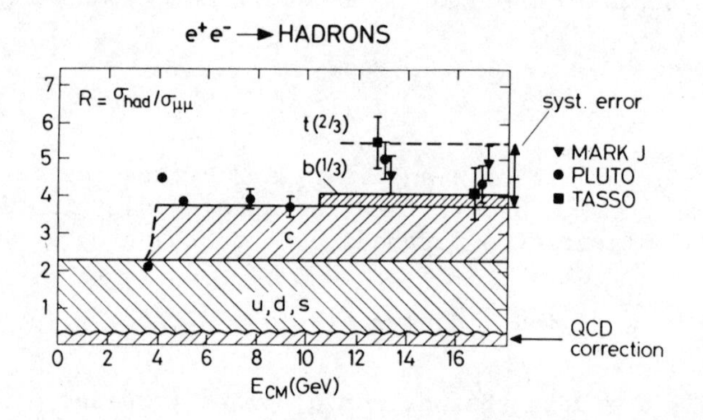

Fig. 7. First measurements of R at PETRA. Earlier measurements by PLUTO at lower energy are included for comparison. Contributions of the various quark flavors to R are indicated. No contribution from $e^+e^- \rightarrow \tau^+\tau^-$. (Ref. 38)

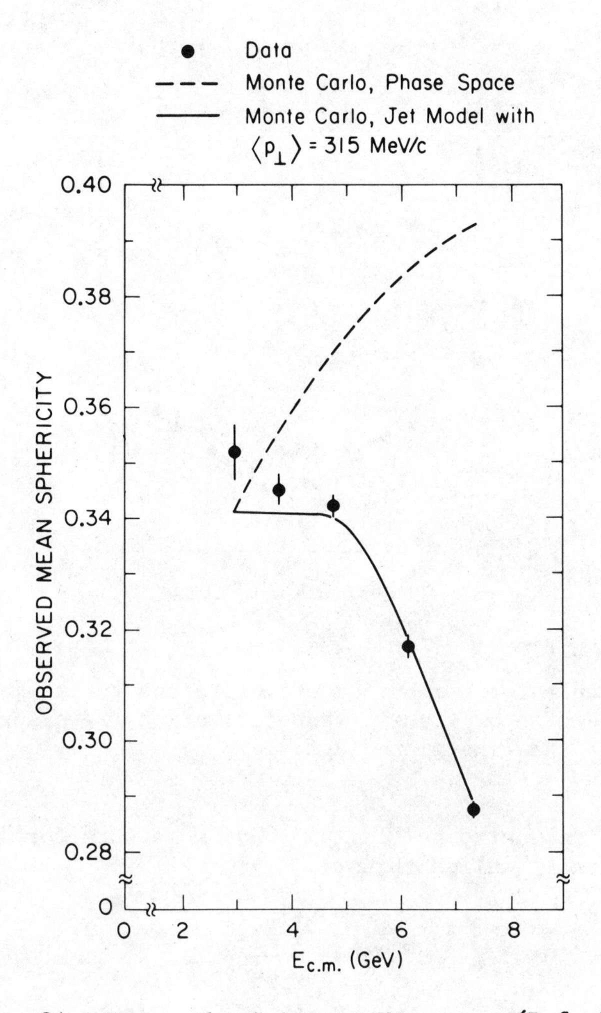

Fig. 8a. Mean sphericity vs CM energy (Ref. 18).

a quantity called sphericity S is determined. It is defined[15]

$$S = \frac{3}{2} \frac{\text{Min}(\Sigma p_{ti}^2)}{\Sigma p_i^2} \tag{13}$$

p_{ti} is the transverse momentum of the i-th particle with respect to a chosen jet axis, p_i is its momentum. The best estimate for the jet axis minimizes the Σp_{it}^2. In the limit of isotropic events with high multiplicity $S \to 1$. On the contrary, for jet-like events and high energy, $S \to 0$.

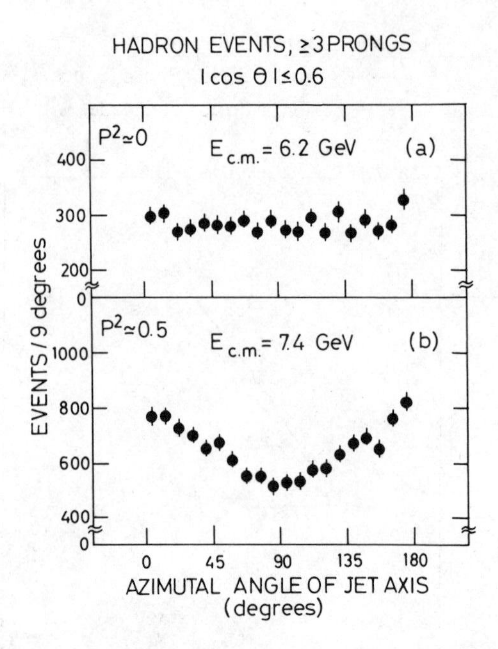

Fig. 8b. Azimuthal distribution of the reconstructed jet axis, P = degree of transverse polarization, zero angle coincides with the ring plane (Ref. 18).

There is another possibility to have a measure for "jet-ness." This is a quantity called thrust:

$$T = \max \frac{\sum_i |p_{li}|}{\sum_i p_i} \tag{14}$$

p_{li} = longitudinal momentum of i-th particle. In the limit of isotropic events with high multiplicity $T \to 1/2$, and $T \to 1$ for jet-like events.

A third measure is spherocity:[17]

$$S_o = \left(\frac{4}{\pi}\right)^2 \frac{\mathrm{Min}(\Sigma|p_{ti}|)^2}{(\Sigma|p_i|)^2} \tag{15}$$

This measure is not affected by infrared divergences in the theory as much as sphericity S. However, in experiment it turns out that the computation of the minimum of S_o leads to practical difficulties.

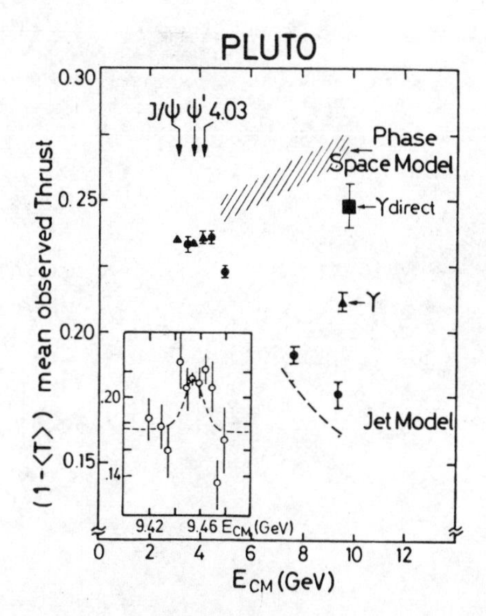

Fig. 9. Mean thrust $\langle T \rangle$ vs CM energy. Value of $\langle T \rangle$ at Υ resonance is also shown. Υ direct: background subtracted $\langle T \rangle$-value on Υ-resonance. Insert: total cross section (Ref. 19). See Section 8 for explanation of Υ.

The first evidence for jet structure in e^+e^- collisions came from the SLAC-LBL group[18] (Fig. 8a, b).

The aximuthal angular distribution and the dependence of S on CM energy go as expected for quarks with spin 1/2. At higher energies jet structures become very apparent, as e.g., in the recent work of the PLUTO detector at DORIS[19] (Figs. 9, 10) and as shown in recent PETRA experiments.[20,21]

Figure 11 shows that the angular distribution of the jet axis is in agreement with the expectations from a quark-jet model ($1 + \cos^2\theta$).

Figure 12 shows, that the transverse momentum with respect to the jet axis is constant or only very slowly rising, whereas the longitudinal momentum of the jet particles increases with beam energy.[19] Figure 13 shows, that the energy flow of neutral particles (mostly π°) follows the jet axis as determined by the charged particles.[19]

Of course, in all these investigations the true jet axis is not known. At energies of 9.4 GeV the axis as determined by sphericity or thrust can show deviations from the true jet axis. The

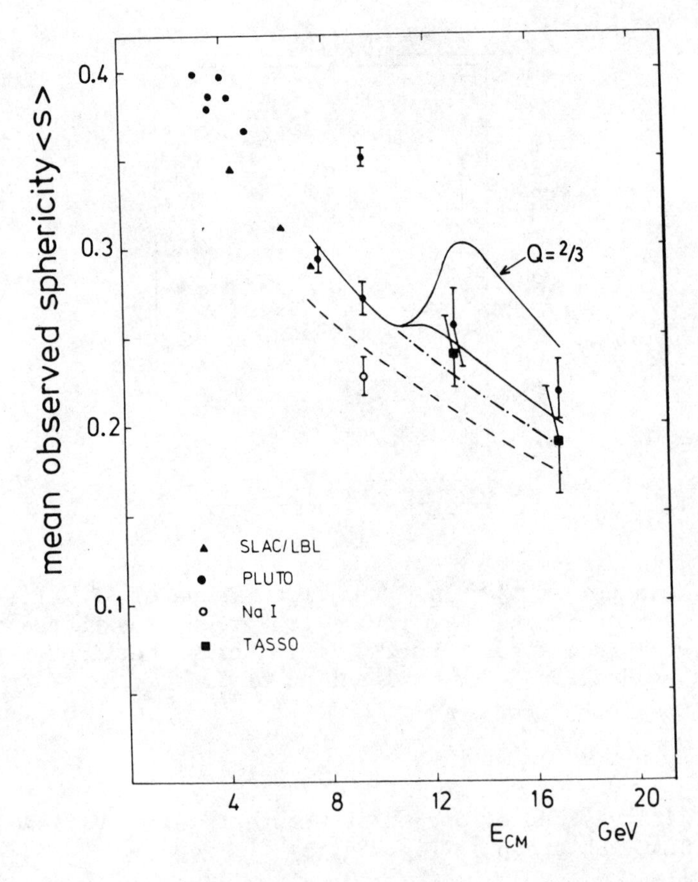

Fig. 10. Mean sphericity <S> vs CM energy. Dashed line: two-jet model (Feynman-Field). Dash-dotted and full lines: inclusion of c- and b-flavors. Contribution t-quark is indicated (Ref. 38).

Monte Carlo study of Fig. 14 shows, that the error in most cases is <20°.

4. Inclusive Distributions

Consider the process $e^+e^- \rightarrow h$ + anything, h stands for hadron.

Call q the four-momentum of the virtual photon,

$$q^2 = s$$

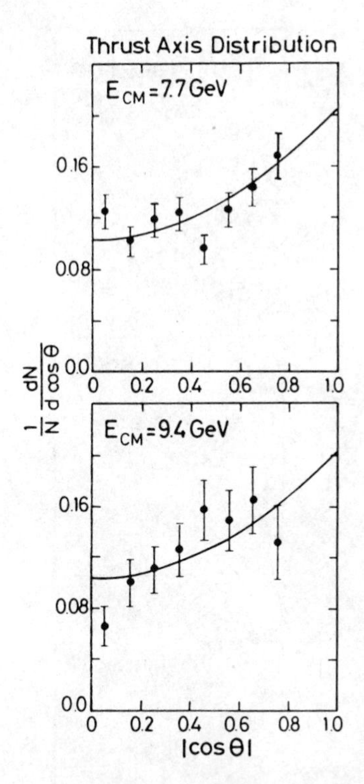

Fig. 11. Angular distribution of thrust axis at 7.7 GeV and 9.4 GeV CM energy. Curves are $1 + \cos^2\Theta$ distribution (Ref. 19).

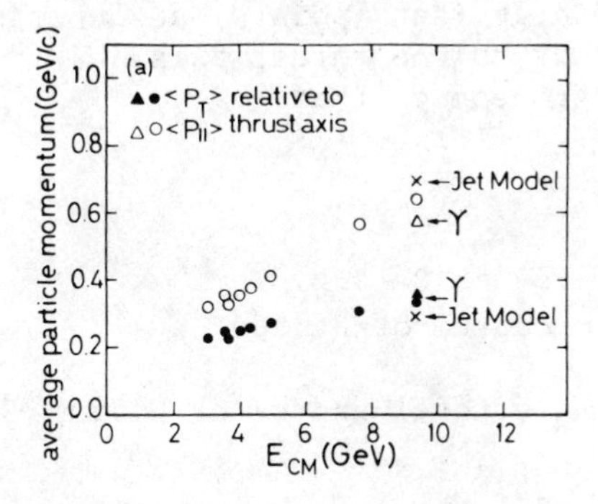

Fig. 12. Mean observed parallel and transverse momentum as function of CM energy. Also included values for Υ resonance (Ref. 19).

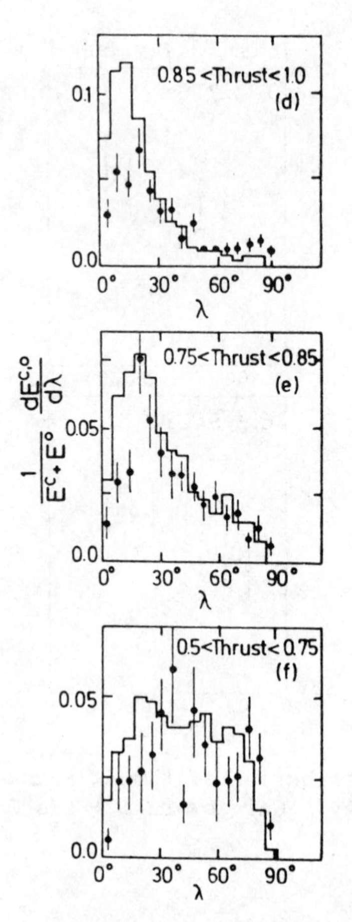

Fig. 13. Angular distribution of charged energy (histograms) and neutral energy (data points) as function of the angle λ with respect to the thrust axis. E_o = neutral energy, E_c = charged energy (Ref. 19).

Call p = four momentum of hadron

E_h = total energy of hadron

$$x = \frac{2p \cdot q}{s} = \frac{2E_h}{\sqrt{s}} = \text{scaling variable} \quad (16)$$

Then, if scaling holds, the cross section $d\sigma/dx$, normalized to the total cross section σ_T can only depend on x:

$$\frac{d\sigma}{dx} = \sigma_T \cdot \phi(x) \tag{17}$$

If we put $\sigma_T \sim 1/s$, then

$$\frac{sd\sigma}{dx} \sim \phi(x) \tag{18}$$

In the quark picture, one interprets this law by a process, in which a $\bar{q}q$-pair is formed, and each quark of the pair fragments into hadrons, described by a fragmentation function $D_q^h(x)$.

$$\frac{d\sigma}{dx}(e^+e^- \to q\bar{q} \to h) = \sigma_{q\bar{q}} 2D_q^h(x) = \frac{8\pi\alpha^2}{s} Q_f^2 D_q^h(x) \tag{19}$$

From Eq. (17) we have

$$\int \frac{d\sigma}{dx}\, dx = \sigma_T \int_{x_{min} = 2m/\sqrt{s}}^{1} \phi(x)\, dx = \sigma_T \cdot \langle n \rangle \tag{20}$$

where $\langle n \rangle$ = average number of produced particles.

Therefore, if exact scaling were true for all x, we should have

$$\langle n \rangle = \int_{x_{min} = 2m/\sqrt{s}}^{1} \phi(x)\, dx \tag{21}$$

$$\frac{d\langle n \rangle}{d\ln\sqrt{s}} = x_{min} \cdot \phi(x_{min}) \tag{22}$$

$x_{min} = 2m/\sqrt{s}$, m = hadron mass

Therefore, if $\phi(x) \sim (1/x)$ for small values of x, we expect a dependence of multiplicity on energy like

$$\langle n(s) \rangle = c_1 + c_2 \ln s \tag{23}$$

Figure 15 shows a compilation[9,25-28] for the average number of charged particles $\langle n_s \rangle$. The charged multiplicity rises faster[29] than the logarithmic law of Eq. (23).

Figure 16 shows the scaling cross section $s(d\sigma/dx_p)$ as a function of the variable $x_p = 2p/\sqrt{s}$. This variable is chosen, because in some experiments only the momentum p of the secondary particles can be measured, not their energy E, for lack of particle identification. For small values, there can be substantial differences between x_p and x. The Fig. 16 shows, that at large x_p values there are discrepancies of unknown origin between laboratories.

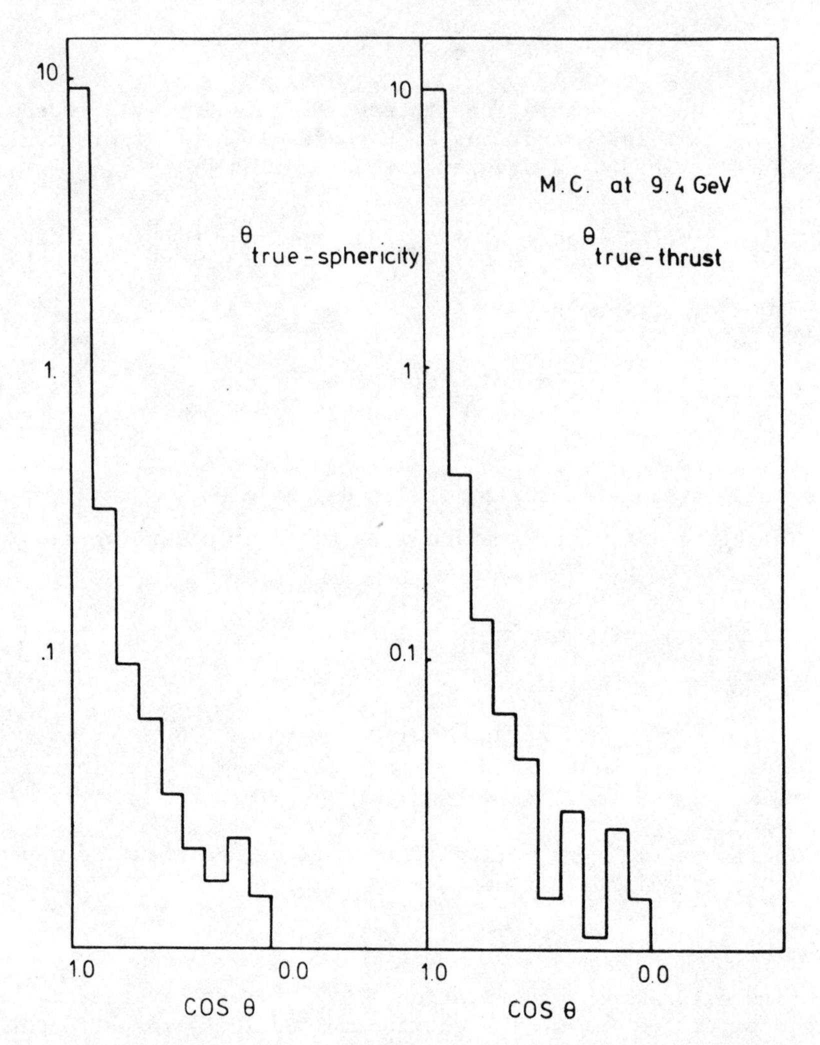

Fig. 14. Distribution of angle between generated quark and the axis defined by sphericity and thrust. Monte Carlo calculation by the PLUTO group, Ref. 38.

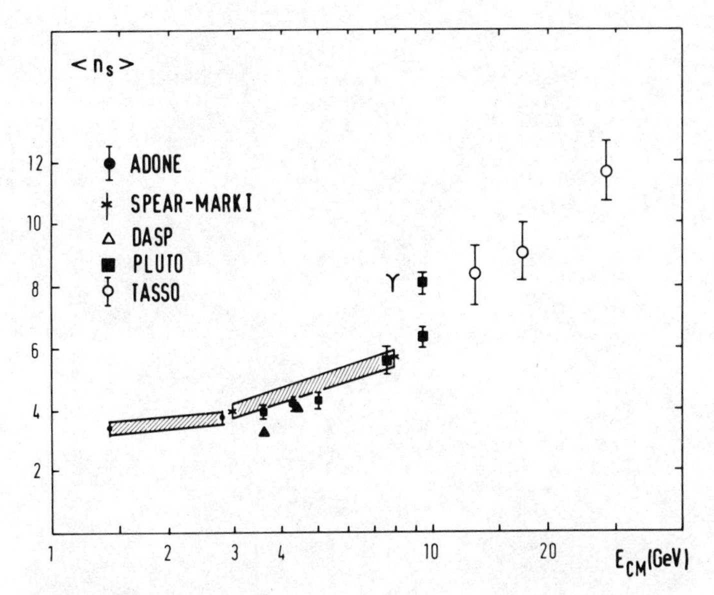

Fig. 15. Average number of charged particles $\langle n_s \rangle$ as a function of CM energy. The bands at low energy mark a series of measurements at ADONE[9] and SPEAR[28]. The points above 3 GeV have been corrected for contamination by $e^+e^- \to \tau^+\tau^-$. References: DASP[27], TASSO[26]. The point marked Υ is taken on the Υ resonance, nonresonant background subtracted (PLUTO[25,28]).

Figure 17 shows a comparison of kaon data. Figure 18 shows the data of the DASP detector, giving a complete survey of the inclusive spectra of pions and kaons. The data show scaling for $x > 0.4$, except in a CM energy region $4.0 < \sqrt{s} < 5$ GeV.

A fit to the scaling cross section of the form

$$\frac{s}{\beta} \cdot \frac{d\sigma}{dx} = a \exp(-bx) \tag{24}$$

leads to

$$a \sim 24 \text{ nb GeV}^2$$
$$b \sim 8.5$$

Below $x \sim 0.4$ one observes a rise in the x-distribution $\phi(x)$ when crossing the charm threshold, which of course violates scaling. This rise can be attributed to the onset of charm production, which leads to an increase of the pion and kaon yield at small x-values.

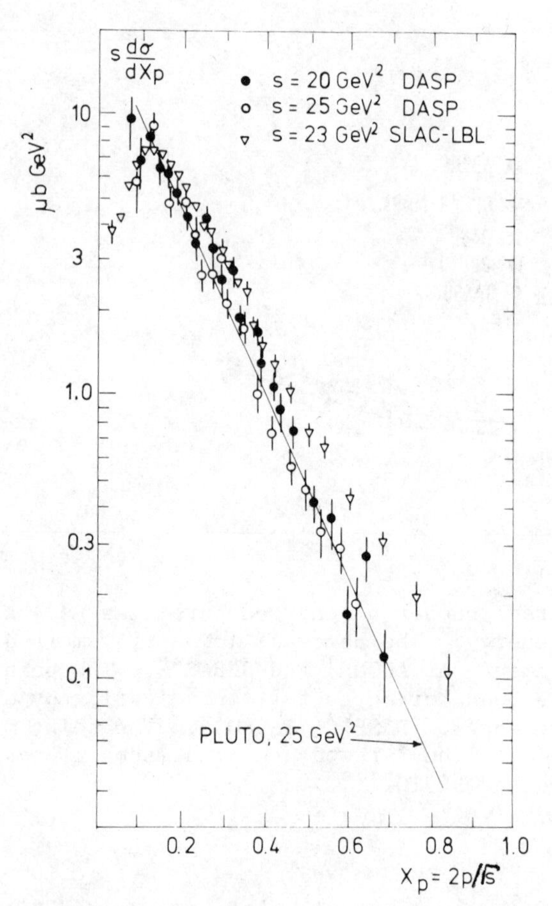

Fig. 16. Differential cross section $s(d\sigma/dx_p)$ of charged hadrons as function of variable $x_p = 2p/\sqrt{s}$. p = momentum of hadron, $\sqrt{s}$ = CM energy. DASP[30], SLAC-LBL[24], PLUTO[31,36].

The rise in the kaon yield is larger than the rise in the pion-yield, in accordance with expectations from the GIM mechanism.[35]

The data of Ref. 30 also show the remarkable fact, that the functions $\phi(x)$ for the different particles $\pi^\pm$, $K^\pm$, p, $\bar{p}$ are quite similar. This is shown in Fig. 19. The effect has been seen before by the SLAC-LBL group.[33]

Figure 20 shows evidence for scaling for the DASP data in comparison between $\sqrt{s}$ = 3.6 GeV (below charm threshold) and at $\sqrt{s}$ = 5 GeV (above the charm resonance region). Previous demonstrations of scaling have come from the SLAC-LBL group. There are new data on inclusive K_S^o and D production, which do not seem to follow the universal line for $\pi^\pm$, $K^\pm$.

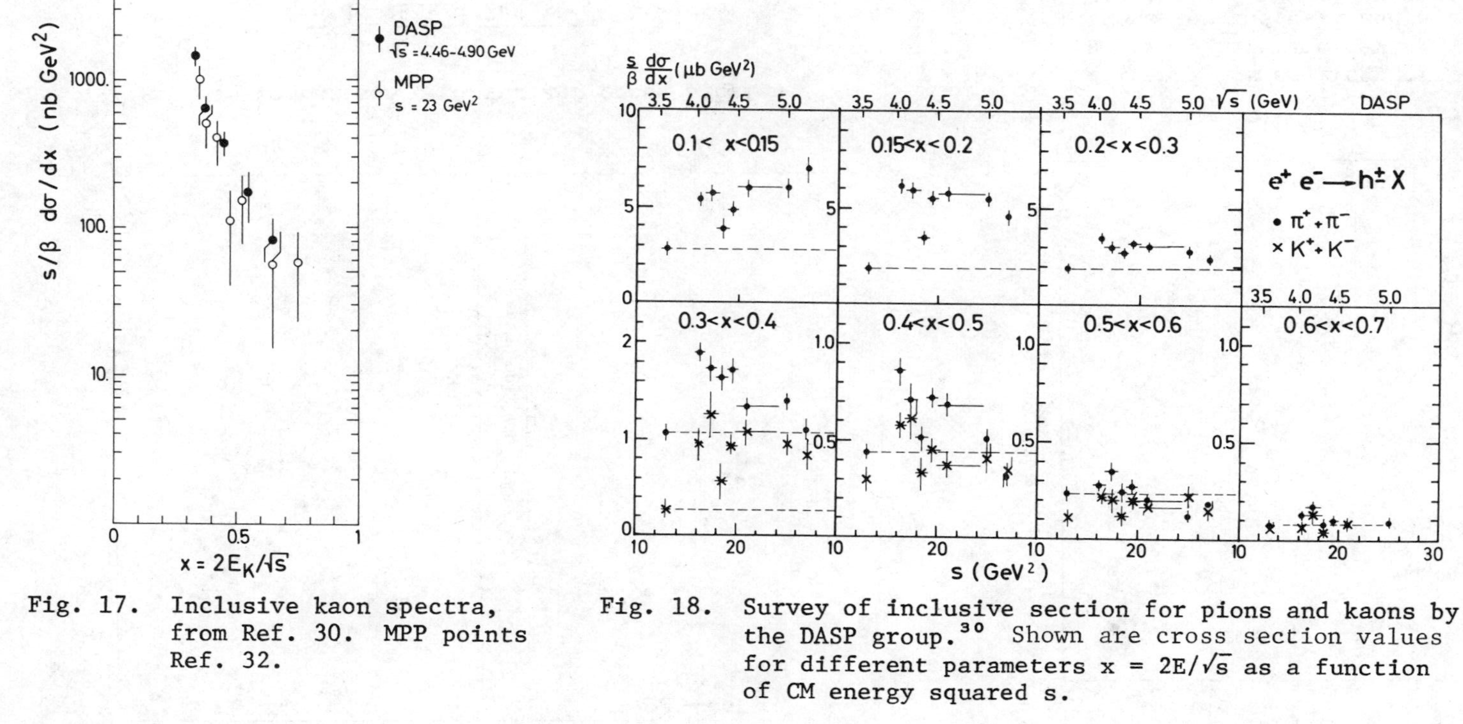

Fig. 17. Inclusive kaon spectra, from Ref. 30. MPP points Ref. 32.

Fig. 18. Survey of inclusive section for pions and kaons by the DASP group.[30] Shown are cross section values for different parameters $x = 2E/\sqrt{s}$ as a function of CM energy squared s.

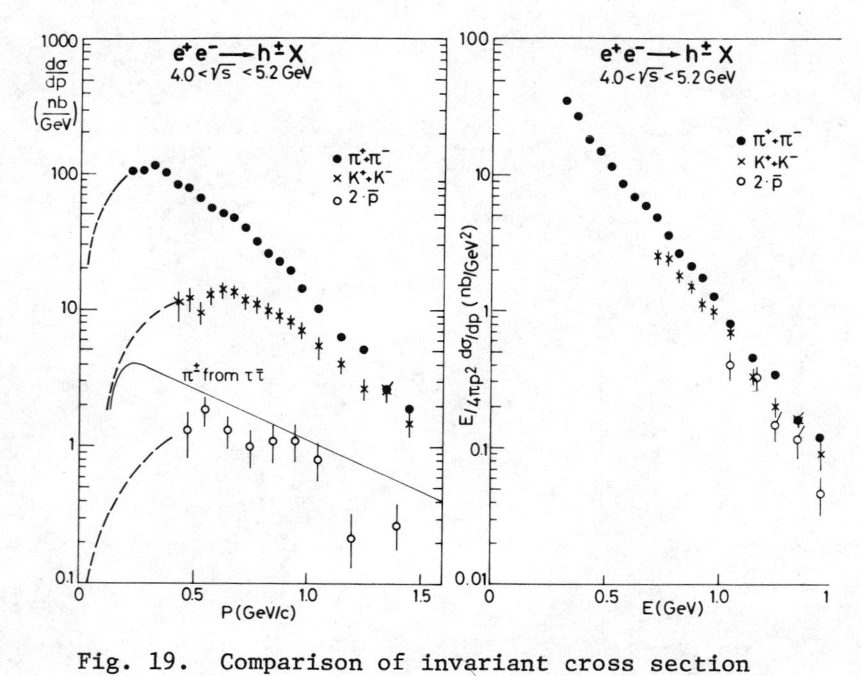

Fig. 19. Comparison of invariant cross section

$$4\pi \frac{E d\sigma}{p^2 dp} = \frac{s d\sigma}{\beta dx} \cdot \frac{1}{2\pi E \cdot s^{3/2}}$$

for pions, kaons and (anti)protons.

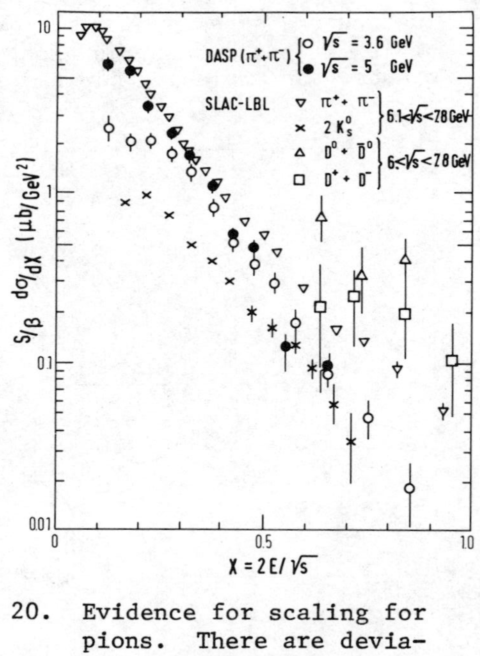

Fig. 20. Evidence for scaling for pions. There are deviations for $x < 0.25$ and a discrepancy between DASP[30] and SPEAR[34] data for $x > 0.5$. Also included are data for K^0_S and D-mesons (Ref. 34).

Table 2. List of Narrow $J^{PC} = 1^{--}$ States

Particle	Mass MeV	Total width MeV	Quark content	Γ_e keV	Threshold for allowed decay
ϕ	1019.6 ± 0.2	4.1 ± 0.2	$s\,\bar{s}$	1.3 ± 0.15	K^+K^-: 987 MeV
J/Ψ	3097 ± 2	0.067 ± 0.012	$c\,\bar{c}$	4.8 ± 0.6	$D^0\bar{D}^0$: 3727 MeV
Ψ'	3686 ± 3	0.228 ± 0.056	$c\,\bar{c}$	2.1 ± 0.3	3727 MeV
Ψ''(3770)	3767 ± 4[44]	24 ± 5	$c\,\bar{c}$	0.13 ± 0.05	3727 MeV
Υ	9460 ± 10	$0.03 < \Gamma < 17$	$b\,\bar{b}$	1.3 ± 0.2	10560 MeV (?)
Υ'	10016 ± 20	<20	$b\,\bar{b}$	0.33 ± 0.1	10560 MeV (?)

A word of caution: The DASP data and probably also the data of other experiments have a contribution of hadrons from $e^+e^- \to \tau^+\tau^- \to$ hadrons. This should not be serious for kaons, and not for pions at momenta $p < 0.6$ GeV. However, for $p \sim 1.5$ GeV the τ-contribution to the pion spectrum could be $\sim 25\%$.

PLUTO has measured the inclusive ρ^0-yield.[36] Its $\phi(x)$-distribution is somewhat above the one of π^+ or π^-.

5. Bound States of Quarks — the $J^{PC} = 1^{--}$ States

A quark and its antiquark may be bound together. If the mass of the system is below the threshold for decay into a pair of mesons with the appropriate flavour, the decay is inhibited and the width of the state will be small.

We shall treat first states with $J^{PC} = 1^{--}$. They couple directly to one photon. Table 2 lists the narrow states seen so far.[37–44]

The widths of J/Ψ, Ψ', Υ, Υ' are too small to be measured directly. However, with the assumption that the reaction proceeds via the formation of a narrow intermediate 1^{--} state, they can be determined indirectly. If $(q\bar{q})$ stands for 1^{--} $q\bar{q}$ bound state, the process can be regarded as follows:

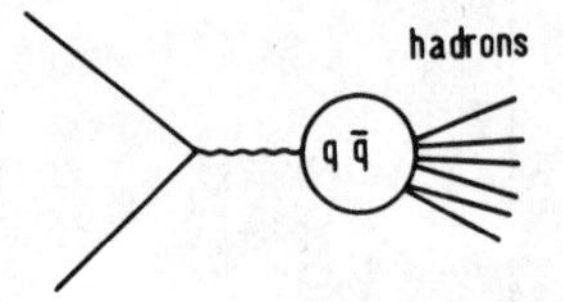

and the cross section $\sigma_h(e^+e^- \to (q\bar{q}) \to \text{hadrons})$ is

$$\sigma_h = \pi \cdot \frac{2J+1}{M^2} \cdot \frac{\Gamma_e \cdot \Gamma_h}{(E_{CM}-M)^2 + \Gamma^2/4} \qquad (25)$$

where

Γ = total width of $q\bar{q}$ state;
Γ_e = width for the decay $(q\bar{q}) \to e^+e^-$;
Γ_h = width for $(q\bar{q}) \to$ hadrons;
M = mass of $q\bar{q}$ state;
E_{CM} = CM energy $\cong 2 \times$ beam energy.

If we integrate Eq. (25) over the resonance, we obtain with $J = 1$

$$\int_{\text{res.}} \sigma_h \, dE_{CM} = \frac{6\pi^2}{M^2} \cdot \frac{\Gamma_h \cdot \Gamma_e}{\Gamma} \tag{26}$$

This quantity can be measured. It yields Γ_e, because $\Gamma_e \ll \Gamma$ and hence $\Gamma_h \approx \Gamma$.

A measurement of $\sigma(e^+e^- \to (q\bar{q}) \to \mu^+\mu^-)$ yields in analogy to Eq. (26).

$$\int_{\text{res.}} \sigma(e^+e^- \to (q\bar{q}) \to \mu^+\mu^-) \, dE_{CM} = \frac{6\pi^2}{M^2} \cdot \frac{\Gamma_e \cdot \Gamma_\mu}{\Gamma} \tag{27}$$

Γ_μ = width for the decay $(q\bar{q}) \to \mu^+\mu^-$.

If we put $\Gamma_\mu = \Gamma_e$($e - \mu$ universality) then from a measurement of the hadronic cross section and a measurement of the decay of the resonance into $\mu^+\mu^-$, one can determine Γ_e and Γ_h from Eq. (26) and Eq. (27).

The same thing can be done for the Υ. Figures 21 and 22 show the Υ and Υ'[39-43,42,43] as seen at DORIS. They allow a precise measurement of the masses and a not so precise measurement of the widths Γ_e (see Table 2).

The total width of the Υ is not yet well known. It requires measurement of the width

$$\Gamma(\Upsilon \to \mu^+\mu^-) = \Gamma_\mu$$

The experiments yield (Table 3).

Table 3

$B_{\mu\mu} = \Gamma_\mu/\Gamma$ %	
2.2 ± 2.0	PLUTO[41]
2.5 ± 2.1	DASP 2[43]
$1.0 \pm {3.4 \atop 1.0}$	DES-HD 2[42]
$2.0 {+2.2 \atop -1.2}$	average

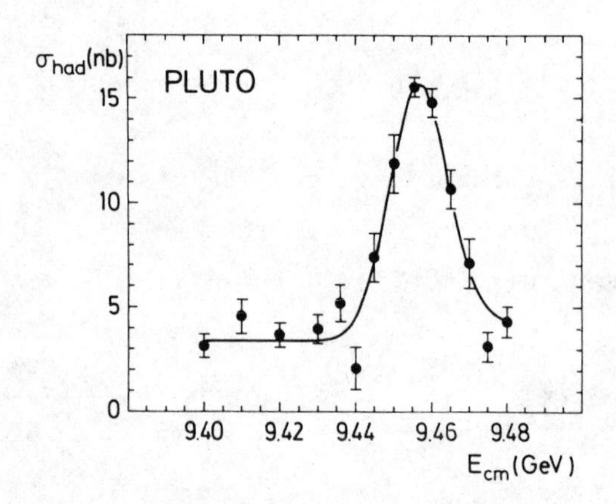

Fig. 21. Hadronic cross section as function of CM energy at Υ resonance.[41]

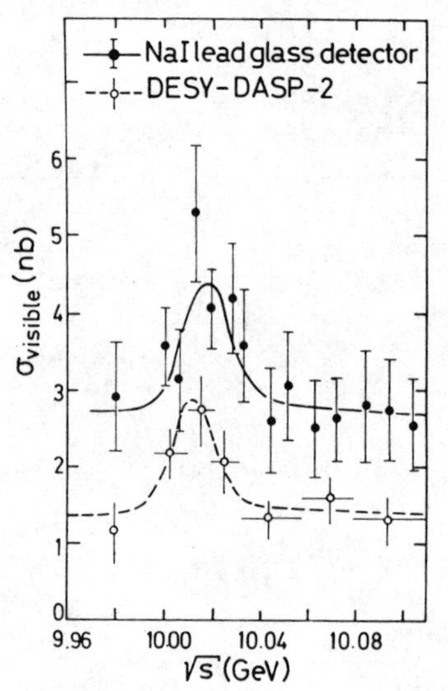

Fig. 22. Visible cross section for annihilation into hadrons as function of CM energy $\sqrt{s}$ at the Υ' resonance (Ref. 42, 43).

From this we have $\Gamma = \Gamma_\mu/B$ and $\Gamma = 65^{+100}_{-34}$ keV which establishes a lower limit for Γ.

The value expected from QCD (see Eq. (35)) is $\Gamma \approx 60$ keV.

From Table 2 we see, that Γ_e for the ground states $\phi(s\,\bar{s})$ and $J/\Psi(c\,\bar{c})$ seems to follow the rule[45-48]

$$\frac{\Gamma_e(J/\Psi)}{Q_c^2} \approx \frac{\Gamma_e(\phi)}{Q_s^2} \tag{28}$$

where Q_c, Q_s = charge of the c,s-quark. In a quarkonium-model this would imply from Eq. (31):

$$\frac{1}{16\pi\alpha^2} \cdot \frac{\Gamma_e}{Q_{c,s}^2} = \frac{|\Psi(0)|^2}{M^2} \approx \text{const} \tag{29}$$

This relation is approximately valid for a wide class of potential models for quarkonium. If we believe, that it also holds for the Υ, this argument favours, because of $\Gamma_e(\Upsilon) = 1.3$ keV, a charge $|Q_b| = 1/3$ over $|Q_b| = 2/3$.

The argument can be considerably sharpened, if the known width Γ_e for the Υ' is used in addition. One gets rather unambiguously[49]

$$|Q_b| = 1/3$$

(see Fig. 23).

The measurements on the Υ and Υ' lead then to the following conclusions:

1) The mass values of Υ and Υ' can be used as input for fits to the mass distribution of muon pairs in the Υ region as measured at the original discovery and subsequently.[50] This leads with a high probability to the existence of a third Υ state Υ' at 10410 MeV.

2) The charge of the quark making the Υ is very likely $\pm 1/3$, therefore one identifies this quark with the bottom (b-) quark with charge $Q_b = -1/3$.

6. Decay of Vector Charmonium States

The states shown in Table 2 have C-parity C = −1, and are therefore coupled to an odd number of photons or gluons (actually, a $q\bar{q}$ − 1^{--} state can decay into any number of gluons $\geqslant 3$).[51] Figure 24 shows the two main ways, in first order, how such a state can decay into hadrons.[52] Diagram (a) proceeds via a one photon intermediate state and leads via quark pair production to a two jet hadron event. Its partial width is

$$\Gamma_a \approx \Gamma_e \cdot R \tag{30}$$

with R ≈ 4 from Eq. (10).

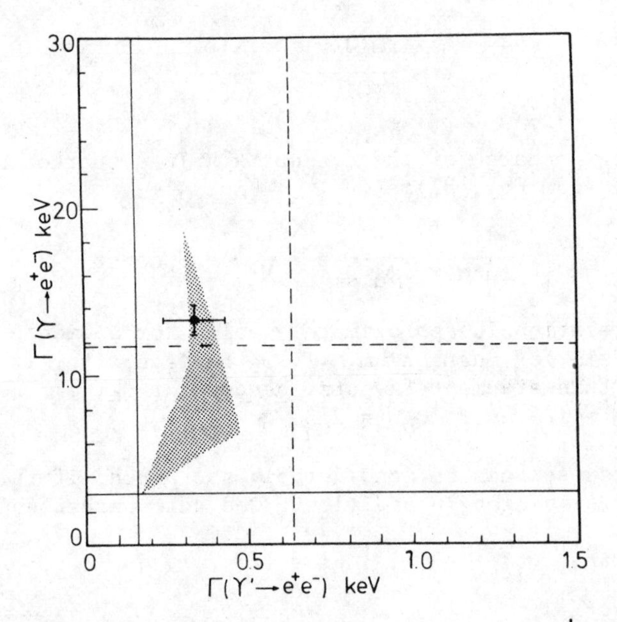

Fig. 23. Decay width $\Upsilon \to e^+e^-$ vs decay width $\Upsilon' \to e^+e^-$. The figure shows lower limits for $\Gamma(\Upsilon \to e^+e^-)$ and $\Gamma(\Upsilon' \to e^+e^-)$ for quark charge 1/3 (full lines) and 2/3 (dashed lines). Experimental point lies in the shaded area given by a series of reasonable potential models for charges 1/3.

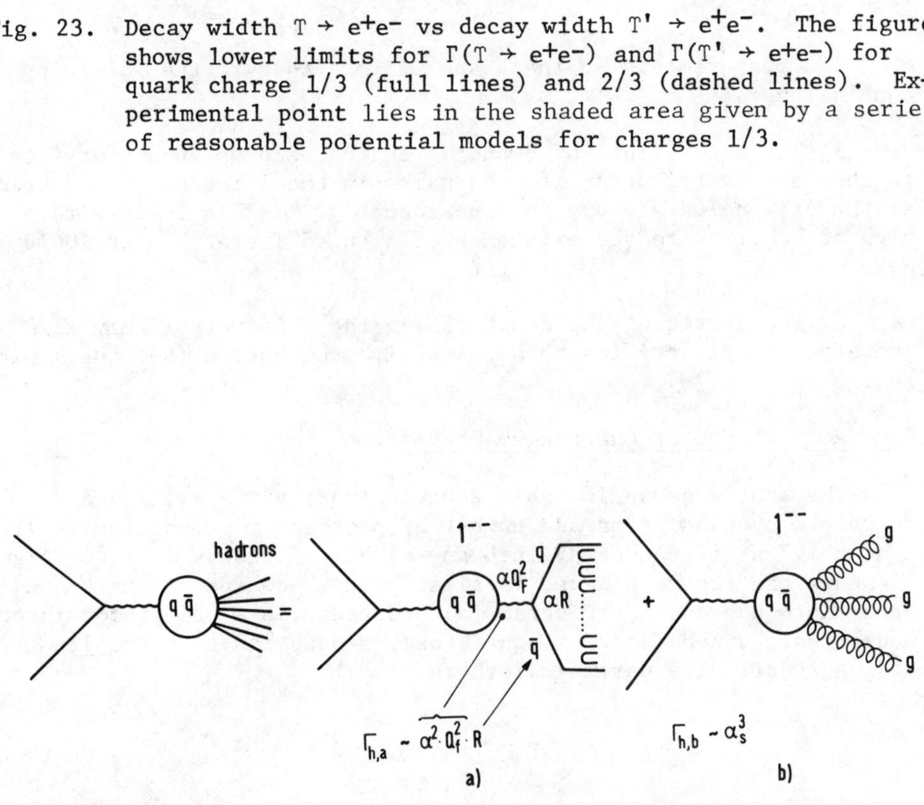

hadrons $q\bar{q}$ = 1^{--} $q\bar{q}$ αQ_F^2 q αR $\bar{q}$ + 1^{--} $q\bar{q}$ g g g

$\Gamma_{h,a} \sim \alpha^2 \cdot Q_f^2 \cdot R$ $\Gamma_{h,b} \sim \alpha_s^3$

a) b)

Fig. 24

Diagram (b) describes a decay via three vector gluons, which results in a three-hadron-jet structure at sufficiently high energy. The width for process (b) can be computed from analogy with three-photon positronium decay, substituting the electromagnetic coupling constant α by the quark-gluon coupling "constant" α_s.

Positronium-like calculations yield[53-55]

$$\Gamma_\mu = \Gamma_e = \Gamma((q\bar{q}) \to \gamma \to e^+e^-) = 16\pi \frac{\alpha^2 Q_q^2}{M^2} |\Psi(0)|^2 \quad (31)$$

Q_q = charge of quark forming the $q\bar{q}$ state,
M = mass of $q\bar{q}$ state,
$\Gamma_b = \Gamma((q\bar{q}) \to 3\,\text{gluons}) \to$

$$= \frac{160(\pi^2 - 9)}{81} \cdot \frac{\alpha_s^3}{M^2} \cdot |\psi(0)|^2 \quad (32)$$

$\Psi(0)$ is the $q\bar{q}$ wave function at the origin. α_s is given by Eq. (1).

The decay width into hadrons is then

$$\Gamma_h = \Gamma_a + \Gamma_b = R\Gamma_e + \Gamma_b \quad (33)$$

If we form Γ_h/Γ_e, the unknown value of $|\Psi(0)|$ drops out, and we have

$$\frac{\Gamma_h}{\Gamma_e} = R + \frac{\Gamma_b}{\Gamma_e} = R + \frac{10(\pi^2 - 9)\alpha_s^3}{81\cdot\pi\cdot\alpha^2\cdot Q_q^2} \quad (34)$$

Equation (34) can be applied to the J/ψ ($c\bar{c}$)-decay: We have $\Gamma_e = 4.8$ keV, $\Gamma = 67$ keV, $\Gamma_h = \Gamma - 2\Gamma_e = 57$ keV, $R = 2.3$ $Q_f = 2/3$. Inserting these values into Eq. (34) we get $\alpha_s(Q^2 = 9.6\ \text{GeV}^2) = 0.19$.

This value of α_s can be compared with the value of α_s given by Eq. (1). Using $Q^2 = M^2_{J/\psi}$, $N_b = 3$, $\Lambda = 0.4$ GeV, we get $\alpha_s \approx 0.34$.

These calculations are probably not very reliable, because of the use of perturbation theory. Still we can try to predict the jet unknown total decay width of the Υ: From Eq. (34) we can compute

$$\Gamma_h = \Gamma_e\left(R + \frac{10(\pi^2 - 9)\alpha_s^3}{81\cdot\pi\alpha^2\cdot Q_q^2}\right) \quad (35)$$

Using $\Gamma_e = 1.3$ keV, $R = 4.5$, $Q_q = -1/3$, and $\alpha_s = 0.19$ yields

$$\Gamma = \Gamma_h + \Gamma_e + \Gamma_\mu + \Gamma_\tau \approx 60\ \text{keV}$$

We also see that

$$\frac{\Gamma_b}{\Gamma_a} = \frac{\Gamma(3g - \text{jets})}{\Gamma(2q - \text{jets})} = \frac{10(\pi^2 - 9)\alpha_s^3}{81 \cdot \pi \cdot \alpha^2 Q^2 \cdot R} \cong 9 \tag{36}$$

Therefore we expect the Υ decays predominantly into three gluons. This should lead to deviations from the two-jet structure and a more isotropic distribution of the secondary hadrons, i.e., to large values of the sphericity S, at the Υ mass. These predictions are indeed borne out experimentally.

7. Charm-Spectroscopy

The $c\bar{c}$ states can be classified according to the oribital angular momentum ℓ between the two quarks (S state $\ell = 0$, P state $\ell = 1$, D state $\ell = 2$, etc.) and according to the total spin S of q and $\bar{q}$ (spins parallel, S = 1, or antiparallel, S = 0).

The states have parity $P = (-1)^{\ell+1}$.

S-states, $\ell = 0$, Spin S = 1: They have total spin J = 1, P = −1, C = −1. They are vector particles, identified with the J/Ψ (total quantum number n = 1) and with the Ψ' (n = 2).

S-states, $\ell = 0$, Spin S = 0: They have $J^{PC} = 0^{-+}$. These states have not been unambiguously identified so far.

P-states, $\ell = 1$, Spin S = 1: We can combine $\ell = 1$, S = 1 to form the following states: $J^{PC} = 2^{++}, 1^{++}, 0^{++}$. These states are identified with the charmonium states shown in Fig. 25.

The quantum number of the state 0^{++} has been determined by its decay properties. The quantum number assignment for the 1^{++} and the 2^{++} is not completely unique, but very likely (see eg., G. Wolf ERICE Lectures 1978). The photon transitions $\Psi' \to \gamma\chi$, $\chi \to \gamma$, J/Ψ are E1 transitions in first order.

The width of the electronic dipole transition is[56]

$$\Gamma(q\bar{q} \to \gamma q\bar{q}) = \frac{4}{3}\alpha Q_f^2 k^3 \cdot |x_{fi}|^2 \tag{37}$$

where k = photon wave number, x_{fi} = matrix element of the dipole operator. This matrix element can be evaluated, if one makes an assumption about the $q\bar{q}$-potential. The "standard potential" normally used is[56]

$$V(r) = -\frac{4}{3}\frac{\alpha_s}{r} + ar \tag{38}$$

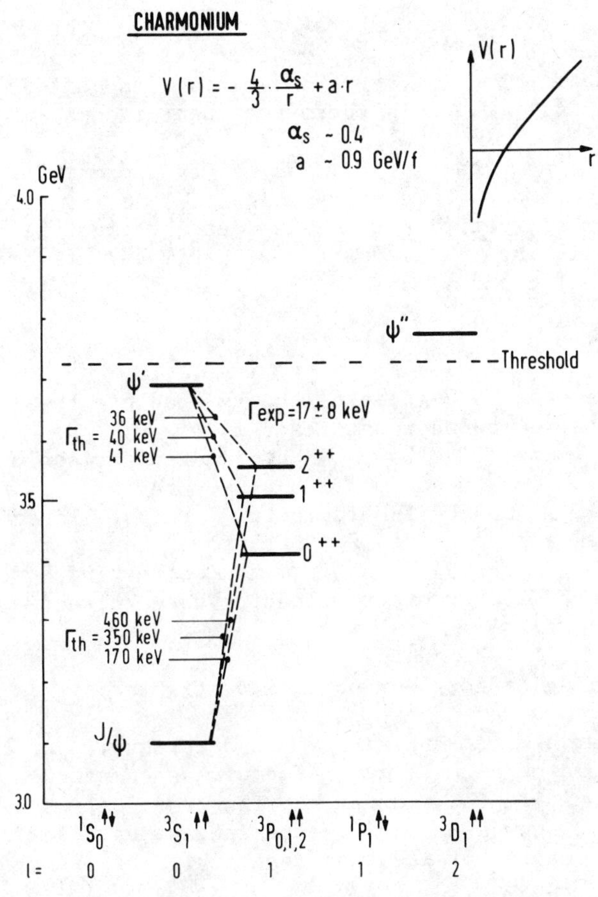

Fig. 25. Charmonium states. The figure indicates the experimentally verified states with their spectroscopic notation and the widths of their photon transitions (from Ref. 51).

with $\alpha_s \approx 0.4$ for charmonium and $a \approx 0.90$ GeV/f.

The first term is a Coulomb-like attraction from one gluon exchange. The second term increases indefinitely at large distances and provides for the confining force. With this standard potential the photon width can be computed and the result agrees to within a factor two with the experiment.[57]

P-states, $\ell = 1$, Spin $S = 0$: This is a $J^{PC} = 1^{--}$ state. It has so far not been unambiguously identified.

D-states, $\ell = 2$, Spin $S = 1$: We can combine $\ell = 2$ and $S = 1$ to total spin $J = 1, 2, 3$. Here we only treat the state $J^{PC} = 1^{--}$.

These are the same quantum numbers as the photon, the state can therefore be made by $e^+e^- \to \gamma_{Virt} \to {}^3D_1$. Its mass can be predicted from the standard potential Eq. (38). It agrees quite closely with the state Ψ'' (3771), which is therefore identified as ${}^3D_1$.

As we see, the potential Eq. (38) gives approximate answers to some simple questions. However, it has grave deficiencies. It lacks deeper theoretical justification. It does not give the correct $\Upsilon - \Upsilon'$ mass splitting. It lacks a spin-orbit part to describe the mass splitting of the P states.

8. Test for Gluons from Quarkonium States

As we have seen in Section 6, theory predicts that the hadronic decays of the narrow bound $q\bar{q}$ states J/ψ, ψ', Υ, Υ' proceed mainly through three gluons (Fig. 24). A detailed comparison of the Υ and Υ' decays with a three-gluon model has been carried out by the PLUTO,[59] DASP II,[60] and DESY-Heidelberg[61] groups. The distributions from PLUTO have been corrected for background coming from decays via $e^+e^- \to \Upsilon \to \gamma \to q\bar{q}$ (Fig. 24a) and from nonresonant background. After corrections about 1200 events remained. These corrected distributions are marked Υ_{dir}.

The following effects have been investigated:

1) Average value of sphericity S and thrust T: On the Υ resonance there should be a three jet structure as compared to the two-jet structure outside the resonance. Therefore on resonance $\langle S \rangle$ and $1-\langle T \rangle$ should show a clear increase indicating a deviation from the two-jet structure outside the resonance. This is clearly seen in the data of Fig. 26a, b.

2) Distribution of sphericity S and thrust T on the resonance: The distribution of S and T can be compared with various theoretical models. A two jet model and an isotropic phase space model disagree with the experimental distributions, whereas a three-gluon model agrees with the data (Fig. 27).

3) Since the three gluons must lie in one plane, one expects a "pancake-like" structure for the momentum distribution of the events. To analyze this, one forms the following quantities, which may be considered a generalized sphericity.

Define a quantity

$$T_{\alpha\beta} = \sum_i (\delta_{\alpha\beta} p_i^2 - p_{i\alpha} p_{i\beta}) \qquad (39)$$

$p_{i\alpha}$ = α-component of i-th particle,
α, β = x, y, z or 1, 2, 3 coordinate.
$p_i^2 = p_{i1}^2 + p_{i2}^2 + p_{i3}^2$.

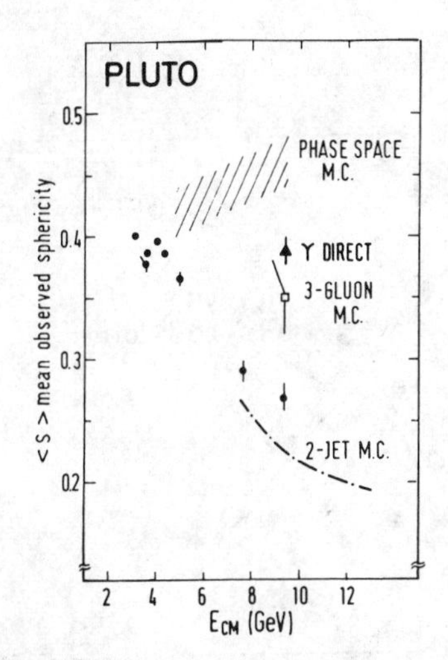

Fig. 26a. Mean sphericity <S> as function of CM energy, showing data at the Υ resonance (background subtracted) compared with the prediction from a three-jet Monte Carlo model and with data around the resonance (Ref. 59).

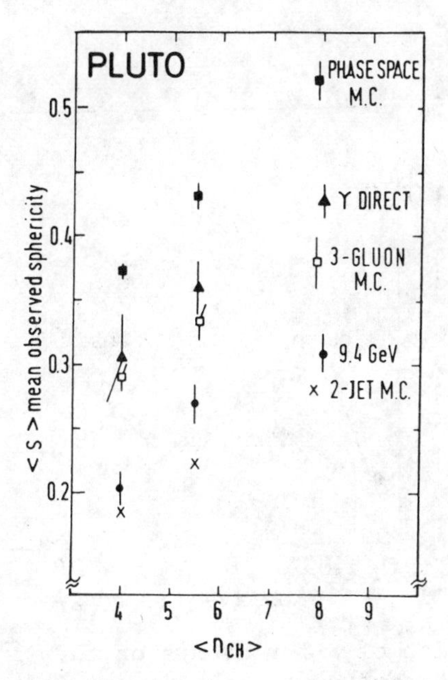

Fig. 26b. Mean sphericity <S> for events on the Υ resonance (Υ direct, background subtracted) and for two-jet-events outside the resonance (9.4 GeV points). One sees, that Υ-events have high <S> independent of the charged multiplicity n_{ch} (Ref. 59).

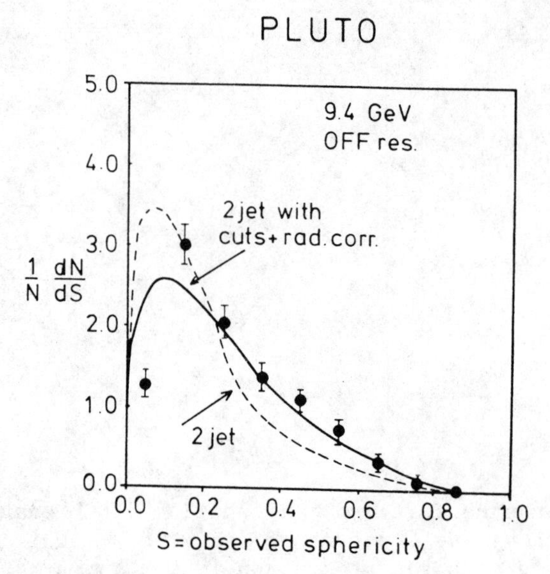

Fig. 27a. Distribution of sphericity S for events outside the Υ resonance. It agrees with a realistic two-jet model (Ref. 59).

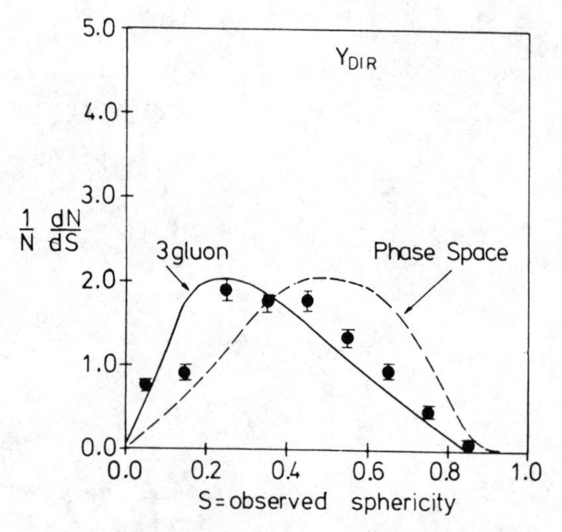

Fig. 27b. Distribution of S for events on the Υ resonance (background subtracted distribution). It agrees with a three-jet model, disagrees with phase space and with a two-jet model (compare Fig. 27a) (Ref. 59).

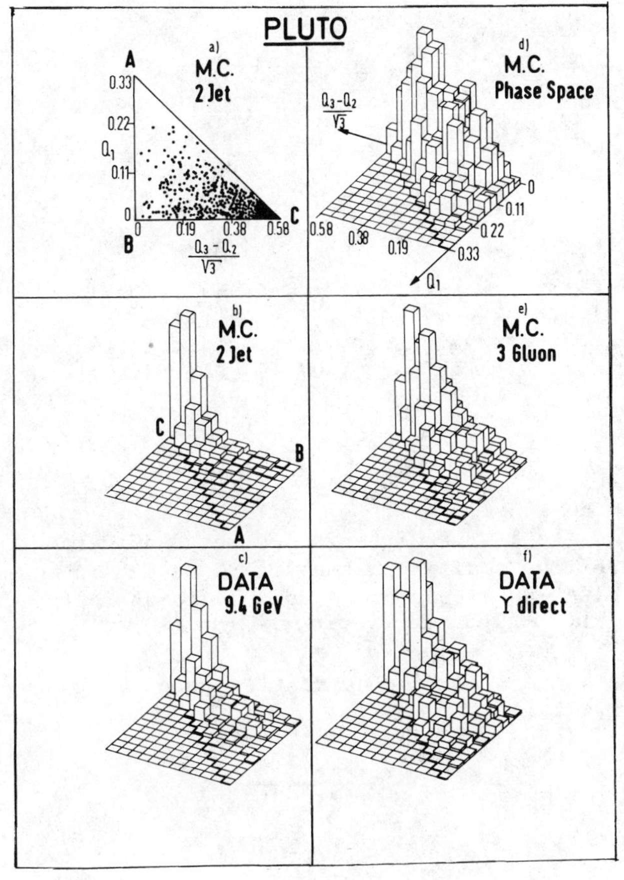

Fig. 28

Find the Eigenvalues λ_1, λ_2, λ_3 of the tensor $T_{\alpha\beta}$ and order them $\lambda_1 \geqslant \lambda_2 \geqslant \lambda_3$.

The values λ_1, λ_2, λ_3 correspond to the three main axis in the momentum space of the event. The direction belonging to λ_3 points towards the minimum value of $T_{\alpha\beta}$. We have with the new directions 1, 2, 3 associated with λ_1, λ_2, λ_3:

$$\frac{\lambda_3}{\Sigma p_i^2} = \frac{\Sigma(p_i^2 - p_{i3}^2)}{\Sigma p_i^2} = \frac{\Sigma(p_{i1}^2 + p_{i2}^2)}{\Sigma p_i^2} = \frac{\Sigma p_{it}^2}{\Sigma p_i^2} = \frac{2}{3}\,S \qquad (40)$$

p_{it} = transverse momentum with respect to the 3-axis. Since λ_3 is the smallest eigen-value, $\lambda_3/\Sigma p_i^2$ can be identified with sphericity (Eq. (13)).

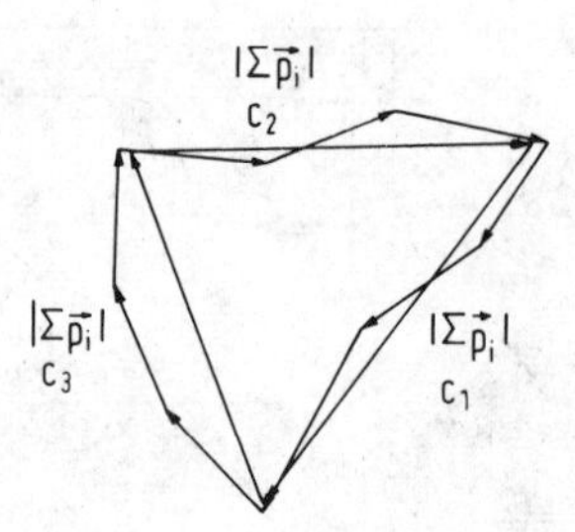

Fig. 29. Definition of triplicity. The momenta of particles (arrows) are ordered into three classes (C_1, C_2, C_3) such that the circumference of the triangle C_1, C_2, C_3 is a maximum. The arrows of C_1, C_2, C_3 give the axis of the three jets.

In contrast to λ_3 the direction corresponding to λ_1 points in the direction of the smallest extent of the event in momentum space. In the three gluon-decay picture it should point in the direction of the normal to the 3-gluon decay plane.

We form now the following quantities relative to the axis defined by λ_1, λ_2, λ_3:

$$Q_k = 1 - \frac{2\lambda_k}{\lambda_1 + \lambda_2 + \lambda_3} = \frac{\Sigma p_{i,k}^2}{\Sigma_i p_i^2} \qquad (41)$$

$$Q_1 \leqslant Q_2 \leqslant Q_3 \qquad (42)$$

$$Q_i + Q_2 + Q_3 = 1$$

Figure 28 shows the combined distributions[59] in Q_1 vs $(Q_2 - Q_3)/\sqrt{3}$ for the data on and off the T resonance, compared with the three models: 2 jets, 3 jets and phase space. Off resonance the data agree with the two-jet model. On resonance the distribution is quite different. It disagrees with a two-jet model. It agrees well with the three gluon model. The phase space model gives a worse description of the data.

4) Triplicity: With this method one tries to find the three-jet directions for each event.[62,63] One forms a quantity, called triplicity T_3. To find it one groups the particles into three non-empty classes C_1, C_2, C_3 such that

$$T_3 = \frac{1}{\Sigma|\vec{p}_i|} \cdot \mathrm{Max}\left[\Big|\sum_{C_1} \vec{p}_i\Big| + \Big|\sum_{C_2} \vec{p}_i\Big| + \Big|\sum_{C_3} \vec{p}_i\Big| \right] \qquad (43)$$

Ordering convention:

$$|\underset{C_1}{\Sigma \vec{p}_i}| \geq |\underset{C_2}{\Sigma \vec{p}_i}| \geq |\underset{C_3}{\Sigma \vec{p}_i}|$$

$T_3 = 1$ for a perfect three-jet, $T_3 = 3\sqrt{3}/8$ for a spherical event. The particles in the three classes C_1, C_2, C_3 are identified with the three gluon jets, the sums $\underset{C_i}{\Sigma \vec{p}_i}$ define the direction of the gluon momenta (Fig. 29). We call

θ_1 = direction between jet 2 and jet 3,
θ_2 = direction between jet 1 and jet 3,
θ_3 = direction between jet 1 and jet 2, and

$$x_1 = \frac{2 \sin \theta_1}{\sin \theta_1 + \sin \theta_2 + \sin \theta_3} \approx \frac{2E_1}{E_{CM}} \tag{44}$$

where E_1 = energy of gluon jet No. 1.

The distributions in triplicity T_3, in x_1, x_3, θ_1, θ_3 can be compared on and off resonance with a three gluon model, a two-jet model and with phase space. Off resonance the two-jet model describes the data. On resonance only the three gluon model describes the data, the other two models are excluded (Figs. 30, 31, 32).

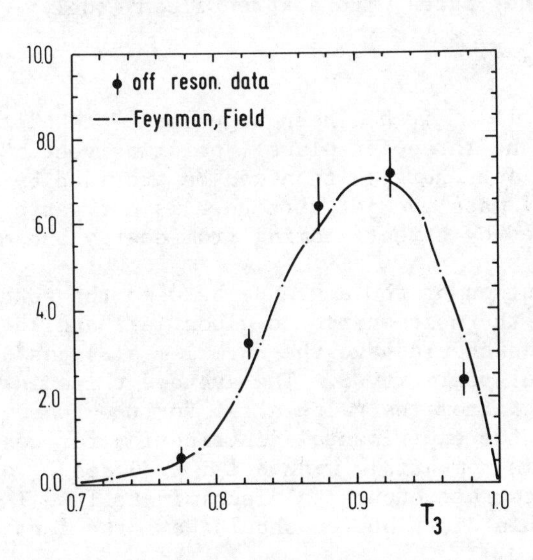

Fig. 30. Distribution of triplicity T_3 for events outside the Υ resonance. They show a two-jet structure and agree with the prediction from a Feynman-Field two-jet model (Ref. 63).

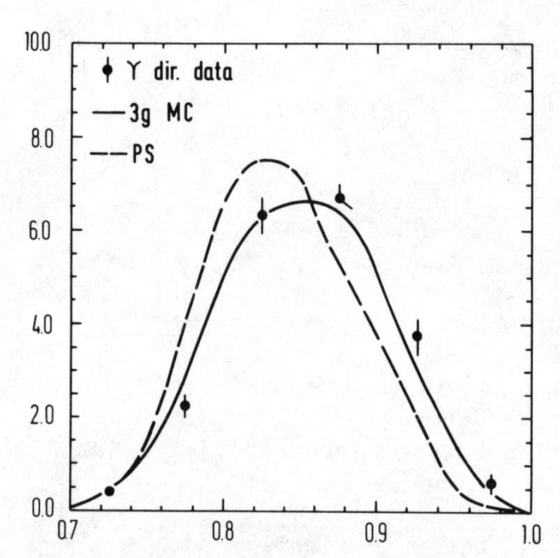

Fig. 31. Distribution of triplicity for events from the ϒ-resonance. The data disagree with expectations from a phase-space model or a two-jet model (compare Fig. 30). They agree with a three-jet model (Ref. 63).

The tests mentioned in (1)-(4) show clearly, that on the ϒ resonance the data are in contradiction with a two-jet or a phase space model. They agree with a three gluon model.

Further tests:

5) Energy flow: it has been suggested[64] to plot the azimuthal energy flow in the three-jet plane ("pointing vector"). This test requires higher jet energies than can be provided by the ϒ decay, because at the ϒ mass the jet structure is not yet pronounced enough and can be masked by effects coming from energy and momentum balance.

6) Distribution of the angle θ_S between the sphericity axis (= direction of the most energetic gluon jet) and the beam axis: This distribution should have the form $1 + a(S) \cos^2\theta_S$ with a(S) depending on the sphericity S. The average value is $\langle a \rangle = 0.39$, clearly different from the value a = 1 for the two quark jet model. Figure 33 shows the experimental distribution for cos θ_S on the ϒ resonance. It is compatible with a $1 + 0.39 \cos^2\theta_S$ distribution, but it is not accurate enough to discriminate from $1 + \cos^2\theta$. Note, however, that this distribution should have the form[66]

$$1 - \cos^2\theta_S$$

if the gluons were scalar, not vector particles. This is clearly excluded by the data.

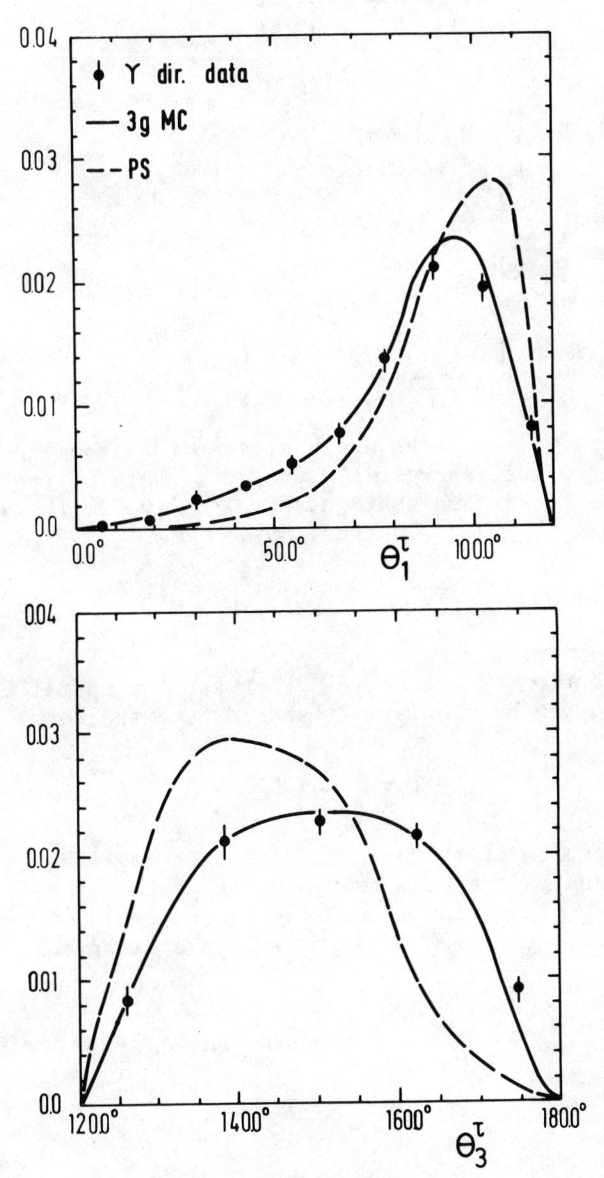

Fig. 32. Distribution of the angles θ_1 and θ_3 between gluon jets (see text) for events from the Υ resonance. The distributions agree with a three gluon jet model, disagree with phase space (Ref. 63).

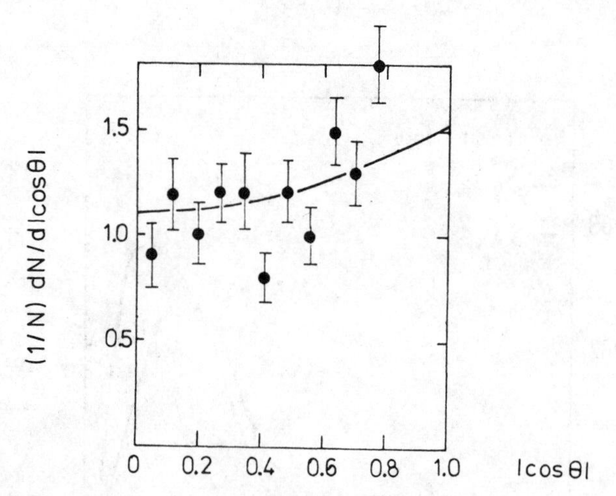

Fig. 33. Distribution of the angle between the incoming e^+e^- beam and the most energetic gluon jet. Data on the Υ resonance, background subtracted. The line is the distribution $1 - 0.39 \cos^2\theta$ (Ref. 59).

7) Distribution of the angle θ_n between the normal to the gluon decay plane and the beam axis: It should have the form[67]

$$1 - \frac{1}{3}\cos^2\theta_n$$

8) Distribution of the azimuth angle χ = angle between: a) the plane defined by the beam axis and the most energetic gluon, and b) the three-gluon plane. Calling θ_S the angle of the most energetic gluon with the beam axis, we expect a distribution of the form[68]

$$\frac{d\sigma}{d\cos\theta_S d\chi} \; 1 + a(S)\cos^2\theta_S +$$

$$\frac{1 - \langle a(S)\rangle}{4}\sin^2\theta_S \cos 2\chi \approx$$

$$1 + 0.39\cos^2\theta_S + 0.15\sin^2\theta_S \cos 2\chi \tag{45}$$

9) Test for the three gluons on the J/Ψ:

The decay $J/\Psi \rightarrow \gamma f$ provides the following check.

The branching ratio for the decay

$$B_{\gamma f} = \frac{\Gamma(J/\Psi \to f\gamma)}{\Gamma(J/\Psi \to \text{hadrons})}$$

has been measured. The result is

$$B_{\gamma f} = (2.0 \pm 0.3)\cdot 10^{-3} \text{ PLUTO}^{69} = (0.9\text{–}1.5)\cdot 10^{-3} \text{ DASP}^{70}$$

This may be compared with the branching ratio for

$$J/\Psi \to f\omega:^{71} \quad B_{\omega f} = (4.0 \pm 1.4)\cdot 10^{-3}$$

Vector meson dominance would predict $B_{\gamma f} \approx \alpha B_{\omega f}$ (Fig. 34b). Since $B_{\gamma f}$ is two orders of magnitude bigger than this prediction, there must be a different decay mechanism for the γf final state. Such a mechanism is provided by a model in which the photon is coupled to the $c\bar{c}$ of the J/Ψ and the 2^{++} meson is coupled to the $c\bar{c}$ via a two-gluon exchange (Fig. 34a).

The production and subsequent decay of the f can be described by three independent helicity amplitudes A_0, A_1, A_2.

Reference 72 presents the experimental result for the ratios. $x = A_1/A_0$ and $y = A_2/A_0$. The most likely value for x, y agrees with a prediction by M. Krammer,[73] but also with a model based on tensor meson dominance.[74]

10) Three-jet analysis according to Wu and Zobernig:[75] This method finds the three-jet axis in a plane by grouping the particles into three classes such that each group minimizes the sum of squares of transverse momenta with respect to the appropriate jet axis chosen for each group.

11) Kaon yield: Gluons couple equally to all flavors. Therefore, the gluon jets should contain (in the limit of very high energy) equal amounts of all flavors. At finite energies mass corrections are required. At the Υ strange particles should easily be produced because of their small mass, and so the cross section for

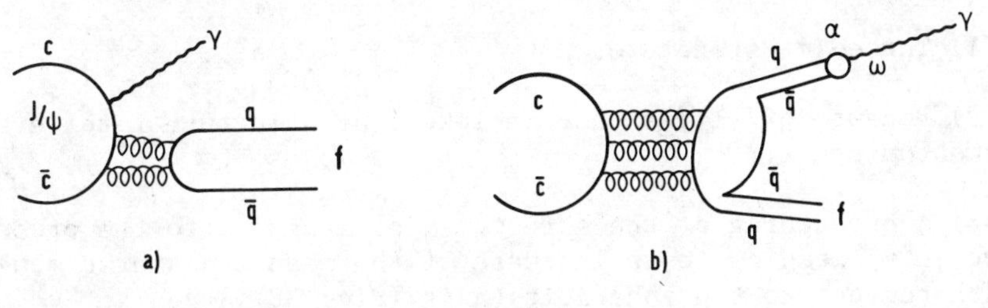

Fig. 34

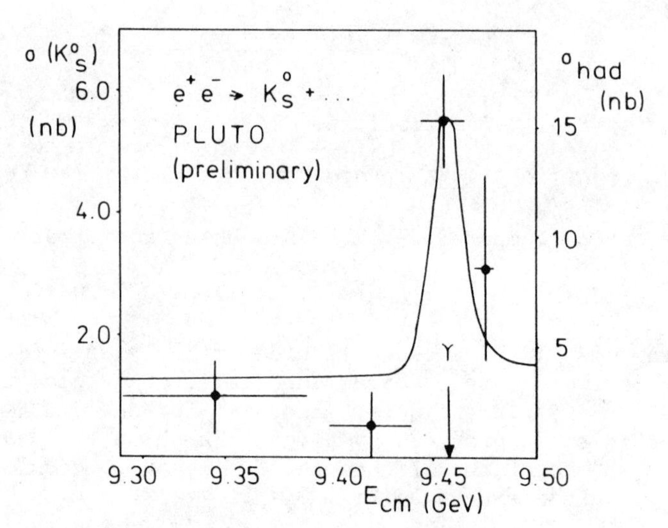

Fig. 35. Inclusive K^o_S yield in the Υ region. The total cross section is shown for comparison (Ref. 38).

strange particles should increase like the cross section of the other hadrons at resonance. This is indeed the case, as Fig. 35 shows for the K^o_S yield.[76] Likewise the $K^\pm$ yield seems to follow the pion yield on the Υ resonance[77] (Fig. 36).

9. Gluon Bremsstrahlung

In first approximation the annihilation of e^+e^- into hadron proceeds via the two quark jet mechanism (Fig. 37a).

Radiation effects like in QED occur if one of the quarks radiates a gluon (Fig. 37b). The probability is of order α_S, as compared to order α in QED. The gluon fragments in a similar way as the quarks. Therefore, at high energies, one expects to see the following effects due to the bremsstrahlung of hard gluons.[78]

1) Three-jet structure.

2) Because of (1), a pancake-like flat structure of the events in momentum space.

3) A broadening of the effective jet cone due to the presence of two jets, leading to an increase of the mean transverse momentum p_t with respect to the sphericity axis (Fig. 38).

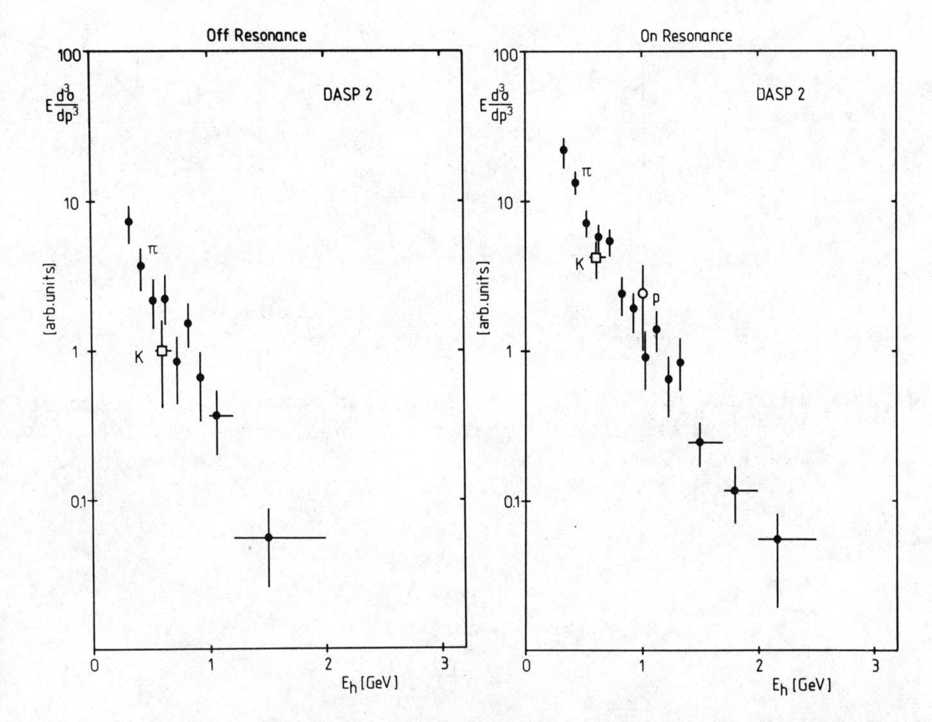

Fig. 36. Invariant cross section $E(d^3\sigma/dp^3)$ vs hadron energy E_h off (left picture) and on (right picture) the Υ resonance. Comparison of pions (full points) and kaons (open points) (Ref. 77).

(a) (b)

Fig. 37

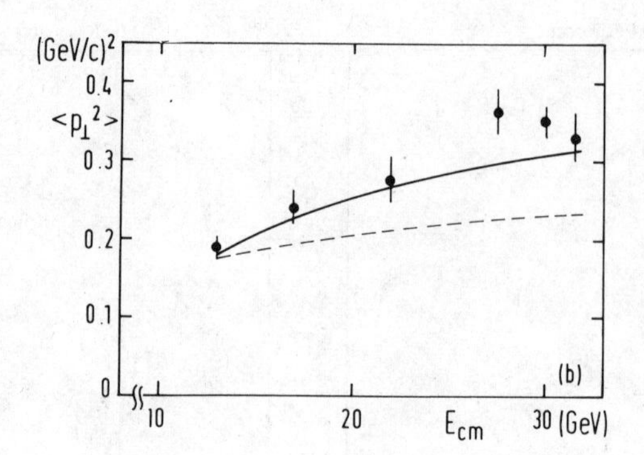

Fig. 38. Increase of $\langle p^2_\perp \rangle$ ($p_\perp$ = transverse momentum with respect to jet axis) with CM energy. Dashed line = two-jet model, full line = two jets with gluon bremsstrahlung. PLUTO group, DESY 79/57.

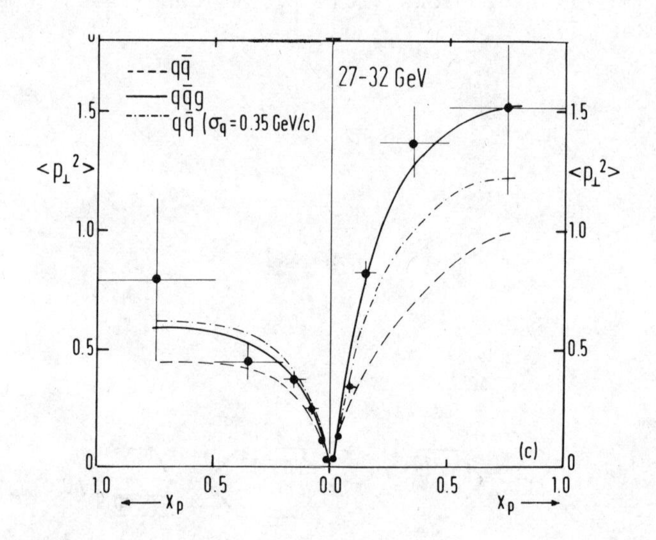

Fig. 39. $\langle p^2_\perp \rangle$ as function of $x_p = 2p/\sqrt{s}$, $\sqrt{s}$ = CM energy = 27-32 GeV dashed line: two-jet model with $\langle p_\perp \rangle$ = 247 MeV/c, dot-dashed line: same, with $\langle p_\perp \rangle$ = 350 MeV/c full line: two jets with gluon bremsstrahlung PLUTO group, DESY 79/57.

4) A quantitive analysis of the effect under 3) shows, that the mean transverse momentum $\langle p_t \rangle$ becomes a strong function of $x_\ell = 2p_\ell/E_{CM}$ (p_ℓ = longitudinal momentum). For $x_\ell \sim 0.4\text{-}0.5$ $\langle p_t \rangle$ rises rapidly with CM energy. This is a very striking effect (Fig. 39).

5) The increase of $\langle p_t \rangle$ with CM energy is due to the occurrence of asymmetric jets: There must be two jet events, having one narrow and one broad jet.[80]

6) There must be characteristic changes in the distributions of thrust and sphericity with CM energy. Also the thrust/sphericity axis no longer has a $1 + \cos^2\theta_S$ distribution, but a distribution of the form $1 + a \cos^2\theta_S$ with $a < 1$.

7) The azimuth angle χ between the $q\bar{q}g$ production plane and the beam axis has a distribution of the form[81]

$$\frac{d\sigma}{d \cos \theta_S \, d\chi} \sim 1 + 0.44 \cos^2\theta_S +$$

$$0.14 \sin^2\theta_S \cos 2\chi + 0.10 \sin^2\theta_S \cos \chi \qquad (46)$$

for thrust $T = 0.75$, θ_S = angle between thrust axis and beam axis.

It must be emphasized, that in principle effects 3), 4), 6) can also occur qualitatively if one crosses a new flavor threshold, and therefore care must be taken in the interpretation.[82]

10. The Heavy Lepton τ

There is now a consistent body of experimental evidence for the τ, all supporting the view, that it is a heavy sequential lepton, i.e., a lepton identical to the e or μ, only differing in mass.[83,1] Below is a table of its properties, taken from a recent compilation of G. Flügge.[1] The following comments are maybe in order:

1) mass: The most accurate data come from DELCO. It is now clear that the τ and the D have different masses.

2) spin: The threshold behavior of the cross section excludes explicity spin 0, 1, 3/2, agrees with spin 1/2. The best check of the pointlike structure of the τ comes from a measurement by PLUTO. They find (for spin 1/2): $\sigma_{\tau\tau}$ ($\sqrt{s}$ = 9.4 GeV) = $(0.94 \pm 0.25)\sigma_{QED}$. At even higher energy the MARK J group has recently also checked the agreement of the cross section $e^+e^- \to \tau^+\tau^-$ with QED.[84]

3) Leptonic decays: A very simple model gives (Fig. 40)

$$B(e\nu\nu) = \frac{\Gamma(\tau^- \to e^-\nu_\tau\bar{\nu}_e)}{\Gamma(\tau^- \to \text{all})} \sim \frac{1}{5}$$

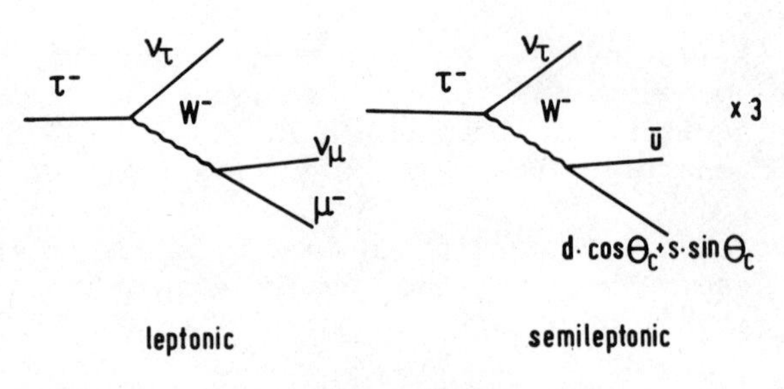

Fig. 40

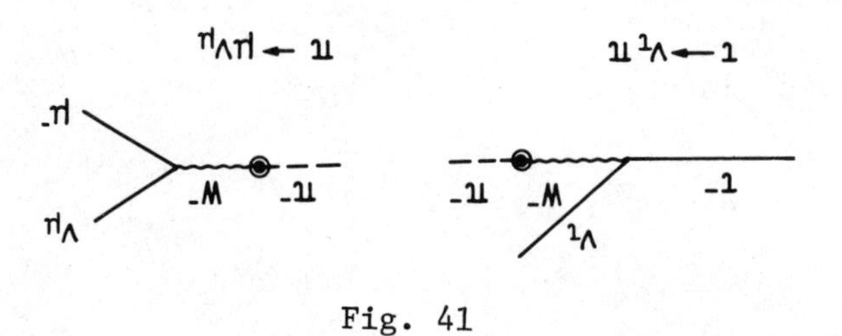

Fig. 41

Simplest version of τ decay. The W^- from the $\tau\ \nu_\tau$ W-vertex couples with all pairs of fermions with about equal strength and therefore each diagram has about the same weight. Final states with $u\bar{d}$ lead to hadrons. The factor 3 is for color. The branching ratio for $\tau^- \rightarrow e^-\nu\nu$ is therefore

$$B(e^-\nu\nu) = \frac{\Gamma(e^-\nu\nu)}{\Gamma(e^-\nu\nu) + \Gamma(\mu\ \nu\nu) + 3\cdot\Gamma(ud\nu)} \approx \frac{1}{5}$$

The experimental values for $B(e\nu\nu)$ and $B(\mu\nu\nu)$ are in accord with theory and with e/μ universality.

4) Semihadronic decays: Predictions for the decays $\nu\rho$ and νA_1 follow from CVC and PCAC. They agree with experiment.

5) Decays $\tau \rightarrow \pi\nu$: Its branching ratio relative to the leptonic decay can be computed unambiguously, because the $W - \pi$ coupling is known (Fig. 41).

This important check is also o.k. with experiment.

6) Strangeness: decay into kaons should be suppressed by $\tan^2\theta_C$ (θ_C = Cabibbo angle). This is consistent with experiment.

7) Unusual decays without neutrinos: the following limits are known for the branching ratios:[1]

B ($\tau \to$ 3 charged particles) <1% PLUTO
B ($\tau \to$ 3 charged leptons) <0.6% SLAC-LBL
B ($\tau \to e\gamma$) <2.6% SLAC-LBL
B ($\tau \to \mu\gamma$) <0.35% MARK II
B ($\tau \to \gamma$) <0.8% MARK II

8) Weak coupling: The best data come from a measurement of the electron spectrum of τ decays by DELCO.[85] The shape of the spectrum can be characterized by the Michel parameter ρ. The experimental value is $\rho = 0.72 \pm 0.15$. It agrees with the expectation $\rho = 0.75$ for V-A, it excludes V + A ($\rho = 0$) and makes very unlikely pure V or A ($\rho = 0.375$).

9) Lifetime of τ: Only an upper limit can be given. This is a pity.

10) τ-neutrino ν_τ: It has not been directly observed. It cannot be identical with ν_μ or $\bar{\nu}_\mu$ otherwise τ would be produced in ν_μ interactions. It cannot be identical with $\bar{\nu}_e$, because this would lead to $B_e \neq B_\mu$. It is not excluded but unlikely that $\nu_\tau \equiv \nu_e$.

11. Two Photon Physics

The simplest approach to two photon physics is through the Weizsäcker-Williams (W.W.-) approximation, where one approximates the electromagnetic field of an electron by a beam of (almost) real photons. For e^+e^- collisions one has then a picture, in which the two photon beams collide. This opens the way to study γ-γ collisions.[86-88] It must be emphasized that the W.W.-approximation becomes very rapidly very bad, if one goes to angles $\gg m_e/E$. It is therefore better to start from the Feynman diagrams (Fig. 42).

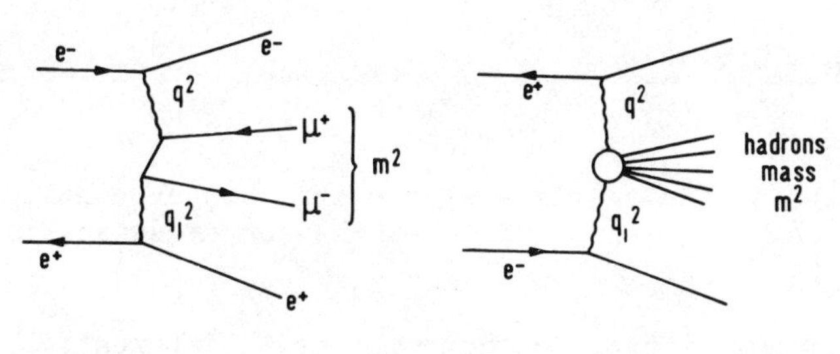

Fig. 42

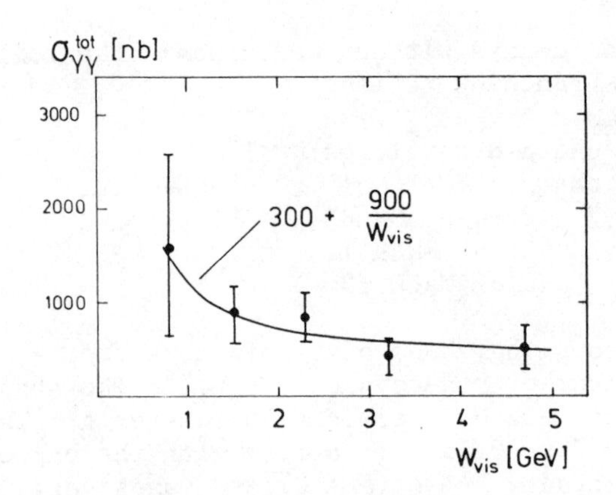

Fig. 43. Total cross section for $\gamma + \gamma \to$ hadrons vs total visible energy of hadrons, PLUTO group (Ref. 38).

Still the W.W.-approximation can offer an intuitive insight into the order of magnitude of the effects: The number of photons in one of the beams is given by

$$N(k)dk = \frac{2\alpha}{\pi} \cdot \ln \left(\frac{2E}{m_e}\right) \cdot \frac{dk}{k} \tag{47}$$

k = photon energy, E = beam energy

The cross section for ee → ee + X is then given by

$$\sigma(eeX) \cong \int N(k_1)dk_1 \int N(k_2)dk_2 \; \sigma_{\gamma\gamma}(X) \tag{48}$$

where $\sigma_{\gamma\gamma}(X)$ is the cross section $\gamma + \gamma \to X$ at the CM energy $W \approx \sqrt{4\, k_1 k_2}$.

For the production of a hadronic state with mass m_H we have approximately

$$\frac{d\sigma(ee \to eeX(m_H))}{dm^2_H} \cong \left(\frac{2\alpha}{\pi}\right)^2 \left[\ln\left(\frac{2E}{m_e}\right)\right]^2 \cdot \frac{\sigma_{\gamma\gamma}(m^2_H)}{m^2_H} \cdot \ln\left(\frac{4E^2}{m_H^2}\right) \tag{49}$$

This cross section rises with energy and eventually becomes very important compared to the annihilation cross section which drops like E^{-2}.

The following investigations may be of interest:

1) Total cross section $\sigma_{\gamma\gamma}(\gamma + \gamma \rightarrow \text{hadrons}) = \sigma_T(\gamma\gamma)$: Here the main part of the cross section may come from the hadron-like nature of the photon. According to vectormeson dominance this part of the photon is described by a mixture of vector mesons. Then the $\gamma\gamma$ cross section follows approximately from factorization arguments:

$$\sigma_T(\gamma\gamma) \approx \frac{\sigma_T{}^2(\ p)}{\sigma_T(pp)} \approx 300\ \text{nb}. \tag{50}$$

A cross section of this order has been seen by the PLUTO group at PETRA[38] (Fig. 43).

2) Production of States with C = + 1: Two photon physics opens a way for the study of these states, which are coupled to two photons. A very important measurement has recently been carried out by the SLAC-LBL group.[89] They have measured the cross section $\sigma(e^+e^- \rightarrow e^+e^-\eta')$ through the decay mode $\eta' \rightarrow \rho\gamma$. Using the branching ratio $B(\eta' \rightarrow \rho\gamma) = 0.298 \pm 0.017$, they find for the width $\Gamma(\eta' \rightarrow \gamma\gamma) = 5.9 \pm 1.6$ KeV ± 20% systematics.

With $B(\eta' \rightarrow \gamma\gamma) = 0.0197 \pm 0.0026$, one obtains a total width for the η':

$$\Gamma(\eta') = 0.3 \pm 0.09\ \text{MeV}.$$

This is in agreement with a recent direct measurement of the η' width,[90] which yields $\Gamma(\eta') = 0.28 \pm 0.10$ MeV.

These measurements are important because they are in accord with a model for the η' made out of Gell-Mann-Zweig fractionally charged quarks, under the assumption that the η' is a SU3 singlet with only small octet mixing and that the singlet and octet decay constants are equal. The model predicts $\Gamma(\eta' \rightarrow \gamma\gamma) = 6$ KeV, which is o.k.[91]

A model with integer charged Han-Nambu quarks is excluded, since it predicts

$$\Gamma(\eta' \rightarrow \gamma\gamma) \approx 26\ \text{KeV}.$$

3) Jet production[92]: The coupling of the photons to a $q\bar{q}$ pair may lead to two quark jets at large angles (Fig. 44).

The occurrence of such jets is a clear prediction of the quark model. It offers a very important way to check the charges of the quarks.[91] This can be done by measuring the ratio of the cross sections $\gamma\gamma \rightarrow \mu^+\mu^-$ and $\gamma\gamma \rightarrow q\bar{q}$, which depend on Q_f^4, Q_f = charge of quark. So we have

$$R_{\gamma\gamma} = \frac{\sigma(e^+e^- \to q\bar{q} \to e^+e^- + 2 \text{ jets})}{\sigma(e^+e^- \to e^+e^-\mu^+\mu^-)} = 3\cdot\sum_f Q_f^4 \qquad (51)$$

where the sum goes over all quark flavor charges.

4) Photon structure function: If one of the two photons is far off the mass shell, one can regard the process as deep inelastic scattering off an electron or photon, and one can then measure the photon structure function. The important point is, that the photon structure function can be computed[93,94] from QCD and therefore a direct check of QCD is possible.

Measurements of $\gamma\gamma$ processes will be difficult, experimentally and from their theoretical interpretation. For example, the WW approximation will only hold in a limited kinematical range. For example, an investigation by C. Carimalo et al. shows,[95] that the WW approximation holds better than 10% only if, for both photons, $\varepsilon < 0.1$, $y > 0.4$ with $y = k/E$, $\varepsilon = q/(E\sqrt{yy'})$. This points towards the importance of tagging the electron under very small angles. Otherwise, the q^2 of the virtual photon will become big, and one has to consider the q^2-dependence of the $\gamma\gamma$-cross section as an additional complication.

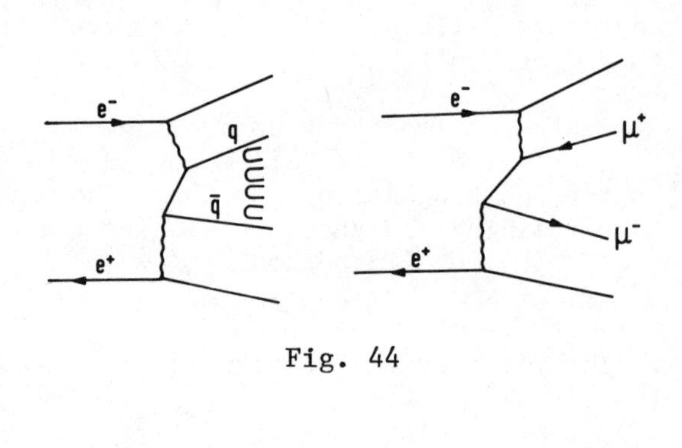

Fig. 44

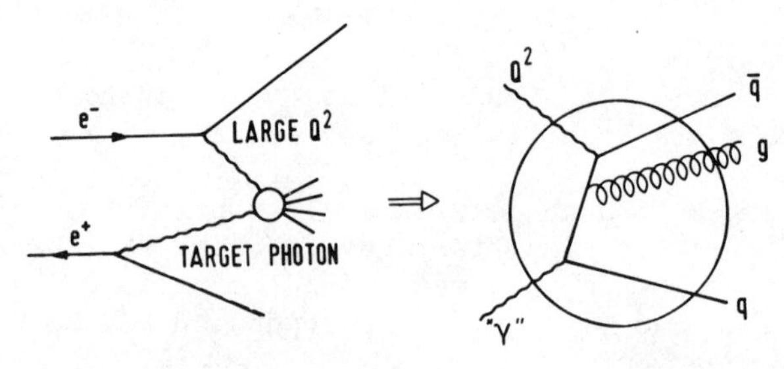

Fig. 45

Finally, radiative corrections still have to be computed.

ACKNOWLEDGEMENTS

It is a pleasure to acknowledge the very efficient help of Dr. G. Mikenberg and of Miss G. Wolter in preparing this report.

REFERENCES

1. G. Flügge, Z. Physik C, Particles and Fields 1, 121 (1979).
2. M. L. Perl, Proc. Int. Symp. on Lepton and Photon Int., Hamburg (1977). G. J. Feldman, SLAC-PUB-2138.
3. H. Harari, Phys. Rep. 42C, 235 (1978). See also Review by A. Zichichi, Lecture at 4th General Conf. of the EPS, York (1978).
4. H. Fritzsch and K. H. Streng, Phys. Lett., 77B, 299 (1978).
5. D. J. Gross and J. Wilczek, Phys. Rev. Lett., 30, 1343 (1973). H. D. Politzer, Phys. Rev. Let., 30, 1346 (1973). H. Georgi and H. D. Politzer, Phys. Rev., D 14, 1829 (1976). In Eq. (1) Q^2 = "the characteristic Q^2 of the interaction."
6. B. A. Gordon et al., Phys. Rev. Lett., 41, 615 (1978). P. C. Bosetti et al., Nucl. Phys., B 142, 1 (1978). J. G. H. de Groot et al., Phys. Lett., 82B, 292, 456 (1979).
7. A. De Rujula and H. Georgi, Phys. Rev., D13, 1296 (1976). The corrections to next higher order in α_S have been computed by M. Dine and J. Sapirstein, Phys. Rev. Lett., 43, 668 (1979). See this paper for more references to Eq. (10).
8. PLUTO group, presentation by U. Timm, DESY 77/52.
9. C. Bacci et al., Phys. Lett., 86B, 234 (1979). G. P. Murtas, Recent work at Frascati, Proc. Int. Conf. on High Energy Phys., Tokyo (1978). J. Perez-Y-Jorba, Recent Work by DCI, ibid.
10. DASP Collaboration DESY 78/18, Phys. Lett., 76B, 361 (1978).
11. J. Siegrist et al., Phys. Rev. Lett., 36, 700 (1976).
12. J. Burmester et al., Phys. Lett., 66B, 395 (1977).
13. Preliminary results have been communicated by J. Kirkby and by E. Bloom, Int. Conf. on Lepton and Photon Int. at high Energy, FNAL (1979).
14. PLUTO group, DESY 79/11.
15. J. D. Bjorken and S. D. Brodsky, Phys. Rev., D1, 1416 (1970).
16. E. Fahri, Phys. Rev. Lett., 39, 1587 (1977), see also S. Brandt et al., Phys. Lett., 12, 57 (1964).
17. H. Georgi and M. Machacek, Phys. Rev. Lett., 39, 1237 (1978), see also A. De Rujula et al., Nucl. Phys. B., 138, 387 (1978).
18. R. F. Schwitters et al., Phys. Rev. Lett., 35, 1230 (1975). G. Hanson et al., Phys. Rev. Lett., 35, 1609 (1975).
19. Ch. Berger et al., Phys. Lett., 78B, 176 (1978), DESY 78/39.
20. PLUTO group, DESY 79/11.

21. G. Wolf, DESY 79/41.
22. S. D. Drell, D. Levy, and T. M. Yan, Phys. Rev., 187, 2159 (1969), D1, 1035, 1617, 2402 (1970).
23. J. D. Bjorken, Phys. Rev., 179, 1547 (1969).
24. G. Hanson, Talk given at the 1976 Tblisi Conf., SLAC-PUB-1814 (1976).
25. PLUTO Collaboration: Ch. Berger et al., Phys. Lett., 81B, 410 (1979).
26. TASSO Collaboration: R. Brandelik et al., Phys. Lett., 83B, 261 (1979).
27. DASP Collaboration: R. Brandelik et al., Nucl. Phys., B 148, 189 (1979).
28. SLAC-LBL: R. F. Schwitters, 1975 Stanford Conf.
29. This was pointed out by G. Wolf, DESY 79/41.
30. R. Brandelik et al., Nucl. Phys., B 148, 189 (1979), Phys. Lett., 67B, 358 (1977).
31. PLUTO Collaboration: A. Bäcker, Thesis Gesamthochschule Siegen (1977) and Ref. 36.
32. T. Atwood et al., Phys. Rev. Lett., 35, 704 (1975). H. F. W. Sadrozinski, Proc. Int. Symp. on Lepton and Photon Int., Hamburg (1977).
33. G. Berdin et al., Nucl. Phys., B 120, 45 (1977). See also Review by J. Perez-Y-Jorba and F. M. Renard, Phys. Reports, 31C, 1 (1977), and by G. J. Feldman and M. L. Perl, Phys. Reports, 33C, 285 (1977).
34. P. A. Rapidis et al., Phys. Lett., 84B, 509 (1979).
35. S. L. Glashow, J. Iliopoulos and L. Maiani, Phys. Rev., D2, 1285 (1970).
36. G. Knies, Int. Symp. on Lepton and Photon Int., Hamburg (1977).
37. Particle Data Group, Phys. Lett., 75B (1978).
38. Review by G. Flügge, DESY 79/26.
39. Ch. Berger et al., Phys. Lett., 76B, 243 (1979).
40. C. W. Darden et al., Phys. Lett., 76B, 246 (1979).
41. Ch. Berger et al., DESY 79/19, Zt. Physik C, Particles and Fields, 1, 343 (1979).
42. J. K. Bienlein et al., Phys. Lett., 78B, 360 (1978).
43. G. W. Darden et al., Contribution to the Int. Conf. on High Energy Physics, Tokyo (1978), page 793, presentation by G. Flügge.
44. SLAC-LBL group, Contribution to the Int. Conf. on Lepton and Photon Interactions at High Energies, FNAL (1979).
45. D. R. Yennie, Phys. Rev. Lett., 34, 219 (1975).
46. F. E. Close, D. M. Scott, and D. Sivers, Phys. Lett., 62B, 213 (1976).
47. T. Walsh, DESY 76/13.
48. G. J. Gounaris, Phys. Lett., 72B, 91 (1977). M. Grecco, Phys. Lett., 77B, 84 (1978).
49. J. L. Rosner, C. Quigg, and H. B. Thacker, Phys. Lett., 74B, 350 (1978).

50. S. W. Herb et al., Phys. Rev. Lett., 39, 252 (1977). W. R. Innes et al., Phys. Rev. Lett., 39, 1240 (1977). K. Ueno et al., Phys. Rev. Lett., 42, 486 (1979).
51. M. Krammer and H. Krasemann DESY 79/20.
52. R. Barbieri and R. Gatto, Phys. Lett., 74B, 225 (1978), contains references to earlier work.
53. H. Pietschmann and W. Thirring, Phys. Lett., 21, 713 (1966). R. Van Royen and V. F. Weisskopf, Nuovo Cim., 50, 617 (1967), ibid., 51, 583 (1967).
54. L. B. Okun and M. B. Voloshin, Preprint ITEP 152, Moscow (1976). V. A. Novikov et al., Phys. Rep., 41C (1978).
55. A. Ore and J. L. Powell, Phys. Rev., 75, 1696 (1949), for a Review see also Ref. 51.
56. See Review Ref. 51.
57. This factor two discrepancy may be a difficulty of the model: C. Quigg, Int. Conf. on Lepton and Photon Int. at High Energy, FNAL (1979).
58. K. Gottfried, Review talk at the Int. Symp. on Lepton and Photon Int. at High Energies, Hamburg (1977).
59. Ch. Berger et al., DESY 78/71 and Phys. Lett., B82, 449 (1979).
60. C. W. Darden et al., DESY Internal Report F 15-78/01.
61. J. K. Bienlein et al., DESY 79/39.
62. S. Brandt and H. D. Dahmen, Z. Physik C, Particles and Fields, 1, 61 (1979).
63. PLUTO Collaboration: Presentation by S. Brandt to the Int. EPS Conf. on High Energy Phys., Geneva (1979), DESY 79/43.
64. A. De. Rujula et al., Nucl. Phys., B138, 387 (1978).
65. K. Koller, H. Krasemann, and T. F. Walsh, Z. Physik C, Particles and Fields, 1, 71 (1979).
66. K. Koller and H. Krasemann, DESY 79/52.
67. M. Krammer and H. Krasemann, DESY 78/66.
68. Ref. 65. Contains a printing error in Eq. (15). See K. Koller and T. Walsh, DESY 78/16, Eq. (10').
69. G. Alexander et al., Phys. Lett., 72B, 493 (1978).
70. R. Brandelik et al., Phys. Lett., 76B, 361 (1978).
71. J. Burmester et al., Phys. Lett., 72B, 135 (1978).
72. G. Alexander et al., Phys. Lett., 76B, 652 (1978).
73. M. Krammer, Phys. Lett., 74B, 361 (1978).
74. W. Gampp and H. Genz, Phys. Lett., 76B, 319 (1978).
75. S. L. Wu and G. Zobernig, Z. Phys. C. Particles and Fields, 2, 107 (1979).
76. PLUTO Collaboration, shown in Ref. 38.
77. C. W. Darden et al., Phys. Lett., 80B, 419 (1979).
78. J. Ellis et al., Nucl. Phys., B111, 253 (1976), ibid. B130, 516 (1977). A. de Rujula et al., Nucl. Phys., B138, 387 (1978). T. A. De Grand, Yee Jack Ng and S. H. Tye, Phys. Rev., D16, 3251 (1977). P. Hoyer et al., DESY 78/21.
79. G. Kramer and G. Schierholz, Phys. Lett., 82B, 108 (1979).
80. P. Hoyer et al., Ref. 78.

81. G. Kramer, G. Schierholz, and J. Willrodt, Phys. Lett., 79B, 249 (1978). Phys. Lett., 80B, 433 (1979), Erratum.
82. A. Ali et al., Z. Physik C, Particles, and Fields, 1, 203 (1979).
83. Review by G. J. Feldman, Int. Conf. on High Energy Physics, Tokyo (1978), see also Ref. 1.
84. MARK-J Group, Talk given by H. Newman at the Int. Conf. on Lepton and Photon Int. at High Energy, FNAL (1979).
85. W. Bacino et al., Phys. Rev. Lett., 42, 6, 749 (1979).
86. F. E. Low, Phys. Rev., 120, 582 (1960). F. Calogero and C. Zemach, Phys. Rev., 120, 1860 (1960).
87. A. Jaccarini et al., Lett. Nuovo Cim., 4, 933 (1970). N. Arteaga-Romero et al., Phys. Rev., D3, 1569 (1971).
88. S. J. Brodsky, T. Konoshita, and H. Terazawa, Phys. Rev. Lett., 25, 972 (1970), and Phys. Rev., D4, 1532 (1971).
89. G. S. Abrams et al., Phys. Rev. Lett., 43, 475 (1979).
90. D. M. Binnie et al., Phys. Lett., 83B, 141 (1979).
91. H. Suura, T. F. Walsh, and B. L. Young, Nuovo Cim. Lett., 4, 505 (1972).
92. S. J. Brodsky et al., Phys. Rev. Lett., 41, 672 (1978).
93. S. J. Brodsky, SLAC-PUB-2240.
94. E. Witten, Nucl. Phys., B120, 189 (1977).
95. C. Carimalo, P. Kessler and J. Parisi, Laboratoire de Physique Corpusculaire, Preprint 79-06.

DISCUSSIONS

CHAIRMAN: Prof. E. Lohrmann

Scientific Secretaries: H. Lierl, G. Mikenberg, and F. Vannuci

DISCUSSION 1

- Ting:

Is there any claim for a structure in R in the region E_m = 1.6 GeV?

- Lohrmann:

Some topological cross sections show some structure in that region. It is, however, of limited statistical accuracy.

- Ting:

Is there any confirmation of the resonance with M = 1.1 GeV found at DESY through the interference effect with the reaction $\gamma p \rightarrow e^+e^-p$?

- Bertolucci:

There is no e^+e^- machine working in that energy region, except for VEP-2M and it has not yet produced results.

- Vannucci:

On the basis of the naive quark model would not one expect R = 1.7 at $E_m \lesssim$ 1 GeV?

- Lohrmann:

It is very hard to believe that the naive quark model should work at such low energies. Moreover, QCD corrections become extremely large and hard to calculate for $E_{cm} \lesssim 1$ GeV.

- Vannucci:

It is unfair to compare the SLAC-LBL R measurements in the region $4.0 \leqslant E \leqslant 4.3$ GeV with the PLUTO and DASP data, since the former has not assumed resonances at 4.03 and 4.16 GeV in its radiative corrections, while the other experiments have.

- Mikenberg:

The structure at $E_{cm} = 4.16$ GeV appears in the DASP data before applying radiative corrections. Those corrections only enhance the effect.

- Jaffe:

R measurements usually present point to point agreements that are much better than what one would expect in term of their statistical errors. Could one have the data presented with only its statistical error together with a statement on its systematic error?

-Lohrmann:

The errors usually shown are statistical errors only. On top of these errors one has to add a normalization uncertainty, which is 10% to 15%.

DISCUSSION 2

- Paffuti:

Is the Gribov-Lipatov relation between the proton structure function as measured in deep inelastic electron scattering and the p scaling function measured in e^+e^- satisfied?

- Lohrmann:

The DASP collaboration finds that the $\bar{p}$ yield obtained in e^+e^- is a factor 2-3 higher than what can be inferred from the deep inelastic scattering results. The authors suggest an explanation due to contributions from decays of N* states (see Ref. 30).

- Preparata:

What happened to the X(2.83) and the 3.45 states?

- Lohrmann:

The crystal ball experiment at SLAC is casting serious doubts on the existence of the 2.8 state. This experiment will continue taking data and a more definite answer will be given in the future. The evidence for the state at 3.45 GeV is now very weak. Probably it does not exist.

- Aubert:

There is another discrepancy between the crystal ball and the DASP results in so far as the decay $J/\psi \rightarrow \gamma\eta'$ which is found to be 3 times higher in the SLAC experiment.

- Gavela Legazpi:

What is the situation about the 3 jet analysis of the Υ decays?

- Lohrmann:

The data are compatible with all tests for 3 jet decays. However, they do not offer compelling proof for a 3 jet decay of the Υ.

- Gavela Legazpi:

There is some contradiction between different theoretical predictions of the D lifetime. What does the statement of experimental agreement with calculation mean?

- Lohrmann:

The present experimental results are not good enough to discriminate between different calculations. There is agreement in the order of magnitude.

- Di Liberto:

In the emulsion experiment WA17 at CERN we found four charged charm decays (1 baryon) for which the lifetime is in the range $.9-7 \cdot 10^{-13}$ sec.

- Shimada:

Is there any charge correlation between jets observed in e^+e^- reactions?

- Lohrmann:

The effects have been calculated by K. Koller, H. Krasemann, P. Zerwas, and T. Walsh, Zt. Physik, C1 (1979), 71. They may not be easy to observe because one needs large statistics.

- Vannucci:

Can not the apparent scaling violation at small x in inclusive hadron production be attributed to the opening of the charm threshold?

- Lohrmann:

The DASP collaboration has shown that, in fact, the scaling violation can be attributed to the crossing of the charm threshold.

- Koh:

Does the sphericity distribution at the Υ mass agree with the theoretical prediction of the process $\Upsilon \rightarrow$ gluons?

- Lohrmann:

The PLUTO data are consistent with a Monte Carlo calculation of the process $T \to 3$ gluons, with the gluons fragmenting subsequently into hadrons.

- Dydak:

How do you know how gluons fragment?

- Lohrmann:

It is assumed that the fragmentation functions for quarks and gluons are similar. This assumption can in principle be tested by looking at the distribution of p_{out}, the momentum component normal to the 3-gluon plane. It seems that the assumption is not too badly wrong.

- Mikenberg:

Scaling violations are expected to be seen at $x_p \gtrsim .1$, since those measurements were done at $E_{cm} \sim 4$ GeV and therefore this region in x_p corresponds to $p \sim 200$ MeV, comparable to meson masses. Above 4.5 GeV the DASP collaboration finds scaling for different particle types down to $x \geqslant .1$.

DISCUSSION 3

- Adkins:

What is the gluon fragmentation function that was used for the model calculation of $T \to 3$ gluons?

- Lohrmann:

The model used the same fragmentation functions as for quarks, namely the Feynman-Field precription.

- Preparata:

The fragmentation of the gluon should not be taken the same as that of quarks because quarks need to pick up a pair from the vacuum, while gluons need two such pairs, thus leading to a softer fragmentation function for the gluons. Moreover, the last quarks in the quark fragmentation chain must be put ad hoc in order to conserve energy, momentum and flavor in a Monte Carlo event. Since most of the information is carried by the high x particles this procedure is adequate for a comparison with experiment. This, however, is not acceptable for a gluon fragmentation.

- Aubert:

What is the triplicity angular distribution at the Υ and how does it compare with the 3 gluon decay?

- Lohrmann:

The comparison has been made and there is very good agreement with the 3 gluon Monte Carlo.

- Vannucci:

Is there an upper limit for the cascade decay of the Υ' into the Υ?

- Lohrmann:

The number of events collected at the Υ' is too small for such studies.

- Ginsparg:

Will the energy flow analysis improve for a heavier quark system decaying into 3 jets?

- Lohrmann:

Phase space shows the same mimicking of a 3 jet structure in the energy flow distribution even for a quarkonium of 30 GeV. However, more refined analysis like the one of Wu and Zobernig should certainly show the 3 gluon structure at those energies.

- Nicolaidis:

Is the Weizsäcker Williams approximation valid for the experimental set-up for the detection of $\gamma\gamma$ processes at PETRA?

- Lohrmann:

Carimalo, Kessler, and Parisi (Ref. 95) have checked the validity of this approximation, and the data here presented are in the region where this approximation is shown to be valid, at least within the present experimental accuracy.

- Louis:

What fraction of the toponium decays via 3 gluons into hadrons?

- Lohrmann:

If one believes in QCD predictions a smaller branching ratio for the decay into 3 gluons is expected for the toponium as compared to the Υ case, due to higher charge of the t quark.

- Oliensis:

Can one observe the states of ortho-τ onium ($\tau^+\tau^-$ bound state) produced in e^+e^- just below the $\tau^+\tau^-$ production threshold?

- Zichichi:

It is a very interesting idea. One could think of detecting this state into its 3 photon decay.

- Paton:

Is beam polarization possible at high energies?

- Lohrmann:

At PETRA the polarization time is expected to be of the order of 30 minutes at 15 GeV beam energy. There are, however, depolarizing effects which may be difficult to control.

- Proudfoot:

The value of Λ (80 MeV) obtained in the Υ decay is considerably lower than that observed in deep inelastic scattering.

- Lohrmann:

The value of Λ = 80 MeV is a rough approximation, because the expression for the 3 gluon decay may be quite inaccurate.

- Ginsparg:

Is there any chance of unambiguously identifying the ν_τ?

- Dydak:

Such an experiment has been attempted at CERN by dumping 10^{18} protons on a target. To identify the ν_τ one has to observe the hadronic decay of the τ. This is hard to disentangle from neutral current events in a high mass detector and therefore one has to use a bubble chamber where the event rate is extremely small.

- Preparata:

What subjects have still to be cleared up in the SPEAR-DORIS energy range?

- Lohrmann:

A better R measurement in the 4 GeV region should be performed in order to iron out the discrepancies between the various measurements. Charmed baryons should be observed in e^+e^- collisions and there is more work to be done in order to complete the spectrum of charmonium.

FIRST RESULTS FROM PETRA

Samuel C. C. Ting

Massachusetts Institute of Technology

Cambridge, Massachusetts

PETRA

PETRA[1,2] is a e^+e^- storage ring located in Hamburg, Germany. It has a circumference of 2.3 km (see Table 1 and Figure 1). The bending radius of 256 m, and it has 8 straight sections. Four of the straight sections are already occupied by detectors; two others, the north and the south, are used for rf halls; the remaining two will be used for future experiments. The injection energy is 6.5 GeV, and the maximum design energy is 19 GeV per beam.

To accelerate the beams from injection to a higher energy, and to compensate for energy loss due to synchrotron radiation, four pairs of klystrons are used; each pair supplies 1.2 MW, totaling 4.8 MW. The energy is transferred to the beams by 32 rf cavities. 32 more cavities will be installed in the near future. The rf frequency is 500 MHz.

The vacuum pipe has water cooling on one side to remove the heat from synchrotron radiation, and distributed pumps of the other side to provide a vacuum of $\sim 10^{-19}$ mbar.

The focusing structure is FODO: focusing, nonfocusing, defocusing, nonfocusing. Before an interaction region, the final bending is done by two bending magnets as shown in Figure 2. The

Table 1. Some properties of PETRA

ENERGY	5 - 19 GeV
MAX. LUMINOSITY (10 m Interaction Region)	$10^{32} cm^{-2} s^{-1}$ (Naive) Now $10^{30} cm^{-2}$/sec !!!
CIRCUMFERENCE	2304 m
BENDING RADIUS	256 m
FREE LENGTH FOR EXPERIMENTS	15 m
FOCUSING STRUCTURE	F O D O
Qx / Qz AT 23 GeV	27.14 / 23.11
β - VALUES AT INTERACTION POINTS	3.00 / .15 m
rf - FREQUENCY	500 MHz
rf - POWER	4.8 MW
LENGTH OF rf-STRUCTURE	96 m (64 cavities)
SINGLE BUNCH CURRENT (MAX.)	20 mA
NUMBER OF BUNCHES	2 x (1 to 4)
INJECTION ENERGY	7 GeV
AUTHORIZATION	OCTOBER 20, 1975
FIRST STORED BEAMS	SEPTEMBER 1978
BEGIN OF EXPERIMENTS	APRIL 1979
COSTS	< 100 MDM

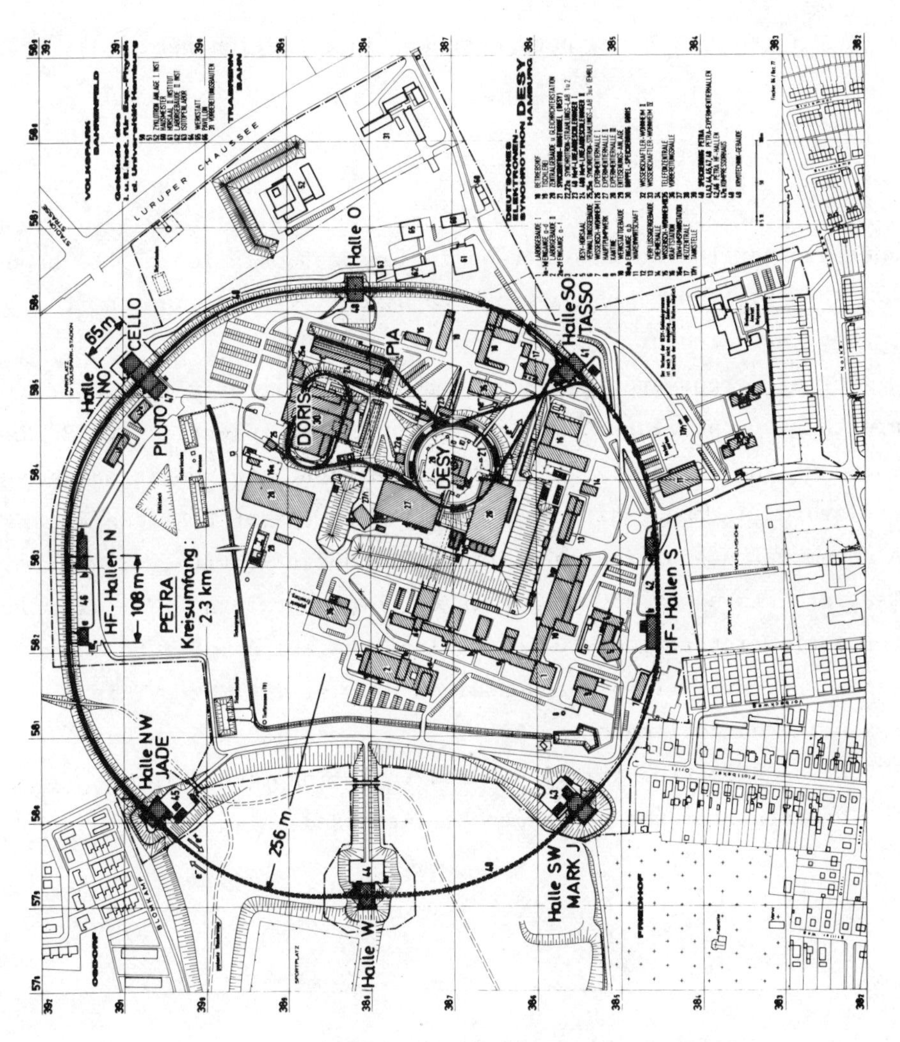

Fig. 1. Layout of PETRA.

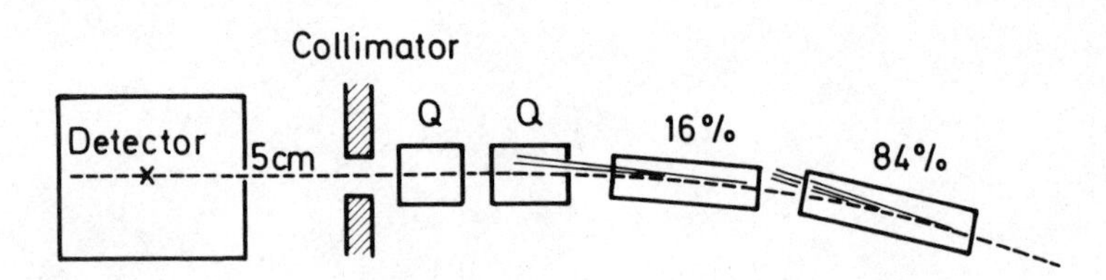

Fig. 2. PETRA magnets near the intersection region.

magnet closest to the interaction region does only 16% of the bending. This is to prevent most of the synchtrotron radiation from entering the detector.

The electrons are injected from LINAC I directly into DESY where they are accelerated to 6.5 GeV and then injected into PETRA. Because of the intensity of the positron LINAC, the positrons are first stored in the Positron-Intensity-Accumulator PIA[3] (see Fig. 3). PIA accumulates positron by using a 10.4 MHz rf-cavity. 20 linac pulses are stored at 450 MeV, each of them containing 10^9 positrons. After a time interval of 60 ms for longitudinal damping to 80 cm, a second rf cavity is powered additionally at 125 MHz. This compresses the bunch length to 25 cm. The bunch is then injected into DESY for acceleration to 6.5 GeV and injection into PETRA.

The rate of interaction can be expressed as

$$r = L\,\sigma$$

where σ is the interaction cross section

$L = N_e N_p f/A$

where N_e = number of e^-

N_p = number of e^+

f = number of crossings per second

A = area of the beams at interaction point.

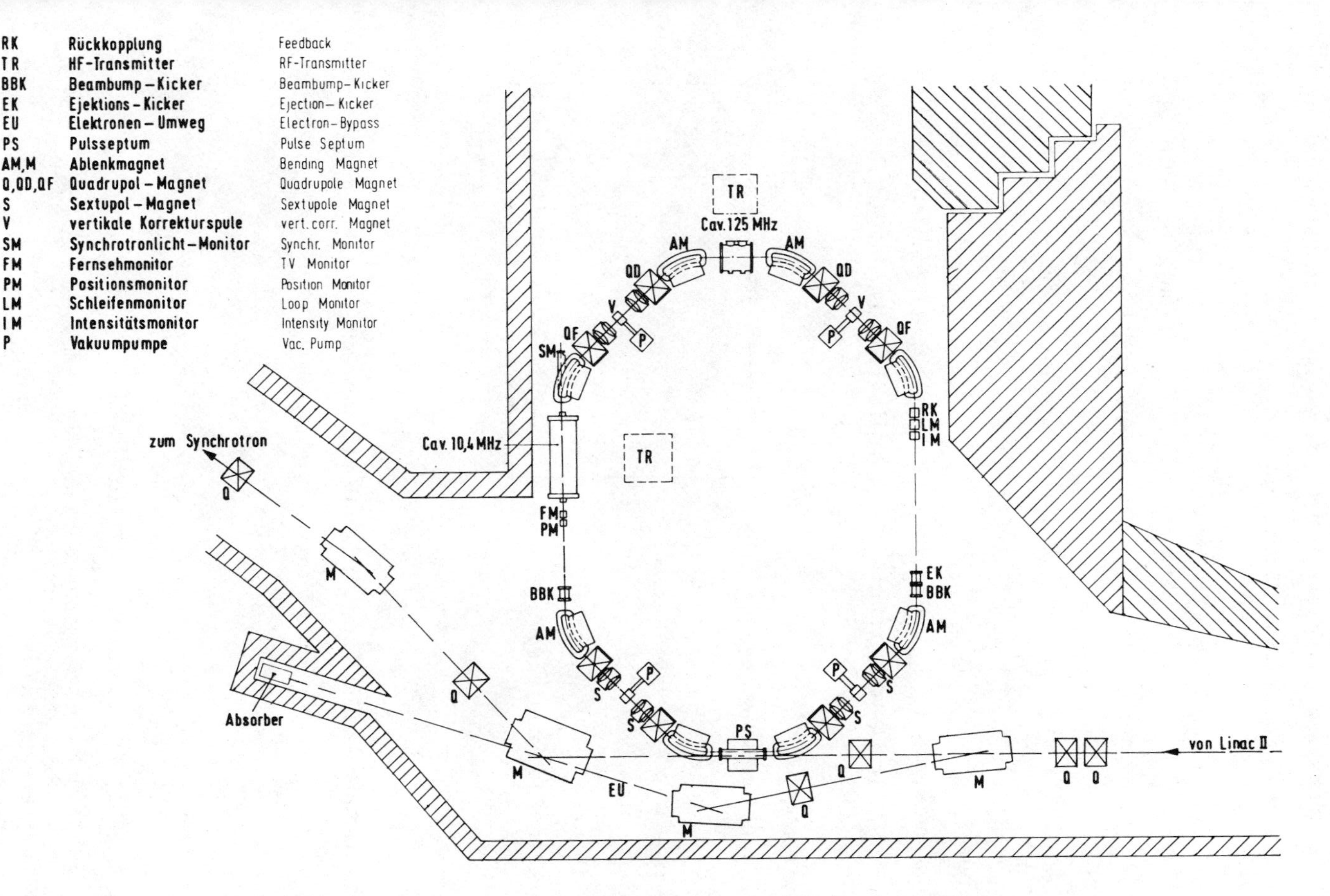

Fig. 3. P.I.A. (Positron Intensity Accumulator).

Each bunch makes 1.3×10^5 turns per second, and presently, PETRA is operating under the two bunch modes, each bunch having about 2×10^{11} particles. Thus,

$$L \cong 1 \times 10^{30} cm^{-2}/sec.$$

At C.M. energy of 30 GeV, the $e^+e^- \to$ hadron cross section is about $\frac{1}{2}$ nb. This means that there is one event in about every 30 minutes.

To understand the luminosity better, let us consider the physics of the storage ring.[4] The coordinates are defined as follows:

s : direction along the beam

x: the radial direction, and

z : direction perpendicular to the orbit plane.

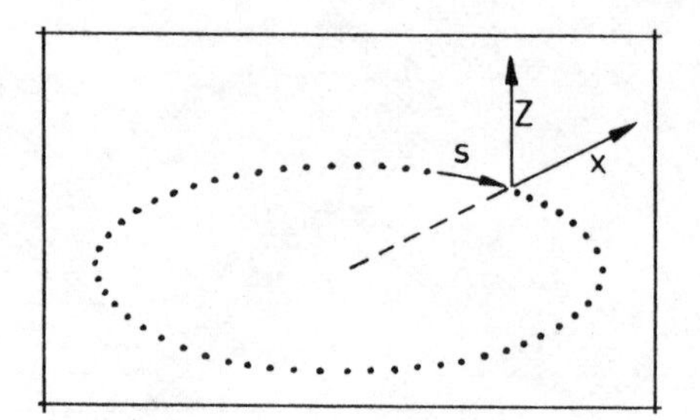

For x, the deviation from an ideal orbit,

$$\chi(s) = \alpha\sqrt{\beta(s)}\ \cos(\phi(s)-\theta)$$

where $\alpha^2 = \varepsilon$, the emittance. It is a constant of a particular beam optics. It denotes the "phase space" for the beam in the $x - x'$ plane, where $x' = dx/ds$.

The value of x in a particular revolution around the orbit is shown below.

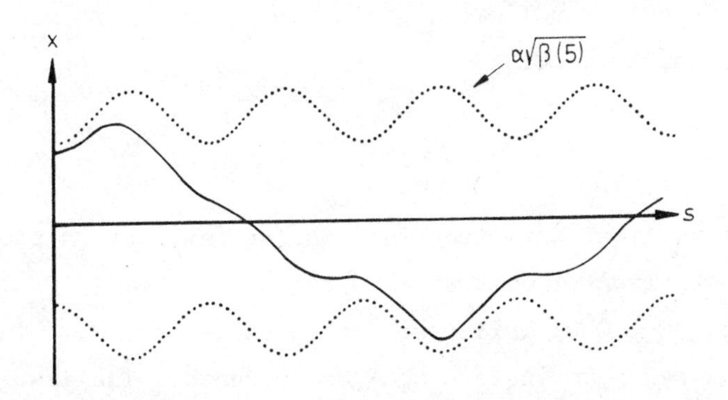

The envelope is $\alpha\sqrt{\beta(s)}$. So, to get the maximum luminosity, one should have as small a β value as possible at the interaction point.

For a particle in the storage ring, the phase $\phi(s)$ is advanced by $2\pi\oint\frac{ds}{\beta}$. The quantity $Q \equiv \oint\frac{ds}{\beta}$ is called the betatron number.

If at s = 0, the orbit of a particle receives a disturbance, and the slope x' changes to x' + Δx', then the orbit becomes:

$$\chi(s) = \frac{\Delta x'\sqrt{\beta(o)}}{2\sin\pi Q}\sqrt{\beta(s)}\cos\{\phi(s)-\pi Q\}.$$

Therefore, if Q is an integer, any disturbance on the orbit causes the orbit to be lost. This is known as a resonance.

Similarly, there is a betatron number for the z direction: Q_z. The tune shift, ΔQ, is the range in Q for a beam. It is clearly limited because of the resonances which a stored beam must avoid. Since L is inversely proportional to the area of the beams at the interaction point, β_x and β_z should be at their minimum values there.

Beam-beam interaction introduces a range of ΔQ; it also increases with decreasing area at the interaction point. Operating the storage ring simultaneously at the horizontal and vertical beam-beam interaction limits, and assuming that both are given by the same tune shift ΔQ,

$$L \propto \frac{(\Delta Q)^2}{\beta_z} \varepsilon_x ,$$

where β_z = vertical amplitude function at interaction point,

ε_x = horizontal emittance.

At present, $\Delta Q = 0.02$ while the design value is 0.06. The bunch number is 2 per beam instead of 4 as intended. The design maximum current of 20 mA per beam is also not reached. The resulting luminosity is a factor of 50 lower than the design.

There is a technique to improve the luminosity by decreasing β_z. In the experimental hall, between the quadrupoles

$$\beta(s) = \beta_o \left\{ 1 + \frac{(s-s_o)^2}{\beta_o^2} \right\} .$$

where β_o is the value of β at the interaction point.

This is shown in the figure for two values of β_o. It can be seen that β_o is limited by the aperture of the focusing quadrupoles, and how close they are positioned to the interaction point.

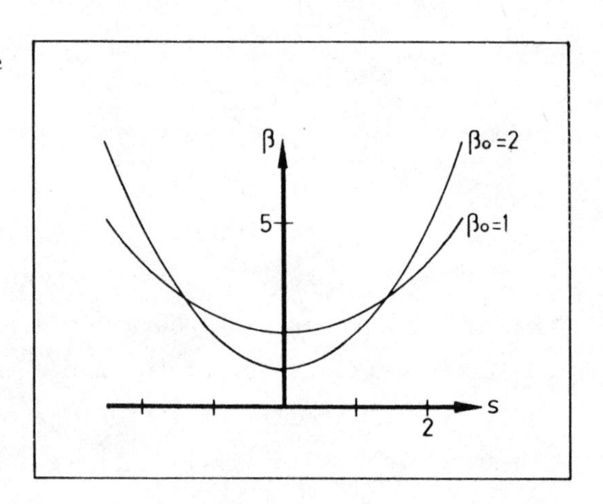

By putting two focusing quadrupoles near the interaction point while matching the boundary conditions, one can improve the luminosity without disturbing the rest of the storage ring. This technique is known as low-β insertion.

In July this year, an attempt was made to measure the beam polarization. A circularly polarized argon laser beam was reflected by mirrors into the electron beam and back-scattered. It is detected down-stream by a shower counter telescope. If the electron beam is polarized, an angular dependence of the reflected photon is expected. The reflected light has been observed, and the results are still being evaluated.

The following table shows the hopes and expectations for the near future:[5]

1979 Installation of 32 more cavities.

1980 Maximum energy of 2 x 19 GeV.
Low-β insertion.

1981 Two more detectors for polarized beams.
Increase of rf-installation (max. energy 2 x 23 GeV).

MARK-J

The MARK-J detector[6] is shown in Figures 4 and 5. It is a detector designed to measure and distinguish hadrons, electrons, neutral particles, and muons. It covers a solid angle of $\phi = 2\pi$ and $\theta = 9^{\circ}$ to 171° (θ is the polar angle and ϕ is the azimuthal angle). The detector is symmetrical in both ϕ and θ directions. The physics objectives of this detector are:

1) to measure the total hadron cross section,
2) to measure the interference effects between weak and electromagnetic interactions by studying the charge asymmetry in the reaction $e^+e^- \to \mu^+\mu^-$ with an accuracy of 1%,
3) to study various quantum electrodynamic processes,
4) to measure μe and μ hadron events and to study asymmetries associated with these events which can come from weak interaction effects, and

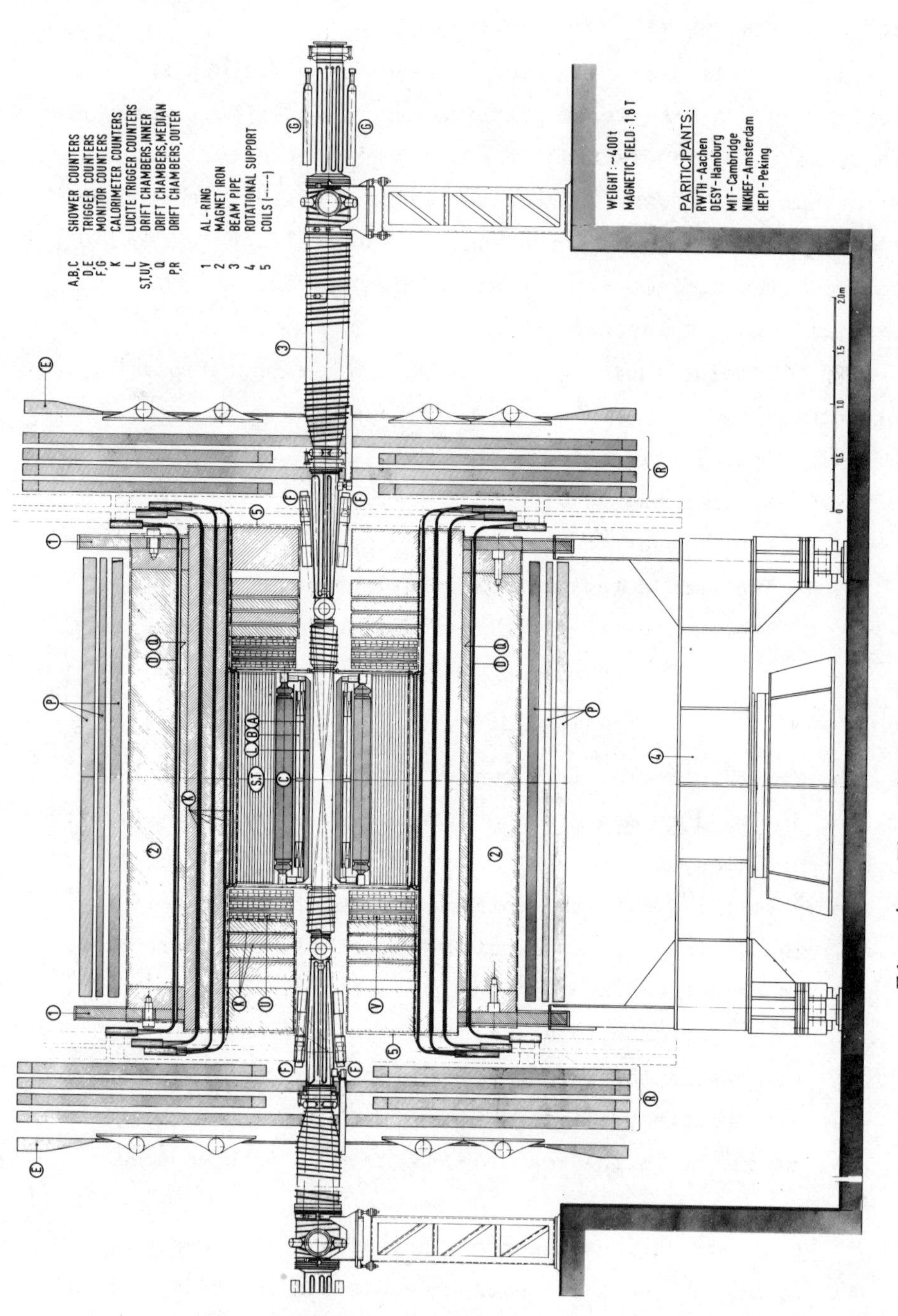

Fig. 4. The MARK J detector in a side view.

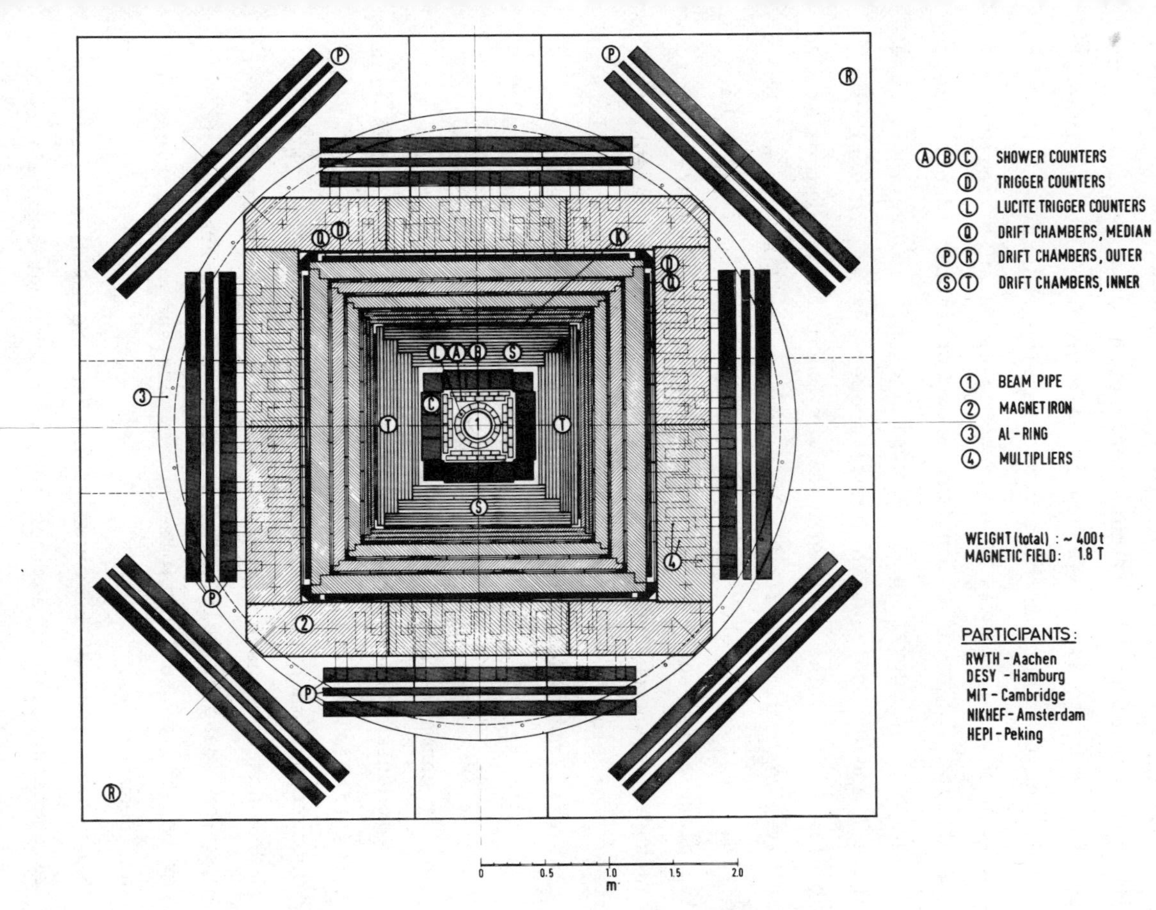

Fig. 5. The MARK J detector in an end view.

5) to search for new quarks and heavy leptons.

As shown in Figures 4, 5, the particles leaving the interaction region pass through a ring of 32 lucite Cerenkov counters, L, sixteen on each side of the interaction point. These counters are used to distinguish charged particles from neutrals and are insensitive to synchrotron radiation. These counters cover down to an angle of $\theta = 9^o$ to 171°. The twenty A counters have one tube at each end, are made out of three radiation lengths of lead sandwiched with 5 mm of scintillators, and cover an angle region of $\theta = 12^o$ to 168°. The 24 B counters are constructed identically to the A counters and cover an angle region of $\theta = 16^o$ to 164°. The counters A and B enable us to locate shower maxima in various θ and ϕ directions. The sixteen C shower counters consist of twelve layers (twelve radiation lengths) of lead-scintillator sandwich also with one tube at each end. The drift chambers S and T have twelve planes each and U and V have ten planes. They are used to sample hadron showers and measure the original muon angles with a spatial resolution $\simeq$ 400 μm. The hadron calorimeter K consists of 192 counters sandwiched with magnetized iron to measure hadron showers. The magnetic field in the iron is toroidal and its value is 17 kG. Finally, in the outermost layer, we have the P and R drift chambers which are used to measure muon exit angles and thus momenta. These chambers have ten to sixteen planes each. The Q drift chambers right next to the D counters have two planes. They measure the muon tracks in the bending direction. The 32 D and 16 E hodoscopes have dimensions of 30cm x 4.5m x 1cm and 80cm x 4.5m x 1cm, respectively. The D and E counters are used to trigger on single- and multiple-muon events and to reject cosmic rays and beam sprays.

Since the study of E.M.-weak interference involves a very sensitive measurement of asymmetries, any inherent asymmetries of the detector could create errors. In order to nullify their effects in the θ and ϕ directions, the supporting structure was made so that the detector can rotate $\pm 90^{\circ}$ in ϕ and 180° in θ. The total weight of the detector is about 500 tons. The detector has been surveyed and it was found that the axis of the ϕ rotation is reproducible to 100 microns.

One quarter of the complete assembly was tested and calibrated at a muon, electron, and pion beam at CERN. The gain of all the counters was adjusted according to the test results.

The luminosity monitor consists of two arrays of twenty-eight lead glass counters[7] each with dimensions of 8cm x 8cm x 70cm located 5.8 meters from the interaction point. They are designed to measure the $e^+e^- \rightarrow e^+e^-$ reaction at small angles. Scintillators in front of the lead glass define the acceptance and the lead glass counters measure the angle and energy of the electron pairs. In addtion, the luminosity was also monitored with the L, A, B, C hodoscopes which measure the forward angle Bhabha scattering ($\theta = 11.5^{\circ}$ to 26°). A very loose trigger was used which collects candidates for electron pairs, single-muon events, muon pairs, and hadron events. For electron pairs we require that the opposite quadrants of A and B counters be in coincidence and that each quadrant has a minimum energy of 0.5 GeV. For single muon events we require at least two A counters, two B counters and one D counter triggered. For muon pairs we require at least two A counters in coincidence with a pair of opposite-quadrant D counters. For hadrons we require at least four A counters and three B counters, and each quadrant A, B, C to be in coincidence with the opposite quadrant and at least have two pairs of the opposite quadrant triggered.

Cosmic ray and accidental events were mostly rejected by requiring the event trigger to be in coincidence with the beam bunch signal to $\pm$ 15 ns. The trigger rate is typically 5 per second. A microprocessor is used to require that the S and T chambers have at least three counts for hadron triggers. This reduces the tape writing rate by a factor of 3.

The total energy of each interaction and directions of a particle or groups of particles was computed from the time and pulse height information of the shower counters and calorimeter counters. From the difference in time between the two phototubes of each shower counter and from the ratio of their pulse heights, we obtain two measurements of the position along the beam direct- tion at which the particles struck the counter. The algorithms used were developed from analysis of test beam data which was accumulated for incident electrons and pions between 0.5 and 10 GeV. The azimuthal position was determined by the finely segmented shower counters. This method enables us to determine the θ and ϕ angles to $< 5^\circ$ for e or γ and $< 15^\circ$ for muons or hadrons.

TEST OF QED

Historically, the original work using the e^+e^- reaction to study the validity of QED and to search for heavy leptons was started by the Bologna-CERN Group at the ADONE colliding beam machine. They were the first to study three reactions:[8,9,10,11,12]

$$e^+e^- \to e^+e^-$$
$$e^+e^- \to \mu^+\mu^-$$
$$e^+ + e^- \to e^- + \mu^+ + \ldots \text{ (to search for heavy leptons)}$$

The results for Bhabha scattering obtained are based on the analysis of 12,827 events, in the s-range 1.44 - 9.0 GeV^2. Figure 6 shows the final results. The systematic uncertainties in the acceptances and efficiencies are estimated to be less or equal to ± 5% at each energy.

If the cross section for reaction is written in the form

$$\sigma = As^n$$

where $s = (2E_{beam})^2$, the result of the best fit gives

$$\frac{A_{exp}}{A_{th}} = 1.00 \pm 0.02, \; n = -(0.99 \pm 0.02).$$

Bhabha scattering is the cleanest QED process and the data reported

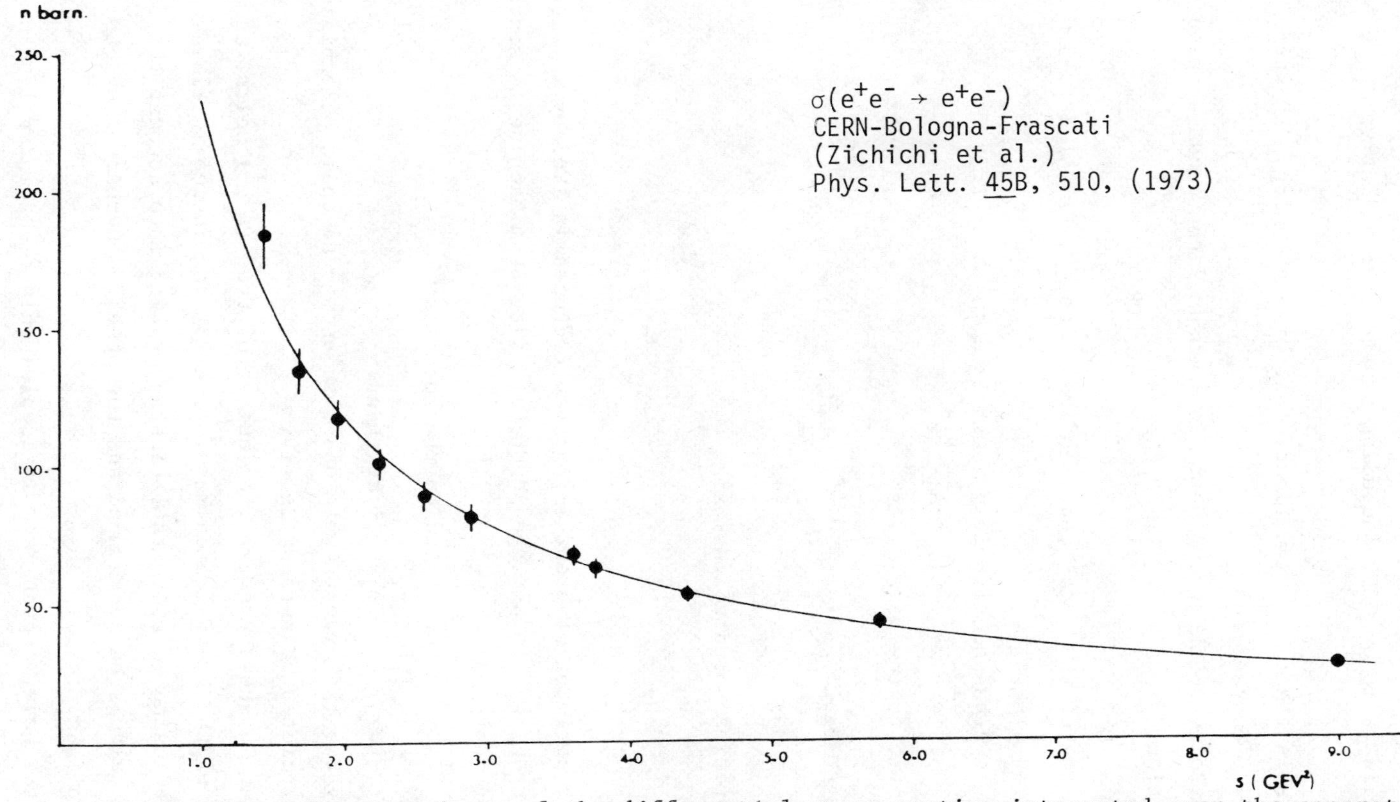

Fig. 6. Experimental s-dependence of the differential cross-section integrated over the apparatus for the process $e^+ + e^- \rightarrow e^\mp + e^\pm$. The line is the best fit to the data.

constitute the highest statistical sample so far collected and analyzed in terms of QED, including first-order radiative corrections.

This is the first significant test of high-energy QED performed on $e^+e^- \rightarrow e^+e^-$ reactions.

The Bologna-CERN group have also measured the cross-section of the reaction

$$e^+e^- \rightarrow \mu^\pm\mu^\mp$$

at 14 values of the total center-of-mass energy, from 1.2 to 3.0 GeV, as shown in Figure 7.

If the cross-section is parametrized in the form

$$\sigma(e^+e^- \rightarrow \mu^+\mu^-) = A \cdot s^{-n},$$

the best fit to the data gives

$$A_{exp} = 15.87 \pm 0.48$$

$$n_{exp} = 1.000 \pm 0.012$$

the theoretical predictions being

$$A^{QED}_{theor} = 15.70$$

$$n = 1.000$$

An experiment of great importance proposed in 1967 by A. Zichichi and his group was to use the ADONE Colliding Beam Machine to search for heavy leptons.

The search concentrated on the reaction

$$e^+ + e^- \rightarrow e^\pm \mu^\mp + \text{anything}.$$

Only two (μe) events were observed and these could be explained by the predicted background level.

Figure 8 shows the expected number of true ($\mu^\pm e^\mp$) pairs, as a function of the heavy lepton mass. The 95% confidence level was calculated using Poisson statistics and taking into account the fact that the calculated background and the observed number of events were both equal to 2.

If the heavy lepton mass is coupled only to ordinary leptons

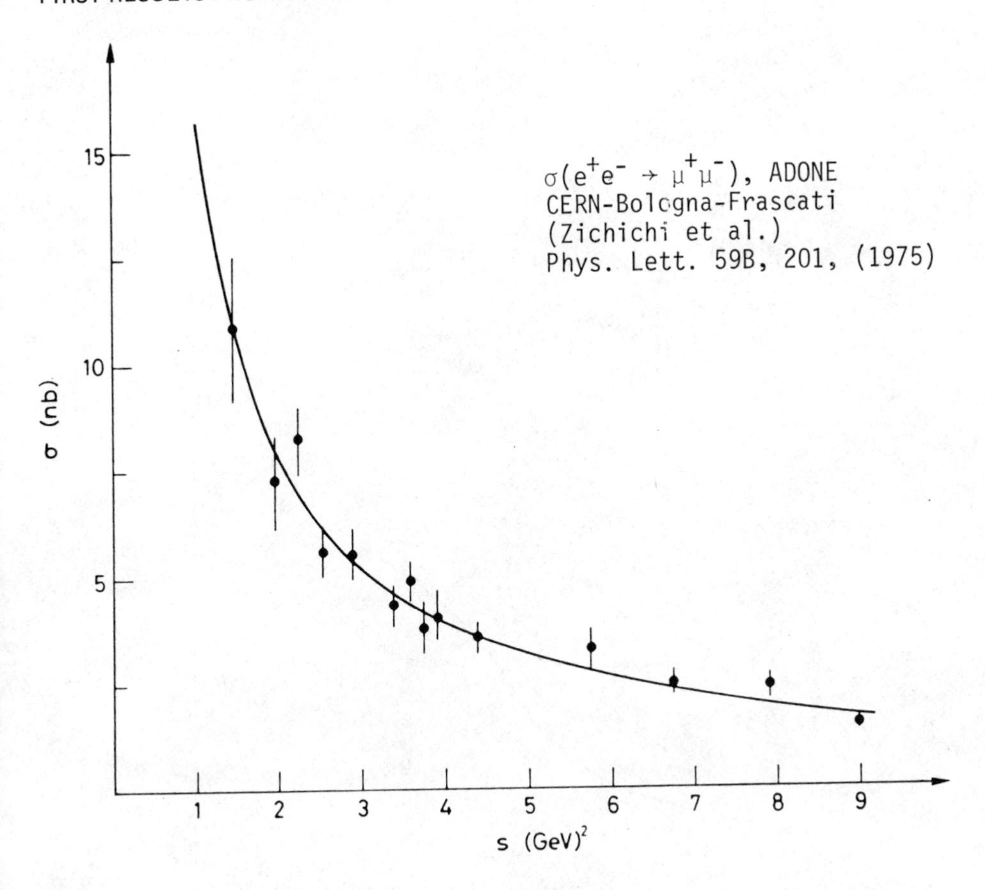

Fig. 7. The values of $\sigma(e^+e^- \to \mu^+\mu^-)$ integrated over the experimental apparatus at the 14 values of the time-like four momentum transfer s. First order radiative corrections have been applied to the data. The solid line is the QED prediction.

(with the universal weak coupling constant), the 95% confidence level for the mass is

$$m_{HL} \geq 1.45 \text{ GeV}.$$

If the heavy lepton is universally coupled to both ordinary leptons and hadrons, the 95% confidence level for the mass is

$$m_{HL} \geq 1.0 \text{ GeV}.$$

The mass limits apply to any type of heavy lepton.

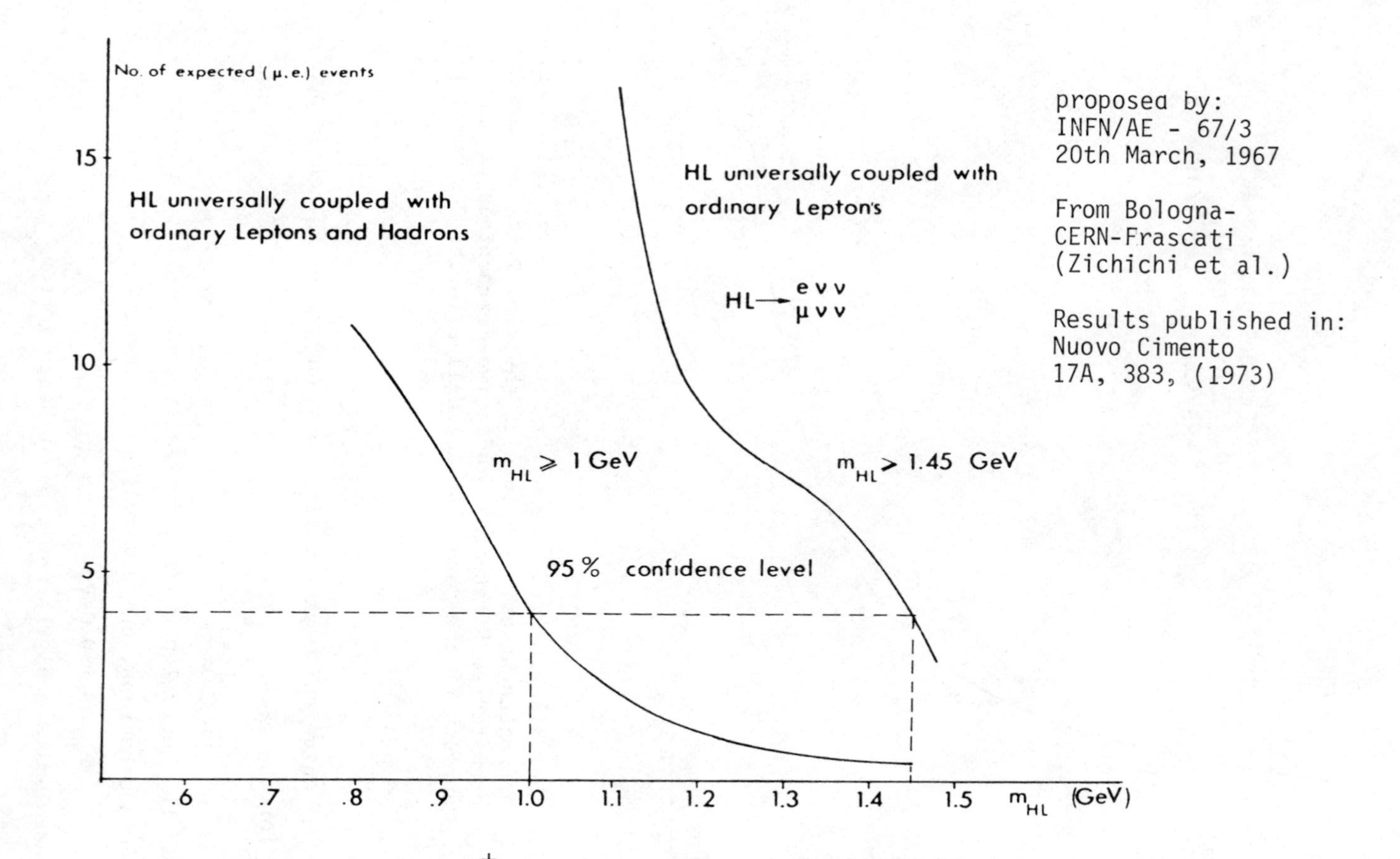

Fig. 8. Showing the number of ($\mu^{\pm} e^{\mp}$) pairs versus m_{HL} for two types of universal weak couplings of the heavy leptons. The dotted lines indicate the 95% confidence levels for m_{HL}.

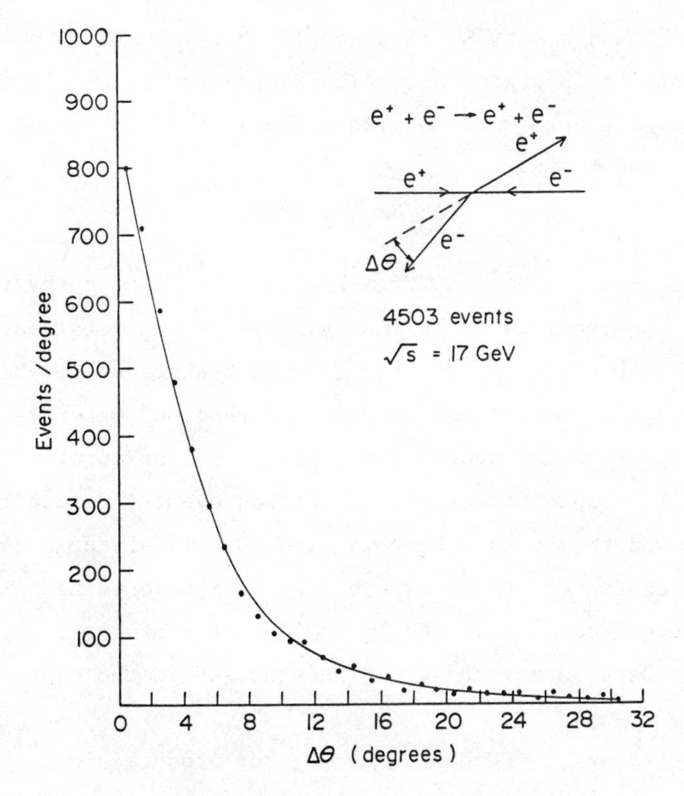

Fig. 9. The measured acollinearity.

At PETRA the MARK J group has measured the Bhabha reaction to much smaller distances.

In Figure 9 we show the measurement of the acoplanarity angle $\Delta\phi$ in the detector with large-angle Bhabha scattering. For each particle, or particles emitted within a cone of 20° which were not separable in the counters, the vector momentum was computed from pulse height and counter position information using the position of the interaction region (which is checked to ±2cm by fitting tracks in the chambers for $e^+e^- \to \mu^+\mu^-$ events). Photons emitted close to either electron are included in the fitted electron momentum.

The Bhabha events are identified by requiring two back-to-back showers which are collinear to within 20° in ϕ and θ and with a measured total shower energy greater than 8 GeV. Because there are few events near the 20° cut in the θ acolinearity spectrum shown in Figure 9 and a very similar acoplanarity spectrum in ϕ, we conclude that the background to elastic scattering events is negligible. The test beam data mentioned above yield a resolution on the energy sum of

$$\Delta E/E = 12\%/\sqrt{E}.$$

To eliminate most background from hadron jets, the energy in the K counters was required to be less than 7% of the total energy. Because the QED test is most sensitive to background in the large angle region, all events having θ larger than 60° were scanned on graphic displays which showed the distribution of counter hits. On the basis of a Monte Carlo study of hadron events, we conclude that the background from this source is less than 1% of these events.

The acceptance for $e^+e^- \to e^+e^-$ was computed using a Monte Carlo technique and is defined by the geometry of the first shower counter A. Both energy and acceptance losses in the corners were found to be small.

The first order QED photon propagator produces an s^{-1} dependence in the $e^+e^- \to e^+e^-$ cross section. The quantity $s\, d\sigma/d\theta$ vs θ is independent of s. The distribution is plotted for the data at $\sqrt{s}$ = 13, 17, and 27.4 GeV in Figure 10. Excellent agreement with QED predictions is seen. To express this agreement analytically, we compare our data with the QED cross section in the following form[13] (since charge is not distinguished here):

$$\frac{d\sigma}{d\Omega} = \frac{\alpha^2}{2s}\left\{ \frac{q'^4 + s^2}{q^4}|F_s|^2 \frac{2q'^4}{q^2 s}\mathrm{Re}(F_s F_T{}^*) + \frac{q'^4 + q^4}{s^2}|F_T|^2 \right.$$

$$\left. + \frac{q^4 + s^2}{q'^4}|F'_s|^2 + \frac{2q^4}{q'^2 s}\mathrm{Re}(F'_s F_T{}^*) + \frac{q'^4 + q^4}{s^2}|F_T|^2 \right\}\{1 + C(\theta)\},$$

where

$$F_2 = 1 \mp q^2/(q^2 - \Lambda_s{}^2)$$

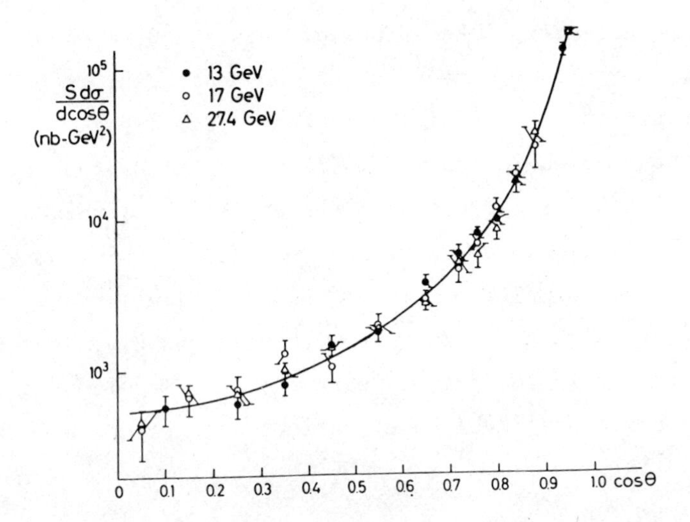

Fig. 10. The differential cross section $s\frac{d\sigma}{d\cos\sigma}$ for $e^+e^-\to e^+e^-$ at $\sqrt{s}$ of 13, 17, and 27.4 GeV.

is the form factor of the spacelike photon,

$$F_s' = 1 \mp q'^2/(q'^2 - \Lambda_s^2)$$

$$F_T = 1 \mp s/(s - \Lambda_T^2)$$

is the form factor if the timelike photon $q^2 = -s\cos^2(\theta/2)$, $q' = -s\sin^2(\theta/2)$, and Λ is the cut-off parameter in the modified photon-propagator model and $C(\theta)$ is the radiative correction term as a function of θ.

The radiative correction to the e^+e^- elastic process was calculated using a modified program from Berends[14] which included the contribution of the heavy-lepton (τ) loop and the hadronic vacuum polarization. The inclusion of these two effects changes the radiative correction from minus a few percent, as was commonly used previously, to +4.6% at θ = 14% and +1.3% at $\theta = 90^o$ for $\sqrt{s}$=17 GeV. It is slightly smaller at $\sqrt{s}$ = 13 GeV.

Electron positron pairs are generated according to the above equation in a Monte Carlo program. Each electron is then traced through the detector. The effect of measured θ, ϕ resolutions are included. A χ^2 fit was made, using the Monte Carlo generated angular distribution, to all of 13, 17 and 27 GeV data. The normalization was treated in two ways: (1) the total Monte Carlo events in the region $0.9 < \cos\theta < 0.98$ was set equal to the total number of measured events in the same region. (2) The minimum χ^2 fit to the entire data sample determines the normalization. The two methods agree with each other to within 3% and give essentially the same result in the cut-off parameter Λ. The lower limits of Λ at 95% confidence level under various assumptions are shown in Table II.[15]

HADRON FINAL STATES

To eliminate the bulk of the beam gas background, which is mainly low energy and one sided, we first require that the total energy deposited in the calorimeter be greater than half of the total center of mass energy, and that the computed total $P_{//}$ and total P_t each be less than 50% of the observed energy. One part in a thousand of the raw events pass the above criteria, and pictures of each surviving event were scanned by physicists on a videoscreen to assure that the counter tracks are reasonably fitted. We further demand at least one track in the drift chambers pointing

Table 2. Cut-off parameters in GeV for photon form factors from Bhabha scattering.

Λ	$1-\frac{q^2}{q^2-\Lambda^2}$	$1+\frac{q^2}{q^2-\Lambda^2}$
Λ_S	52	55
Λ_T	60	52
$\Lambda_S=\Lambda_T$	65	64

back to the interaction region to distinguish hadronic events from beam gas events. The interaction region is defined by $e^+e^- \rightarrow \mu^+\mu^-$ events to $\sigma = 2$ cm. To discriminate against events of electromagnetic origin such as $ee \rightarrow ee$, $ee \rightarrow ee\gamma$, and so forth, we have accepted three types of events with different shower properties which are as follows:

(1) Two narrow showers penetrating into the third and fourth layers of the hadron calorimeter counters (K). There is a total of 33 radiation lengths from the interaction region. To discriminate against e^+e^- and $\gamma\gamma$ final states, the total energy in the K calorimeter counters is required to be greater than 7% of the total shower energy in A+B+C.

(2) Two broad showers penetrating into the first or second layer of the K calorimeter counters with energy greater than 1% of the total A, B and C shower energy.

(3) Three or more broad showers with tracks in the drift chambers are also considered.

The energy spectrum of the events passing the cut is shown as the non-hatched area of Figure 11. The energy spectrum of beam gas events, which are defined by a chamber track pointing at least 15 cm from the intersection point, is shown as the hatched area in Figure 11. A comparison of these two spectra (which were taken simultaneously) shows that the real hadron events can be readily separated from beam gas events by an energy cut. Such an energy cut also reduces the contamination from two photon processes and $e^+e^- \rightarrow \tau^+\tau^-$ events which yield hadrons in the final state.

A Monte Carlo program was used to compute the acceptance for

$$e^+e^- \rightarrow \text{hadrons} \qquad (1)$$

and to determine the contribution to our hadron event sample from the two photon process

$$e^+e^- \rightarrow \text{hadrons} + e^+e^- \qquad (2)$$

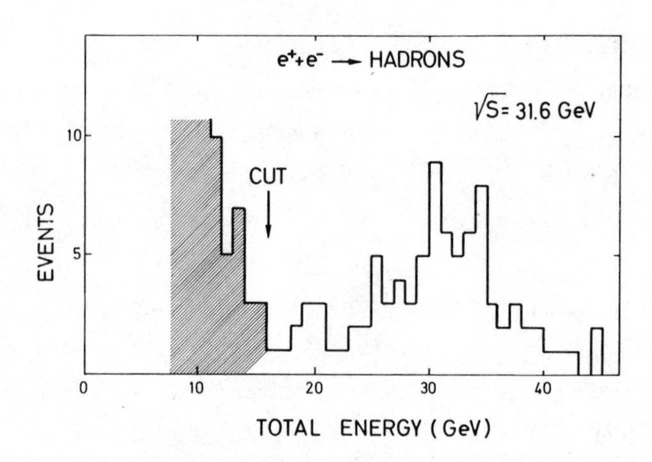

Fig. 11. Nonhatched area: energy spectrum of the events passing the criteria as defined in the text.
Hatched area: energy spectrum of the beam-gas events.

and

$$e^+e^- \to \tau^+\tau^- \qquad (3).$$

The Monte Carlo program generates two jets for reaction (1) according to the Feynman-Field ansatz,[16] which includes not only u,d, and s quarks but also the contribution of c (charm) and b (bottom) quarks. The branching ratios for the decays of the D and F mesons rely either on available experimental data or otherwise on isospin-statistical models.[17] The branching ratios for the decay modes of B mesons (from $b\bar{b}$) and T mesons, which are yet to be observed, are based largely on the work of Ali et al.[18,19] in the framework of the Kobayashi-Maskawa six-quark model.[20] The probability of producing each quark flavor is taken to be proportional to the square of the quark charge q, with $q_c = 2/3$ and $q_b = -1/3$. Acceptances for reaction (1) were computed using u,d,s,c, and b quarks only, with the mass of the B mesons taken to be in the range of 5-6 GeV, expected if the Υ is a $b\bar{b}$ state. The production of top quark pairs was only considered in connection with the jet analysis of the events (described later), where the T meson was taken to be 9-14 GeV. Above this threshold the $t\bar{t}$ production probability was

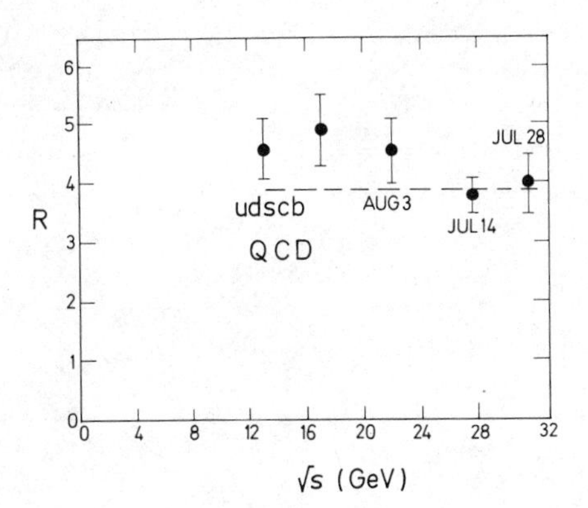

Fig. 12. The total relative hadronic cross section $R = \frac{\sigma(e^+e^- \to \text{hadrons})}{\sigma(e^+e^- \to \mu^+\mu^-)}$ at all energies by the MARK J Group. Statistical errors only shown. Additional systematic error 15%. $R = R_o(1+\alpha_s(s)/\pi)$ $R_o = 3 \sum_f Q_q^2$ $\alpha_s = \frac{12\pi}{(33-2N_f)} \frac{1}{\ln s/\Lambda^2}$

assumed to be 4/15 of the total cross section; i.e. proportional to the square of the quark charges so that threshold effects were ignored. To improve the accuracy of the model computation for reaction (1), we incorporated initial radiative effects in the computer program.[21,22]

The generation of events for reaction (2) was performed in the equivalent photon approximation with multipions only in the final states.[23] Simulated events for reaction (3) were produced according to the "standard model", where the τ is considered as a sequential heavy lepton, as indicated by available data.[24]

The total cross section for $e^+e^- \to$ hadrons, as expected in terms of

$$R = \sigma(e^+e^- \to \text{hadrons}/\sigma(e^+e^- \to \mu^+\mu^-),$$

is shown in Figure 12. Systematic errors due to model uncertainties for the acceptance of reaction (1) are 10% and the uncertainties in evaluating the reactions (2) and (3) are limited by the lack of

Table 3. R Measurement at 31.6 GeV. July, 1979. MARK J Group.

$R = \sigma(e^+e^- \rightarrow \text{hadron}) / \sigma(e^+e^- \rightarrow \mu^+\mu^-)$

$\sqrt{s}$	31.6 GeV
$\int Ldt\ (nb^{-1})$	243
EVENTS	88
ACCEPTANCE	83%
R ± STATISTICAL ERROR	4.0 ± 0.5
SYSTEMATIC ERROR	0.6

SUBTRACTIONS INCLUDED

$\tau^+ \tau^-$	$\Delta R = -0.3$
(2γ)	$\Delta R = -0.04$
RADIATIVE CORRECTIONS	- 8%

experimental data on high energy high multiplicity states. Table III gives an example of R measurement at 31.6 GeV.

JET ANALYSIS

A jet analysis[25] of the hadronic events was performed using the spatial distribution of the energy deposited in the detector. For each counter hit, a vector p^i is constructed, whose direction is

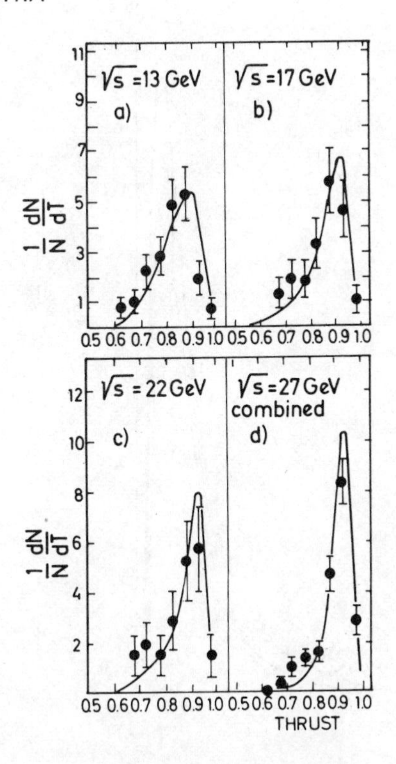

Fig. 13. Thrust distribution observed at $\sqrt{s}$ = a) 13, b) 17, c) 22, d) 27 GeV. The solid line is the quark model prediction for u, d, s, c and b quarks with no gluon emission.

given by the position of the signal in the counter, and magnitude by the corresponding deposited energy. The thrust parameter T and the spherocity parameter S' are then defined as:

$$T = \max \left[\Sigma |p^i_{//}| / \Sigma p^i \right] \qquad S' = \left(\frac{4}{\pi} \right)^2 \min \left[\Sigma p^i_t / \Sigma p^i \right]^2$$

where $p^i_{//}$ (p^i_t) is the parallel (perpendicular) component of p^i along a given axis, and the maximum (minimum) is found by varying the direction of this axis. The sums are taken over all counter hits.

The normalized thrust distributions $\frac{1}{N} \frac{dN}{dT}$ for 13, 17, 22 and the combination of 27.4 and 27.7 (labeled 27 GeV combined), shown in Figure 13 along with the Monte Carlo predictions. The indivi-

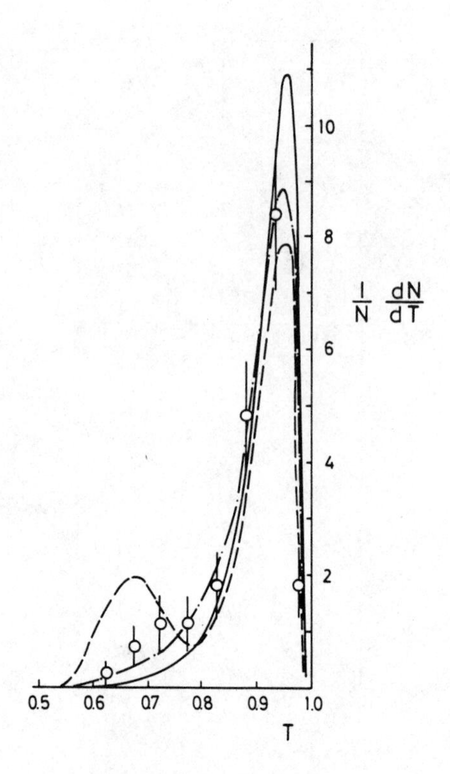

Fig. 14. The thrust distribution N^{-1} dN/dt. The solid curve is the Monte Carlo prediction based on u, d, s, c and b quarks. The dash-dotted curve includes gluon contributions. The dashed curve has included the t quark contributions. The data are inconsistent with t contribution but agree with the gluon emission model.

dual distributions at 27.4 and 27.7 GeV are in agreement with each other.

As expected for production of final states with two jets of particles, the distributions peak at high T. The data are consistent with Monte Carlo distributions which include u,d,s,c, and b quarks.

A jet analysis is carried out for the 31.6 GeV data. The results are summarized in Figure 14 which is the normalized thrust distribution $\frac{1}{N}\frac{dN}{dT}$. The solid curve is the Monte Carlo prediction based on u,d,s,c, and b quarks. The dash-dot curve includes gluon

Table 4. Thrust and spherocity measured by the MARK J Group.

$\sqrt{s}$	13	17	27.5	31.6
<T>	0.82±0.01	0.85±0.01	0.88±0.01	0.88±0.01
<S>	0.32±0.03	0.24±0.03	0.18±0.03	0.20±0.02

contribution. The dotted curve has included the hypothetical charge 2/3 t quark contributions assuming a t mesons mass of 14 GeV. The $t\bar{t}$ production probability was assumed to be 4/15 of the total cross section; i.e. proportional to the square of the quark charges so that threshold effects were ignored. The data is incosistent with the production of t-quarks, and is in good agreement with $q\bar{q}$ with gluon emission. A charge 1/3 quark, however, cannot be ruled out by the present data.

A similar analysis is carried out in terms of spherocity, and the distributions yield the same conclusion. The average thrust and spherocity values for the 13, 17, 27.5 and 31.6 GeV data are summarized in Table 4. The energy dependences of these quantities are smooth, showing no steps which would have appeared at new flavor thresholds.

The possible effect of gluon emission is studied in greater detail. The characteristic features of hard non-collinear gluon emission in $e^+e^- \to q\bar{q}g$ are illustrated in the figure below. Because of momentum conservation the momenta of the three particles have to be coplanar. For events where the gluon is sufficiently energetic, and at large angles with respect to both the quark and anti-quark, the observed hadron jets also tend to be in a recognizable plane.

This is shown in the upper part of the figure where a view down onto the event plane shows three distinct jets; distinct because the fragmentation products of the quarks and gluons have limited P_t with respect to the original directions of the partons. The lower part of the figure shows a view looking towards the edge of the event

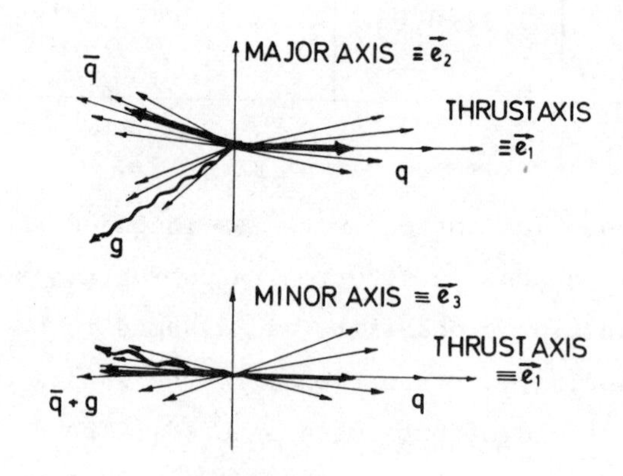

plane, which results in an apparent 2-jet structure. The figure thus demonstrates that hard non-collinear gluon emission is characterized by planar events which may be used to reveal a 3-jet structure once the event plane is determined.

The spatial energy distribution is described in terms of three orthogonal axes called the thrust, major and minor axes. The axes and the projected energy flow along each axis T_{thrust}, F_{major} and F_{minor} are determined as follows:

(1) The thrust axis, $\vec{e}_1$, is defined as the direction along which the projected energy flow is maximized. The thrust, T_{thrust}, and $\vec{e}_1$ are given by

$$T_{thrust} = \max \frac{\Sigma_i \left|\vec{E}^i \cdot \vec{e}_1\right|}{\Sigma_i \left|\vec{E}^i\right|}$$

where $\vec{E}^i$ is the energy flow detected by a counter as described above and $\Sigma_i|\vec{E}^i|$ is the total visible energy of the event (E_{vis}).

(2) To investigate the energy distribution in the plane perpendicular to the thrust axis, a second direction, $\vec{e}_2$, is defined as perpendicular to $\vec{e}_1$. It is the direction along which the projected energy flow in that plane is maximized. The quantity F_{major} and $\vec{e}_2$ are given by

$$F_{major} = \max \frac{\Sigma_i |\vec{E}^i \cdot \vec{e}_2|}{E_{vis}}; \; \vec{e}_2 \perp \vec{e}_1$$

(3) The third axis, $\vec{e}_3$, is orthogonal to both the thrust and the major axes. It is found that the absolute sum of the projected energy flow along this direction, called F_{minor}, is very close to the minimum of the projected energy flow along any axis, i.e.,

$$F_{minor} = \frac{\Sigma_i |\vec{E}^i \cdot \vec{e}_3|}{E_{vis}} \simeq \min \frac{\Sigma_i |\vec{E}^i \cdot \vec{e}|}{E_{vis}}$$

If hadrons were produced according to phase-space or a $q\bar{q}$ two-jet distribution, then the energy distribution in the plane as defined by the major and minor axes would be isotropic, and the difference between F_{major} and F_{minor} would be small. Alternatively, if hadrons were produced via three-body intermediate states such as $q\bar{q}g$, and if each of the three bodies fragments into a jet or particles with $\langle P_t^h \rangle \sim 325$ MeV, the energy distribution of these events would be oblate (P_t^h refers to the final state hadrons). Following the suggestion of H. Georgi, the quantity oblateness, O, is defined as

$$O = F_{major} - F_{minor}$$

The oblateness is $\sim P_t^{gluon} / \sqrt{s}$ for three-jet final states and is approximately zero for final states coming from a two-jet distribution.

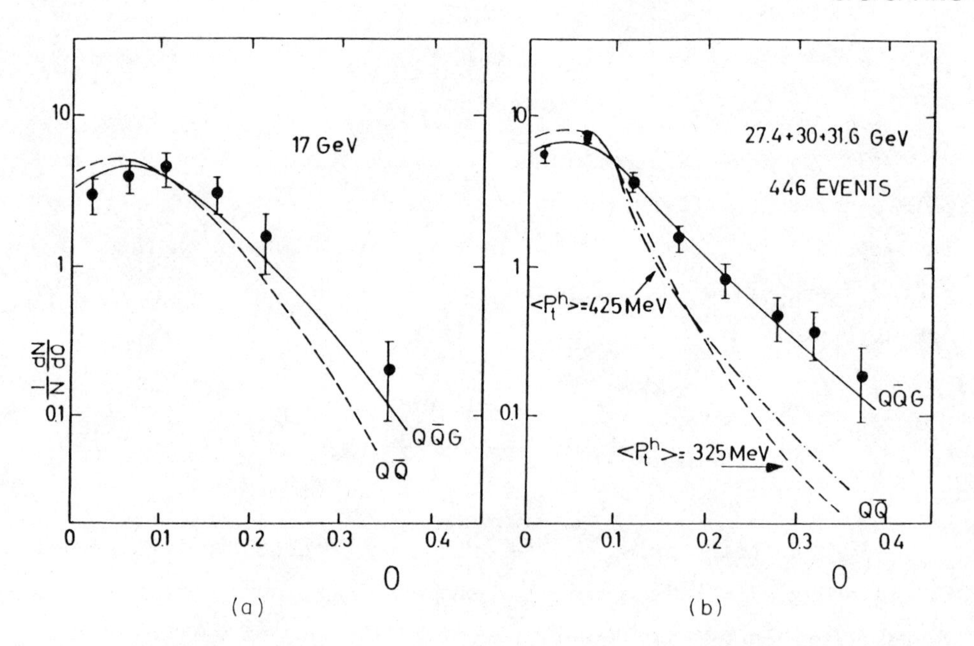

Fig. 15. Differential oblateness distribution at a) $\sqrt{s}$ + 17 GeV and b) at high energies (combined data) compared to the predictions of QCD (solid line) and quark model (dashed lines).

Figure 15a shows the event distribution as a function of oblateness for the data at $\sqrt{s}$ = 17 GeV where the gluon emission effect is expected to be small. The data indeed agree with both models, although the prediction with gluons is still preferred.

Figure 15b shows the event distribution as a function of oblateness for part of the data at 27.4 $\leq$ s $\leq$ 31.6 GeV as compared with the predictions of $q\bar{q}g$ and $q\bar{q}$ models. Again, in the $q\bar{q}$ model we use both $\langle P_t^h \rangle$ = 325 MeV and $\langle P_t^h \rangle$ = 425 MeV. The data have more oblate events than the $q\bar{q}$ model predicts, but they agree with the $q\bar{q}g$ model very well. The MARK J data shows the observed planar events are in good agreement with gluon emission. The 3-jet structure analysis is still being carried out.

TASSO

The end view of TASSO[26] is shown in Figure 16. Around the beam pipe are four scintillator counters and cylindrical proportional

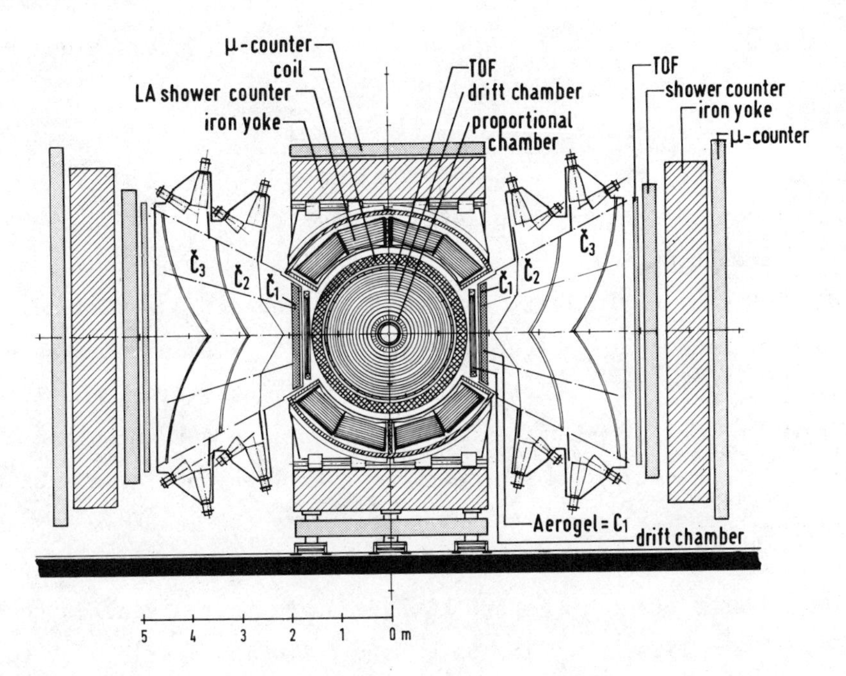

Fig. 16. End view of TASSO.

chambers. From 36.6 to 122.2 cm in radius are 15 layers of drift chambers. There are then 48 time of flight counters each viewed from both ends. A large magnetic solenoid produces 0.5 T of magnetic field inside a volume 4.4 m long and 1.35 m in radius. The liquid argon shower counters have not yet been installed. Conventional lead scintillators are used instead. Outside the iron yoke are four layers of proportional tubes to measure muons.

The end cap consists of time of flight counters and shower counters. There are also small angle detectors consisting of scintillators, lead glass counters and proportional chambers to monitor luminosity. Compensating magnets are needed to correct for the effect of the solenoidal field on the beam particles.

Two hadron spectrometer arms are designed to distinguish π, k and p. They cover a θ range of 90° and a ϕ range of 60°. Plane drift chambers are followed by three layers of Cerenkov counters. The first uses aerogel, the second freon, and the third CO_2. Outside, there are time of flight counters, shower counters, muon - filter and finally muon chambers.

The trigger consists of beam gate and time of flight coincidence, and three tracks in the drift chamber as detected by an online microprocessor, or two coplanar tracks in the drift chamber. The microprocessor does the pattern recognition within four μsec.

The charge multiplicity for the hadron events are studied. In Figure 17a the average charge multiplicity $\langle n_{ch} \rangle$ is plotted as a function of $\sqrt{s}$.[27,28,29] Although the data are preliminary since most of them have not yet been published and corrections for acceptance, for photons converting in the beam pipe, etc. may not always have been made in the same way, they suggest that the multiplicity above $\sim$ 10 GeV is rising (logarithmically) faster than at lower energies.

The dashed curve in Figure 17a gives the energy dependence of $\langle n_{ch} \rangle$ for pp[30] collisions. The pp multiplicity is lower by 0.5 to 1 units but has almost the same behaviour with energy. A good fit to the e^+e^- data is obtained with the function (see solid curve)

$$\langle n_{ch} \rangle = 2 + 0.2 \ln s + 0.18 (\ln s)^2 .$$

Scaling behaviour is e.g. expected from the hypothesis of quark fragmentation: at energies large enough that particle masses can be neglected, the number of hadrons h produced by a quark q with fractional energy x, $D_q^h(x)$, is independent of s. This leads to

$$\frac{d\sigma}{dx} (e^+e^- \to q\bar{q} \to h) = \sigma_{q\bar{q}} \cdot 2D_q^h(x) = \frac{8\pi\alpha^2}{s} e_q^2 D_q^h(x).$$

At PETRA energies, inclusive cross sections have been measured for the sum of all charged particles. Since the mass of the particle is not known, the scaling variable used is $x_p = p/E_{beam}$ and the quantity sd/dx_p is measured.

Figure 17b displays the data from TASSO measured at energies of 13, 17 and 27.4 GeV together with measurements from SLAC-LBL[31] at

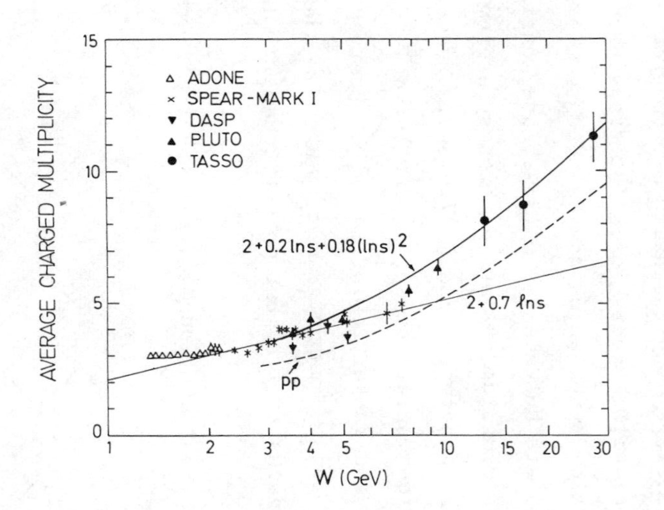

Fig. 17a. Average charge multiplicity from TASSO experiment and from measurements at lower energies. The dashed line shows the result for pp collisions. The solid curve shows the result of the fit.

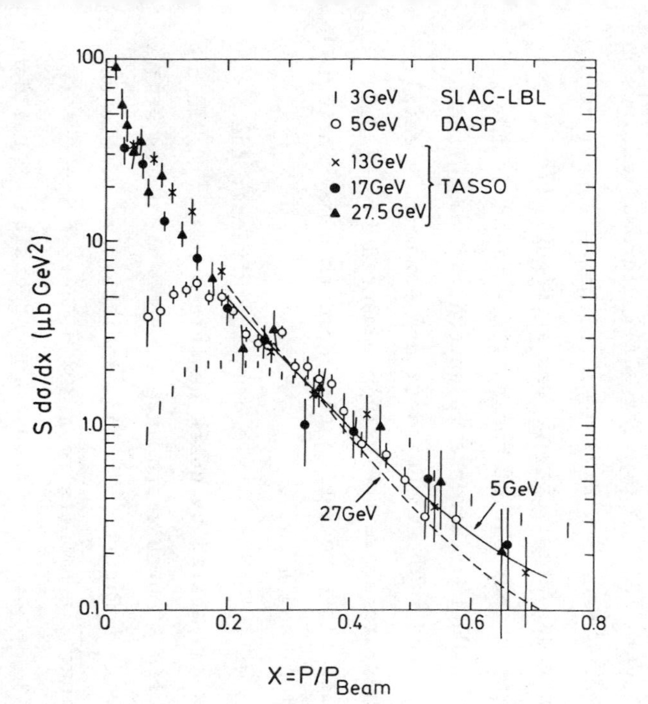

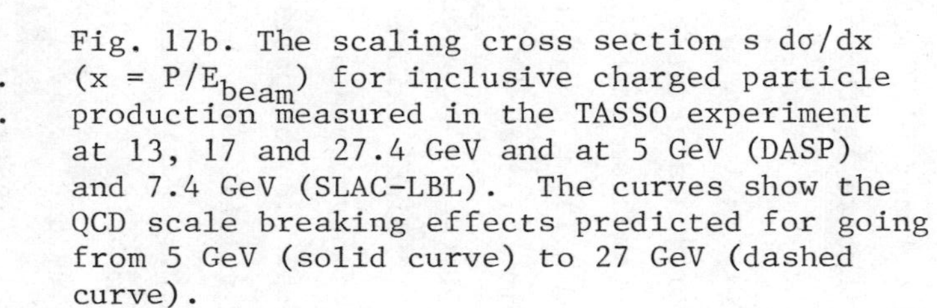

Fig. 17b. The scaling cross section s dσ/dx ($x = P/E_{beam}$) for inclusive charged particle production measured in the TASSO experiment at 13, 17 and 27.4 GeV and at 5 GeV (DASP) and 7.4 GeV (SLAC-LBL). The curves show the QCD scale breaking effects predicted for going from 5 GeV (solid curve) to 27 GeV (dashed curve).

3 GeV and DASP[29] at 5 GeV. At $x > 0.2$ the scaling cross sections are found to be the same between 5 and 27.4 GeV within errors (∿20-30%). The rise of the charged multiplicity we saw in Figure 17a is related to the dramatic increase of the particle yield at low x; for instance at $x = 0.06$ the increase is an order of magnitude going from 5 to 27.4 GeV. The 13 GeV data are somewhat special in that for $x > 0.2$ they are above the values measured for 17 GeV. Since 13 GeV is still close to the $b\bar{b}$ threshold this may indicate copious $B\bar{B}$ production (and decay).

Gluon emission will lead to scale breaking effects: the primary momentum is now shared by quark and gluon resulting in a depletion of particles at high x and an excess of particles at low x values. The curves in Figure 17b indicate the size of the expected scale breaking.[32] It amounts to a 30% effect at $x = 0.6$ comparing 5 and 27.4 GeV cross section. The precision of the data does not allow us to test the predicted charge.

Figure 19 shows the energy dependence of the average $p_{//}$ and p_T values. The average longitudinal momentum grows almost linearly in accordance with expectation. The transverse momentum shows a rapid rise below 5 GeV which may be due to the increase in phase space. The data between 6 and 13 GeV are consistent with a constant $\langle p_T \rangle$. The measurements at the highest energy, 27.4 GeV, show that $\langle p_T \rangle$ is rising between 17 and 27.4 GeV. Measurement errors can lead to widening of the $\langle p_T \rangle$ distribution. However, the increase in $\langle p_T \rangle$ and in particular in $\langle p_T^2 \rangle$, observed by TASSO cannot be accounted for by instrumental biases or by the larger phas space which allows quarks to fragment more frequently into heavier particles (kaons).

In the e^+e^- - hadrons continuum, first order perturbation QCD involves the graphs

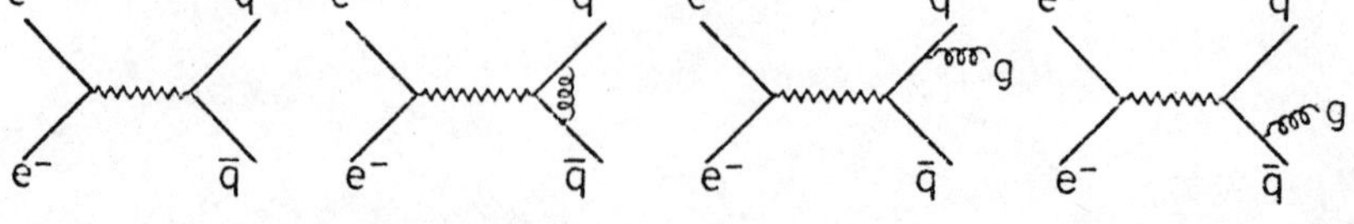

e^+ q e^+ q e^+ q g e^+ q g

e^- $\bar{q}$ e^- $\bar{q}$ e^- $\bar{q}$ e^- $\bar{q}$

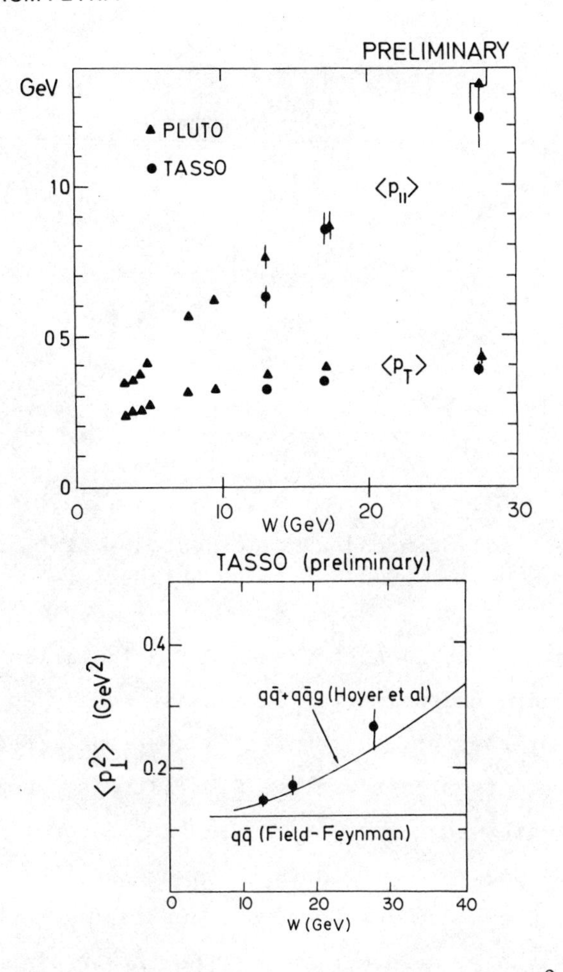

Fig. 18. The energy dependence of $\langle p_{//} \rangle$, $\langle p_T \rangle$ and $\langle p_{\perp}^2 \rangle$ relative to the thrust axis for charged particles and compared with model predictions.

They give

$$\langle p_T^2 \rangle q/\text{jet axis} \simeq \frac{1}{3\pi} \alpha_s(Q^2) \cdot Q^2 = \text{const} \cdot Q^2/\log Q^2$$

and therefore large effects are predicted[34,35] at the highest energy at which data have so far been taken at PETRA, namely Q = 27.4 GeV, $Q^2=W^2=750\ \text{GeV}^2$.

In Figure 18 the evidence for an increase of the overall $\langle p^2 \rangle$ with W is presented.

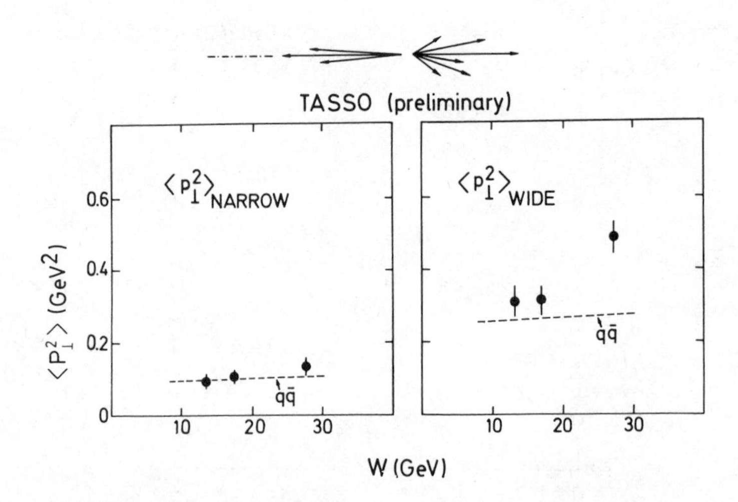

Fig. 19. One-sided jet broadening, expected in 1st order perturbative QCD, is appearing at the highest PETRA energy.

Comparison with the prediction from a QCD calculation extended to include q and g hadronization in a least model-dependent way, presented by Hoyer et al.,[35] demonstrates that the effects are of the size expected from perturbative QCD. It was checked with Monte Carlo studies that resolution or acceptance effects of the detector or the analysis procedure have no significant effect on these results. One-sided jet broadening is tested in Figure 19. The jet axis was here determined from the thrust of one the particles in the "narrow" jet. There is of course a trivial difference between the $\langle p_T^2 \rangle$ on the "narrow" and "wide" side, due to the built-in bias in the selection of the two sides. The effect of this bias can be seen from the curves labelled $q\bar{q}$ showing the effect of an identical selection of 2-jet Field-Feynman events. There remains a clear indication in the data of a true-one-sided jet broadening at the highest W. It would appear difficult to find a plausible explanation for such an effect in terms of some peculiarity of the hadronization process in a pure $q\bar{q}$ picture.

At the European Physical Society Meeting P. Söding of TASSO has also reported seeing a few 3-jet events. The origin of these 3-jet events has not yet been determined.

PLUTO

The PLUTO group is the first group to show the upsilon decay into three gluons.

The PLUTO detector[36] is shown in Figure 20. The central detector consists of thirteen layers of proportional wires. Two layers of lead are positioned after the seventh layer to convert neutrals. The central detector is surrounded by lead-scintillator arrays. Wire chambers are used to locate the shower origin. The angular coverage is 96% of 4π.

A superconducting coil produces a magnetic field of 1.7 T inside a volume 1.55 m long and 1.4 m in diameter. Two layers of proportional tubes outside the iron yoke cover 82% of 4π. An extra iron housing is used to filter hadrons. The outermost multi-layer of drift chambers cover 83% of 4π. There are two small angle detectors one covering $15^o > \theta > 4^o$ and the other $4^o > \theta > 1.3^o$. Compensating coils are also used to correct for the effects of the solenoidal field on the beam particles.

The trigger conditions are beam crossing in coincidence with two coplanar or more than two tracks detected by the online wire logic for the inner detector, or, energy > 3 GeV in the central shower counter or forward spectrometers, or, energy in the forward spectrometer together with 1 GeV shower energy or $\geq$ 1 track in the central detector.

The acceptance factor ε of hadron events is obtained from a Monte Carlo study using the Feynman-Field model (with u,d, and s quarks) in a realistic simulation of the detector. The result is $\varepsilon = 0.72$ (average). For the determination of the total cross section additional corrections are necessary for radiation effects ($\sim$ 10%), and for the estimated contribution by two photon exchange events derived from a Monte Carlo study. A systematic error of 20% is mainly due to uncertainties in the luminosity determination and in the acceptance calculation. The R values are shown in Figure 21 together with values measured by PLUTO below 10 GeV.[37] The QCD expectation $R = 3\, \Sigma\, Q_i^2 (1 + \alpha_s/\pi)$ is also shown. In the

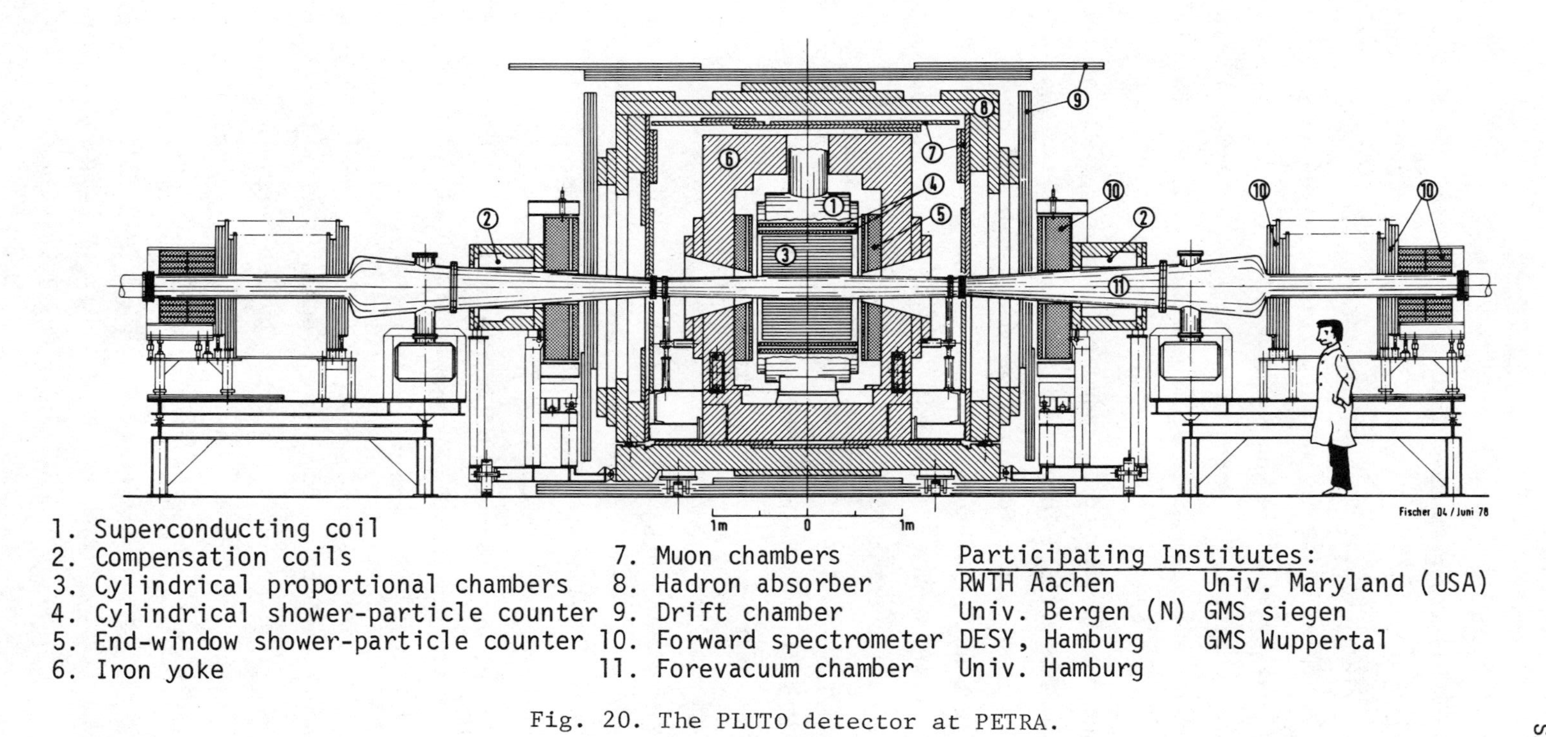

Fig. 20. The PLUTO detector at PETRA.

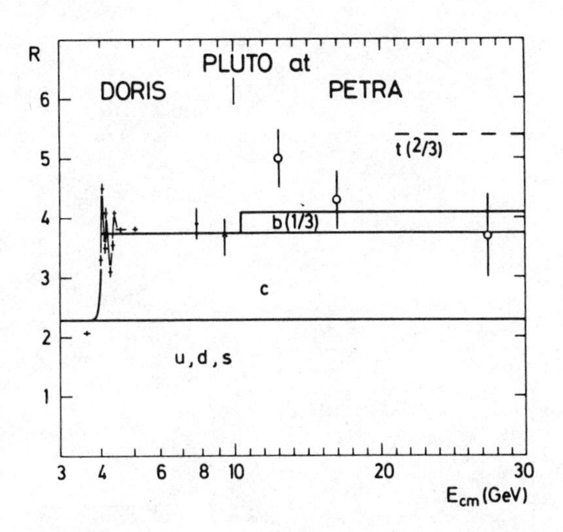

Fig. 21. Measurements by the PLUTO group from both DORIS and PETRA.

asymptotic regions the total cross section are saturated by the contributions from the u,d,s, and c quarks below 10 GeV, the higher energy data allow for a small increase due to a charge 1/3 quark (b). The 27.4 GeV data do not show evidence for an increase of R due to a potential new charge 2/3 quark (t).

A jet analysis was carried out in terms of thrust, T. Figure 22a shows the observed mean values of <1-T> between 7.7 and 27.4 GeV, showing a clear decrease with increasing energy. The distributions of the observed thrust are shown in Figure 22 b-f for the different energies. Also shown are curves from a Monte Carlo study based on the Feynman-Field model of quark parton jets, with full simulation of the detector and radiative corrections. The distributions generally follow the expected behavior, using u,d, and s quarks only. Also shown in Figure 22a is the dependence of <1-T> , if $b\bar{b}$ pair production and decay is included.[38] In the thrust distributions at higher energies the additional contributions for the c and b quarks are included. At 27.4 GeV in addition the expectation is shown, if a new heavy quark (2/3 charge) is added, assuming a threshold a few GeV below 27.4

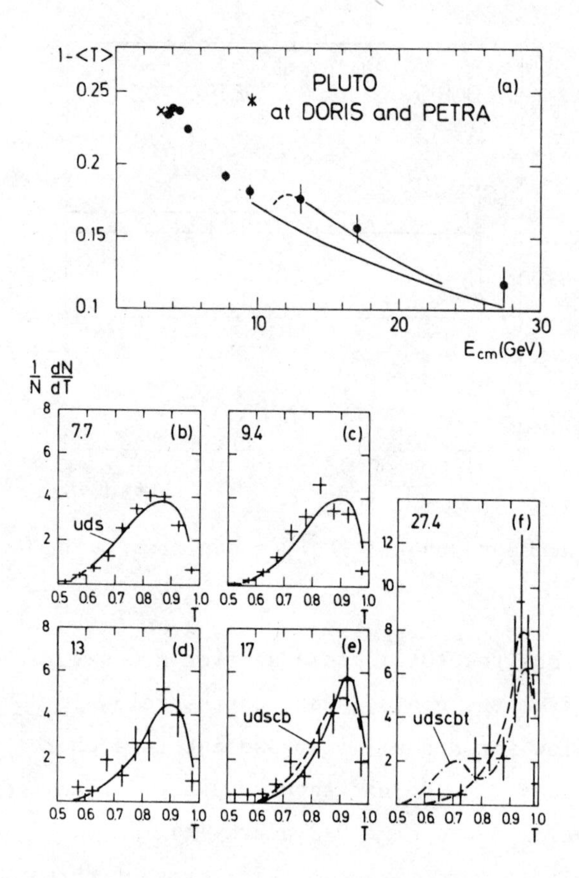

Fig. 22. PLUTO results on thrust measurements from DORIS and PETRA.

GeV. No evidence is found for this additional contribution, which results in values of T around 0.7.

Figures 23, 24, and 25 show the "seagull" effect with p_T plotted against x_L at 13, 17 and 27 GeV.[39] For comparison, Monte Carlo calculations with u, d, and s quarks are shown. A u-d-s with gluon calculation is also shown with the 27 GeV data. Again gluon emission explains the data better.

JADE

Figure 26 shows the JADE detector.[40] Innermost are the beam pipe counters. From 20 to 80 cm radii are three layers of jet chambers which are 2.36 m long. They are divided into sectors of

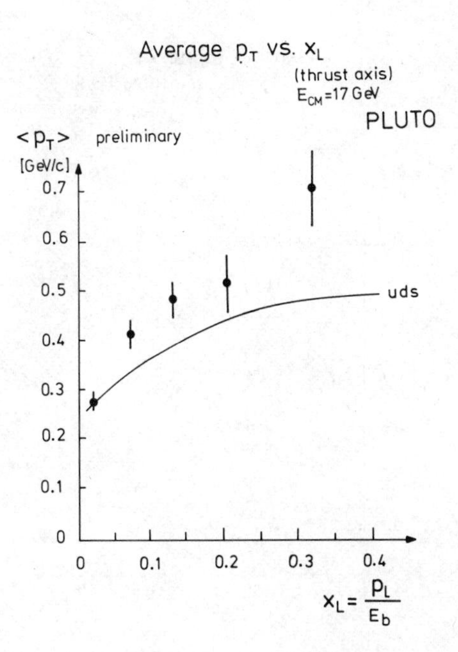

Fig. 23. Study of the "seagull" effect from the PLUTO Group at 13 GeV.

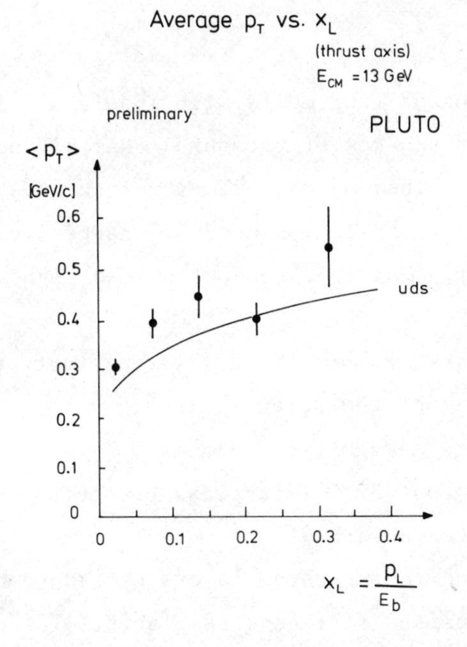

Fig. 24. Study of the "seagull" effect from the PLUTO Group at 17 GeV.

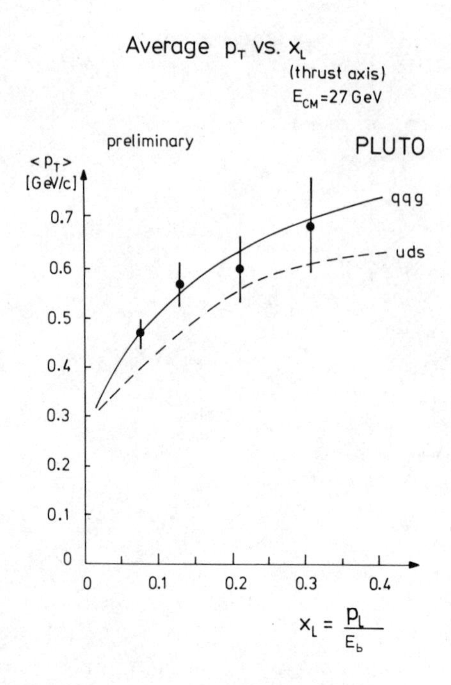

Fig. 25. Study of the "seagull" effect from the PLUTO Group at 27GeV.

15° each consisting of four cells with 16 wires. The jet chambers are under four atmospheres of pressure. Radial and azimuthal coordinates are obtained by measurements accurate to 150m per point, and the axial coordinate is measured by charge division to an accuracy of 1.6 cm. Multiple readout allows eight particles per signal wire.

Outside the jet chambers is a layer of forty two time of flight counters. The coil of the solenoid is 3.5 m long and 2 m in diameter. It produces a magnetic field of 0.5 T. The inner detector is surrounded by about 2900 lead-glass counters covering 93% of 4π.

The muon filter covers 94% of 4π. It consists of 4 planes of drift chambers sandwiching three layers of iron-concrete. A double tagging systen consists of lead-glass, scintillator and drift chambers.

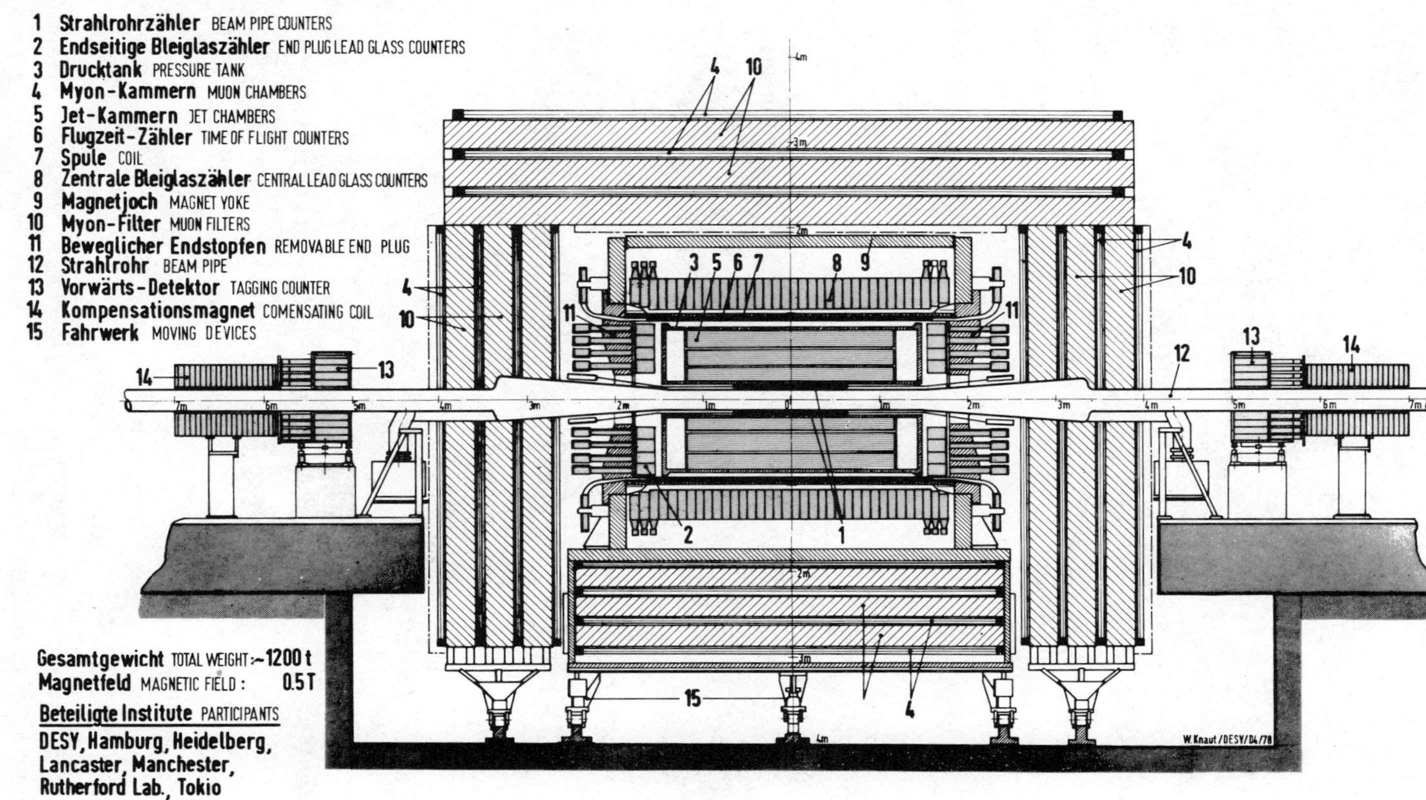

Fig. 26. The JADE detector at PETRA.

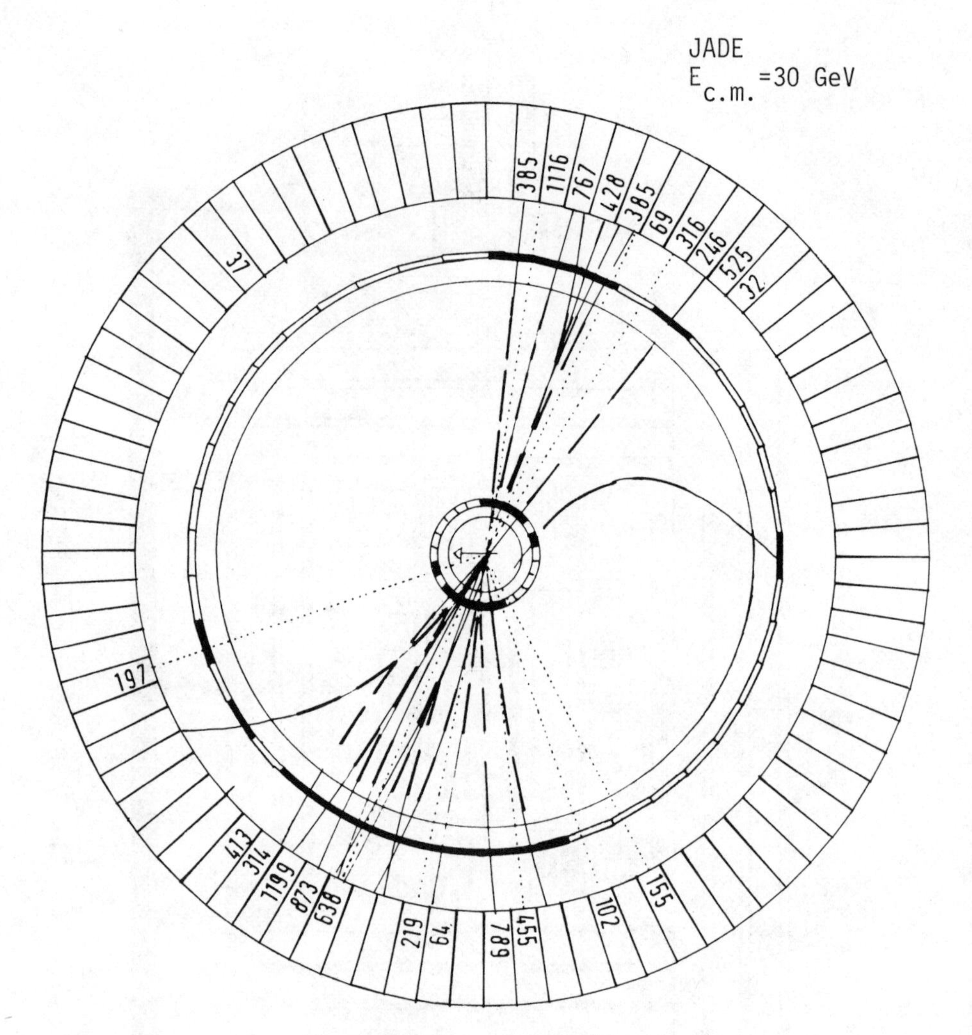

Fig. 27. A typical hadronic event seen at the JADE detector.

Finally, there are compensating coils to correct for the effects of the solenoidal magnetic field on the beam particles.

This detector is just starting to take data. Figure 27 shows one of the events.

ACKNOWLEDGEMENTS

I wish to thank Professor A. Zichichi for his kind hospitality during my stay at his school. I wish also to thank Professor Min

Chen of MARK J, Professor G. Wolf of TASSO, Professor R. Felst of JADE and Professor U. Timm of PLUTO for discussions, and I wish especially to thank Dr. D. Fong for his help in preparing this manuscript.

REFERENCES

1) PETRA Proposal (updated version), DESY, Hamburg, (February, 1976).

2) G. Voss, The 19 GeV e^+e^- Storage Ring, Internal Report, DESY M/79/16.

3) A. Febel and G. Hemmie, "PIA, the Positron Intensity Accumulator for the PETRA Injection", Internal Report, DESY M/79/13.

4) M. Sands, Proceedings of the International School of Physics "Enrico Fermi": Physics with Intersecting Storage Rings, Varenna on Lake Como, 1969 (Academic Press, New York and London, 1971) p. 257.

5) G. Voss, Private Communication.

6) U. Becker et al., "A Simple Detector to Measure e^+e^- Reactions at High Energy", Proposal to PETRA Research Committee (March 1976).

7) P.D. Luckey et al., Proceedings of the International Symposium on Electron and Photon Interactions at High Energies, Hamburg, 1965, (Springer, Berlin 1965), Vol. II, p. 397.

8) M. Bernardini et al., Phys. Lett. 45B, 510 (1973).

9) V. Alles-Borelli et al., Phys. Lett. 59B, 201 (1975).

10) M. Bernardini et al., INFN/AE-67/3, Proposal submitted 20th March, 1967.

11) V. Alles-Borelli et al., Nuovo Cimento Letters 4, 1145 (1970).

12) M. Bernardini et al., Nuovo Cimento 17A, 383 (1973).

See also S. Orito et al., Phys. Lett. 48B, 165 (1974).

13) S.D. Drell, Ann. Phys. (N.Y.) 4, 79 (1958), T.D. Lee and G.C. Wick, Phys. Rev. D2, 1033 (1970).

14) S.A. Berends et al., Phys. Lett. 63B, 432 (1976).

15) D.P. Barber et al., Phys. Rev. Lett. 42, 1110 (1979).

16) R.D. Field and R.P. Fenyman, Nuclear Physics B136, 1 (1978).

17) M.K. Gaillard, B.W. Lee and J.L. Rosner, Rev. Mod. Phys. 47, 277 (1975).
C. Quigg and J.L. Rosner, Phys. Rev. D17, 239 (1978).

18) A. Ali, J.G. Koerner, G.K. Kramer, and J. Wilrodt, DESY Report 78/51 and 78/67 (1978), unpublished.

19) A. Ali, Z. Phys. C, Particles and Fields, 1, 25 (1979).
A. Ali and E. Pietarinen, DESY-Report 79/12 (1979), unpublished.

20) M. Kobayashi and T. Maskawa, Prog. Theor. Phys. 49, 652 (1973).

21) G. Bonneau and F. Martin, Nuclear Physics B27, 381 (1971).

22) Y.S. Tsai, Rev. Mod. Phys. 46, 815 (1974).
L.W. Mo and Y.S. Tsai, Rev. Mod. Phys. 41, 205 (1969).

23) H. Terazawa, Rev. Mod. Phys. 45, 615 (1973).
We use $\sigma(\gamma\gamma\text{-multipion}) = |240+270/W^2|$ nb. W is the energy of the two-photon system.

24) Y.S. Tsai, Phys. Rev. D4, 2821 (1971).
H.B. Thacker and J.J. Sakurai, Phys. Lett. 36B, 103 (1971).
K. Fujikawa and N. Kawamoto, Phys. Rev. D14, 59 (1976).
M.L. Perl, in Proceedings of the 1977 Symposium on Lepton and Photon Interactions at High Energies, edited by F. Gutbrod, (DESY, Hamburg 1977), p. 145.

25) D.P. Barber et al., Phys. Rev. Lett. 42, 1113 (1979).
D.P. Barber et al., Phys. Rev. Lett. 43, 901 (1979).
D.P. Barber et al., Phys. Lett. 85B, 469 (1979).
and M. Chen, Private Communication.

26) TASSO Collaboration, R. Brandelik et al., Phys. Lett. 83B, 201 (1979).
G. Wolf, High Energy Trends in e^+e^- Physics, DESY 79/41.
P. Söding, Invited Talk at the EPS Conference, Geneva, 1979.

27) PLUTO Collaboration, Ch. Berger et al., Phys. Lett. 81B, 410, (1979).

28) TASSO Collaboration, R. Brandelik et al., Phys. Lett. 83B, 261 (1979).

29) DASP Collaboration, R. Brandelik et al., Nucl. Phys. B148, 189 (1979).

30) E. Albini, P. Capiluppi, G. Giacomelli and A.M. Rossi, Nuovo Cimento 32A, 101 (1976).

31) R. Schwitters, Rapperteur Talk, 1975 Stanford Conference, p.5.

32) R. Baier, J. Engels and B. Petersin, University of Bielefeld, Report BI-TP 79/10 (1979).
W.R. Frazer and J.F. Gunion, University of California, Report UCP-78-5 (1978).

33) PLUTO Collaboration, Ch. Berger et al., Phys. Lett. B78, 176 (1978).

34) A. De Rujula, J. Ellis, E.G. Floratos and M.K. Gaillard, Nucl. Phys. B138, 387 (1978).
G. Kramer and G. Schierholz, Phys. Lett. 82B, 108 (1979).

35) P. Hoyer, P. Osland, H.G. Sander, T.F. Walsh and P.M. Zerwes, DESY 79/21 (1979).

36) PLUTO Collaboration, Contribution to the EPS Conference, Geneva 1979.

37) PLUTO Collaboration, J. Burmester et al., Phys. Lett. 66B, 395 (1977).
PLUTO Collaboration, Ch. Berger et al., Z.Physik C, 1, 343 (1979).

38) A. Ali et al., DESY Report 79/16 (1979).

39) PLUTO Collaboration, Preliminary Results, Private Communication. I wish to thank Dr. U. Timm for discussion of his results with me in July 1979.

40) R. Felst, Private Communication.

DEEP INELASTIC PHENOMENA

J.J. Aubert

LAPP, Annecy le Vieux, France, and

CERN, Geneva, Switzerland

ABSTRACT

Deep inelastic phenomena are reviewed and a few typical historical experiments are described to show the continuity in this field. Scaling, scaling violation, and new recent measurements are described. Experimental difficulties are pointed out and care is taken to show how to compare the data with theoretical prediction. A comparison of Drell-Yan pair production with lp deep inelastic scattering is made.

1. HISTORY OF THE ELECTRON-ATOM AND ELECTRON-NUCLEUS SCATTERING[1)]

It is not the purpose of this lecture to give a complete description of all the past experiments on electron scattering; the

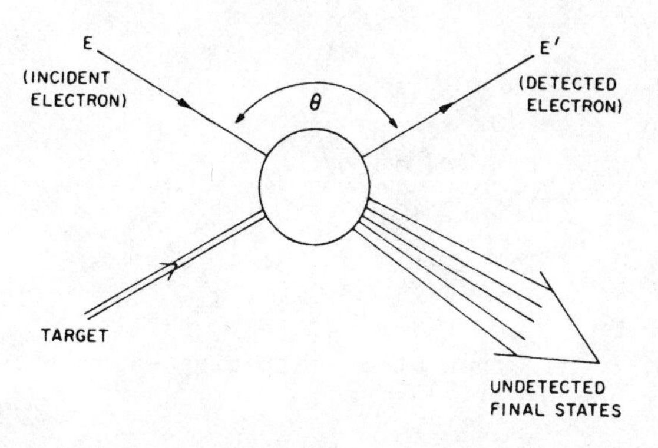

lecture will be restricted to the study of two characteristic experiments which have shown the corpuscular structure of matter. Emphasis will be laid on the continuity of these experiments.

1.1 Electron-Atom Scattering - Mohr and Nicoll's Experiment, 1932 [2)]

The layout of this experiment is shown in Fig. 1. The electron source consists of a tungsten wire and of a collimator which fixes the emission angle. The electrons interact with the gas inside the chamber. Scattered electrons are then collected through an adjustable potential and the collimators S which define the scattered energy; the angle is defined by the orientation of the source. The current measured in the Faraday cup is proportional to the scattering cross-section at a given value of θ and E'.

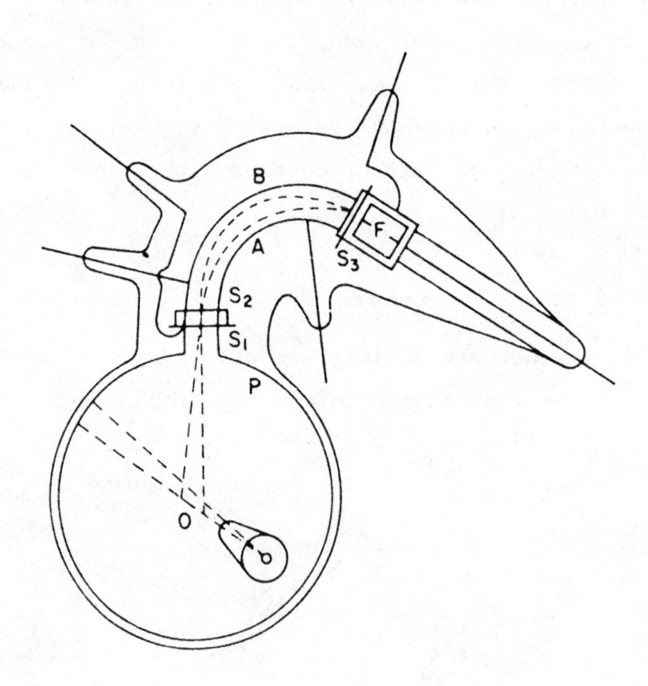

Fig. 1 Electron-atom scattering experiment of Mohr and Nicoll[2)]

Figure 2 displays the scattering cross-section (arbitrary units) as a function of the energy loss ($\nu = E - E'$) at a fixed angle θ. One sees the elastic peak at $\nu = 0$ as well as other structures. These peaks correspond to the excitation level of the given atom: a) mercury, b) helium, or c) argon. (They are indeed the same as those determined with the emission spectra.)

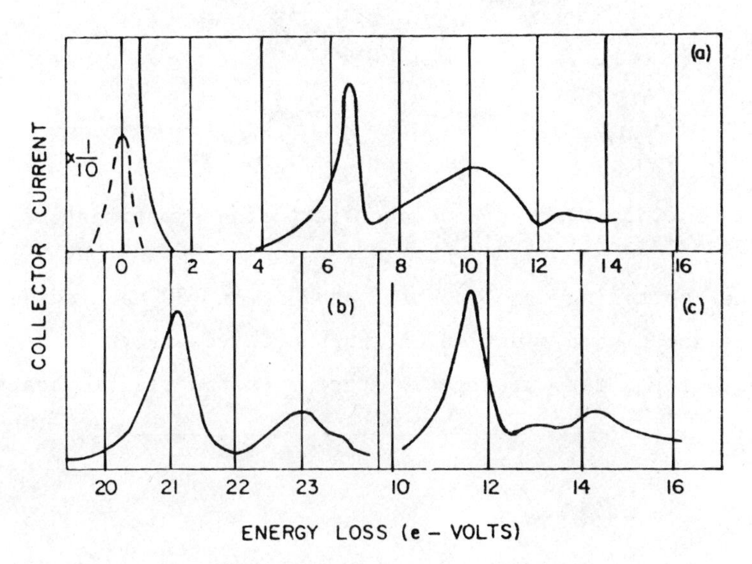

Fig. 2 Experimental results of electron-atom scattering[2] in a) mercury, b) helium, c) argon

This experiment reveals the structure of atoms with their different excitation levels. Notice the energy scale, which is a few electronvolts.

1.2 Electron-Nucleus Scattering - Hofstadter's Experiment, 1950 [3]

The electron source is now a linear accelerator (see Fig. 3). A collimated electron beam is transported to a scattering chamber, where electrons will interact with the target nucleus.

The measurement principle is the same as before; scattered electrons are measured through a spectrometer, which determines the flux at fixed angle θ and energy E'.

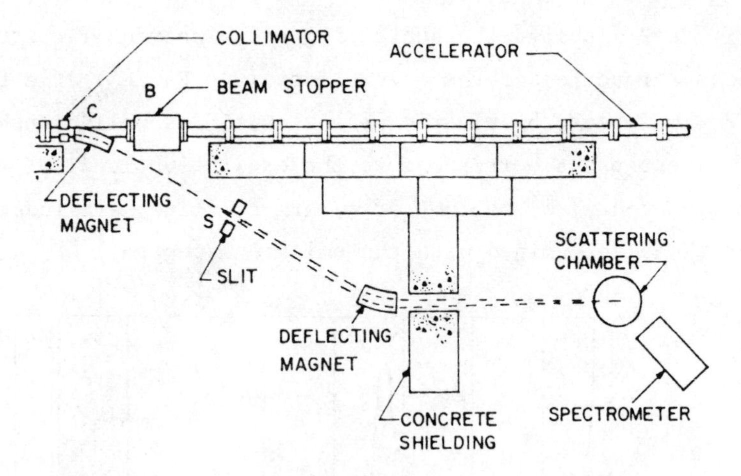

Fig. 3 Electron-nucleus scattering experiment of Hofstadter et al.[3]

The scattering cross-section at 90° for 150 MeV incident electrons is shown as a function of the scattered electron energy (E′) in Fig. 4 for a carbon target. Apart from the elastic scattering,

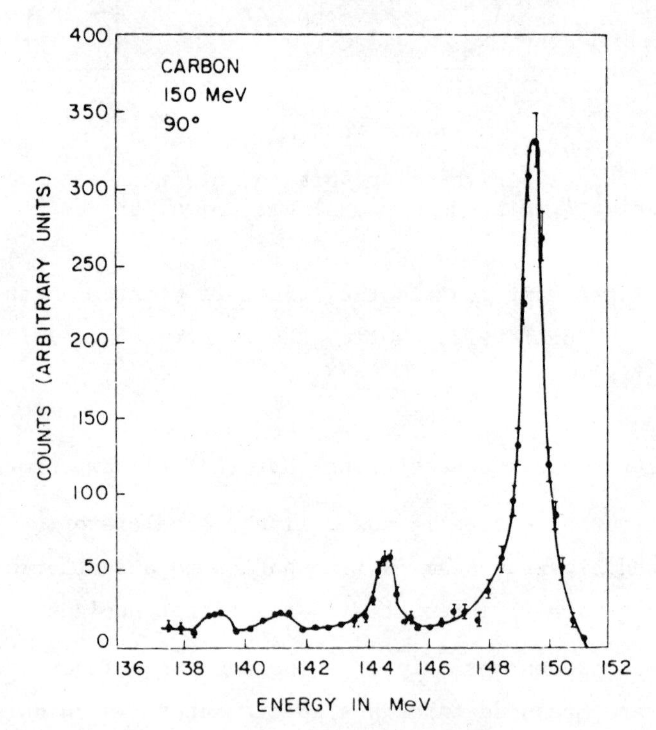

Fig. 4 Electron-C scattering: differential cross-section[3]

one observes the different excitation levels of the C nucleus. In Fig. 5 the scattering cross-section e^{4}He is displayed at 400 MeV incident electron energy and 45° scattering angle. There is no other narrow peak than that of the elastic scattering on the nucleus. At the position of the elastic scattering on the nucleon there is a broad peak, which is interpreted as the scattering on each nucleon smeared by the Fermi motion of nucleons inside the nucleus.

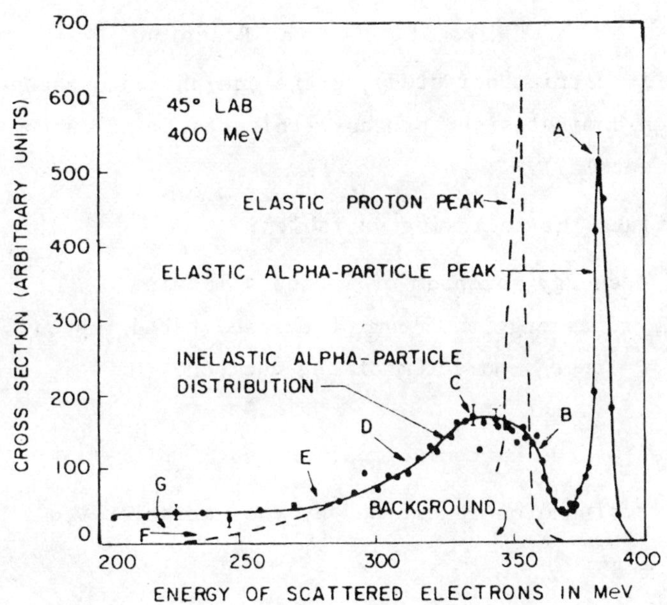

Fig. 5 Electron-^{4}He scattering: differential cross-section[3)]

We have seen that low-energy (a few eV) electron scattering shows the structure of the atoms and that medium energy (a few hundred MeV) reveals the nucleus structure. Let us see what has been learnt from electron-nucleon scattering with a few GeV electron probe.

2. ELECTRON-NUCLEON SCATTERING -- SCALING AND PARTON MODEL

We will describe the phenomenology of electron-nucleon scattering experiments carried out from 1960 to 1970.

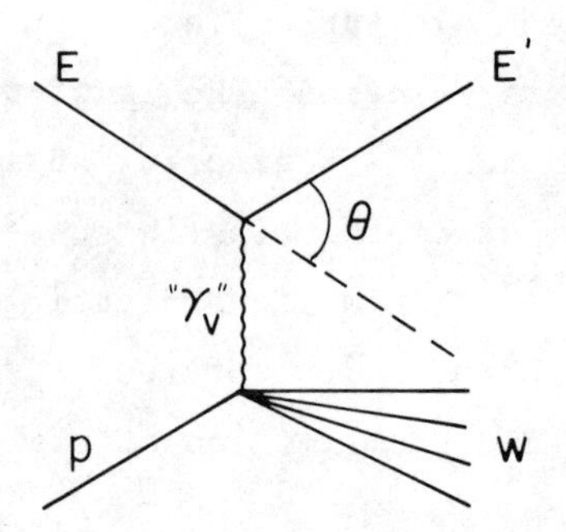

2.1 Kinematics of the Reaction

We shall restrict our study to the one-photon exchange cross-section which dominates the process (this statement can be verified experimentally).

One defines the following variables:

$k(E,\vec{k})$ = energy momentum of incident lepton

$k'(E',\vec{k}')$ = energy momentum of the scattered lepton

$p(E_p,\vec{p})$ = energy momentum of the nucleon, $(M,0)$ in the laboratory

W = mass of the hadronic final state.

The virtual photon is defined by two variables ν, q^2:

$$\nu = E - E'$$

$$q^2 = (k - k')^2 = -2(EE' - |\vec{k}||\vec{k}'| \cos\theta - m^2).$$

m is the mass of the lepton; if one neglects m $(E \simeq |\vec{k}|)$, then

$$q^2 = -4EE' \sin^2 \theta/2$$

$$W^2 = q^2 + M^2 + 2M\nu .$$

Then one defines

$$\omega = \frac{2p\cdot q}{q^2} = \frac{2M\nu}{q^2}$$

$$x = \frac{1}{\omega} .$$

On a plot q^2,ν one sees the lines of constant W, constant x, and constant θ.

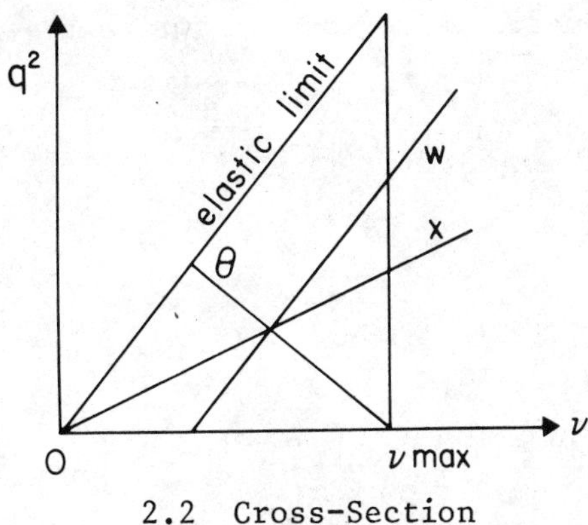

2.2 Cross-Section

The Lagrangian of the interaction is

$$H_i = e\, A^{\mu}_{(x)} \left[J^{lept}_{\mu(x)} + J^{had}_{\mu(x)} \right]$$

where

A^{μ} is the photon field,

J^{lept}_{μ} is the lepton electromagnetic current for a point Dirac particle,

J^{had}_{μ} is the hadron electromagnetic current; it is unknown.

The matrix transition can then be calculated with Feynman graph rules

$$T_{xi} = -e^2 \frac{J^{lept}_{\mu} J^{had}_{\mu}}{q^2} ,$$

where

$J^{lept}_{\mu} = \bar{u}(k',\lambda')\gamma^{\mu}u(k,\lambda)$ (μ = spinor of lepton, λ = lepton spin),

$J^{had}_{\mu} = \langle x | J_{\mu} | \text{nucleon} \rangle$.

If one detects only the scattered muon, the cross-section is obtained when one sums over all final states x. Furthermore, if incident particles are not polarized and one does not measure the scattered lepton polarization, one then has to sum over all the helicity of the outgoing lepton and average on the initial polarization state.

Up to a kinematic factor one can write the cross-section:

$$\frac{d\sigma}{d^3\vec{k}'} \approx L^{\mu\nu} W_{\mu\nu} ,$$

where

$$L^{\mu\nu} = \frac{1}{2} \sum_{\text{spin}} J_{\mu}^{\text{lept}} J_{\nu}^{\text{lept}}$$ is the known leptonic tensor ,

$$L^{\mu\nu} \approx 2 \left[2k^{\mu}k^{\nu} - k^{\mu}q^{\nu} - q^{\mu}k^{\nu} + g^{\mu\nu} \frac{q^2}{2} \right] ,$$

and $W_{\mu\nu}$ is the unknown hadronic tensor.

To ensure the Lorentz invariance of the matrix element, the general form for $W_{\mu\nu}$ is a second-rank tensor, which depends only on two variables, p and q:

$$W_{\mu\nu} = Ag_{\mu\nu} + \frac{B}{M^2} p_{\mu}p_{\nu} + \frac{C}{M^2} q_{\mu}q_{\nu} + \frac{D}{M^2} (q_{\mu}p_{\nu} + q_{\nu}p_{\mu}) +$$

$$+ \frac{E}{M^2} (q_{\mu}p_{\nu} - q_{\nu}p_{\mu}) + \frac{F}{M^2} \varepsilon_{\mu\nu\alpha\beta} \, p^{\alpha}q^{\beta} ,$$

where A, B, C, D, E, and F are scalar functions of ν and q^2, $\varepsilon_{\mu\nu\alpha\beta}$ is the total antisymmetric tensor, $g_{\mu\nu}$ is the metric ($g_{00} = 1$,

$g_{11} = g_{22} = g_{33} = 0.1$, $g_{\mu\nu} = 0$ if $\mu \neq \nu$).

If one asks that parity is conserved, then F = 0 (see Appendix). If one also applies the current conservation $q^{\mu}J_{\mu} = 0$, then $q^{\mu}W_{\mu\nu} = q^{\nu}W_{\mu\nu} = 0$, so one can then write $W_{\mu\nu}$ with the following form (see Appendix):

$$W_{\mu\nu} = -W_1 \left[g_{\mu\nu} - \frac{q_{\mu}q_{\nu}}{q^2} \right] + \frac{W^2}{M^2} \left[p_{\mu}p_{\nu} + q_{\mu}q_{\nu} \frac{(q\cdot p)^2}{q^4} - \frac{q\cdot p}{q^2} (q_{\mu}p_{\nu} + q_{\nu}p_{\mu}) \right.$$

W_1 and W_2 are scalar functions of ν and q^2 only. They are called the nucleon structure functions.

One then calculates the cross-section

$$\frac{d^2\sigma}{dq^2 d\nu} = \frac{4\pi\alpha^2}{q^4}\frac{E'}{E}\left[2\sin^2\frac{\theta}{2}W_1(q^2,\nu) + \cos^2\frac{\theta}{2}W_2(q^2,\nu)\right]. \quad (1)$$

It is worth noticing

- that the photon propagator introduces a $1/q^4$ dependence;
- that the measurement of the differential cross-section gives only a combination of W_1 and W_2;
- that the cross-section can be written also in the form

$$\frac{d^2\sigma}{dq^2 d\nu} = \text{Mott cross-section} \times \left[W_2 + 2\tan^2\frac{\theta}{2}W_1\right]$$

(the Mott cross-section corresponds to the electron scattering on a point-like structure of spin 0, and the structure function measures the departure from a point-like structure);

- that by the analogy of the virtual photon with a real photon one can define a cross-section corresponding to the longitudinal (or transverse) polarization of the virtual photon

$$\frac{d^2\sigma}{dq^2 d\nu} = \Gamma(\sigma_L + \varepsilon\sigma_T),$$

where

Γ is the flux of the virtual photon,
σ_L is the cross-section for longitudinal polarization,
σ_T is the cross-section for transverse polarization,
ε is the polarization term,

$$\Gamma = \frac{e^2}{4\pi E^2 m}\left[\frac{2M\nu}{-q^2} - 1\right]\frac{1}{1-\varepsilon},$$

$$\sigma_L = \frac{4\pi^2\alpha}{\sqrt{\nu^2 - q^2}}\left[W_2\left(1 + \frac{\nu^2}{q^2}\right) - W_1\right],$$

$$\sigma_T = \frac{4\pi^2\alpha}{\sqrt{\nu^2 - q^2}} W_1 ,$$

$$\varepsilon = \frac{1}{1 + 2\left(1 - \frac{\nu^2}{q^2}\right) \tan^2 \frac{\theta}{2}} .$$

2.3 Electron-Nucleon Scattering -- SLAC Results

The experimental set-up used at SLAC with the 30 GeV linear electron accelerator is very similar to that used at lower energy for probing the structure of the nucleus, except for the size which is scaled up (Fig. 6). They have been using two different spectrometers [which have been described copiously*)], but the measurement principle is the same; one determines the scattering probability at fixed values of the scattering angle and of the scattered energy.

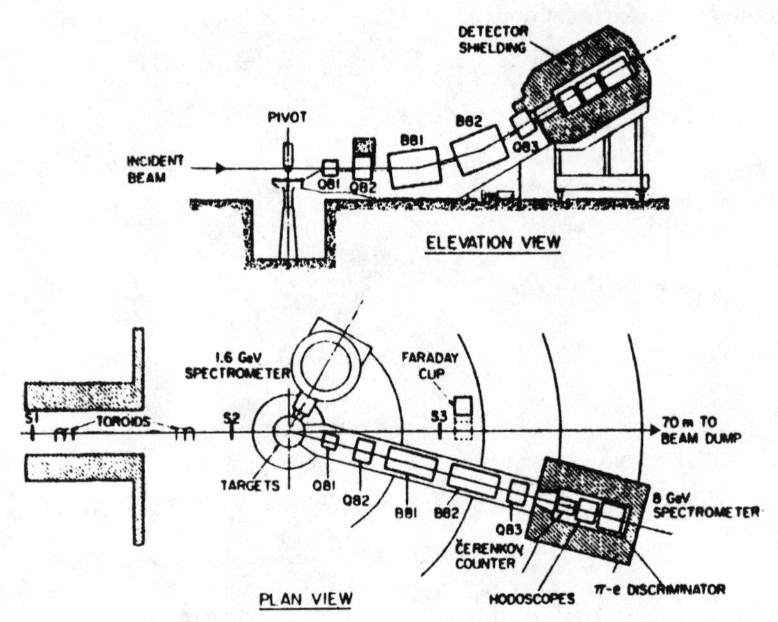

Fig. 6 Experimental set-up used in electron-nucleon scattering at SLAC: one example[4]

---

*) They have used a 20 GeV and an 8 GeV spectrometer; a description of the 8 GeV spectrometer can be found in a paper by Bodek et al.[4].

One example of these measurements is shown in Fig. 7 for incident electron energy of 4.88 GeV at a scattered angle of 10° as a function of the scattered electron. Except the elastic peak and three resolved resonances, there is a broad continuum. It is tempting to compare this continuum with the broad peak observed in the e^4He scattering (Fig. 5). This peak was explained as a quasi-elastic scattering on the He constituents broadened by the Fermi motion. One can speculate about the constituents of the nucleon, which will produce also a broad peak in the electron-nucleon scattering. It is clear that results shown in Fig. 7 are not a clear indication of nucleon constituents; nevertheless let us see what are the implications of such constituents (partons).

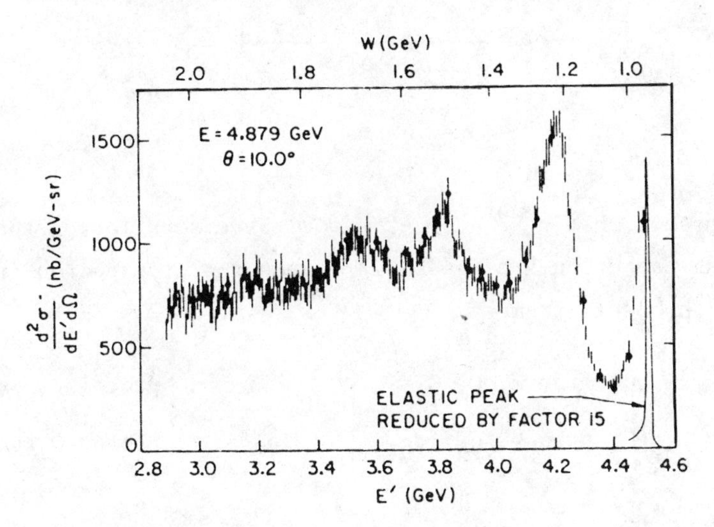

Fig. 7 Electron-proton scattering: one cross-section curve

Parton model[5] and scaling. Nucleons are built of partons, and in the electron-nucleon scattering the virtual photon interacts punctually with a parton. The Breit frame is the frame where the virtual photon energy is null, the z axis is the proton direction and the Lorentz boost is given by $\gamma v = -Q/2Mx$. In this frame, before the interaction, the kinematics are the following

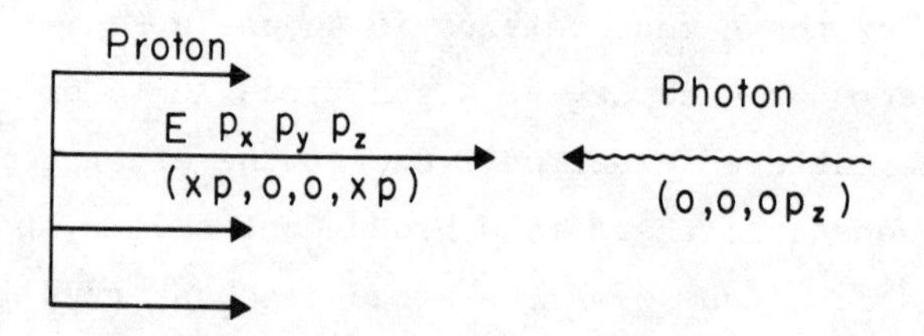

where

p is the nucleon momentum,

x is the fraction of the proton momentum carried by the parton.

One then has an interaction: photon + parton → parton. After the interaction one has the following kinematics.

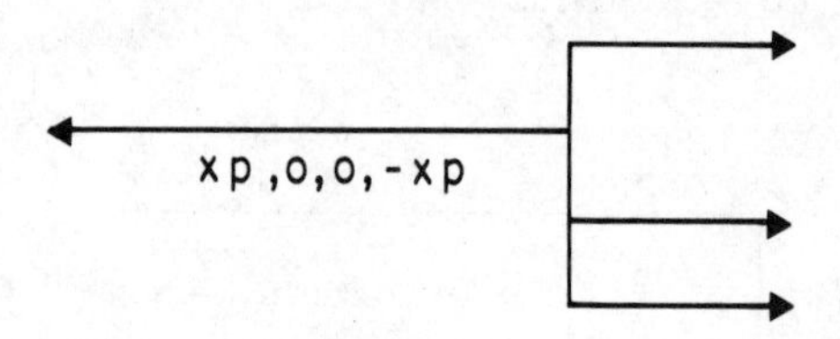

One can show that $p_z^{photon} = -2xp$ for the absorption of the photon, then one can calculate the scalar product proton·photon in the lab. and in the Breit frame

$$\text{Laboratory} \begin{cases} \text{proton } M,\ 0,\ 0,\ 0 \\ \text{photon } \nu,\ q_x,\ q_y,\ q_z \end{cases} \qquad \text{Breit} \begin{cases} \text{proton } p,\ 0,\ 0,\ p \\ \text{photon } 0,\ 0,\ 0,\ -2px \end{cases}$$

$$M\nu = +2p^2x$$
$$-q^2 = 4p^2x^2$$

so

$$x = \frac{Q^2}{2M\nu} .$$

In the Breit frame one can neglect the transverse momentum of the parton ($xp >> p_T$), so one can treat the parton as a free particle. The interaction of the virtual photon with the nucleon will then be an incoherent sum of the interaction photon-parton, the parton being punctual,

$$\frac{d^2\sigma}{dq^2 d\nu} = \sum_i \int dx\ f_i(x)\ \frac{d\sigma_x}{dq^2 d\nu}\ ep \to ep_i\ ,$$

where $f_i(x)$ is the probability of finding the parton i,

$$\left(\frac{d^2\sigma}{dq^2 d\nu}\right) ep_i \to ep_i = Q_i^2\ \frac{4\pi\alpha^2}{q^4}\ \frac{E'}{E}\ \left(2\ \sin^2\frac{\theta}{2}\ V_1 + \cos^2\frac{\theta}{2}\ V_2\right)\ ,$$

Q_i is the charge of the parton i.

For a spin 0 parton

$$V_1 = 0, \quad V_2 = \frac{x}{\nu}\ \delta\left(x - \frac{|q^2|}{2M\nu}\right)\ .$$

For a spin ½ parton:

$$V_1 = \frac{1}{2m}\ \delta\left(x - \frac{|q^2|}{2M\nu}\right)\ ,$$

$$V_2 = \frac{x}{\nu}\ \delta\left(x - \frac{|q^2|}{2M\nu}\right)\ .$$

So $\nu W_2(q^2,\nu) = \sum_i Q_i^2 x f_i(x) = F_2(x)$ is a function of x only, and

$M_p W_1 = 0$ for a parton of spin 0

$M_p W_1 = F_2(x)/2x$ for a parton of spin ½

(one notices that for a spin ½ parton one has $F_2(x) = 2xF_1(x)$, the Gallan-Gross relation).

So the structure functions νW_2 and $M_p W_1$ will depend only on one dimensionless variable $x = Q^2/2M\nu$, x being the fraction of the proton momentum carried out by the parton. This property has been called the scaling prediction.

Experimentally if one measures the structure function as a function of x for different values of ν and Q^2, one can verify this scaling prediction (the same x can be reached with different Q^2 values). Indeed these tests have been done. Figure 8 shows SLAC results[6)] in the early 70's. Within experimental error, scaling is very well verified (for $Q^2 > 1$ GeV/c^2, $W > 2$ GeV/c^2) in perfect agreement with the parton model.

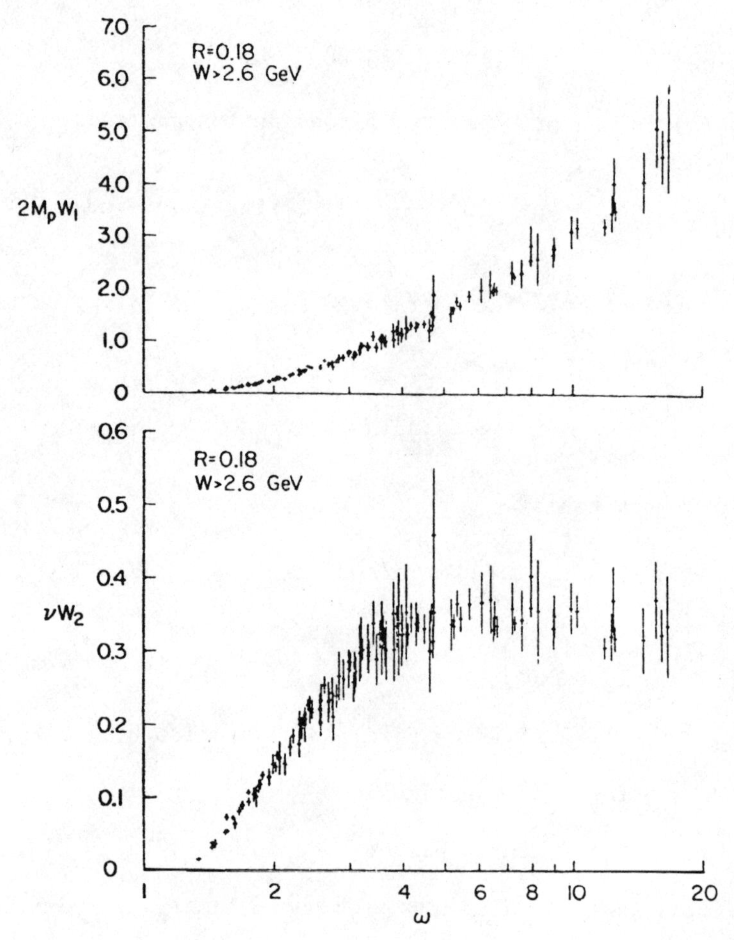

Fig. 8 Experimental evidence for scaling in electron-proton scattering: SLAC, early 1970[6])

To appreciate the quality of this work, one has to notice that the structure functions computed at a fixed value are derived from cross-sections which vary over three orders of magnitude.

2.4 Scaling Violation in Electron-Nucleon Scattering

More accurate experiments have been progressing at SLAC and FNAL since 1974 and they have all shown discrepancies with exact scaling; only a description of the Cornell-Michigan-La Jolla-Princeton[7]) experiment, which has been specifically set up for an accurate test of scaling, will be given here.

If scaling is correct, then

$$E_0 \frac{d^2\sigma}{dxdy} = \frac{4\pi\alpha^2}{2Mx^2y^2}\left[F_2(x)(1 - y) + 2xF_1(x)\frac{y^2}{2}\right]$$

where

$$x = \frac{Q^2}{2M\nu} ,$$

$$y = \frac{\nu}{E_0} ,$$

E_0 = incident muon energy ,

$F_2(x) = \nu W_2$,

$F_1(x) = M_p W_1$.

According to the authors of Ref. 7: "The test of scaling is made by comparing distributions of kinematic quantities at two incident muon energies 56.3 and 150 GeV, using a large-aperture spectrometer which changes with energy so as to keep ω acceptance and resolution constant. The realization of the scaling geometries is shown in Fig. 9. Longitudinal distances scale as $\sqrt{E_0}$; q^2 and E'

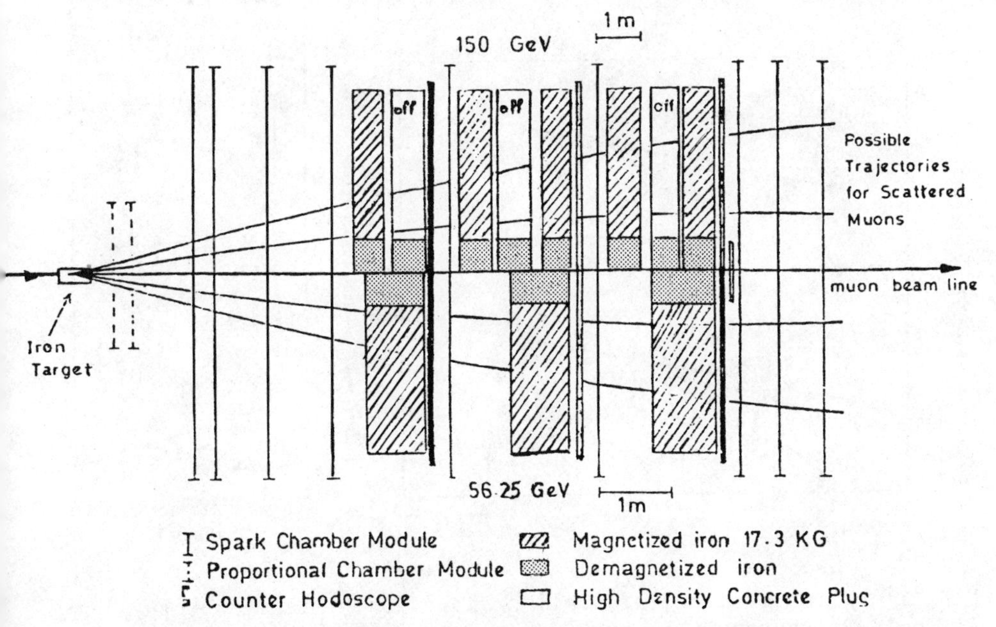

Fig. 9 Michigan-Cornell-La Jolla-Princeton Collaboration experimental set-up[7])

scale as E_0. The counting rate scales as E_0^{-1} and is compensated by scaling the target material (233 g/cm^2 at 56.3 to 622 g/cm^2 at 150 GeV). Relative momentum resolution of 14% is kept constant by using three degaussed magnets as extra scattering material in the 150 GeV configuration. The total magnetic field integral scales as $\sqrt{E_0}$ ($\langle p_T \rangle$ = 1.3 or 2.2 GeV/c) and is known to 1% with an uncertainty of 0.5% in its variation over the radius". Their results, first published in 1974 [7] (Fig. 10) show a departure from the

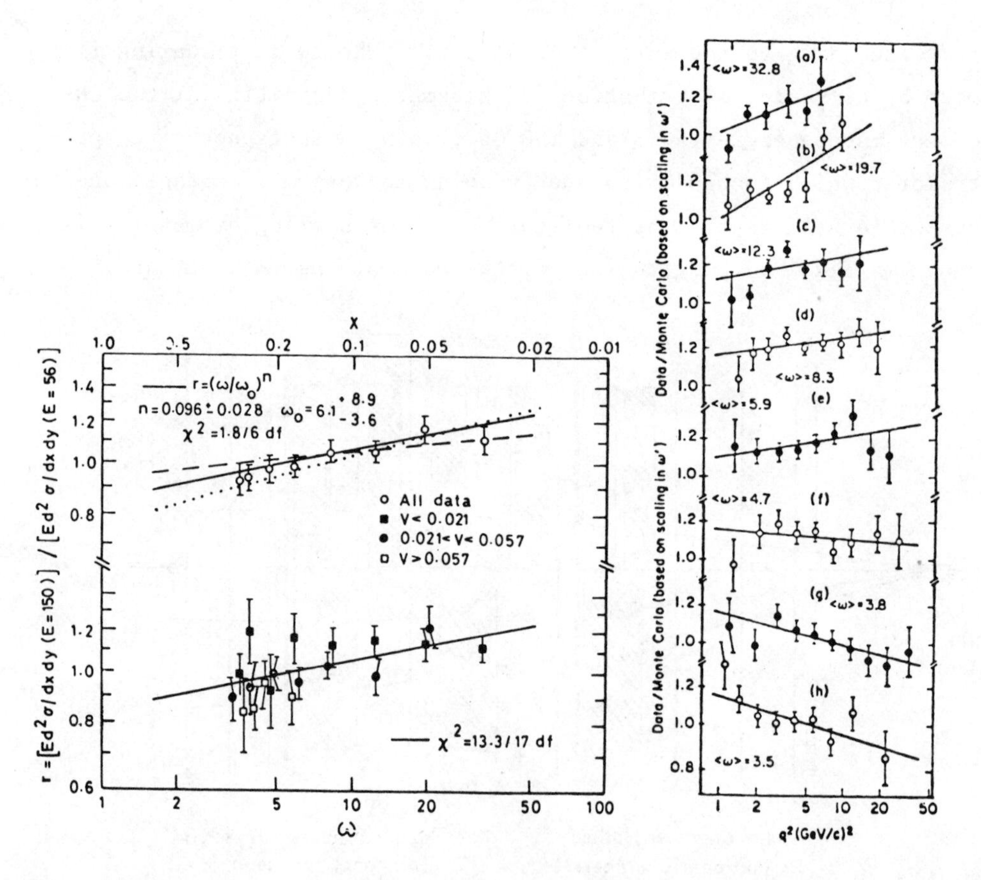

Fig. 10 First evidence of scaling violation: Michigan-Cornell-La Jolla-Princeton Collaboration[7])

precocious scaling observed before at SLAC. For $\omega > 5$, the yield of data increases as a function of Q^2 and for $\omega < 5$, there is a decrease compared with the scaling prediction. This experiment has been done on an iron target. These scaling violations have also been observed since then in more precise experiments at SLAC[4,8] and at FNAL[9] on H_2 and D_2 targets.

2.5 A Possible Explanation of the Scaling Violation

We have first to notice that the observed scaling violation is only a 20% maximum deviation when the cross-section is varying over three orders of magnitude. So the effect is small.

There have been many explanations for the scaling violation; it is not intended to review them or to explain the details of them here. A few of them will just be classified to introduce the next generation of deep-inelastic μp scattering.

2.5.1 QCD. For deep understanding of QCD, see De Rújula's talk at this school; let us just make a few simple and elementary remarks here.

In the parton model we have seen that $F_2(x)$ is related to the parton charge

$$F_2(x) = \sum_i Q_i^2 x f_i(x)\, dx \; .$$

We define

$$M_2 = \int_0^1 F_2(x)\, dx \; .$$

If $f_i(x) = f(x)$ for all partons, then

$$\int_0^1 x f(x) = \langle x \rangle = \frac{1}{\langle N \rangle} \; ,$$

where $\langle N \rangle$ is the total number of partons (neutral + charged). So

$$M_2 = \frac{\sum_i Q_i^2}{\langle N \rangle} .$$

If one supposes that the proton is built of three valence quarks (charge $+2/3$, $+2/3$, $-1/3$), then

$$M_2^{\text{proton}} = \frac{\left(\frac{2}{3}\right)^2 + \left(\frac{2}{3}\right)^2 + \left(-\frac{1}{3}\right)^2}{3} = 0.3333 .$$

Experimentally one finds $M_2 = 0.17$. So M_2 is not saturated and we have to add to the valence quark something which is called a gluon and which will carry 50% of the momentum.

In QCD, with 4 quarks and 3 colours in the limit $Q^2 \to \infty$, one predicts $M_2 = 0.12$. The experimental situation[9] is shown in Fig. 11.

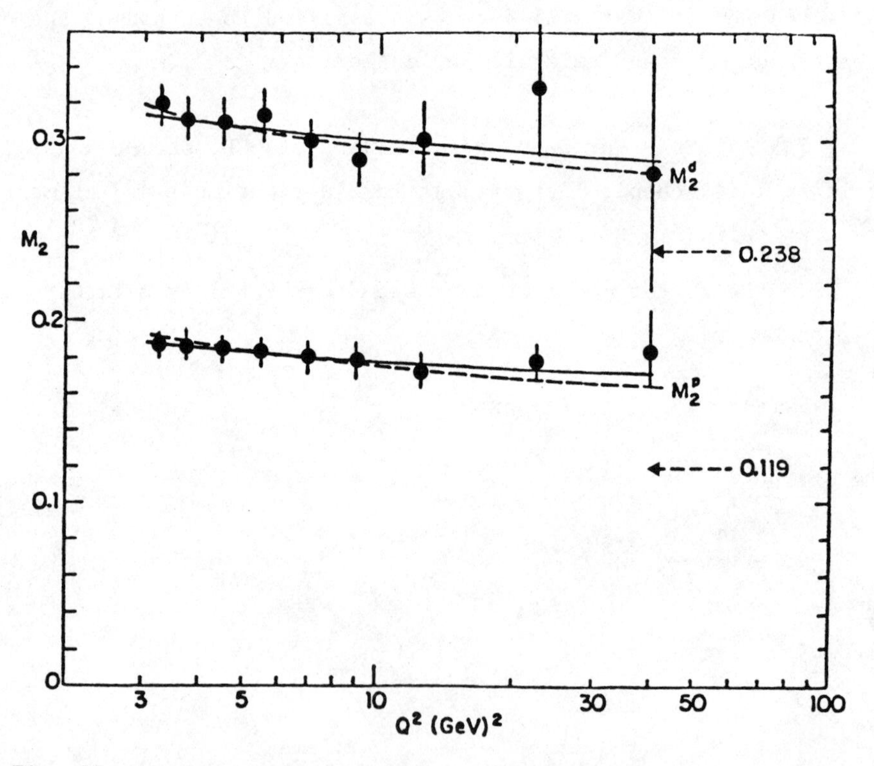

Fig. 11 Energy-momentum sum rule determined by the Chicago-Harvard-Illinois-Oxford Collaboration[9]

Now if quarks (partons) can radiate gluons, one has then to modify the calculation made with the simple parton model. By a simple argument one sees that gluon radiation will change the value of the observed $x(x' < x)$, and so there will be a scaling violation. The magnitude of this scaling violation will induce correction in $\log q^2$.

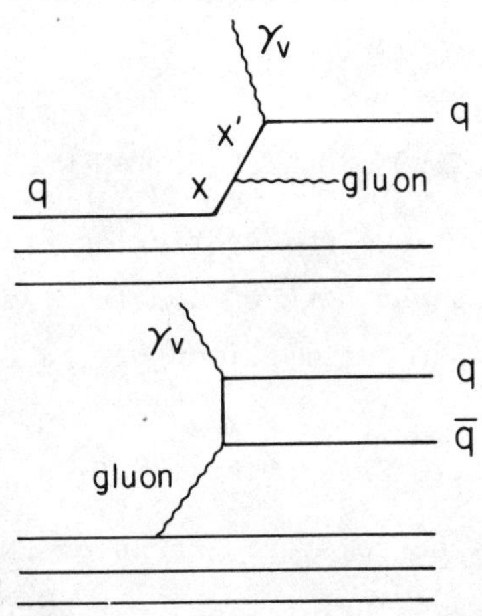

2.5.2 Pre-asymptotic régime. In the parton model the quark mass has been neglected as well as the transverse momentum of the quark. These effects have been calculated[10] using a ξ variable instead of x,

$$\xi = x \frac{2}{1 + \sqrt{1 + 4x^2m^2/Q^2}} ,$$

and it seems that only 30% of the scaling violation can be explained.

It is also clear that a new particle like charm or beauty will introduce new threshold and some increase of structure function when one crosses a new threshold; estimates have been made and again they contribute to the observed scaling violation but are not responsible for all the effect.

To explain the residual scaling violation, some authors[11] have attempted to reproduce the experimental results using a parametrization for the quark (antiquark) fragmentation function which takes into account the Regge behaviour near the limit $x \to 0$ and $x \to 1$. With six free parameters they reproduce the actual data. The scaling violation goes as $\sqrt{Q^2}$.

Using only information from deep inelastic scattering, discrimination from the different theories will come from the high Q^2 régime of the structure function (logarithmic or power law Q^2 behaviour, etc.).

3. MUON-NUCLEON SCATTERING: NEW EXPERIMENT

There are three muon-nucleon scattering experiments currently producing new results in the high Q^2 domain. After describing their main features we will make a comparison of the different results.

3.1 Experimental Set-Up

3.1.1 Multimuon spectrometer: Berkeley-FNAL-Princeton[12] (BFP). According to the authors of Ref. 12, "The spectrometer magnet (Fig. 12) serving also as a target and hadron absorber, reaches

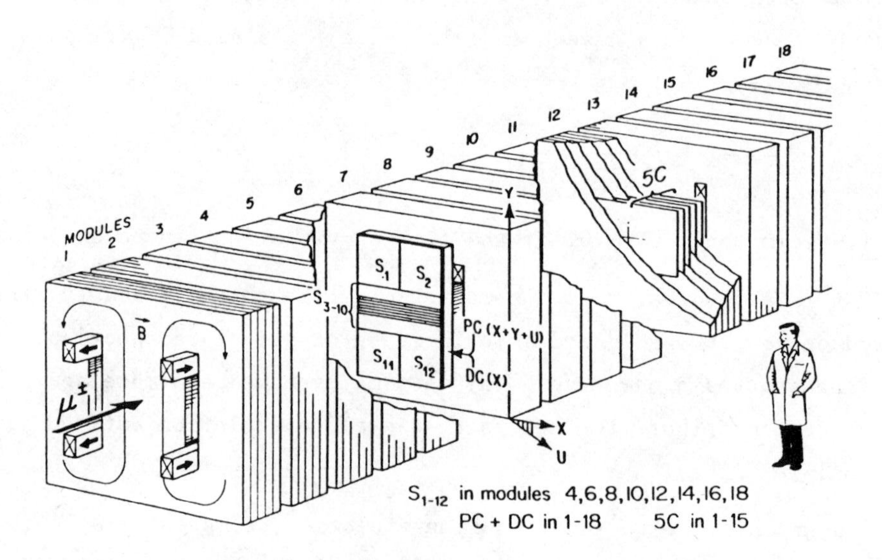

Fig. 12 Berkeley-Fermilab-Princeton Collaboration multimuon spectrometer[12]

19.7 kG within a 1.8 × 1 × 16 m^3 fiducial volume. Over the central 1.4 × 1 × 16 m^3, the magnetic field is uniform to 3% and mapped to 0.2%. Eighteen pairs of proportional chambers (PCs) and drift chambers (DCs), fully sensitive over 1.8 × 1 m^2, determine muon momenta typically to 8%. The PCs register coordinates at 30° (u) and 90° (y) to the bend direction (x) by means of 0.5 cm wide cathode strips. Banks of trigger scintillators (S_1-S_{12}) occupy 8 of the 18 magnet modules. Interleaved with the 10 cm thick magnet plates in modules 1-15 are 75 calorimeter scintillators resolving hadron energy E_{had} with r.m.s. uncertainty 1.5 $E_{had}^{\frac{1}{2}}$ (GeV). Not shown upstream of module 1 are one PC and DC, 63 beam scintillators, 8 beam PCs, and 94 scintillators sensitive to accidental beam and halo muons." Their single arm triggers request one muon in three consecutive counters outside the beam region. Their detection efficiency is shown in Fig. 13.

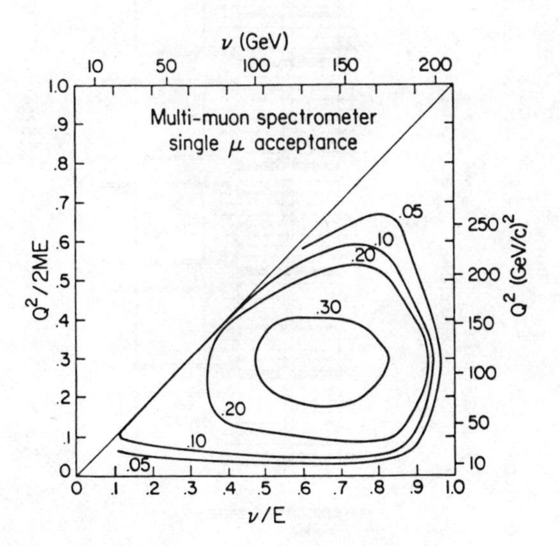

Fig. 13 Berkeley-Fermilab-Princeton Collaboration multimuon spectrometer: acceptance.

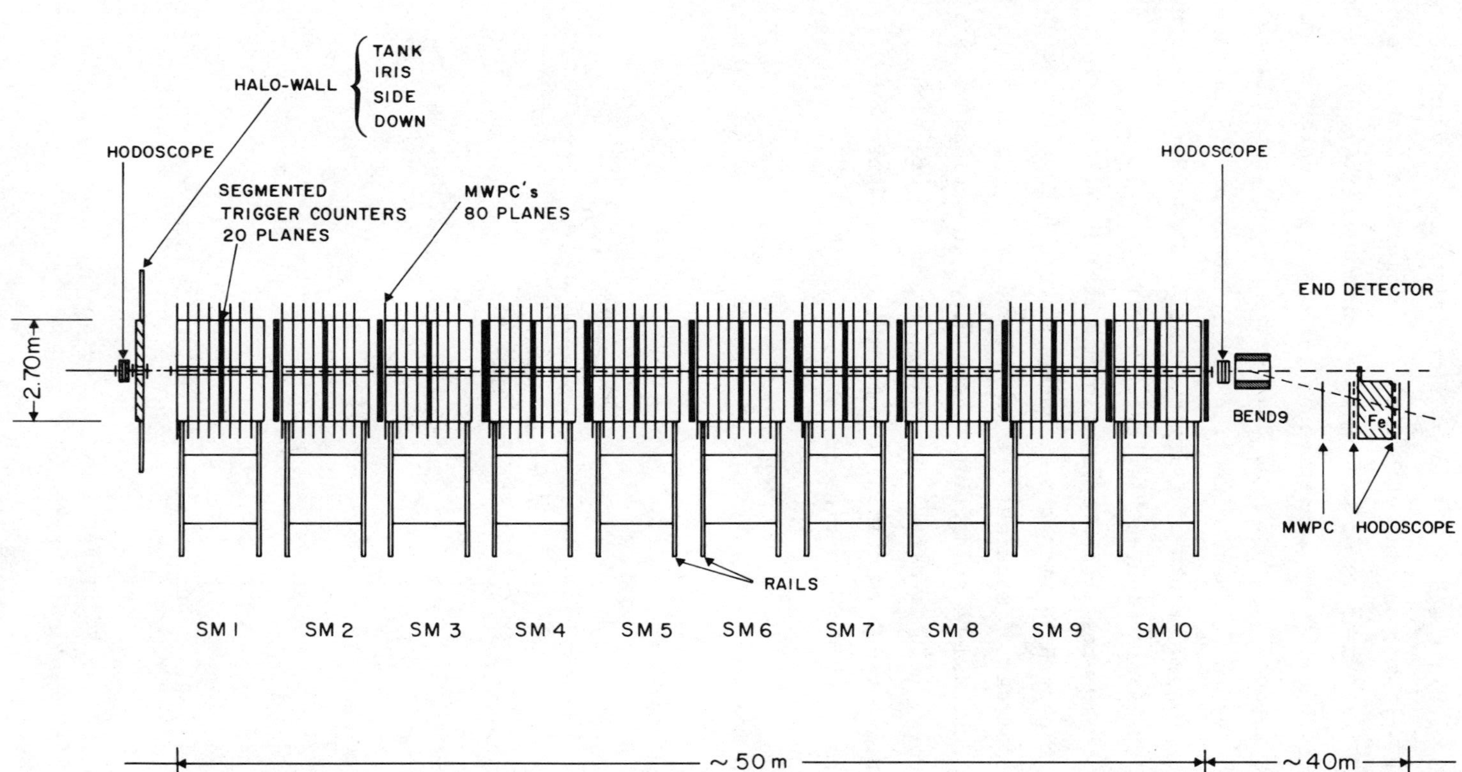

Fig. 14 Bologna-CERN-Dubna-Munich-Saclay Collaboration spectrometer[13]

3.1.2 Bologna-CERN-Dubna-Munich-Saclay Spectrometer[13] (BCDMS). The set-up of this experiment has two main features (Fig. 14):

- a 40 m long carbon target using the property of the radiation length for the muon which is $(m_\mu/m_e)^4$ larger than for the electron;
- scattered muons are analysed inside a toroidal field-focusing spectrometer. Neglecting the energy loss inside iron the trajectory is a helix and the radius of the helix Δ is proportional to Q^2/Q^2_{max}. It is a useful property for triggering on large Q^2 events. The over-all detector consists of 10 modules; each module is a sandwich of a toroidal iron magnet, a proportional chamber, and triggering counter rings. At this stage they have no calorimeter to measure the hadronic energy.

3.1.3 European Muon Collaboration[14] (EMC). The forward spectrometer used in this experiment (Fig. 15) is a conventional spectrometer with an air gap magnet, and large drift chambers before and after it measure particle direction and momentum. The muons are filtered by a calorimeter and a 2 m magnetized iron absorber. At the end of the experiment there are drift chambers which constrain the muon identification. A set of five hodoscope planes allows a trigger on large-angle scattered muons coming from the target.

A 2 kg/cm^2 iron target was used, iron being interleaved with scintillator to achieve hadron calorimetry with a resolution $\sigma_E/E = 0.4/\sqrt{E}$. Obviously the resolution achieved on momentum and angle determination is much better in such a spectrometer ($\Delta p/p = 10^{-4}$ p in GeV, $\sigma_\theta = 0.3$ mrad).

Above the angular trigger cut, the acceptance is large and uniform up to very large Q^2 and is easier to handle than for vetoing trigger.

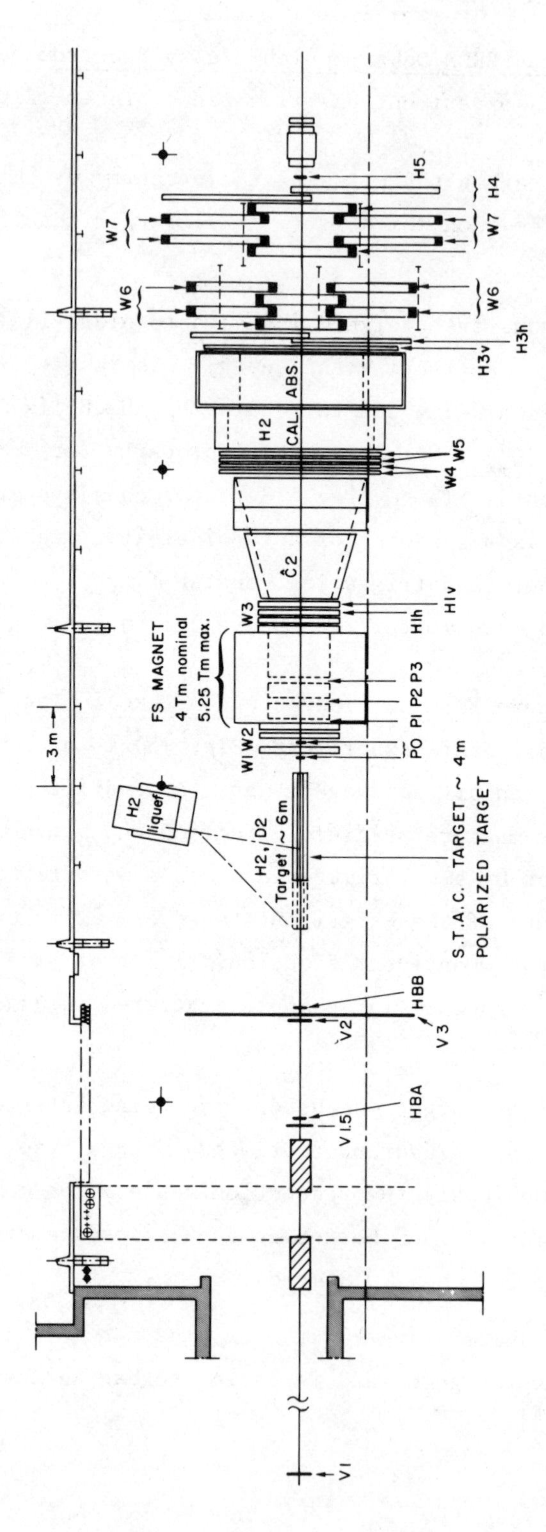

Fig. 15 European Muon Collaboration forward spectrometer[14])

3.2 Results and Comparison with Previous Experiments

3.2.1 Luminosity of the different experiments. The comparison of the different experiments is resumed in Table 1.

Table 1

	BFP	BCDMS	EMC
Target	Iron	Carbon	Iron
E_0 (GeV/c^2)	213 (4×10^{11} μ^+) 100 (5×10^9 μ)	280 (10^{11} μ)	280 (8×10^{10} μ) 250 (5×10^{11} μ)
$\mathcal{L}$ (cm^2/s)	1.5×10^{39}	5×10^{38}	2×10^{38} at 280 5×10^{38} at 250
% data presented	17	20	15

3.2.2 Results on nucleus. All data presented here are preliminary and are shown as they were presented in July 1979 at the EPS conference by the groups in the parallel session and by E.G. Gabathuler in a rapporteur's talk. Their absolute normalization is accurate to 10-15% only at this stage. Figure 16 shows

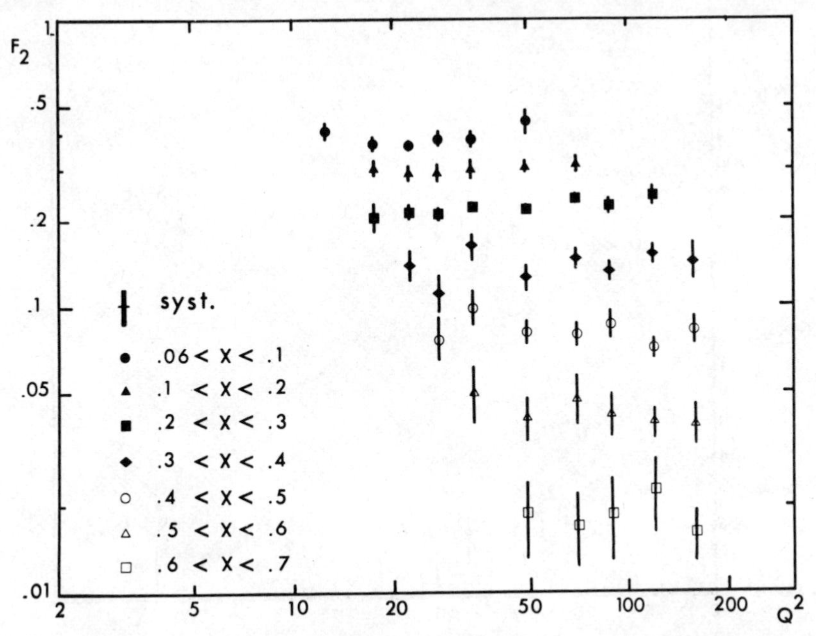

Fig. 16 Structure function $F_2(x,Q^2)$ as determined by the European Muon Collaboration[14])

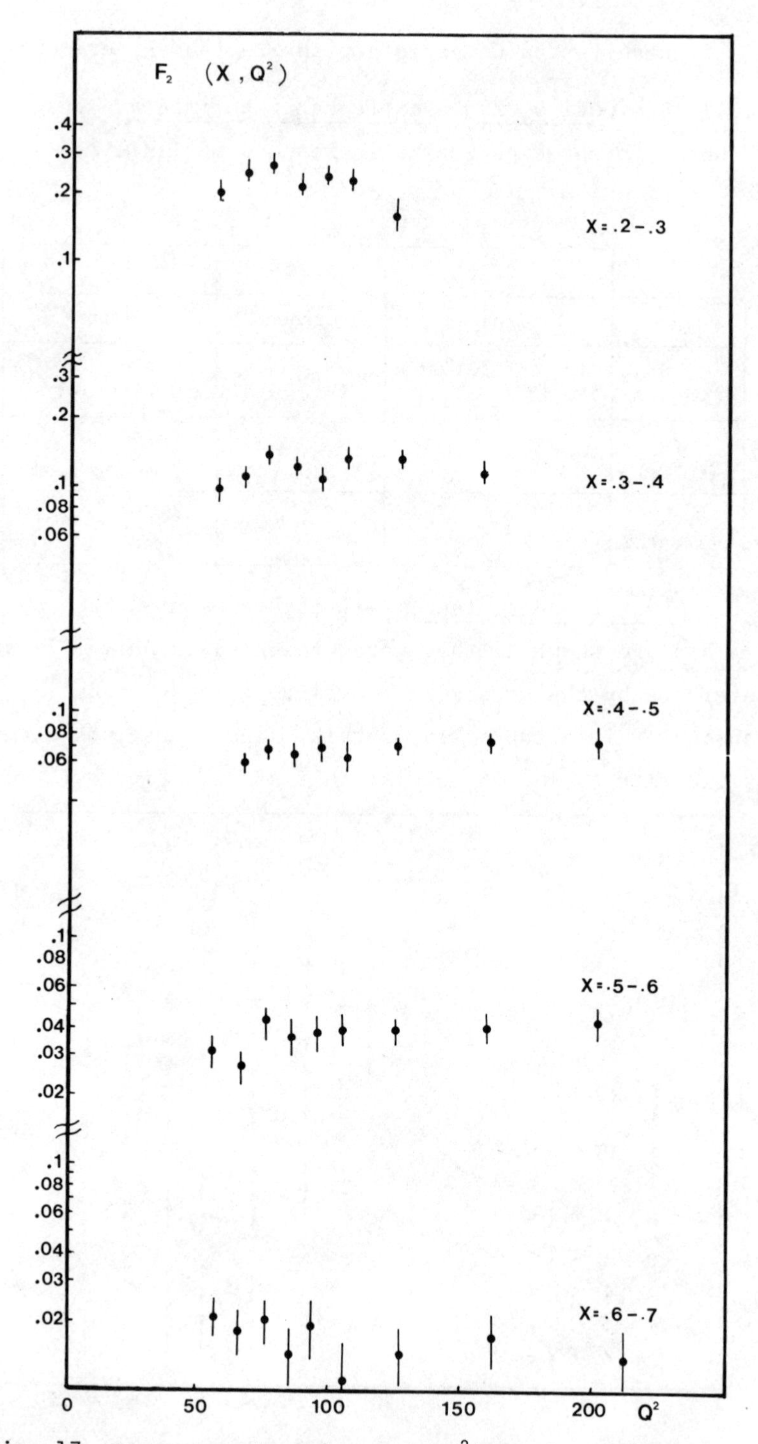

Fig. 17 Structure function $F_2(x,Q^2)$ as determined by the Bologna-CERN-Dubna-Munich-Saclay Collaboration

the results of $F_2(x,Q^2)$ obtained by the EMC and calculated per nucleon for different bins of x. Note that Fermi motion has been taken into account. At a fixed value of x the data is rather Q^2-independent over the whole range of Q^2 (from 15 GeV/c^2 up to 200 GeV/c^2). The BCDMS evaluation of νW_2 is made for the same x binning from Q^2 = 50 GeV/c^2 up to Q^2 = 220 GeV/c^2 (Fig. 17) and Fig. 18 displays the results obtained by the BFP Collaboration.

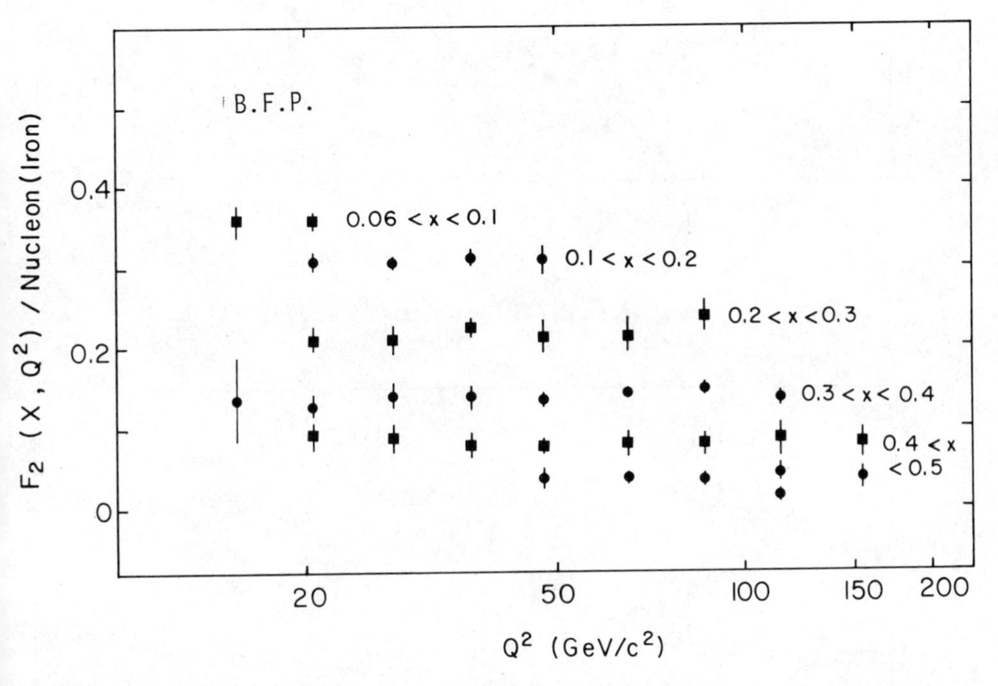

Fig. 18 Structure function $F_2(x,Q^2)$ as determined by the Berkeley-Fermilab-Princeton Collaboration

If one plots the EMC and the BCDMS data in the same way as BFP (Figs. 19 and 20), one can compare them; they all agree perfectly well within the error bar.

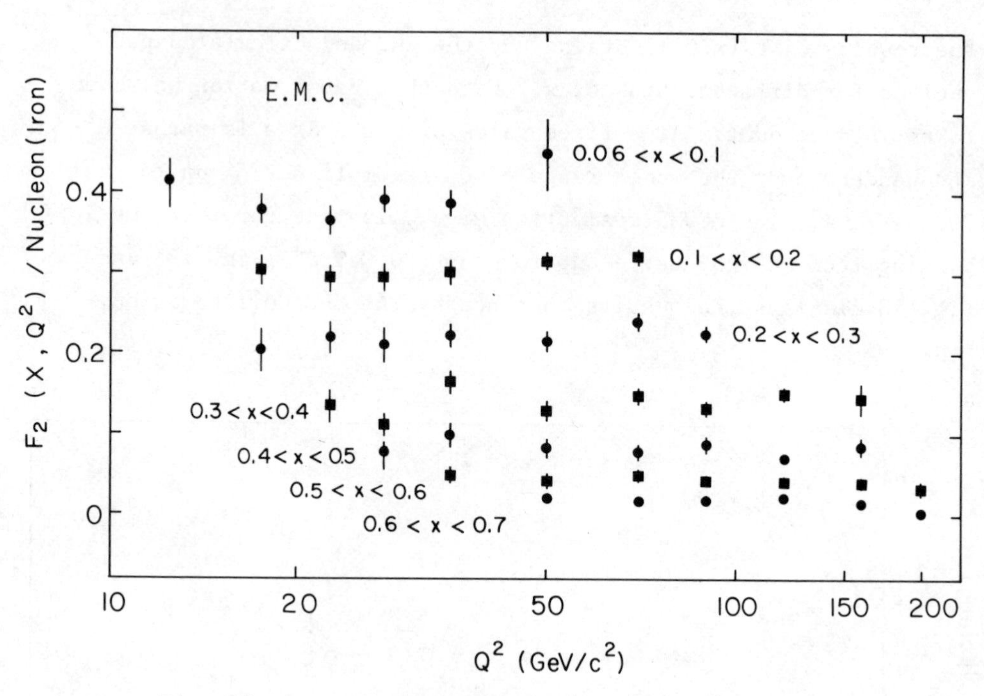

Fig. 19 Same as Fig. 16 but linear scale for F_2

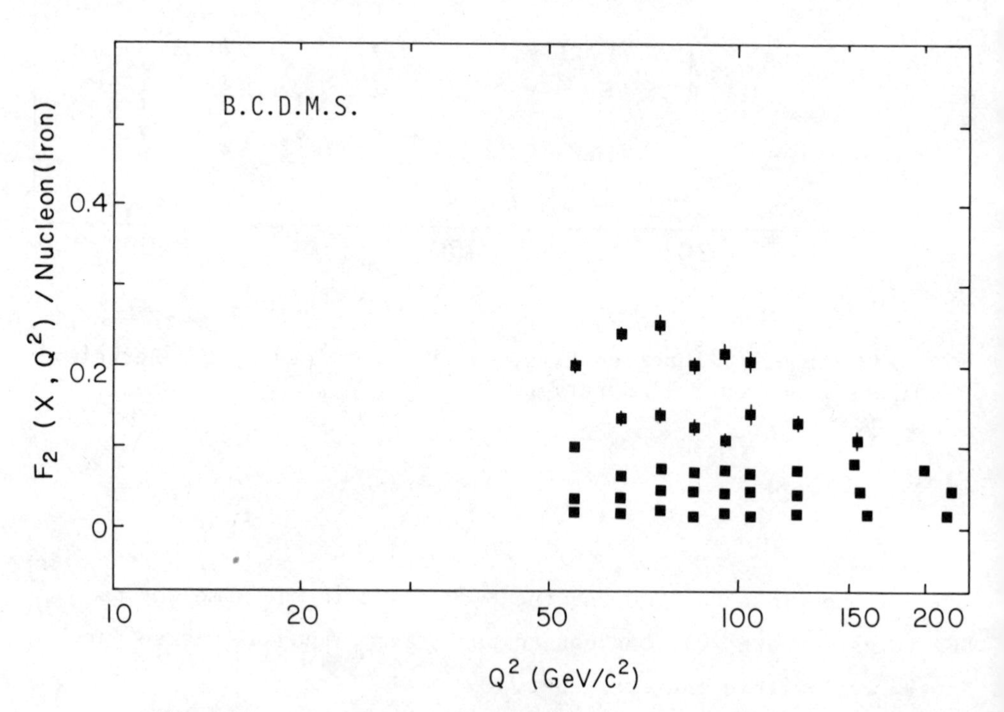

Fig. 20 Same as Fig. 17 but linear scale for F_2

One can also compare the EMC data with the CERN-Dortmund-Heidelberg-Saclay (CDHS) neutrino data[15] (Fig. 21); again the agreement is very good despite the fact that the CDHS Collaboration quote results on the nucleus.

From all these consistent measurements one can conclude that after the small Q^2 value ($Q^2 < 15$) the Q^2 dependence of $F_2(x,Q^2)$ is weak. One will still have to wait until full analysis of these data has been made to draw definite conclusions on a log Q^2 or the power behaviour of F_2. Furthermore, more data at lower Q^2 is needed in the same experiment.

3.2.3 New results on H_2/D_2 data. The EMC is currently running with a 6 m H_2 and D_2 target and 10^{12} μ^+ have been used at 280 GeV on H_2 and on D_2; at 240 GeV, 5×10^{11} μ^+ on H_2 have already been taken.

Results are at a preliminary stage. An example is shown in Fig. 22a of $F_2(x)$ for $30 < Q^2 < 50$ GeV/c^2 for the EMC and in Fig. 22b for the Chicago-Harvard-Illinois-Oxford (CHIO) experiments. Obvious improvement has been made.

4. DETERMINATION OF THE LONGITUDINAL CROSS-SECTION

As we have seen before, the cross-section for electron (muon) scattering can be expressed in terms of a longitudinal cross-section σ_L and a transverse cross-section σ_T:

$$\frac{d^2\sigma}{dQ^2 d\nu} = \Gamma(\sigma_T + \varepsilon\sigma_L) .$$

The longitudinal cross-section is in general expressed in terms of $R = \sigma_L/\sigma_T$, which can be learnt from R determination.

4.1 R Expectation

We have seen that in a simple parton model, for spin ½ constituent $R = 0$. In fact we know that if one cannot neglect the

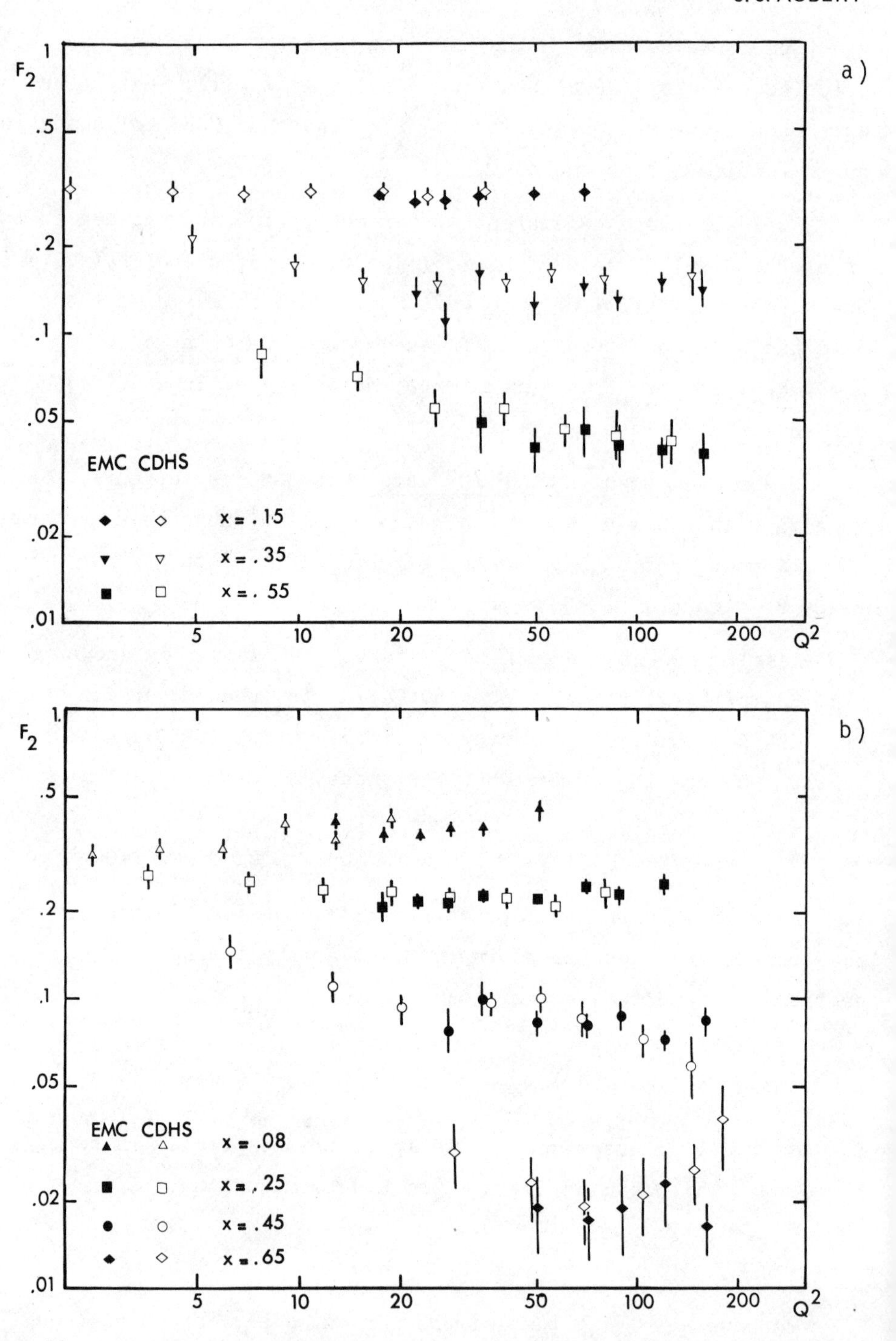

Fig. 21 Comparison of CERN-Dortmund-Heidelberg-Saclay Collaboration structure function with European Muon Collaboration results

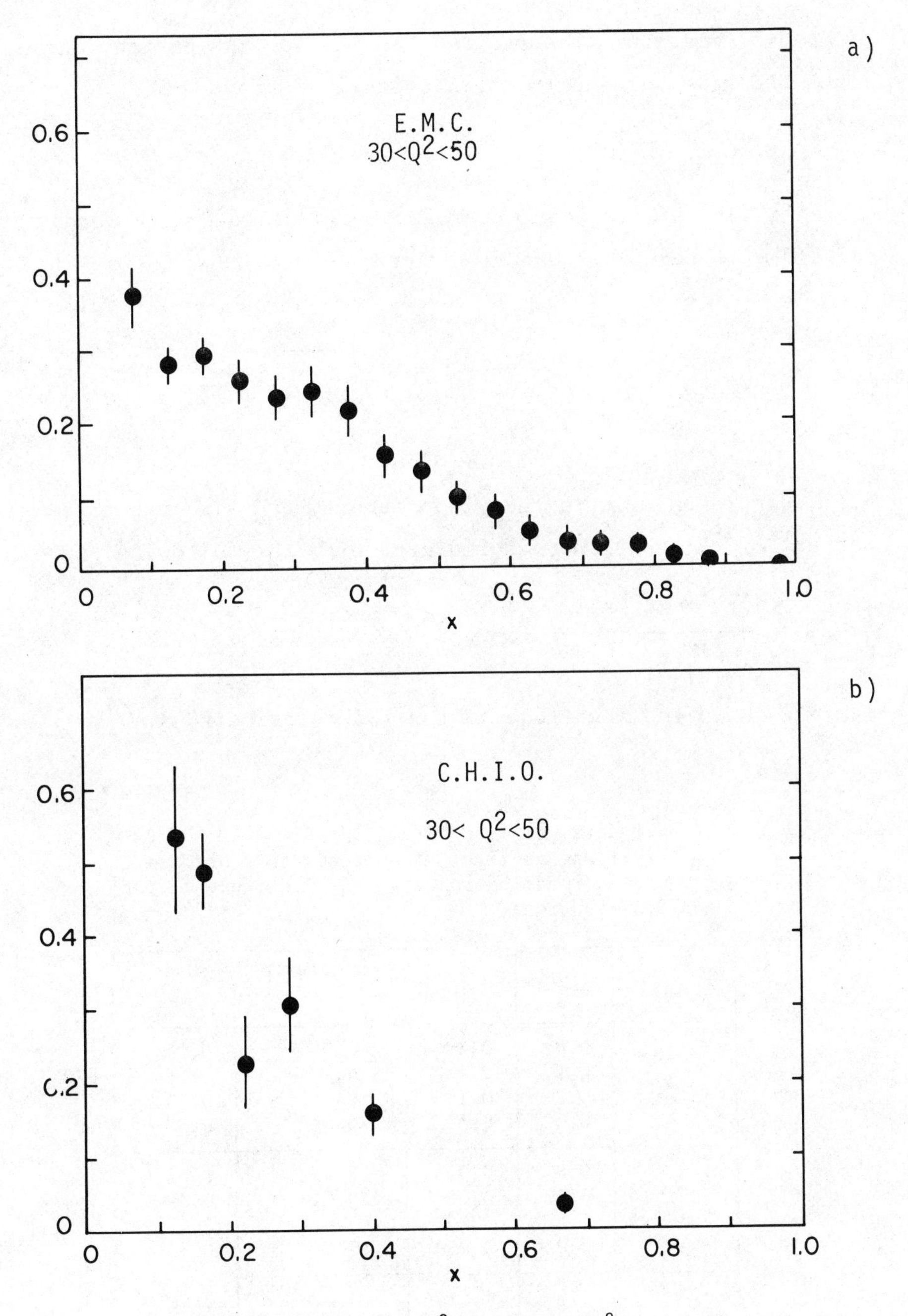

Fig. 22 a) $F_2(x)$ on H_2 for $30 < Q^2 < 50$ GeV/c^2 determined by the European Muon Collaboration

b) $F_2(x)$ on H_2 for $30 < Q^2 < 50$ GeV/c^2 determined by the CHIO Collaboration primordial momentum transfer and the mass of the parton, one can reach the zero value only asymptotically; so

$$R = \frac{4[m_q^2 + p_T^2]}{Q^2} .$$

QCD calculation also introduces another term to R; to leading order R will receive a contribution from

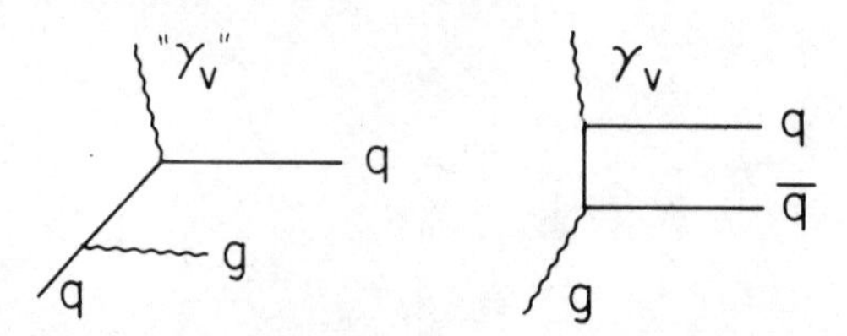

i.e. being sensitive to the gluon density and also the three-gluon vertex[16]. R can then be formulated with the following expression:

$$R = \frac{4[m_q^2 + p_T^2]}{Q^2} + \frac{F(x)}{\log (Q^2/\Lambda^2)} .$$

Table 2 shows the magnitude of the two contributions.

Table 2

Predicted values for $R(x,Q^2) = R^{intrinsic} + R^{QCD}$ for electroproduction. The predictions at Q^2 = = 2 GeV^2 correspond to naive (Q^2-independent) parton distributions.

Q^2 (GeV^2)	x	R	$R^{intrinsic}$	R^{QCD}
2	0.8	0.1	0.08	0.02
	0.5	0.16	0.1	0.06
	0.2	0.19	0.04	0.15
	0.02	0.53	0	0.53
10	0.8	0.025	0.015	0.01
	0.5	0.05	0.02	0.03
	0.2	0.08	0.01	0.07
	0.02	0.23	0	0.23
20	0.8	0.014	0.006	0.008
	0.5	0.035	0.01	0.025
	0.2	0.063	0.003	0.06
	0.02	0.18	0	0.18
100	0.8	0.007	0.002	0.005
	0.5	0.02	0.002	0.018
	0.2	0.04	0	0.04
	0.02	0.12	0	0.12

4.2 Experimental Situation

As we will see, the experimental situation is not what one would like and I think it is worth explaining the difficulties of the experimental determination of R.

4.2.1 <u>Experimental difficulties.</u> To extract W_1 and W_2 (or σ_L and σ_T) from the data, one must measure the differential cross-section at two different incident energies for the same x,Q^2 couple; then $(d^2\sigma/dQ^2dx)/\Gamma = \sigma_T + \varepsilon\sigma_L$.

If one plots on a linear graph $\sigma_T + \varepsilon\sigma_L$ as a function of ε (ε photon polarization, which is a function of x,Q^2, and incident energy E_0) for the two different incident energies, one gets two points and therefore σ_T and σ_L.

It is obvious that to determine accurately σ_L one needs two ε values ε_1 and ε_2 as different as possible ($\varepsilon < 1$).

In Fig. 23 ε is displayed as a function of E_0; ε is close to 1 up to large ν values, where it decreases and goes to zero. But in the same way when $\varepsilon << 1$, the radiative correction is large (so is the uncertitude in their determination).

So already there is a problem; large $\Delta\varepsilon$ means better σ_L separation but more problems with radiative correction.

Furthermore one can see that the comparison implies the absolute determination of the cross-section (running at two different energies) in different spectrometers (SLAC case), at different times in the same spectrometer (at FNAL and CERN). The cross-section at the same x,Q^2 bin is not measured exactly in the same position inside the spectrometer; the beam size is different in the two energy regions, etc. So there is difficulty to handle the absolute cross-section determination which has to be determined with an accuracy of better than 2% to reach a 50% error on σ_L (if $R = 0.2$).

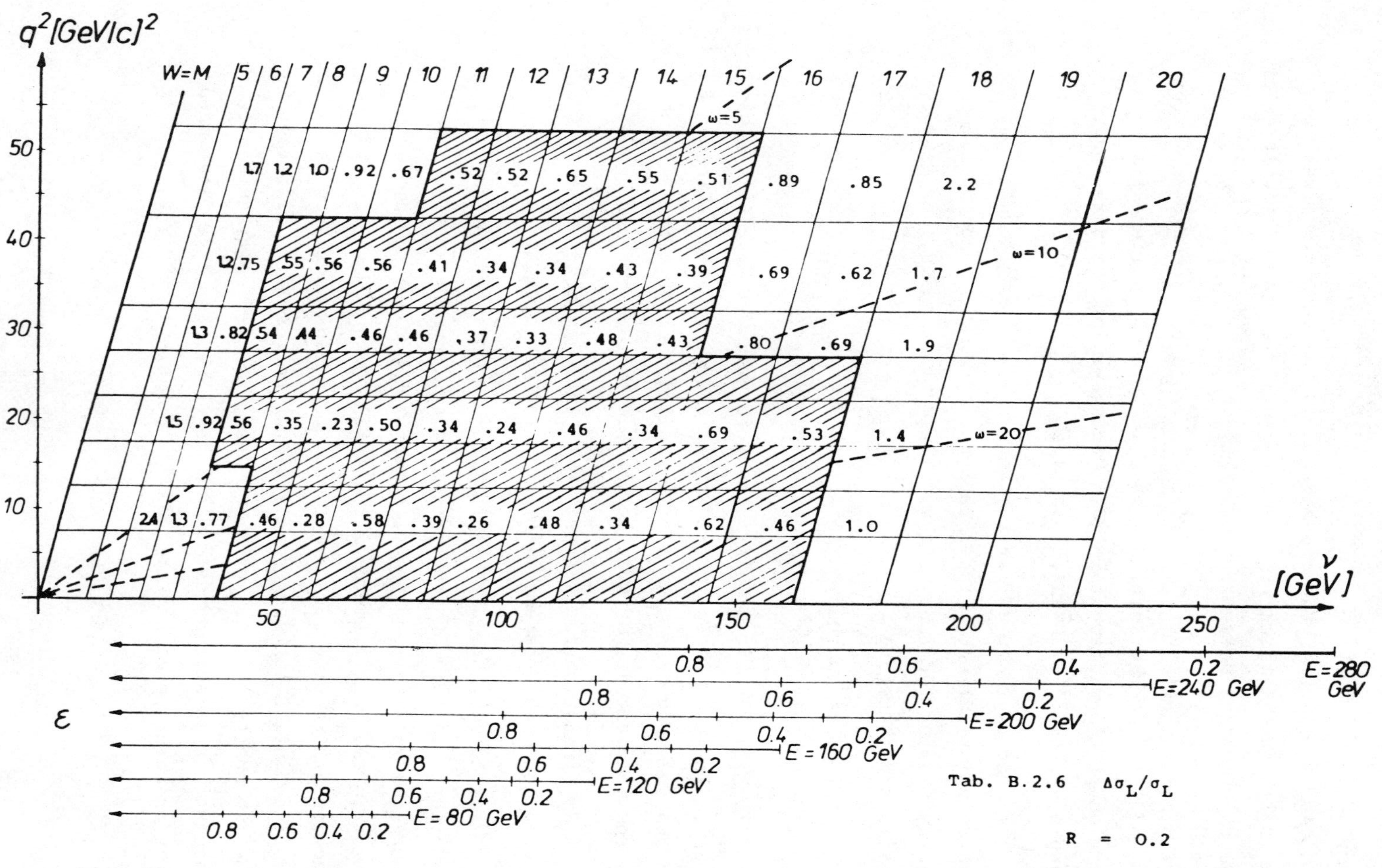

Fig. 23 Prospect on R measurements in the European Muon Collaboration forward spectrometer[14])

In principle, if one makes measurements with three incident energies one can control the magnitude of the systematic error in R determination even if this does not really improve the knowledge of R. Anyway it is easy to see from these remarks that the error on R will be mainly systematic.

4.2.2 Experimental results. At SLAC a very accurate experiment has been pursued for many years on structure function determination[8]. Figure 24 shows their results on the proton. The mean value of $R = 0.21 \pm 0.1$. Figure 25 shows the results obtained by the CHIO Collaboration[9], the mean value being $R = 0.52 {}^{+0.24}_{-0.20}$. The disagreement is not statistically significant and we will also notice that it was measured at very small x at FNAL, but at large x at SLAC.

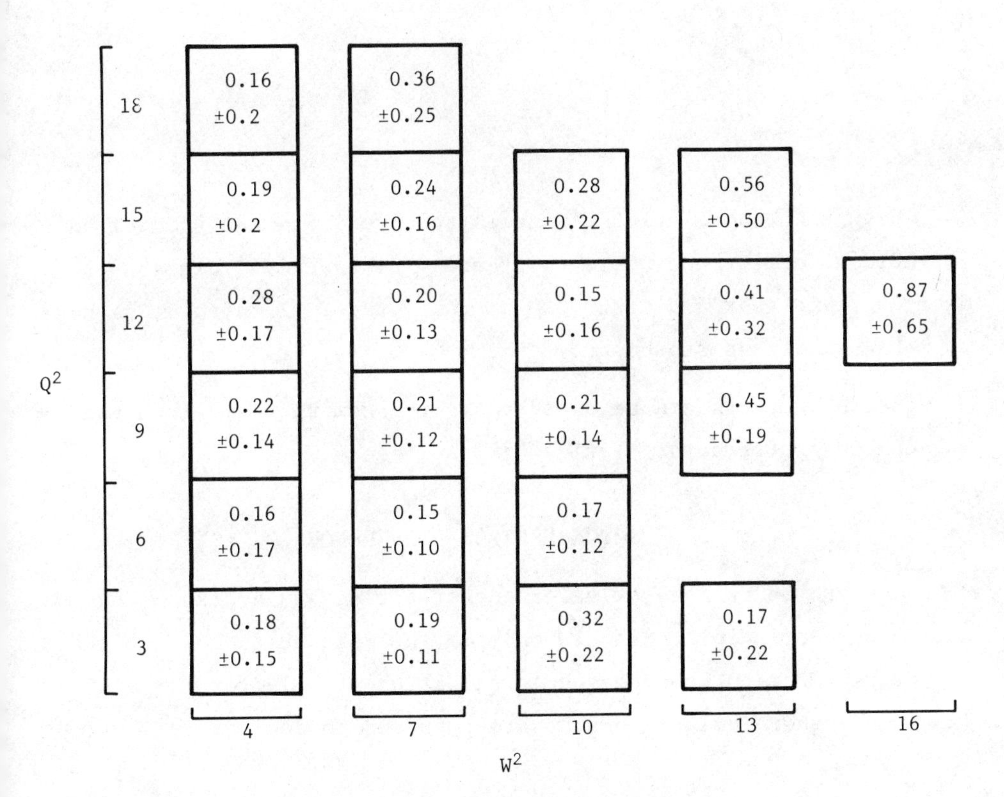

Fig. 24 R determination at SLAC[17]

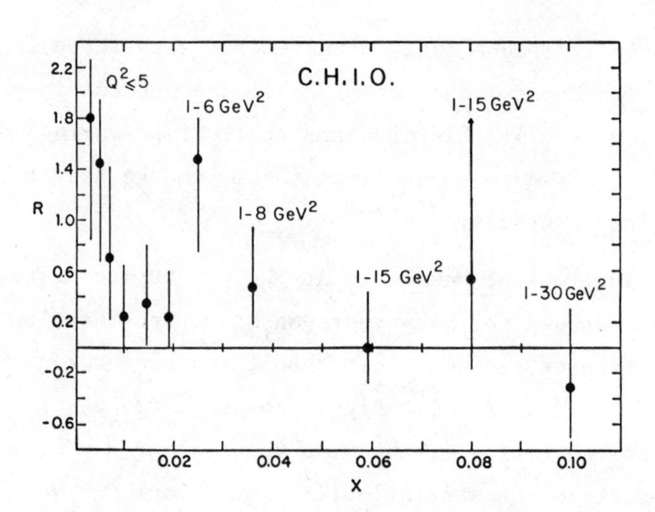

Fig. 25 R determination by the Chicago-Harvard-Illinois-Oxford Collaboration[9]

Figure 26 shows all the results compared with the prediction (primordial effect and QCD). Is there a disagreement between prediction at large x and large Q^2, at this stage? I think it is hard to make any strong statement.

Better results can be expected in the coming years from the two CERN μ experiments; let us wait and see.

5. PROTON-NEUTRON DIFFERENCE

The proton is made of valence quarks plus sea quarks. The neutron, in principle, differs only in the valence quarks, so the structure function is expected to be different only for large x ($x > 0.2$), where valence quarks are supposed to dominate.

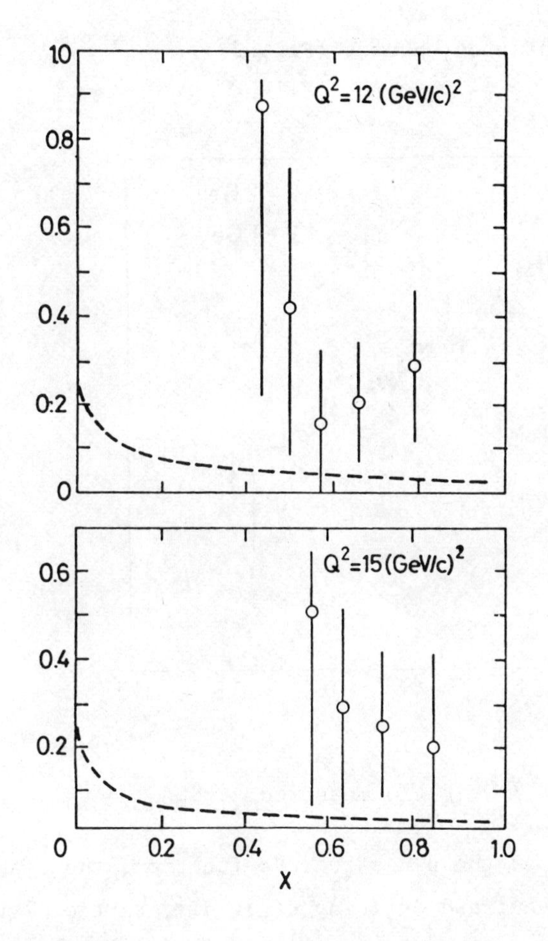

ϕ S.L.A.C. M. iT

ᵻ C.H.I.O.

Fig. 26 R prediction and measurement

On more precise ground, Nachtmann has shown that the structure function ratio is bounded

$$4 \geq \frac{\nu W_2^n}{\nu W_2^p} \geq 0.25 \ .$$

Indeed the data[17] support these bounds (Fig. 27).

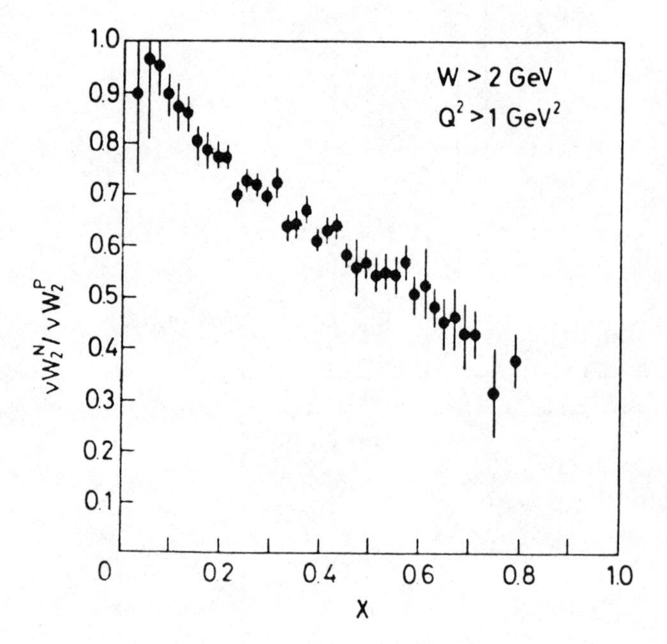

Fig. 27 $\nu W_2^n/\nu W_2^p$ ratio measured at SLAC

If one then calculates the p-n structure function, the sea quark contribution drops out and defining u(x), d(x) as the distribution function of up and down quarks inside the proton one has

$$\nu W_2^p - \nu W_2^n = x \left[\frac{1}{3}\, u(x) - \frac{1}{3}\, d(x)\right] \simeq x\, u(x)\, \frac{1}{3} \ .$$

The data shown in Fig. 28 allow an experimental determination of u(x).

The moment sum rule

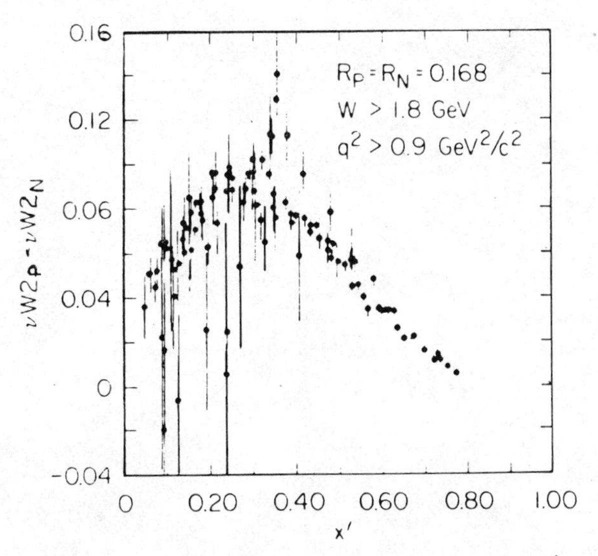

Fig. 28 $\nu W^p - \nu W_2^n$ measured at SLAC[8,17)]

$$M_2^n = \int_0^1 F_2^n(x)\,dx = \frac{\left(\frac{1}{3}\right)^2 + \left(\frac{1}{3}\right)^2 + \left(\frac{2}{3}\right)^2}{3} = \frac{2}{9}$$

is not experimentally satisfied, $M_2^n = 0.12$, as it was for the proton. So partons are not carrying all the nucleon momentum.

Using QCD prediction with 4 flavours and 8 gluons one gets $M_2^n = \sum_i q_i^2/(3f + 2g) = 0.119$ in agreement with the data. One should also notice that the ratio M_2^n/M_2^p predicted by the simple quark model (2/3) is observed in the data $M_2^n/M_2^p = 0.12/0.18$.

6. MOMENT ANALYSIS

As has been explained by De Rújula earlier, the scaling violations predicted by QCD are more simply described in terms of the moments. This will not be repeated here; just a few formulae will be used here when necessary. For example,

$$M_i(n,Q^2) = \int_0^1 x^{n-2} F_i(x,Q^2)\, dx \ .$$

Because there is mass effect (quark mass) it is better to use the Nachtmann moments[18])

$$M_2(n,Q^2) = \int_0^{\xi_{max}} \xi^{n-2} \left[1 - (M^4/Q^4)\xi^4\right](1 + Q^2/\nu^2)(1 + 3\eta_n) \times$$

$$\times F_2(\xi,Q^2)\, d\xi \ ,$$

where M is the nucleon mass and

$$\eta_n = \left[(n + 1)M\nu\xi - (n + 2)Q^2\right]/\left[(n + 2)(n + 3)(\nu^2 + Q^2)\right]$$

$$\xi = \left[(\nu^2 + Q^2)^{1/2} - \nu\right]/M$$

$$\xi_{max} = \left[\frac{1}{2} + \left(\frac{1}{4} + M^2/Q^2\right)^{1/2}\right]^{-1} \ .$$

In μN scattering one has then to distinguish two cases:

i) $F_2^{\mu p} - F_e^{\mu n}$ (identical with $xF_3^{\nu N}$) non-singlet structure function

$$M_{NS}(n,Q^2) = M_{NS}(n, Q_0^2)\, e^{-\lambda_{NS}^n s}$$

where

$$s = \ln\left[\ln(Q^2/\Lambda^2)/\ln(Q_0^2/\Lambda^2)\right] ,$$

Λ = scale parameter not fixed by QCD, and

$$\lambda_{NS}^n = \frac{4}{33 - 2f}\left[1 - \frac{2}{n(n + 1)} + 4\sum_{j=2}^{n} \frac{1}{j}\right]$$

is the anomalous dimension where f is the number of flavours.

ii) F_2^{eN}. There are two additional singlet terms S_+ and S_-:

$$M_2^N(n,Q_0^2) = M_{NS}(n,Q_0^2)\, e^{-\lambda_{NS}^n s} + M_+(n,Q_0^2)\, e^{-\lambda_+^n s} +$$

$$+ M_-(n,Q_0^2)\, e^{-\lambda_+^n s} \ .$$

These relations are correct to leading order in the coupling constant α_s.

Let us turn to the experimental difficulties connected with the determination of the moment. $F_2(x)$ is not measured over all the range of x, so one would have first to mix the results of different experiments and secondly to extrapolate outside measurements.

CHIO data[8]. In their case they use the SLAC data to complete their sample, which extends mainly to the low x region.

They then extrapolate the data using

$$F_2(x = 0) = F_2(x_{min})$$

$$F_2(x = 1) = 0 .$$

They included the resonance and the elastic scattering with the dipole formulae; they use their R value for their point and the SLAC R value for the SLAC point. Then they add a 2.8% systematic uncertainty for different measurements.

Figures 29 and 30 show their input, $F_2(x)$, $x^2F_2(x)$, $x^4F_2(x)$ at $Q^2 = 3.75$ GeV/c^2, and 22.5 GeV/c^2 which is their extreme point. We see that the contribution of the different experiments is changing over the moment degree and that the resonance makes a substantial contribution to higher order moments even up to $Q^2 = 20$ GeV/c^2, so great care has to be taken when one has to quote errors on the moment determination.

In Fig. 31 the Nachtmann proton moment (n = 4, 6, 8, 10) is shown as a function of Q^2, and is compared with the fit in leading order to QCD which shows a good agreement with the data.

Figure 32 shows the Λ determination made for each moment that one would expect from first order, Λ independent of n. Data show a rising value of Λ with the order n, in disagreement with expectation; they point out that second-order effects can simulate this rise and furthermore if they constrained their data to a unique Λ value they would still get a reasonable χ^2.

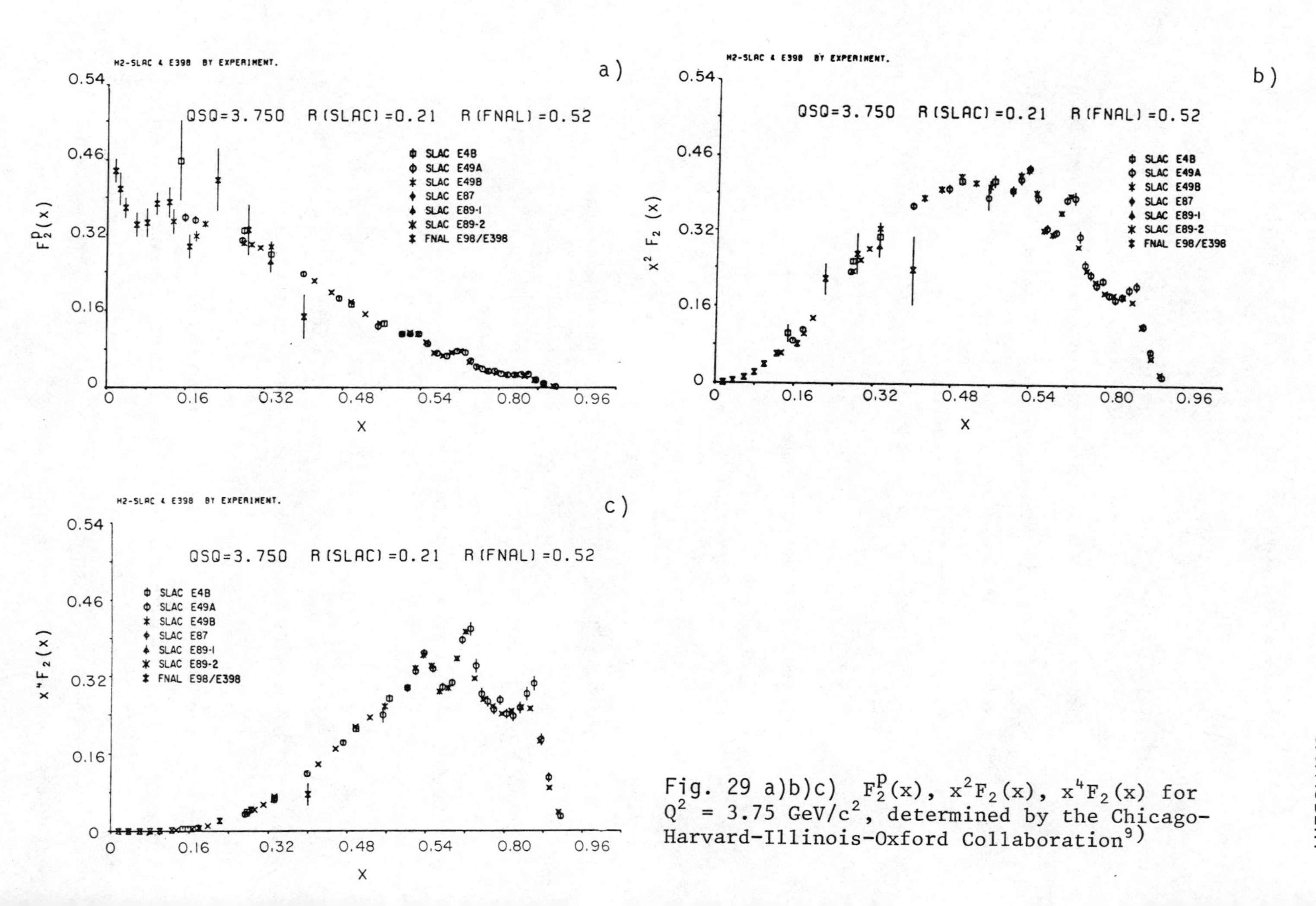

Fig. 29 a)b)c) $F_2^p(x)$, $x^2F_2(x)$, $x^4F_2(x)$ for Q^2 = 3.75 GeV/c^2, determined by the Chicago-Harvard-Illinois-Oxford Collaboration[9])

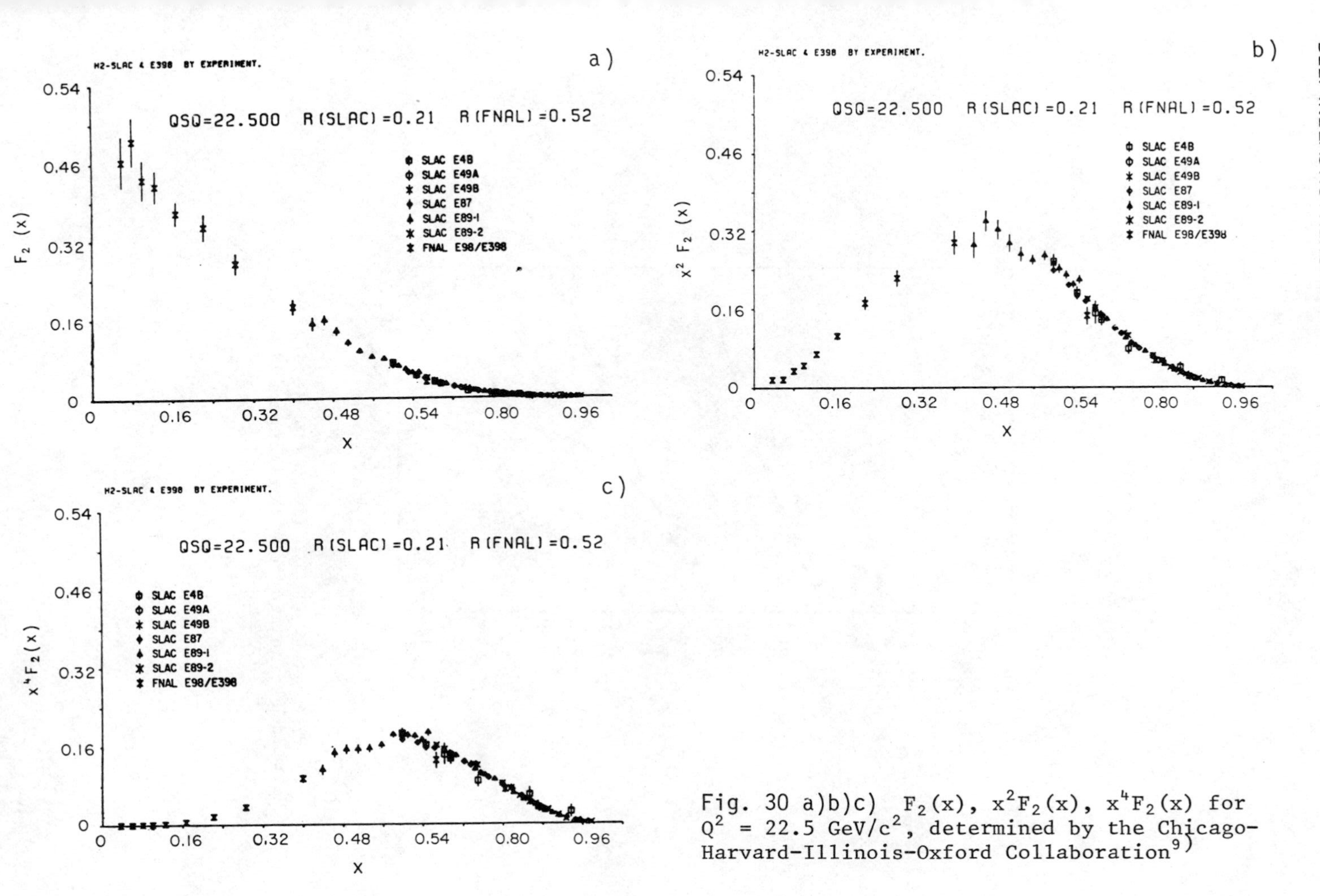

Fig. 30 a)b)c) $F_2(x)$, $x^2F_2(x)$, $x^4F_2(x)$ for $Q^2 = 22.5$ GeV/c^2, determined by the Chicago-Harvard-Illinois-Oxford Collaboration[9]

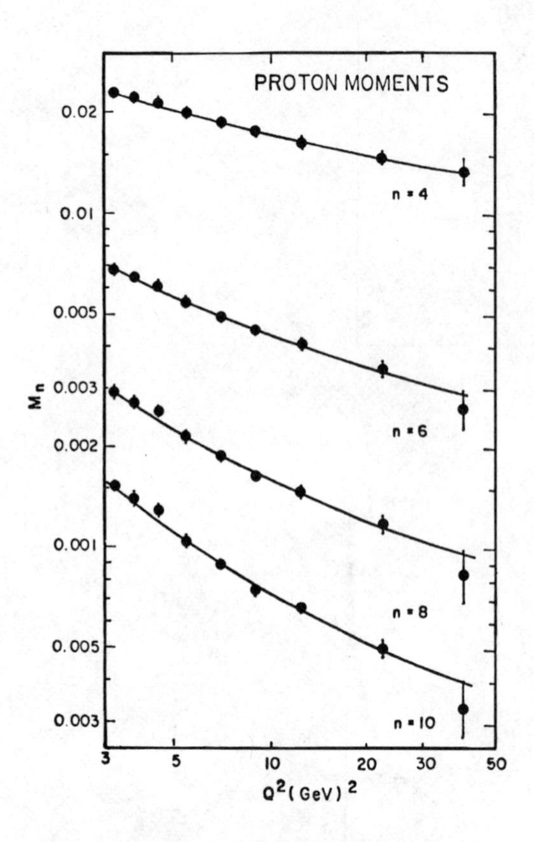

Fig. 31 Nachtmann proton moment determined by the Chicago-Harvard-Illinois-Oxford Collaboration[9])

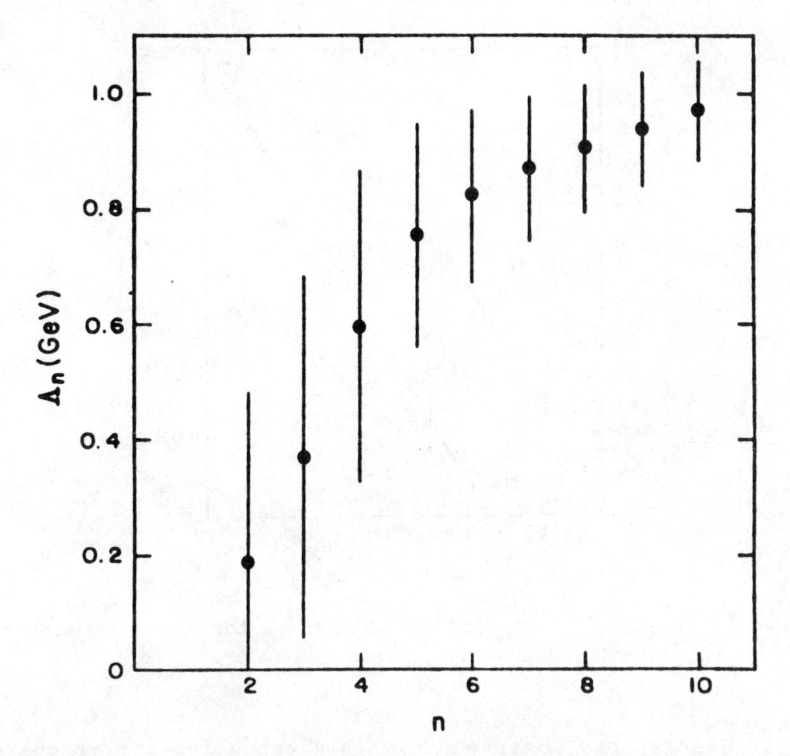

Fig. 32 Λ determination by the Chicago-Harvard-Illinois-Oxford Collaboration[9])

Figure 33 shows log Mn versus log Mm; one expected to have straight lines, the slope being predicted by QCD.

Results and predictions are shown in Table 3.

Table 3 F_2 moment slopes and QCD predictions

Moments	Observed slope	Leading order QCD predicted values (4 flavours) $\lambda_{NS}^{(m)}/\lambda_{NS}^{(n)}$	$\lambda_-^{(m)}/\lambda_-^{(n)}$
d ln M(n = 6)/d ln M(n = 4)	1.62 ± 0.19	1.290	1.310
d ln M(n = 8)/d ln M(n = 4)	2.15 ± 0.24	1.499	1.533
d ln M(n = 10)/d ln M(n = 4)	2.61 ± 0.29	1.662	1.701
d ln M(n = 8)/d ln M(n = 6)	1.33 ± 0.12	1.162	1.166
d ln M(n = 10)/d ln M(n = 6)	1.62 ± 0.14	1.288	1.294
d ln M(n = 10)/d ln M(n = 8)	1.22 ± 0.09	1.109	1.107

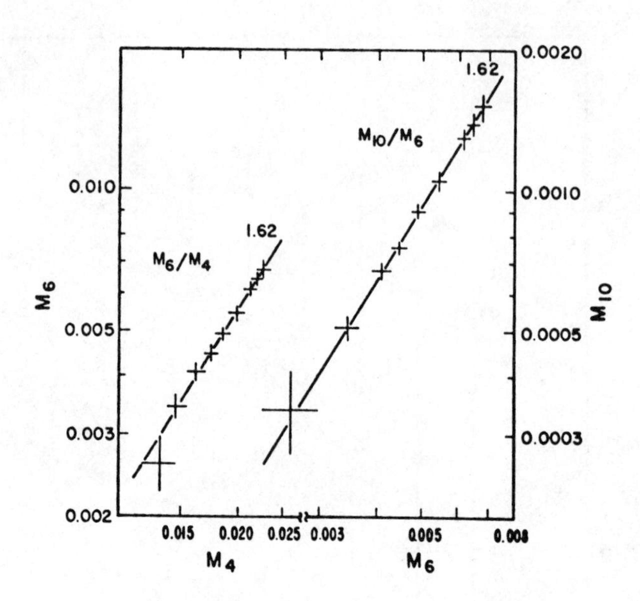

Fig. 33 Log M_n/log M_m determination by the Chicago-Harvard-Illinois-Oxford Collaboration[9]

The discrepancy observed is mainly cancelled out if one imposes Λ to be independent of n.

Then using the function u(x), d(x), s(x), G(x) (for quarks u, d, s, and the gluon), they calculate the quark and gluon content for the different moments. They also tried to include second-order effects (in α_s)[19] and then calculate Λ results as shown in Fig. 32. Apart from the second moment, results are flatter than before (see Table 4). They also calculate the quark and gluon content for the different moments with a single value of Λ (Λ = 459 MeV) (see Table 5). The results are not very different.

From their analysis they conclude that there is an excellent agreement with QCD, when one includes second-order effects. Gluons carry 47% of the momentum. Then using a constraint fit they find $\Lambda = 0.46 \pm 0.11$ and they deduce

Table 4 Moment analysis (including Fermi motion and second-order corrections)

n	$\langle u\rangle_n$	$\langle d\rangle_n$	$\langle s\rangle_n$	$\langle G\rangle_n$	Λ_n (MeV)	χ^2
2	0.3520	0.1960	-	0.4520	90 ± 119	3.1
4	0.0355	0.0126	-	0.0708	397 ± 164	2.6
6	0.0092	0.0021	-	0.0241	472 ± 108	3.8
8	0.0035	0.0004	-	0.0089	500 ± 79	5.1
10	0.0016	-	-	0.0042	531 ± 52	7.1

Table 5 Quark and gluon moments (including second-order and Fermi motion corrections: global fit with common value of Λ

n	$\langle u\rangle_n$	$\langle d\rangle_n$	$\langle s\rangle_n$	$\langle G\rangle_n$	Λ (MeV)	χ^2
2	0.3389	0.1871	-	0.4740	459	10.3
4	0.0353	0.0130	-	0.0883	459	2.7
6	0.0093	0.0021	-	0.0225	459	3.8
8	0.0035	0.0032	-	0.0065	459	5.4
10	0.0016	-	-	0.0019	459	8.6

$$\alpha_s(Q^2 = 3\ \mathrm{GeV/c^2})\ /\ \pi = 0.033 \pm 0.006\ .$$

They claim that the residual dependence of Λ as a function of n might be explained by higher order twist effects.

Here we shall close the discussion on the deep inelastic scattering of nucleons (DIS) without forgetting that we have not looked at structure functions G_1, G_2, measured with a polarized beam on a polarized target, or at the hadronic induced final state.

We shall then discuss phenomena related to DIS and just because of time limitation we shall restrict ourselves to the study of Drell-Yan pair production.

7. DRELL-YAN PAIRS

Most of the information in this section comes from Berger's[20] review on hadroproduction of massive lepton pairs and QCD.

7.1 Drell-Yan Pair Cross-Section

The Drell-Yan (DY) pair process is the annihilation of a pair $q\bar{q}$, going into a $\mu^+\mu^-$ final state; as an example see the pp case.

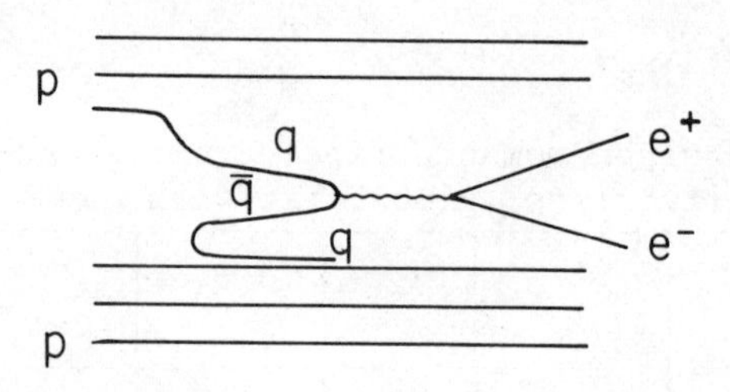

In the parton model, if

- $q_i(x_a, k_{T_a})$ is the probability of finding a quark i in the hadron a,
- x_a the fraction of the nucleon momentum carried by the quark, and
- k_{T_a} the transverse momentum of the quark,

one expresses the cross-section in the following form:

$$\frac{Q^2\, d\sigma^{ab}}{dQ^2 dx_L\, d^2Q_T\, d\cos\theta^*} = \frac{\pi\alpha^2}{2} \int d^2\vec{k}_{T_a}\, dx_a\, d^2\vec{k}_{T_b}\, dx_b$$

$$\times \frac{1}{3} \sum_i e^i \left[q_i(x_a, k_{T_a}) \overline{q_i}(x_b, k_{T_b}) + a \rightleftarrows b \right]$$

$$\times (1 + \cos^2\theta^*)(\vec{Q}_T - \vec{k}_{T_a} - \vec{k}_{T_b})$$

$$\times \delta(x_L - x_a - x_b)\delta(Q^2 - x_a x_b s) ,$$

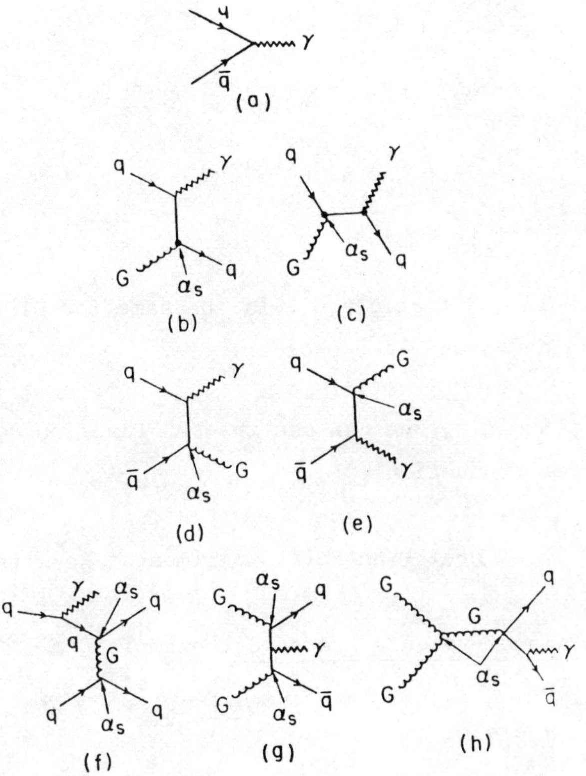

Fig. 34 Drell-Yan pair production: Feynman graph including gluons

where 1/3 represents the three colours and the cross-section is given for the $Q^2 \to \infty$ limit.

With QCD one has to have corrections coming from gluons (see Fig. 34) and the cross-section is

$$\sigma_{ab} \to \gamma^* x = \int q_a(x_a)\bar{q}_b(x_b)\sigma_{q\bar{q}} \to \gamma^* x_1 +$$

$$+ \alpha_s \int q_a(x_a)G_b(x_b) + \bar{q}_a(x_a)G_b(x_b)\sigma_{qG} \to \gamma^* x_2 +$$

$$+ \alpha_s^2 \int q_a(x_a)q_b(x_b)\sigma_{qq} \to \gamma^* x_3 +$$

$$+ \alpha_s^2 \int G_a(x_a)G_b(x_b)\sigma_{GG} \to \gamma^* x_4 +$$

$$+ (a \rightleftarrows b) .$$

One can prove in QCD that $q(x,Q^2)$ is the same for DIS and DY, i.e. for time-like and space-like photons.

So, in principle, one can use the DIS function q,G to compute all the DY pair production.

7.2 Drell-Yan Pair Experimental Results

7.2.1 Valence-valence processes, example π^-N

$$\sigma \propto q_\pi \bar{q}_N + \bar{q}_\pi q_N .$$

If one neglects $\bar{q}_N$, which is small, then $\sigma \propto \bar{q}_\pi q_N$, using the u, d, s, ..., formalism one has

$$\sigma \propto \bar{q}_\pi(x_n,Q^2) \left[\frac{4}{9} x_N U_N(x_N,Q^2) + \frac{1}{9} x_N \bar{d}(x_N,Q^2)\right] .$$

Then using u,$\bar{d}$ from DIS one can determine $\bar{q}_\pi$, or making an over-all fit to the data one can find $\bar{q}$, u, d, and cross-check the universality of u and d.

One also has to find the factorization through the data. There are two nice experiments giving results: that of Chicago-Princeton (CP)[21] at FNAL and Saclay-CERN-Coll. de France-Ec. Poly.-Orsay (SCCDFEPO)[22] at CERN. Figure 35a shows the results of the two experiments, which agree reasonably well. The solid curve shows the results of the structure function using a parametrization of u, d, s, ... All the results agree. Using π^+ and π^- data SCCDFEPO can eliminate the sea contribution giving results shown in Fig. 35b, and the results are again the same. One important question is to see if the absolute cross-section predicted by the theory with the structure function measured in DIS is given by the data. SCCDFEPO have measured this scale factor K using their own G(x), q(x) determination or the CDHS determination. They always find K larger than 1 (K = 1.4 and K = 2.3, respectively); they claim that there is an over-all 35% error on the K measurement.

7.2.2 Valence sea processes.

$$pp \rightarrow \mu^+\mu^- X \quad \text{where} \quad \sigma \propto q_p(x_p)\bar{q}_N(x_N) + (p \rightleftarrows N) \ .$$

One expects to have scaling to within the scaling violation of 20% observed in DIS

$$M^4 \frac{d\sigma}{dM^2} = f\left(\frac{M}{\sqrt{s}}\right) .$$

Data from the Columbia-FNAL-Stony Brook (CFS) Collaboration[23] at FNAL, from the Athens-BNL-CERN-Syracuse-Yale (ABCSY)[24] and CERN-Harvard-Frascati-MIT-Naples-Pisa (CHFNP)[25] Collaborations at the CERN ISR are used to test the scaling; see Figs. 36a,b. The scaling holds within 20-30%, which is expected from DIS measurement.

From the pp DY production one can fit the sea distribution and compare it with what has been measured in lepton-hadron scattering. CFS have compared their data with the CDHS prediction, and using the same hypothesis s = c = 0 they get a factor of 1.3 of discrepancy at x = 0.2, where the two measurements can be compared.

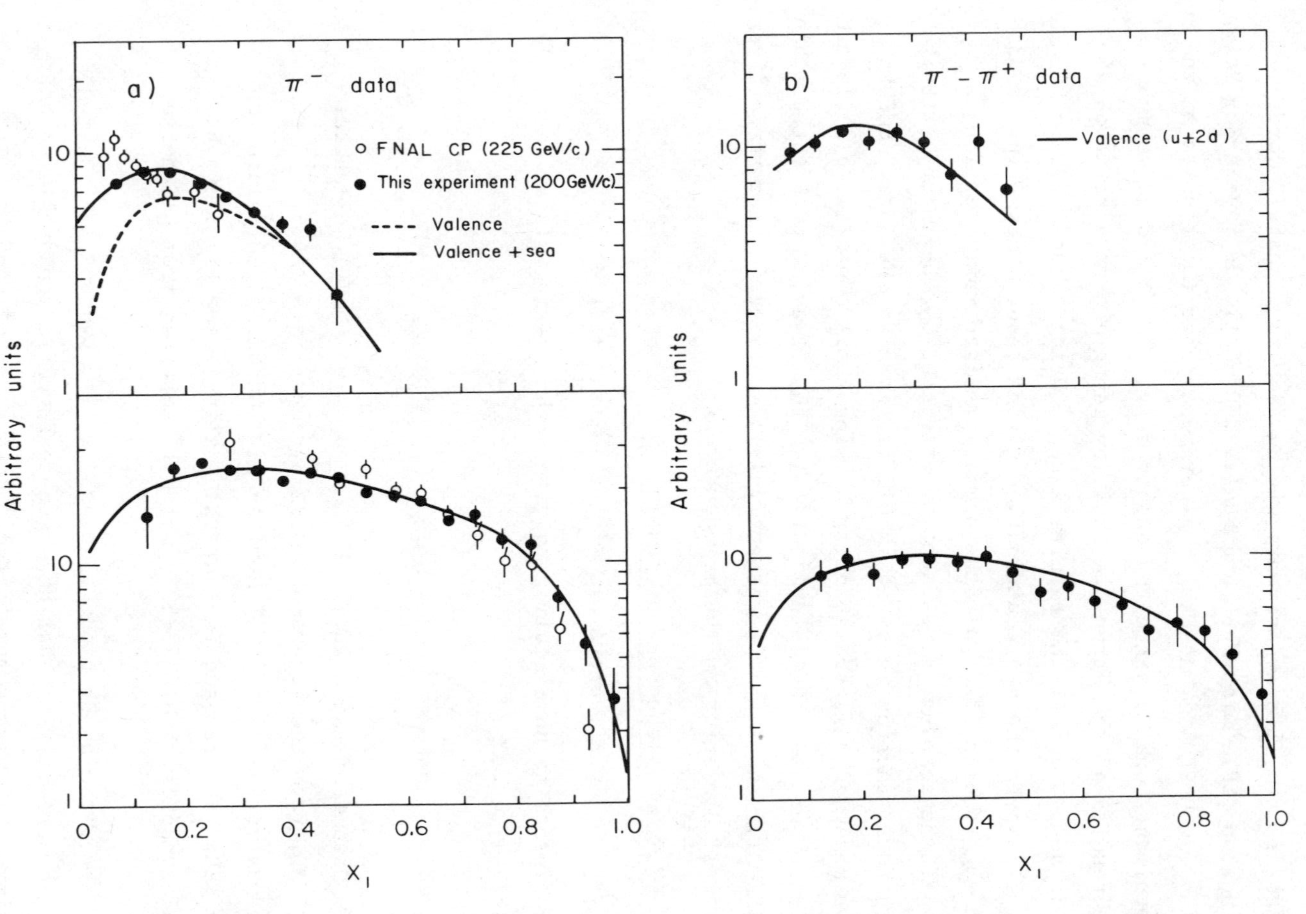

Fig. 35 π^- structure function, a) using the factorization, b) using π^- and π^+ data

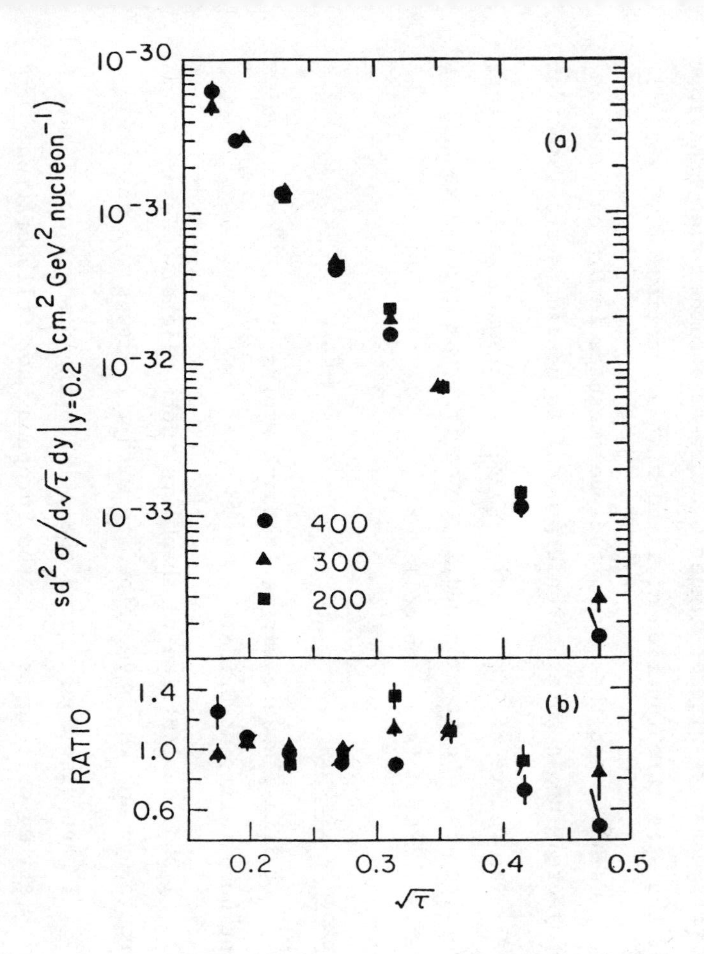

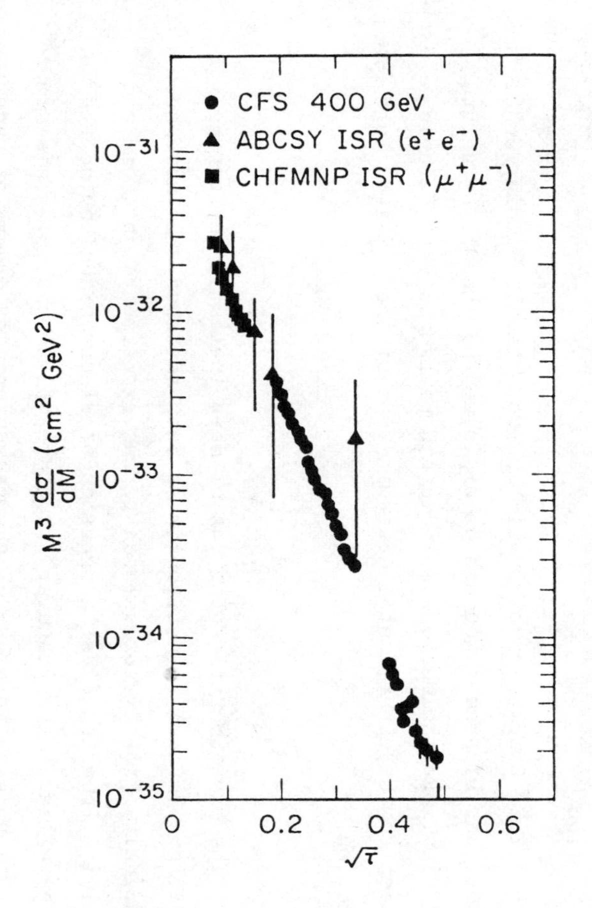

Fig. 36 Test of scaling in Drell-Yan pairs[21-23]

The factor 1.3 might reflect QCD correction to the Drell-Yan production.

7.2.2 Conclusion of Drell-Yan production. Within 20-30% it seems that DY production can be explained by the DIS structure function. Furthermore, π,K-induced μ pairs can provide unique access to the experimental determination of the π,K structure function.

8. CONCLUSION

The experimental situation of deep inelastic scattering for electrons (muons) on nucleons, has been reviewed. We have seen the continuity from historical scattering experiments on atoms and nuclei; from the first generation of electron-nucleon scattering we have learnt the granular structure of the proton. This simple picture holds within 20-30% accuracy over many orders of magnitude of measured cross-sections.

Scaling violation has been observed for moderate Q^2; recent experimental results at higher Q^2 suggest a rather flat behaviour of the structure function at fixed x as a function of Q^2.

We have seen that the structure measured in DIS can then be projected into a pure hadronic process to predict a cross-section.

The future of DIS is still ahead of us, when we get accurate results in μp scattering and later on from the doubler at FNAL and perhaps from a future ep collider.

Acknowledgements

It is a pleasure for me to thank Professor A. Zichichi for the hospitality at Erice.

I should also like to thank Drs. John Oliensis, Tokuzo Shimada, and In-Gyu Koh who have helped me with the discussion session.

I have also appreciated the stimulating discussions I have had with all my colleagues from the European Muon Collaboration.

REFERENCES

1) J.J. Aubert, J. Phys. (France) 39, C3-67 (1978).
G.B. West, Phys. Rep. 18, 263 (1975).

2) E. Mohr et al., Proc. R. Soc. A138, 229, 469 (1932).

3) R. Hofstadter et al., Rev. Mod. Phys. 102, 1586 (1956).
J.H. Fregeau et al., Phys. Rev. 99, 1503 (1955).

4) A. Bodek et al., Experimental studies of the neutron and proton electromagnetic structure function, SLAC-PUB 2248, Jan. 1979, to be published in Phys. Rev. D.

5) J.D. Bjorken, Phys. Rev. 179, 1547 (1969).

6) G. Miller et al., Phys. Rev. D 5, 528 (1972).

7) Y. Watanabe et al., Phys. Rev. Lett. 35, 898 (1975).
D.J. Fox et al., Phys. Rev. Lett. 33, 1504 (1974).

8) S. Stein et al., Phys. Rev. D 12, 1884 (1975).

9) H.L. Anderson et al., Phys. Rev. Lett. 37, 4 (1976).
H.L. Anderson et al., A measurement of the nucleon structure function, FNAL-PUB 79-30, May 1979, to be published in Phys. Rev. D.

10) See, for example,
H. Georgi et al., Phys. Rev. Lett. 36, 1281 (1976); Phys. Rev. D 14, 1829 (1976).
R. Barbieri et al., Phys. Lett. 64B, 171 (1976); Nucl. Phys. B117, 50 (1976).

11) P. Castorina et al., A realistic description of deep inelastic structure functions, Preprint CERN-TH. 2670, May 1979.

12) G. Gollin et al., IEEE Trans. Nucl. Sci. NS26, 59 (1979).

13) Experiment NA4, SPS Proposal P19, 1974.

14) Experiment NA2, SPS Proposal P18, 1974.

15) J. de Groot et al., Z. Phys. C 1, 139 (1979).

16) E. Reya, Deep inelastic structure functions; decisive test of QCD, to be published in Proc. Highly Specialized Seminar on Probing Hadrons with Leptons, Erice, March 1979.

17) A review of SLAC work has been made by R.E. Taylor, Proc. 19th Int. Conf. on High-Energy Physics, Tokyo, Japan, 1978 (Physical Society of Japan, Tokyo, 1979), p. 285.
See also W.B. Attwood, Electron scattering of H_2 and D_2 at 50° and 60° at SLAC, Thesis-185 (1978).

18) See W.A. Bardeen et al., Phys. Rev. D 18, 3998 (1978).

19) E.G. Floratos et al., Phys. Lett. 80B, 269 (1979).

20) E.L. Berger, Hadroproduction of massive lepton pairs and QCD, Invited talk presented at Orbis Scientiae, 1979, Coral Gables, Florida, January 1979, SLAC-PUB 2314 (April 1979).

21) K.J. Anderson et al., Phys. Rev. Lett. 43, 1217 (1979).

22) J. Badier et al., Experimental determination of the pion and the nucleon structure function by measuring high-mass muon pairs produced by pions of 200 and 280 GeV/c on platinium target, Contributed paper to the EPS Int. Conf. on High-Energy Physics, Geneva, Switzerland, 1979, and CERN-EP/79-67 (1979).

23) J.K. Yoh et al., Phys. Rev. Lett. 41, 684 and 1083 (1978).

24) I. Manelli, Proc. 19th Int. Conf. on High-Energy Physics, Tokyo, Japan, 1978 (Physical Society of Japan, Tokyo, 1979), p. 189.

25) F. Vannucci, Contribution to the Karlsruhe Summer Institute, 1978.
U. Becker, Drell-Yan pair measurement at ISR, Contributed paper to the EPS Int. Conf. on High-Energy Physics, July 1979, to be published in the Proceedings.

APPENDIX

$$W_{\mu\nu} = A\, g_{\mu\nu} + \frac{B}{M^2}\, p_\mu p_\nu + \frac{C}{M^2}\, q_\mu q_\nu + \frac{D}{M^2}\,(q_\mu p_\nu + q_\nu p_\mu) +$$

$$+ \frac{E}{M^2}\,(q_\mu p_\nu - q_\nu p_\mu) + \frac{F}{M^2}\,\varepsilon_{\mu\nu\alpha\beta}\, p^\alpha g^\beta \; .$$

1) Parity conservation.

$$W_{\mu\nu} \approx \int d^4y\; e^{iqy}\, \langle p|J_\mu(y)J_\nu(0)|p\rangle \; .$$

If $\mathcal{P}$ is the parity operator one has

$$\mathcal{P}|p\rangle = \eta|p\rangle \quad \text{with} \quad |\eta|^2 = 1$$

$$\mathcal{P}J_\mu(y)\mathcal{P}^{-1} = \eta^\mu J_\mu(\bar{y}) \; ,$$

with

$$\bar{y} = (y_0, -\vec{y}); \quad \text{and } \eta^0 = +1 \quad \eta^\mu = -1 \quad \mu = 1,\, 2,\, 3 \; ,$$

then

$$\langle p|J_\mu(y)J_\nu(0)|p\rangle = \langle p|\mathcal{P}^{-1}\mathcal{P}J_\mu(y)\mathcal{P}^{-1}\mathcal{P}J_\nu(0)\mathcal{P}^{-1}\mathcal{P}|p\rangle =$$

$$= \langle p|\mathcal{P}^{-1}\eta^\mu J_\mu(\bar{y})\eta^\nu J_\nu(0)\mathcal{P}|p\rangle \; ,$$

so

$$W_{\mu\nu}(p,q) = \eta^\mu\eta^\nu W_{\mu\nu}(\bar{p},\bar{q}) \; .$$

In the original $W_{\mu\nu}$ term the four first terms have this transformation with parity except the last pseudo-tensor term

$$F\varepsilon_{\mu\nu\alpha\beta}\, p^\alpha q^\beta = -\eta^\mu\eta^\nu F\varepsilon_{\mu\nu\alpha\beta}\, p^{-\alpha} q^{-\beta} \; ,$$

so parity conservation implies F = 0.

2) Current conservation.

$$q^\mu W_{\mu\nu} = q^\nu W_{\mu\nu} = 0 \; ,$$

so

$$A\, q_\nu + \frac{B}{M^2}(q\cdot p)p_\nu + \frac{C}{M^2}(q^2)q_\nu + \frac{D}{M^2}\left[(q^2)p_\nu + (q\cdot p)q_\nu\right] +$$

$$+ \frac{E}{M^2}\left[(q^2)p_\nu - (q\cdot p)q_\nu\right] = 0 \ ,$$

and

$$A\, q_\mu + \frac{B}{M^2}(q\cdot p)p_\mu + \frac{C}{M^2}(q^2)q_\mu + \frac{D}{M^2}\left[(q^2)p_\mu + (q\cdot p)q_\mu\right] +$$

$$+ \frac{E}{M^2}\left[(q\cdot p)q_\mu - (q^2)p_\mu\right] = 0 \ .$$

These relations have to be satisfied for all p and q values, so coefficients of p_ν and q_ν (or p_μ and q_μ) have to be zero. Then

$$A + \frac{C}{M^2}q^2 + \frac{D}{M^2}(q\cdot p) - \frac{E}{M^2}(q\cdot p) = 0$$

$$\frac{B}{M^2}(q\cdot p) + \frac{D}{M^2}q^2 + \frac{E}{M^2}q^2 = 0$$

$$A + \frac{C}{M^2}q^2 + \frac{D}{M^2}(q\cdot p) + \frac{E}{M^2}(q\cdot p) = 0$$

$$\frac{B}{M^2}(q\cdot p) + \frac{D}{M^2}q^2 - \frac{E}{M^2}q^2 = 0 \ .$$

So

$$E = 0$$

$$D = -B\,\frac{(q\cdot p)}{q^2}$$

$$C = B\,\frac{(q\cdot p)^2}{q^4} - A\,\frac{M^2}{q^2} \ .$$

Then using $A = -W_1$ and $B = W_2$ we derive the well-known relation.

DISCUSSIONS

CHAIRMAN: Prof. J.J. Aubert

Scientific Secretaries: I.G. Koh, J. Oliensis and T. Shimada

DISCUSSION 1

- VANNUCCI:

What is the compatibility of the newer results in muon scattering with the old Fermilab results which found scaling violation? How do they compare to the ν results?

- AUBERT:

Scaling violation is seen by several experiments (E26, SLAC, E98) consistently for $2 < Q^2 < 20$ GeV^2. The new results extend to 200 GeV^2 and give to date essentially nothing below 20 GeV^2. These experiments show no deviation from scaling in this Q^2 range within the error bars, and are consistent with each other. Also the comparison with ν data is consistent. In fact, if you just look at data (ν and μ) above Q^2 = 15 GeV^2 there is no indication of scaling deviation; it all comes from points below Q^2 = 15-20 GeV^2.

- VANNUCCI:

I would also like to ask a more general question. In what ways can μ physics do better than ν, particularly since F_3 gives you an extra (and easy to use) handle?

- AUBERT:

The μ experiments have the advantage of statistics. Also it is probably an order of magnitude easier to analize the μ data than the ν data. The lack of F_3 is compensated by the possibility of extracting the non-singlet $F_2^{\mu p} - F_2^{\mu n}$.

- GINSPARG:

Do you have any predictions from QCD for the data you have just shown us? I am wondering if one expects the scaling and higher twist corrections to be much smaller for μ than for ν scattering.

- AUBERT:

No and no.

- SCHMIDT:

In order to compare the EMC data with the CDHS data on the scaling violations, it is useful to know what acceptance you have at the low Q^2 (< 20 GeV^2) in EMC for the data you have already corrected.

- AUBERT:

The trigger in the EMC experiment selects events with θ larger than θ_{min}. We have run with $\theta \geq 2°, 1°, 0.5°$ triggers simultaneously, although the $\theta = 0.5°$ and $1°$ data rates are too high, so they are scaled down. We have also taken data for $E_{inc.} = 120$ GeV with the H_2 target. There will then be no limitations due to statistics in the low Q^2 data, and only the systematic errors will be of concern.

- KOH:

My question is about this diagram.

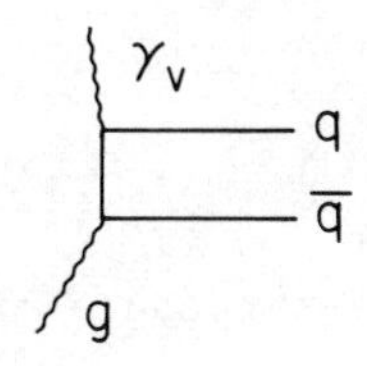

It is counted in the sea quark distribution and also in the scale violations. Isn't this double counting?

- JAFFE:

At one given Q_o^2, there is no way to evaluate the parton distribution. But it is possible to calculate the structure function at another Q^2 value by an integral equation of the form:

$$q(x,Q^2) = \int dx' \, K(x,x',Q^2,Q_o^2)\{q(x',Q_o^2) + \bar{q}(x',Q_o^2) + G(x',Q_o^2)\}$$

where the kernel is computed from the gluon bremsstrahlung and the formation of quark and antiquark pairs from the radiated gluons. So, in fact, it is not double counting.

- VANNUCCI:

There is very large scaling violation in the region $x < .2$ and energies $E > 5$ GeV in $e^+e^- \rightarrow$ hadrons. This is the region you are probing, so why do you get such small scaling violation?

- AUBERT:

The answer must be that we are looking at different processes.

DISCUSSION 2

- LOUIS:

What are the peaks in the structure functions for μp due to? What is the discrepancy between the theory and experiment in the R ratio at high Q^2 due to?

- AUBERT:

The peaks are due to $\mu p \rightarrow \mu \Delta^+$ and the higher resonances. Whether you want to call it a discrepancy is up to you. I should point out that the fact that all the data seem to be above the predictions could be due to a common systematic error of about 1σ. The error bars include an estimate of the systematic error.

- ROMANA:

Can you show where the experiments take place in the (Q^2, ν) plane?

- AUBERT:

The maximum incident energy used for the data is 280 GeV at CERN. Region C is excluded because we don't know how to handle the radiative corrections there. Because of the low beam intensity, FNAL was unable to fill the whole Q^2 region.

- PROUDFOOT:

The errors in the measurement of R seem to be large. Can you explain where they come from? Have you compared $F_2(x,Q^2)$ for H_2

with F_2/nucleon for heavy target in the region of overlap of the data?

- AUBERT:

At the high Q^2, the errors are dominated by statistics and at the low Q^2 by the systematic error. Yes, the data agree where they overlap.

- VANNUCCI:

Comments on the Drell-Yan; the comparison of ISR and Fermilab results is not a check of scaling. The ISR spans $\sqrt{\tau}$ from .05 to .15 while the Fermilab data start at $\sqrt{\tau}$ = .2. They do not overlap and they do not give a test of scaling. For the results, three experiments (Chicago-Priceton, Lezard and Goliath) are inconsistent. In particular, even after correcting for different A dependence, the Goliath results lie a factor 3 higher than the Chicago-Princeton experiments. Also inconsistencies in the structure functions are obtained. Conclusion. The only scaling proven comes from the tests done with the data gathered in a single given experiment. This means rather small Q^2 range. And it is too early to draw firm conclusions about the structure functions.

- AUBERT:

If you take the Chicago-Princeton data and use an A dependence A^1, the data agree with the Lezard experiment.

- TING:

For the last ten years, nobody talks about the elastic scattering. Do you know why?

- AUBERT:

It is not measured now because at high energy it is hard to get good enough resolutions. I think you need a Δp_T < 10 MeV/c .

- TING:

You have some data on μ pair production by μ . I was wondering what is the amount of τ produced by μ . It should be the same where the Q^2 is very large.

- AUBERT:

With our present statistics, we can expect something like few tau events decaying semileptonically, which means events with 3μ in the final states and about 40 or 50 GeV missing energy. We have not yet analyzed our data from this point of view and anyway we cannot contribute much in that field.

- VANNUCCI:

You showed $F_2(x)$ for one Q^2 range. Do you have a comparison at different Q^2 to see a turnover, etc.?

- AUBERT:

Not here.

DIQUARKS

R.T. Van de Walle

University of Nijmegen
Physics Laboratory
Nijmegen, The Netherlands

I. INTRODUCTION

Our lectures will in essence deal with SU(6) violations observed for baryons both in their static and dynamic properties and the possible interpretation of these violations in terms of an SU(6) breaking diquark-quark substructure (*).

We will start with a short review of the known SU(6) baryon problems. This review will be incomplete, both in its content and in its references. Our aim is only to present some background for the topics to be discussed further-on. We will also give an abbreviated 'history' of the notion of diquark, as it has appeared

(*) SU(6) violations should be judged in terms of the quarks involved; thus breakings of the order of 20-30% should not be a source of too much concern if the strange quark is involved, as such breakings already occur at the SU(3) level. For processes involving only u and d quarks (e.g. proton-neutron differences) we can be more severe.

in the literature, either directly or implicitly.

The 'static' successes of SU(6) are well-known; many baryons are succesfully classified using an SU(6) x O(3) symmetry; the μ_p/μ_n-ratio is succesfully predicted, etc. However the dominance of the so-called minimal spectrum i.e. the dominance, of the states $\{56, L^+_{even}\}$ and $\{70, L^-_{odd}\}$ and the absence of solid evidence for the {20} states has remained in essence unexplained [1]. In terms of accomodating SU(6) breaking effects there has always been the incompatibility between the configuration mixings needed to explain the mass levels and those required to explain the decay patterns (**). Also in the static field is the notorious SU(6) violation of the non-zero (negative) charge radius of the neutron; strict SU(6) requires this radius to be zero.

Dynamically the baryon SU(6) successes are less numerous; there is e.g. the usefulness of SU(6) as a decay vertex symmetry for the $\{70, 1^-\}$ and $\{56, 2^+\}$ multiplets [4] and the experimentally well-satisfied spin-density matrix elements relations for many quasi-two-body reactions [5] [6], etc. And also in this area there has been a persistent number of problems. Well-known in this respect is the erroneous SU(6) prediction of the ratio of the nucleon deep inelastic structure functions F_2^{en}/F_2^{ep} measured in e-N scattering; SU(6) requires this ratio to be a constant (equal to or larger than 2/3), experiment yields a ratio approaching 1/4 or 1/3 as x (the Feynman-variable) approaches 1. [7].

Of direct interest for our further discussions are the large violations observed in the SU(6) predictions for the (forward) cross section ratio's of strangeness-exchange reactions such as:

$$K^- p \to M^0 + (\Lambda, \Sigma^0, Y^{*0}_{1385}) \tag{1}$$

(M =neutral nonstrange meson)

(**) See e.g. Ref. 2. Recently there has been progress in interpreting these discrepancies by Isgur and Karl using the mechanism of 'kinematic' mixing i.e. configuration mixing arising from the non-equality of the u, d and s quark masses. (Ref. 3).

These discrepancies were first discussed extensively by Hirsch et al.[8]. More recently the A.C.N.O.-collaboration - using data of a large K^-p bubble-chamber experiment as well as data from a variety of other experiments - has found this effect to occur not only in the total cross sections but also consistently at all (forward) angles in the differential cross section [9]. In a subsequent paper Zralek et al. analysed these effects in terms of an SU(6)-broken diquark-spectator model. [10].

The idea of the existence of diquarks inside (ground-state) baryons was first introduced by Ida and Kobayashi [11] and Lichtenberg and coworkers [12] primarily to explain why nature does not use the full range of multiplet-possibilities contained in SU(6) i.e. restricts itself to the already mentioned minimal spectrum in addition to no {20}'s (see I - 1.3). In this original work diquarks appeared alternatingly as new elementary objects (bosons) or just as two-quark bound states; the formula's used were not sensitive to this distinction. This same feature will be present in our treatment of the diquarks in sections II and IV. However in section III we will explicitly use the bound state interpretation. The latter hypothesis is thus not only the less far-reaching one, it is also the most general-purpose one.

After the work of Lichtenberg et al., the diquarks more or less disappeared from the literature as such but were used implicitly - and embedded in different model assumptions - as x-clusters in an infinite momentum frame description. In this role they were e.g. called 'squeezed wee-quarks' by Feynman [13] and 'nucleon-cores' by Close [14] and used to explain the SU(6) violations observed in the $F_2^{en}(x)/F_2^{ep}(x)$-ratio. The most recent analysis along these lines in the one by Carlitz [15].

The name 'diquark' re-emerged explicitly in the Pavkovic-analysis of the F^{eN}-ratio-problem [16] and in the baryonium work of Chan and Hogaason et al on $qq\bar{q}\bar{q}$ states [17]. However in the latter work diquark was never used to indicate more than states of two quarks with zero relative orbital angular momentum, kept away from the other quarks by a high-L angular momentum barrier.

To explain the SU(6) violating (negative) neutron charge radius Carlitz et al [18] introduced spin-singlet diquarks, bound by an attractive spin-spin interaction of the type suggested by one-gluon-exchange. They showed that using such a force, it was possible to relate the neutron - charge radius quantitatively to the nucleon - Δ mass difference.

Still in the static field there has been the recent work by Isgur et al., [3] reproducing the baryon mass-levels using potential hyperfine terms again inspired by the (SU(6)-violating) spin-spin interaction of OGE; this approach is related to introducing spin-singlet diquarks. Goldstein et al. [19] made a satisfactory overall fit to the baryon spectrum using directly a diquark-quark potential containing the same features.

Several theoretical models, including string and bag models, lead to 'linear' three quark configurations, i.e. quark clustering [20]; to the extent that these models are phenomenological representations of QCD, clustering should also follow from this theory. Actually colour-forces even directly predict the existence of an attractive q-q potential. Within lattice gauge theories arguments have been obtained for dynamical diquark substructure in baryons [21]; similar conclusions have been reached by 'instanton' arguments [22].

A general review of the possible roles of diquarks in the 'static' structure of mesons and both normal and exotic baryons has been given in a recent article by Lichtenberg and Johnson. [23].

Finally in the dynamical field (and as mentioned above) Zralek et al. invoked a SU(6) broken diquark-quark model (coupled to a spectator role for the diquark) to explain the violent SU(6) breakings observed in a large number of quasi two-body reactions [10].

In these lectures we will discuss in some detail three of the problems refered to above. In section II we will show how an SU(6)-broken diquark-quark substructure is capable of explaining the SU(6) violations observed in a large number of strangeness- and charge-exchange quasi-two-body reactions. In section III the same model will be examined as to its possibilities for explaining the nucleon charge radii, and in particular, the neutron charge radius. In section IV this model will be translated into a parton model type description and used for an explanation of the F_2^{en}/F_2^{en} -discrepancy. Section V finally will be reserved for a summary and outlook.

II. SU(6) - BROKEN DIQUARK MODEL FOR MESON-BARYON QUASI TWO-BODY REACTIONS

There are in principle several remedies for the SU(6)-violations observed in quasi-two-body reactions.

An immediate possibility seems to be the breakdown of the additivity assumption used when applying SU(6) - i.e. the symmetric quark model - to these reactions. This possibility however raises the question why additivity works so well so often in many other (baryon as well as meson) cases. Adopting this viewpoint would merely shift the question. Placing the SU(6) breaking inside the baryon wave function, thus appears a more logical starting point.

There are in essence two groups of methods for breaking the SU(6) symmetry of the {56}-multiplet; one can either keep a 3q-symmetry for the internal wave function or abandon it. For simplicity we restrict ourselves to one-parameter SU(6) breaking mechanism.

Configuration mixing is a breaking mechanism which preserves the 3q-symmetry. It usually corresponds to mixing the {56} with the $\{70\} = \{70\}_{M,A} + \{70\}_{M,S}$. The literature contains various claims as to the possibilities of explaining SU(6)-violations with such configuration mixing [24]. In the most recent analyses along this line, configuration mixing is introduced via the SU(6)-violating spin-spin term of one-gluon exchange (OGE) [3]. However for the SU(6)-violations observed in quasi- two-body reactions, configuration mixing is not an adequate explanation. We will come back to the reasons for this impossibility in the conclusions at the end of this section.

Quark-diquark breaking of the ground state baryons is a special case of (one-parameter) SU(6) breaking abandoning the 3q symmetry for the internal variables. It can be shown to be equivalent to the mixing of {56} with $\{70\}_{M,S}$. This is clearly one of the several ways of breaking SU(6) with one parameter. SU(6) breaking associated with an explicitly defined diquark-quark substructure was (re)proposed not only because of its inherent simplicity and its "natural" spectator possibilities, (see further) but also because there are two sources of evidence for such substructure. The first one derives from the observed baryon spectrum; the experimental evidence for the dominance of the so-called minimal spectrum and the absence of the {20} baryons favours a quark-diquark baryon picture (see below). The second one is theoretical, and supplied by the many model indications for diquark clustering inside baryons enumerated in the introduction.

We will now first give a detailed discussion of the model and its assumptions as it was used to explain various sum rules for strangeness - and charge -exchange quasi-two body meson-baryon reactions. However, there are two important remarks which we would like to make already here:

(a) As explained in the introduction the word 'diquark' can have at least two meanings. The essence of the diquark model as used in these lectures is that one avoids the necessity of complete three-quark symmetrization of the wave functions by the assumption that one of the three quarks is spatially further removed from the other two than the latter ones are among themselves. Thus in our discussions 'diquark' never has to mean more than a 'sufficient' clustering of two quarks inside the baryon, sufficient so as to be able to neglect the effects of symmetrizing with respect the third quark when comparing these effects to the (included ones) resulting from symmetrizing in the quarks constituting the so-called diquark. Actually in section III we will make explicit use of the (possible) spatial extension of the diquark.

(b) We will further discuss explicitly the SU(6) violating features entering into the calculations. We would now already like to stress however that it is not the spatial asymmetry inherent in the clustering picture itself which causes this SU(6)-breaking. The breaking really results from the fact that the different possible two-quark states inside the baryon are allowed to take on relative weights differing from those demanded by the symmetric quark model.

1. Assumptions of the Model

Essentially four dynamical assumptions are used in deriving experimentally verifiable relations from the diquark-quark picture in its application to quasi-two-body reactions. The first three are major ones; the fourth one is less crucial. We will discuss these assumptions one by one (including some of their trivial details) in order to present a complete picture.

1.1 First and foremost the diquark model uses the additivity assumption [25]. As in the conventional additive quark model this implies usage of an amplitude for a quasi-two-body interaction (with momentum transfer t) of the form:

$$H(t) = D(t) \, . \, \sum_{a'} C_{a}^{a'} . h_{a'}(t) \tag{2}$$

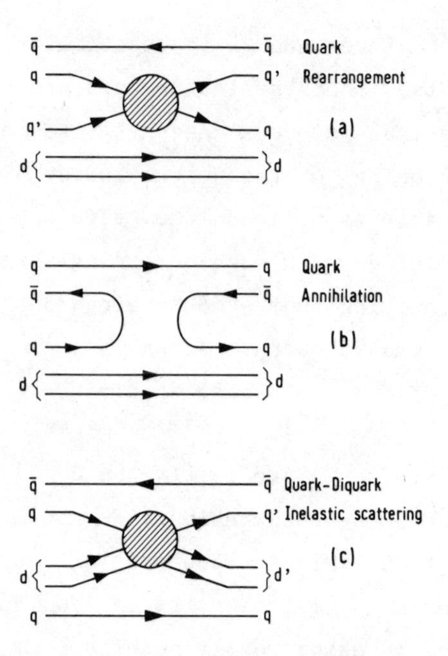

Fig. 1. Quark diagrams for peripheral quasi-two-body scattering: (a) quark rearrangement, (b) quark annihilation-creation, and (c) quark-diquark inelastic scattering.

where a, a' represent the helicity or transversity configuration of the particles and quarks (or diquarks) respectively, $h_{a'}(t)$ represents the quark-quark, quark-diquark or diquark-diquark reaction amplitudes, $C_a^{a'}$ are the coefficients (weight factors) derived from the particle wave functions (as postulated by the model), and D(t) is a form factor which effectively takes into account that the spectator diquark or quark must be dragged away from the colinear (incident) direction by the quark or diquark undergoing the interaction.

1.2 In fig. 1 we show the different types of quark-quark and quark-diquark amplitudes as they contribute (within the additivity model) to peripheral meson-baryon scattering in general and to the two-body reactions considered here in particular. In the usual quark model we only have to consider quark-quark amplitudes; one then deals with two possible types of processes only, namely so-called rearrangement processes and so-called annihilation creation processes. Figs. 1(a) and 1(b) show these two possibilities as they appear in the quark-diquark model. Whether one or both of these diagrams contribute to the strangeness- and charge-exchange processes considered, depends on the specific final state involved. In principle we should also consider quark-diquark amplitudes of the type shown in fig. 1(c). However for the two-body processes studied here (i.e. charge- and strangeness- exchange two-body reactions) these amplitudes must be inelastic and of a rather specific type: they require the substitution of one of the diquark-quarks by one of the quarks from the meson without breaking the diquark structure. This will presumably be a rather improbable process; normally a quark-diquark scattering would probably either be elastic or else break up the diquark structure. In both cases it would not 'feed' the two-body reactions considered here. We will therefore neglect all contributions from diagram 1(c) [*]. This is equivalent to the <u>assumption</u> that the diquark plays a <u>spectator</u> role only. We would like to stress that this inactive role of the diquark is not a general feature of all diquark-model calculations but a specific consequence of the two-body reactions considered here and their (conserved) quantum numbers. (e.g. in the calculations of sections III and IV a diquark-contribution will enter).

1.3 The third important <u>assumption</u> concerns the <u>quark substructure of the baryons.</u> We will asume that - as a result of

[*] An additional (heuristic) reason for not including these contributions in our model calculations is that they would substantially increase the number of parameters used in the model; as we will show later, the model does not need these additional paramters.

the colour q-q interactions - baryons consist of two quarks forming a more or less bound cluster (=the diquark) which in turn interacts with the third quark thus forming a baryon. For the two quarks forming the diquark we will assume zero relative orbital angular momentum. Let us now examine what are the consequences of these assumptions for the diquark and baryon wave functions.

- The diquark wave function

Making the usual Ansatz that quarks belong to the {6} representation of SU(6), the diquark dimensionality must be given by

$$\{6\} \times \{6\} = \{21\} + \{15\}$$

Table I gives the spin-flavour decomposition of both the {21} and the {15}. Note that each of these multiplets contains a flavour-sextet (with a non-strange isotriplet) and a flavour-triplet (with a non-strange isosinglet) but that they combine these states differently with the diquark total spin. In the {21} the I=1 triplet has spin 1 and the I=0 singlet spin 0. In the {15} the opposite pairings I=1, S=0 and I=0, S=1 are observed. It is trivial to derive that in the {21}, the product $\psi_{space} \cdot \psi_J \cdot \psi_I$ is always symmetric in the interchange of the diquark-quarks (both for the triplet and the sextet) while for the {15} multiplet-states it is always anti-symmetric (remember that we assumed ℓ, the diquark relative orbital angular momentum, to be zero).

In terms of colour, a diquark can belong either to a $\{\bar{3}\}$ or a {6} and it must belong to a $\{\bar{3}\}$ if it is to form a colour-singlet with the remaining quark. A $\{\bar{3}\}$ is an anti-symmetric state in the colour-variables of the two quarks forming the diquark (*). However also the total diquark wave

(*) The expectation value of the $\bar{\lambda}_1 . \bar{\lambda}_2$ SU(3)-colour Gell-Mann matrices product in the $\{\bar{3}\}$-state is indeed negative.

Table I

Quantum-numbers of the Diquarks

		I	I_3	Y	Q	Spin S - {21}	Spin S-{15}
FLAVOUR - SEXTET	u u	1	+1	+2/3	+4/3	1 (= s_1)	0
	u d	1	0	+2/3	+1/3	1 (= s_2)	0
	d d	1	−1	+2/3	−2/3	1 (= s_3)	0
	d s	1/2	+1/2	−1/3	+1/3	1 (= s_4)	0
	u s	1/2	−1/2	−1/3	−2/3	1 (= s_5)	0
	s s	0	0	−4/3	−2/3	1 (= s_6)	0
FLAVOUR - TRIPLET	u/d	0	0	+2/3	+1/3	0 (= t_1)	1
	u/s	1/2	+1/2	−1/3	+1/3	0 (= t_2)	1
	d/s	1/2	−1/2	−1/3	−2/3	0 (= t_3)	1

function must be anti-symmetric if it has to refer to a system of two fermions. Ergo: only the {21} multiplet diquark is allowed. The immediate consequence of this fact and the relations:

$$\{15\} \times \{6\} = \{20\} + \{70\}_{M,A}$$

and

$$\{21\} \times \{6\} = \{56\} + \{70\}_{M,S}$$

is that the baryon spectrum should only contain {56} and {70} states and in particular no {20} states, a highly welcome feature as it is supported by the data.

Another consequence of the zero orbital angular momentum assumption for the two quarks forming the diquark is that we immediately have the rule that:

$$\text{Baryon-Parity} = (-1)^L$$

where L is the relative orbital angular momentum of the quark-diquark system (i.e. the total baryon L). This explains the dominances of the 0^+, 1^-, 2^+..., SU(6) multiplets. (*).

The above success of the diquark model assumptions are the ones that - historically - led to its introduction more than 10 years ago. In appreciating them one should of course realize that any reasonable three-quark potential will lead to a spectrum with $\{56,0^+\}$ and $\{70,1^-\}$ as lowest lying states. (Cf. harmonic oscillator). The real test of the diquark-model success therefore lies with the states lying above these ground states (the N=2 level in harmonic oscillator language). However here also the evidence for

(*) To explain the correlations $\{56,L^+_{even}\}$ and $\{70,L^-_{odd}\}$ one needs an additional assumption, namely the assumption of the existence of a sizeable exchange force between quark and diquark of appropriate characteristics.

the minimal spectrum states is substantial. (+) There is at the mass level of, say, the $\{56,2^+\}$, no really compelling evidence for the existence of states others than ones which can be classified inside the minimal spectrum.

- The baryon wave function

With the diquark described above we now construct the most general diquark-quark wave function for all the baryons entering in the analysis (p, n, Λ^0, Σ^0, Δ^0 and Y^{*0}) retaining SU(3) conservation but allowing SU(6) breaking in the coupling of the diquark to the 'odd' quark. Table II gives the resulting wave functions. The symbols s_i and t_i stand for the {21} sextet and triplet diquark states of table I; u, d and s for the individual proton quark states; r_D and r_q are the position vectors with respect to the nucleon CM of the diquark and the odd quark respectively. We will assume - for the time being - that the diquarks are pointlike. (This condition will be relaxed in part II). The fact that the sextet and triplet diquark has a different (i.e. not factorizable) space part is essential; a wave function with a completely factorizable space part could never be SU(6)-violating [26], and vice versa, the fact that the different possible sextet- (respectively triplet-) combinations do carry the same space dependence is a reflection of the fact that SU(6) is not broken inside the diquark. Both ϕ-functions must of course satisfy the (space) normalization conditions:

(+) Leaving out the Ξ^* (and Ω^*) states we have on the order of 20 well established baryon states 'above' the $\{56,0^+\}$ and $\{70,1^-\}$ ones. Approximately half of these can be assigned to a $\{56,2^+\}$. The remaining half can be interpreted as negative parity states either belonging to a $\{70,3^-\}$ or to radial excitations of the $\{70,1^-\}$ and as positive parity states either belonging to a $\{56,4^+\}$ or to radial excitations of the $\{56,0^+\}$ or $\{56,2^+\}$ [1].

$$\int |\phi_{t(s)}(\bar{r}_D, \bar{r}_q)|^2 d^3\bar{r}_D d^3\bar{r}_q = 1 \tag{3}$$

The SU(6) breaking is controlled by an angle Γ ; for $\Gamma = \pi/4$ one regains the unbroken SU(6) wave functions, i.e., the wave functions of the symmetric three-quark model. (*) For any other value of the wave functions in Table II differ from the SU(6) wave functions specifically by the fact that they are symmetrized only in terms of the two quarks inside the diquark and do not contain any symmetry requirements that would result from interchanging the third quark with one of the quarks inside the diquark. Note that the wave functions of decuplet members do not contain Γ ; for the decuplet we cannot break SU(6) without also breaking SU(3).

To respect SU(6) at the diquark-level and to allow its violation in the coupling of the odd quark to the diquark, implies that we assume part of the short-range forces (namely those leading to the diquark clustering) to be SU(6) conserving. In the approach of de Rujula et al., [27] all long-range quark forces are SU(6) conserving and all short-range ones (due to one-gluon-exchange) contain a spin-dependent part which is SU(6) violating. In our model these short-range violating effects are restricted to the short-range quark-diquark interaction. Inclusion of SU(6) violating effects inside the diquark could of course act as 'corrections' to the dominant (attractive) part of the q-q potential influencing diquark structure parameters (such as e.g. its size).

1.4 The final assumption of the model concerns the form factor D(t). We cannot calculate this form factor because we do not know the space part of the wave functions involved. We can make the

(*) This exemplifies that a diquark-substructure assumption is not in itself SU(6) violating.

Table II

Baryon Wave functions in Quark-Diquark model

(SU(3) conserved; SU(6) broken)

$$|p, \pm> = \sin\Gamma \frac{1}{\sqrt{3}} (\sqrt{2}\, s_1 d - s_2 u)\, \phi_s(\bar{r}_D,\bar{r}_q)\, \chi_\pm + \cos\Gamma\, t_1 u\, \phi_t(\bar{r}_D,\bar{r}_q)\, \alpha_\pm$$

$$|n, \pm> = \sin\Gamma \frac{1}{\sqrt{3}} (s_2 d - \sqrt{2}\, s_3 u)\, \phi_s(\bar{r}_D,\bar{r}_q)\, \chi_\pm\ \cos\Gamma\, t_1 d\ \phi_t(\bar{r}_D,\bar{r}_q)\, \alpha_\pm$$

$$|\Lambda, \pm> = \sin\Gamma \frac{1}{\sqrt{2}} (s_4 d - s_5 u)\, \phi_s(\bar{r}_D,\bar{r}_q)\, \chi_\pm + \cos\Gamma \frac{1}{\sqrt{6}}[(t_2 d - t_3 u) + 2t_1 s]\, \phi_t(\bar{r}_D,\bar{r}_q)\, \alpha_\pm$$

$$|\Sigma^0, \pm> = -\sin\Gamma \frac{1}{\sqrt{6}} [2\, s_2 s - (s_5 u + s_4 d)]\, \phi_s(\bar{r}_D,\bar{r}_q)\, \chi_\pm - \cos\Gamma \frac{1}{\sqrt{2}} (t_3 u + t_2 d)\, \phi_t(\bar{r}_D,\bar{r}_q)\, \alpha_\pm$$

$$|Y^{*0}, m> = \frac{1}{\sqrt{3}} (s_5 u + s_4 d + s_2 s)\ \phi_s(\bar{r}_D,\bar{r}_q)\, \chi_m$$

$$|\Delta^0, m> = \frac{1}{\sqrt{3}} (s_3 u + \sqrt{2}\, s_2 d)\ \phi_s(\bar{r}_D,\bar{r}_q)\, \chi_m$$

- Γ = the SU(6) breaking angle
- s_i (t_i) = the internal wave functions of the sextet (triplet) diquark in the {21} multiplet
- ϕ_s (ϕ_t) = the space parts of the quark-diquark system
- $\alpha_\pm$ = the (up and down) single quark spin wave functions
- $\chi_\pm$ (χ_m) = the combined quark-sextet diquark spin wave functions for a total spin $\frac{1}{2}$ (spin $\frac{3}{2}$)

usual assumption that this space part depends both on the total angular momentum inside the baryon (in our case on L, the relative orbital angular momentum between the quark and the diquark) and on the SU(6) representation to which the internal baryon wave function belongs in the limit of unbroken SU(6) ($\Gamma \rightarrow \pi/4$), i.e. $D_L^{56}(t)$, $D_L^{70}(t)$, etc. This leaves us with just one (common) form factor D_0^{56} for all of the $\{\underline{56},0^+\}$ baryons considered here. In principle D(t) should also depend on the diquark-quark configuration considered (coupling to a triplet vs. a sextet diquark) as the corresponding space parts do differ (see above). It can be shown however that (at least in an impulse approximation) this dependence will be very weak [28]. In the present application of our model we will neglect this dependence; this will however not be the case for the applications discussed in part III and IV.

2. Predictions of the Model

The model can now be used to predict relations between the peripheral cross sections and polarizations of the strangeness-exchange reactions

$$A + p \rightarrow M + (\Lambda^0, \Sigma^0, Y^{*0}) \qquad (4)$$

and the charge-exchange reactions

$$A + p \rightarrow M + (n, \Delta^0), \qquad (5)$$

where A and M are mesons satisfying all required conservation laws. We will consider the reactions where A is always a pseudoscalar meson (π or K) and M is either a pseudoscalar or a vector meson. From now on we will suppress all charge indices on the baryon involved and assume we deal with neutral outgoing particles only ($Y = Y^*_{1385}$); $\sigma(\Lambda)$, $\sigma(\Sigma)$, etc. will be used as an abbreviation for the σ (or $d\sigma/dt$) of $Ap \rightarrow M\Lambda$, $Ap \rightarrow M\Sigma$, etc.

Table III shows the different meson-baryon amplitudes contributing to the strangeness-exchange reactions. All amplitudes are expressed in the transversity frame, i.e., with respect to a spin quantization axis perpendicular to the two-body reaction plane. The symbol m is used as a general index to indicate quantum numbers along this axis. In the table we not only distinguish the case of a final-state pseudoscalar meson and vector meson, but for the latter we also distinguish between the three possible 'transversities' with which the meson can be produced ($m = 0, \pm 1$). There exists a correlation between the naturality of the particles exchanged in a two-body process and the transversity state of the meson in the final state. [29] If the produced meson is a pseudoscalar ($M = 0^-$) or a vector meson ($M=1^-$) with $m=0$, the exchanged particle has natural parity (N); for the case $M=1^-$ with $m= \pm 1$ the exchanged object must have unnatural parity (U). (Actually, for $M=1^-$ these statements are strictly speaking valid only in the limit $s \to \infty$).

An analogous table can be set up for the charge exchange reactions. [10].

The amplitudes are expressed in terms of $x = \sin^2\Gamma$ (where Γ is the SU(6)-breaking angle previously defined) and the quantities P_i and V_i. In terms of the quantities in Eq. (2), the x-dependent factors are derived from the $C_a^{a'}$ coefficients and the P_i, V_i are proportional to combinations of the $D(t).h_{a'}(t)$. The P_i and V_i are in fact meson-quark transversity amplitudes; they depend on the meson quark structure and are linear combinations of the quark-quark transversity amplitudes defined by the diagrams Fig. 1(a) and Fig. 1(b). As an example for reactions $K^-p \to \pi^0 + \ldots$ we have

$$P_1 = \langle \pi^0 s_+ | K^- u_+ \rangle$$

$$= \frac{1}{2\sqrt{2}} [\langle u_+ s_+ | s_+ u_+ \rangle + \langle u_- s_+ | s_- u_+ \rangle]$$

Table III

Meson - Baryon transversity amplitudes for strangeness exchange reactions in terms of $x(=\sin^2\Gamma)$ and meson-quark amplitudes P_i, V_i - (see text definitions)

Reactions / Amplitudes	$0^-+p\to 0^-+(\Lambda,\Sigma,Y)$ (NP-exch.)	$0^-+p\to 1^-+(\Lambda,\Sigma,Y)$ m=0 (NP-exch.)	m=+1 (UP-exch.)	m=−1 (UP-exch.)
$p_{1/2}\to\Lambda^0_{1/2}$	$\frac{\sqrt{2}}{\sqrt{3}}(1-x)P_1$	$\frac{\sqrt{2}}{\sqrt{3}}(1-x)V_1$	0	0
$p_{1/2}\to\Lambda^0_{-1/2}$	0	0	$\frac{\sqrt{2}}{\sqrt{3}}(1-x)V_3$	$\frac{\sqrt{2}}{\sqrt{3}}(1-x)V_5$
$p_{-1/2}\to\Lambda^0_{1/2}$	0	0	$\frac{\sqrt{2}}{\sqrt{3}}(1-x)V_4$	$\frac{\sqrt{2}}{\sqrt{3}}(1-x)V_6$
$p_{-1/2}\to\Lambda^0_{-1/2}$	$\frac{\sqrt{2}}{\sqrt{3}}(1-x)P_2$	$\frac{\sqrt{2}}{\sqrt{3}}(1-x)V_2$	0	0
$p_{1/2}\to\Sigma^0_{1/2}$	$\frac{\sqrt{2}}{9}x(2P_2+P_1)$	$\frac{\sqrt{2}}{9}x(2V_2+V_1)$	0	0
$p_{1/2}\to\Sigma^0_{-1/2}$	0	0	$-\frac{\sqrt{2}}{9}xV_3$	$-\frac{\sqrt{2}}{9}xV_5$
$p_{-1/2}\to\Sigma^0_{1/2}$	0	0	$-\frac{\sqrt{2}}{9}xV_4$	$-\frac{\sqrt{2}}{9}xV_6$
$p_{-1/2}\to\Sigma^0_{-1/2}$	$\frac{\sqrt{2}}{9}x(2P_1+P_2)$	$\frac{\sqrt{2}}{9}x(2V_1+V_2)$	0	0
$p_{1/2}\to Y^0_{3/2}$	0	0	$-\frac{\sqrt{2x}}{3\sqrt{3}}V_4$	$-\frac{\sqrt{2x}}{3\sqrt{3}}V_6$
$p_{1/2}\to Y^0_{1/2}$	$\frac{\sqrt{2x}}{9}(P_1-P_2)$	$\frac{\sqrt{2x}}{9}(V_1-V_2)$	0	0
$p_{1/2}\to Y^0_{-1/2}$	0	0	$\frac{\sqrt{2x}}{9}V_3$	$\frac{\sqrt{2x}}{9}V_5$
$p_{1/2}\to Y^0_{-3/2}$	0	0	0	0
$p_{-1/2}\to Y^0_{3/2}$	0	0	0	0
$p_{-1/2}\to Y^0_{1/2}$	0	0	$-\frac{\sqrt{2x}}{9}V_4$	$-\frac{\sqrt{2x}}{9}V_6$
$p_{-1/2}\to Y^0_{-1/2}$	$\frac{\sqrt{2x}}{9}(P_1-P_2)$	$\frac{\sqrt{2x}}{9}(V_1-V_2)$	0	0
$p_{-1/2}\to Y^0_{-3/2}$	0	0	$\frac{\sqrt{2x}}{3\sqrt{3}}V_3$	$\frac{\sqrt{2x}}{3\sqrt{3}}V_5$

and for reactions $K p \to \rho^0 + \ldots$ we have

$$V_1 = \langle \rho^0_0 \, s_+ | K^- u_+ \rangle$$

$$= \frac{1}{2\sqrt{2}} [\langle u_- s_+ | s_- u_+ \rangle - \langle u_+ s_+ | s_+ u_+ \rangle]$$

Note that although we have formulated the meson-quark amplitudes in terms of a particular meson-quark interaction mechanism, our results will be independent of these details and will depend only on the assumptions made at the baryon vertex.

We are now ready to extract predictions from the model by eliminating the unknown amplitudes P_i, V_i. We will write all these predictions in terms of cross sections corrected for kinematic effects resulting from mass differences and use the symbol $\bar{\sigma}$ to indicate these.

Some of the relations obtained do not depend on x; they are relations which the model has in common with unbroken SU(6) [10]. Within the model the validity of these relations is dependent on the assumption that we can neglect the quark-diquark inelastic amplitudes. As it is our aim to explain disagreements with SU(6) predictions in situations where the diquark model <u>does</u> yield predictions different from unbroken SU(6), we will not go into details of these 'common-with-SU(6)' - relations. Let us just mention that in this class belong the many relations between (single and double) spin statistical tensors for weakly decaying particles and/or strongly decaying resonances produced in quasi-two-body reactions [5]; they have been (successfully) tested by many authors [6]. Somewhat related to these 'common-with SU(6)' relations are those that do depend on x but for which the x-dependence is too weak to lead to meaningful differences with strict-SU(6) relations. Examples of such relations are given in ref. 30; again they will not be discussed here.

Instead let us turn to the relations which are (significantly) different from the corresponding SU(6) predictions.

2.1 Predictions for strangeness-exchange reactions:

$$\frac{3(1-x)^2}{x^2}\left[\bar{\sigma}(\Sigma) + 2x\bar{\sigma}(Y)\right] = \bar{\sigma}(\Lambda) \qquad (6)$$

where $x=\sin^2\Gamma$ (as before). This relation replaces the SU(6) relation [31]

$$3\left[\bar{\sigma}(\Sigma) + \bar{\sigma}(Y)\right] = \bar{\sigma}(\Lambda) \qquad (7)$$

and is valid for the natural-parity (NP) exchange ($\bar{\sigma}_N$) and unnatural-parity (UP) exchange ($\bar{\sigma}_U$) cross sections separately. For UP exchange we find in addition

$$\bar{\sigma}_U(\Sigma) : \bar{\sigma}_U(\Sigma) : \bar{\sigma}_U(Y) = 27\ (1-x)^2 : x^2 : 4x \qquad (8)$$

replacing the symmetric quark model prediction [8] [32]

$$\bar{\sigma}_U(\Lambda) : \bar{\sigma}_U(\Sigma) : \bar{\sigma}_U(Y) = 27 : 1 : 8 \qquad (9)$$

For NP exchange we also have the relation

$$P_N^m(\Lambda)\ \bar{\sigma}(\Lambda) = -9\left(\frac{1-x}{x}\right)^2 P_N^m(\Sigma)\bar{\sigma}_N(\Sigma) \qquad (10)$$

instead of the SU(6) relation [9]

$$P_N^m(\Lambda)\ \bar{\sigma}_N(\Lambda) = -9\ P_N^m(\Sigma)\ \bar{\sigma}_N(\Sigma) \qquad (11)$$

where P is the polarization of the hyperon defined as:

$$\vec{P} = \left[\frac{3}{2J^2(J+1)}\right]^{\frac{1}{2}} \langle\vec{J}\rangle$$

with J = spin of the hyperon [P_N^m (Y) = 0 is a prediction common to SU(6) and the diquark model.]

As required all our diquark-model predictions reduce to the corresponding SU(6) ones for $x = 1/2$ ($\Gamma = \pi/4$).

2.2 Predictions for charge-exchange reactions:

For $M = 1^-$ we have that

$$\bar{\sigma}_U(n) = \frac{(9-8x)^2}{16x} \cdot \bar{\sigma}_U(\Delta) \qquad (12)$$

and in general (both for $M=0^-$ and 1^-)

$$\bar{\sigma}(n) \geqslant \frac{(9-8x)^2}{16x} \cdot \bar{\sigma}(\Delta) \qquad (13)$$

instead of the quark-SU(6)-model relation [33]

$$\bar{\sigma}(n) \geqslant \frac{25}{8} \cdot \bar{\sigma}(\Delta) \qquad (14)$$

Relation (13) is obtained by expressing the cross section ratio in terms of the quantity x and the meson-quark transversity amplitudes and minimizing the resulting expression in these variables.

One final remark should be made: All relations derived here, although written down for total cross sections, are in principle only valid for the forward part of the cross section (both total and differential).

3. Comparison with Experiment

In order to check the relations derived in the preceding section it is usually necessary, because of the lack of data, to integrate over (frequently quite large) intervals in t. Sometimes one even has to use total cross sections. Although the model can be expected to be valid only for peripheral interactions, the relative smallness of the backward production cross sections should not unduly influence these comparisons. Where one is able to use peripheral cross sections, integration over equal intervals of $t'=|t-t_{min}|$ has been prefered instead of over equal t-intervals in order to reduce purely kinematical effects in the very forward direction.

Before being able to compare the model predictions with experiment, one must also adopt a prescription for treating the differences in kinematic factors resulting from mass differences. This is a standard difficulty when comparing symmetry predictions with experiment for which there is at present no rigorous solution. Let us first stress that all the comparisons have been performed between reactions starting from the same initial state (both in terms of beam momentum and particle type). Furthermore, although occasionally beam momenta as low as 3 GeV/c have been used, most of the comparisons are at 4 GeV/c or higher; the latter implies a variation in Q values (or C.M. momentum p^*) between the different inelastic final states of at most 25% (or 15%) respectively.

The prescription which is most often used in the literature is the comparison at equal Q-values given by Meshkov et al [34]. Trilling [35] has convincingly pointed out this method can lead to serious and unphysical discrepancies which are avoided if one compares cross sections reduced by phase space and angular momentum barriers at equal s-values. We refer the reader to ref. 10 for full details but just state that in essence the more correct Trilling-prescription was used and that it was verified that the conclusions obtained were qualitatively independent of whether one compares at equal s or equal Q. Actually, within the present model, all differences resulting from using different prescriptions are small and absorbed by small variations in the breaking angle.

Finally, one needs to adopt a value of $x = \sin^2\Gamma$. While making a fit to all available data might seem to be the best method of determining x, in practice such a fit poses severe problems not the least of which is the evaluation of the errors in a large number of data. One therefore has chosen to simply determine x from the cross sections for $K^-p \to \Lambda\rho$ and $K^-p \to Y\rho$ in the high-statistics 4.2 GeV/c K^-p experiment (see table VI). A direct evaluation of eq. (8) for these cross sections yields:

$$x = (0.64 \pm 0.02)$$

or

$$\Gamma = (53 \pm 1.2)^0$$

All the comparisons were made using this value for x.

3.1 Comparison of strangeness-exchange reactions

For the strangeness-exchange reactions there are two types of processes which can be used to test the model, namely

$$K^-p \to M + (\Lambda,\Sigma,Y) \text{ with } M = \pi^0,\ \rho^0, \eta \text{ or } \phi \qquad (15)$$

and

$$\pi^-p \to M + (\Lambda,\Sigma,Y) \quad \text{with } M = K^0,\ K^{*0} \qquad (16)$$

Some of these reactions (e.g. $K^-p \to \pi^0\Sigma^0$) are quite difficult to measure; for such reactions we will use relations such as

$$\sigma(K^-p \to \pi^0\Sigma^0) = \frac{1}{4}\,\sigma(K^-p \to \pi^-\Sigma^+) \qquad (17)$$

For reactions producing an I=0 meson or I=0 baryon these relations follow directly from isospin conservation; otherwise the additional assumption of pure I=1/2 exchange must be made, an assumption already present in our model (and in the usual SU(6) quark model) where only single quark transitions are considered.

Tables IV and V show the tests of the diquark relation (6) vs the SU(6) relation (7) for K^-p and π^-p initial states, respectively. Table VI tests the diquark relation (8) vs the SU(6) relation (9) for K^-p reactions. The agreement between the data and the diquark-model predictions is surprisingly good while in most cases the SU(6) prediction is in strong disagreement with the data.

In the above-mentioned tables we have examined cross sections integrated over large t' intervals. It is however also possible to examine the diquark relations differentially. We will use the high statistics of the 4.2 GeV/c K^-p experiment and concentrate on natural-parity-exchange reactions ($M=\pi^0$) or on the natural-parity-exchange part of the reactions with $M = \rho^0$, ϕ. The most complete method then consists of combining the relations (6) and (7) with (10) and (11) respectively, by performing amplitude fits in each t' interval separately. As may be seen in Table III for each choice of meson there are only two independent natural-parity-exchange meson-quark amplitudes. In each t' bin one can then fit the two amplitude magnitudes and one relative phase to five quantities ($d\sigma_N/dt'$ for Λ, Σ, Y, and P^m_N for Λ, Σ). For $M= \pi^0$ [40] the fitted results are shown in Fig. 2. Solid (broken) lines connect the results of the diquark model [SU(6) model] fits. There is little difference between the polarization fit results of the two models (only the diquark result is shown), which is not surprising since the polarizations are independent of x. Inspecting the $d\sigma/dt'$ distributions, we once more observe that the diquark model is in reasonable agreement with the data while SU(6) is again in violent disagreement. Also shown are the fitted values of the two amplitudes and their relative phase from the diquark fits. The relatively smooth variation of these quantities as a function of t' increases confidence in the fits.

Similar fits were performed for the natural-parity-exchange parts of the reactions with $M=\rho^0$ and $M=\phi$; the results obtained,

Table IV

Test of Diquark-relation (6)

$$\frac{3(1-x)^2}{x^2} \bar{\sigma}(\Sigma) + 2x\, \bar{\sigma}(Y) = \bar{\sigma}(\Lambda)$$

versus SU(6)-relation (7)

$$3\, \bar{\sigma}(\Sigma) + \bar{\sigma}(Y) = \bar{\sigma}(\Lambda)$$

for $K^- p \to M^0 + (\Lambda, \Sigma, Y)$ and $x \simeq 0.64$ (in μb-units)

Reaction (M^0)	$\bar{\sigma}(\Sigma) = \frac{p^*_\Lambda}{p^*_\Sigma}\, \sigma(\Sigma)$	$\bar{\sigma}(Y) = \frac{p^*_\Lambda}{p^*_Y}\, \sigma(Y)$	L.H.S. SU(6) relation	L.H.S. Diquark relation	$\bar{\sigma}(\Lambda) = \sigma(\Lambda)$	Ref.
	p_{lab} = 3 GeV/c (forward hemisphere)					
π^0	108 ± 18	70 ± 20	534 ± 81	193 ± 31	233 ± 30	36
η	37 ± 6	59 ± 12	288 ± 40	106 ± 16	100 ± 26	36
ρ^0	85 ± 7	75 ± 10	480 ± 37	177 ± 14	219 ± 50	36
ρ^0	91 ± 13	84 ± 14	525 ± 57	193 ± 22	160 ± 54	8, 37
ω	94 ± 12	84 ± 23	534 ± 78	197 ± 31	196 ± 36	8,37
φ	51 ± 10	59 ± 15	330 ± 54	123 ± 21	82 ± 20	36
	p_{lab} = 3.9 GeV/c (forward hemisphere)					
ρ^0	26 ± 5	26 ± 2	157 ± 17	57 ± 6	63 ± 9	38
φ	40 ± 8	21 ± 5	183 ± 30	63 ± 10	59 ± 7	38
	p_{lab} = 4.6 GeV/c (forward hemisphere)					
ρ^0	20 ± 3	17 ± 3	110 ± 13	40 ± 4	45 ± 5	38
φ	23 ± 6	19 ± 4	125 ± 21	44 ± 7	41 ± 4	38
	p_{lab} = 14.3 GeV/c (t' < 1.0 GeV^2)					
π^0	5.1 ± 0.5	2.1 ± 0.2	22 ± 2	7.6 ± 0.5	8.4 ± 1.2	39,8
	p_{lab} = 4.2 GeV/c (t' < 1.0 or 1.2 GeV^2)					
π^0 (1)	31 ± 2	20 ± 1	153 ± 7	55 ± 2	61 ± 3	40
ρ^0 (2)	17 ± 2	20 ± 3	111 ± 11	42 ± 2	46 ± 3	42,41
ρ^0 NP-exchange	14 ± 1	4 ± 1	54 ± 4	19 ± 2	24 ± 2	42,41
ρ^0 UP-exchange	3 ± 1	16 ± 2	57 ± 7	23 ± 3	22 ± 2	42,41
φ (1)	28 ± 2	26 ± 2	162 ± 8	60 ± 3	58 ± 3	42,41
φ NP-exchange	23 ± 2	8 ± 1	93 ± 7	32 ± 2	27 ± 2	42,41
φ UP-exchange	4 ± 1	18 ± 2	66 ± 7	26 ± 3	32 ± 2	42,41

(1) t' < 1 GeV^2

(2) t' < 1.2 GeV^2

Table V

Test of Diquark relation (6) versus SU(6) relation (7) for $\pi^- p \rightarrow \binom{K}{K^{*0}} + (\Lambda, \Sigma, Y)$ and $x \simeq 0.64$ (in μb-units)

Reaction	$\bar{\sigma}(\Sigma) = \frac{p^*_\Lambda}{p^*_Y}\sigma(\Sigma)$	$\bar{\sigma}(Y) = \frac{p^*_\Lambda}{p^*_Y}\sigma(Y)$	L.H.S. SU(6) relation	L.H.S. Diquark relation	$\bar{\sigma}(\Lambda)=\sigma(\Lambda)$	Ref.
		p_{lab} = 3.9 GeV/c	(t' < 1.0 GeV^2)			
K^{*0}	32 ± 3	24 ± 3	168 ± 13	61 ± 5	57 ± 4	43
		p_{lab} = 4.5 GeV/c	(t' < 1.0 GeV^2)			
K^0	27 ± 3	14 ± 3	123 ± 13	43 ± 5	50 ± 5	44

Table IV

Test of Diquark relation (8)

$$\bar{\sigma}_U(\Lambda) : \bar{\sigma}_U(\Sigma) : \bar{\sigma}_U(Y) = 27(1-x)^2 : x^2 : 4x$$

versus SU(6) relation (9)

$$\bar{\sigma}_U(\Lambda) : \bar{\sigma}_U(Y) = 27:1:8$$

for $K^- p \rightarrow \binom{\rho^0}{\phi^0} + (\Lambda, \Sigma, Y)$ and $x \simeq 0.64$ (in μb-units)

P_{lab} = 4.2 GeV/c

Reaction	$\bar{\sigma}(\Lambda)$	$\bar{\sigma}(\Sigma)\ f_{dq}$	$\bar{\sigma}(\Sigma) f_{SU(6)}$	$\bar{\sigma}(Y) f_{dq}$	$\bar{\sigma}(Y) f_{SU(6)}$	Ref.
ρ^0 (1)	22 ± 2	26 ± 9	81 ± 27	22 ± 3	54 ± 7	41,42
ϕ (2)	32 ± 2	35 ± 9	108 ± 27	25 ± 3	61 ± 7	41,42

$f_{dq} = 8.8, \frac{8.8}{6.3}$ for Σ and Y respectively

$f_{SU(6)} = 27, \frac{27}{8}$ for Σ and Y respectively

(1) t' < 1.2 GeV^2

(2) t' < 1.0 GeV^2

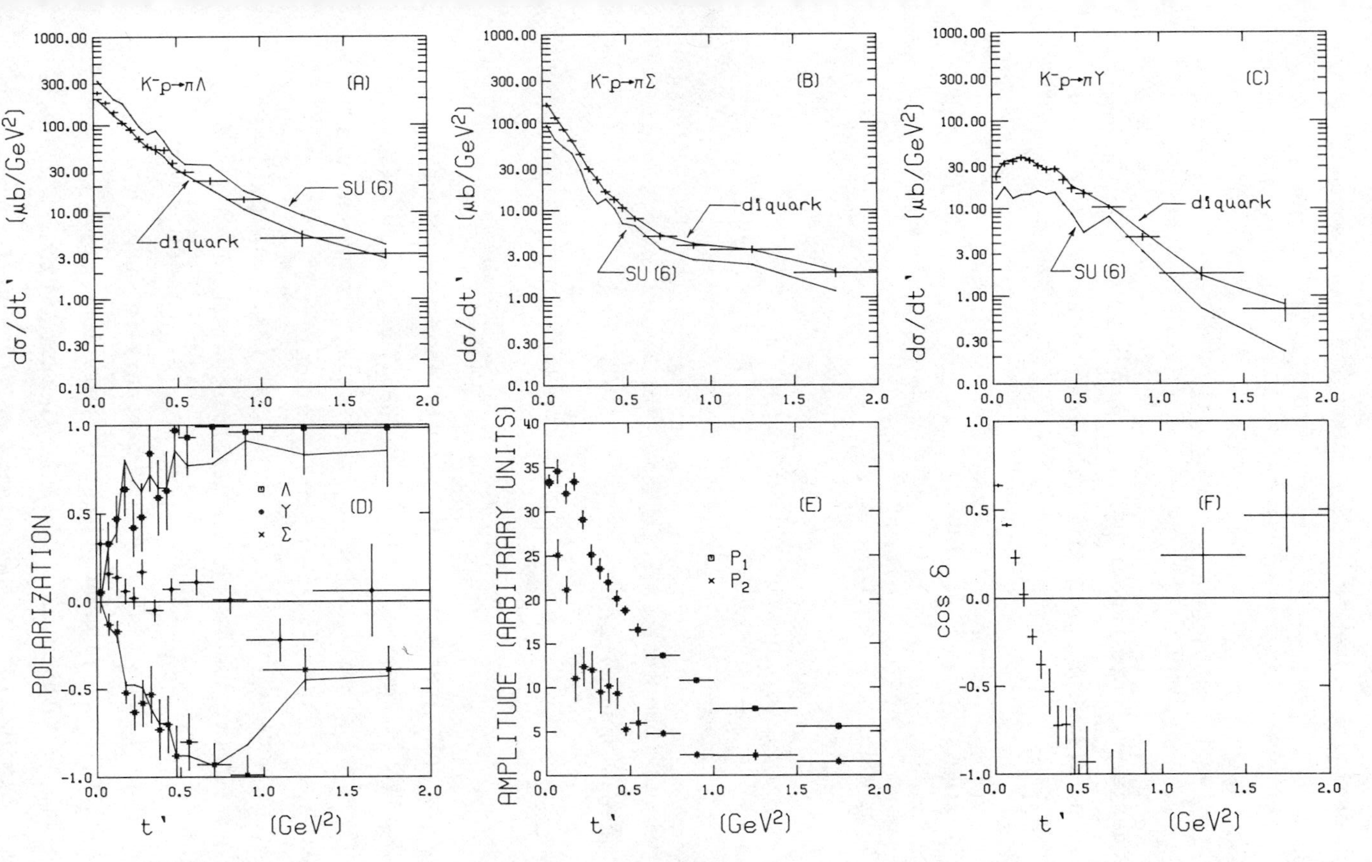

Fig. 2. (a-d) Differential cross sections and polarization for $K^-p \to \pi^0$ (Λ,Σ,Y) at 4.2 GeV/c compared with the results of fits of the diquark or SU(6) quark transversity amplitudes; (e) the magnitudes of these diquark transversity amplitudes and (f) their relative phase as found in the fits.

are comparable, the only difference being that the statistics in these channels is lower and the results therefore somewhat less significant.

3.2 Comparison of Charge-exchange reactions

For the charge-exchange reactions there are three types of reactions which allow a test of relations (12), (13) and (14), namely:

$$K^- p \to M^0 + (n,\Delta) \text{ with } M^0 = \bar{K}^0, \overline{K^{*0}} \qquad (18)$$

$$K^0 p \to M^+ + (n,\Delta) \text{ with } M^+ = K^+, K^{*+} \qquad (19)$$

$$\pi^- p \to M + (n,\Delta) \text{ with } M^0 = \pi^0, \rho^0, \omega, \phi \qquad (20)$$

Here too we will use isospin relations to avoid experimentally difficult reactions, e.g.

$$\sigma(\pi^- p \to \pi^0 \Delta^0) = \frac{1}{3}\sigma(\pi^+ p \to \pi^0 \Delta^{++}) \qquad (21)$$

Most of these relations result from pure isospin conservation. Only those with I=1 mesons and a Δ in the final state [reactions (20)] require the assumption of no I=2 exchange in the t channel (or equivalently, single-quark transitions only).

First we make a test of the equality (12) using the 4.2 GeV/c $K^- p \to K^{*0}(n,\Delta)$ data [45]. The diquark-model prediction for $\bar{\sigma}_U(n)/\bar{\sigma}_U(\Delta)$ is 1.5; the SU(6) value is 3.1. Experimentally we find 1.3 ± 0.4. The diquark model is strongly favoured.

In tables VII we present a test of relations (13) and (14) for a $K^- p$ initial states. The agreement with the diquark model is remarkable. Examination of the forward differential cross sec-

Table VII

Test of Diquark relation (13)

$$\bar{\sigma}(n) \geq \frac{(9-8x)^2}{16x} \cdot \sigma(\Delta)$$

versus the SU(6) relation (14)

$$\bar{\sigma}(n) \geq \frac{25}{8} \cdot \bar{\sigma}(\Delta)$$

for $K^- p \rightarrow \binom{\bar{K}^0}{\bar{K}^{*0}} + (n,\Delta)$ and $x \simeq 0.64$ (in μb-units)

Reaction	$\bar{\sigma}(\Delta) = \sigma(\Delta)$	$\bar{\sigma}(n) = \frac{p^*_\Delta}{p^*_n}\sigma(n)$	$f_{dq}\ \bar{\sigma}(\Delta)$	$f_{SU(6)}\bar{\sigma}(\Delta)$	Ref.
	p_{lab} = 3 GeV/c (forward hemisphere)				
$\bar{K}^0$	210 ± 100	365 ± 31	315 ± 150	656 ± 313	36
$\bar{K}^{*0}$	300 ± 120	596 ± 89	450 ± 180	938 ± 375	36
	p_{lab} = 4.2 GeV/c				
$\bar{K}^0$	147 ± 17	242 ± 14	216 ± 25	459 ± 53	45,46
$\bar{K}^{*0}$	306 ± 25	459 ± 93	459 ± 38	956 ± 78	45
	p_{lab} = 5 GeV/c ($t' < 1.2\ GeV^2$)				
$\bar{K}^0$	86 ± 12	142 ± 12	129 ± 18	269 ± 38	47
	p_{lab} = 8.36 GeV/c ($t' < 1.0\ GeV^2$)				
$\bar{K}^0$	32 ± 4	69 ± 9	48 ± 6	100 ± 13	48
	p_{lab} = 12.8 GeV/c ($t' < 1.0\ GeV^2$)				
$\bar{K}^0$	10.7 ± 0.6	35 ± 2	16 ± 1	33 ± 2	48
	p_{lab} = 13 GeV/c ($t' < 1.2\ GeV^2$)				
$\bar{K}^0$	19 ± 4	39 ± 4	29 ± 6	61 ± 12	49
	p_{lab} = 15.7 GeV/c ($t' < 0.7\ GeV^2$)				
$\bar{K}^0$	13 ± 3	31 ± 4	20 ± 5	41 ± 9	50
			$f_{dq} \simeq 1.50$	$f_{SU(6)} = 25/8$	

tions for the K^-p reactions at 5, 8.4, 12.8 13, and 15.7 GeV/c shows that this conclusion is also valid at each t. A test of these same relations for a K^0p initial state leads to the same conclusions [10].

For the pion-induced reactions we are again able to test directly the equality (12) for the unnatural-parity exchange, using the reactions $\pi^+n \to \omega p$ at 6.95 GeV/c [51] and $\pi^+p \to \omega\Delta^{++}$ at 7.1 GeV/c [52] which are related to reactions (20) by isospin. We have extracted the UP cross section for $0.06<t<0.50$ $(GeV/c)^2$ and obtained:

$$\bar{\sigma}_U(n) = \frac{p^*_\Delta}{p^*_n} \cdot \sigma_U(n) = (9.8 \pm 2.5)\mu b$$

$$\bar{\sigma}_U(\Delta) = \sigma_U(\Delta) = (7.4 \pm 0.5)\mu b$$

The equality (12) is then

$$(9.8 \pm 2.5)\mu b = (11.1 \pm 0.8)\mu b \quad [\text{diquark}]$$

or

$$(9.8 \pm 2.5)\mu b = (23.1 \pm 1.6)\mu b \quad [SU(6)]$$

Again, the diquark model agrees well with the data while SU(6) is in complete disagreement.

In tables VIII and IX we again test relations (13) and (14) but now for pion-induced processes and for $M = \pi^0$ and respectively. (In ref. [10] tests for $M = \rho^0$ and η are also presented). Examination of these tables indicates strong disagree-

Table VIII

Test of Diquark relation (13) versus SU(6) relation (14) for $\pi^- p \to \pi^0 = (n,\Delta)$ and $x \simeq 0.64$ (in μb-units).

p_{lab} (GeV/c)	$\bar{\sigma}(\Delta)=\sigma(\Delta)$ (1)	$\bar{\sigma}(n)=\frac{p^*_\Delta}{p^*_n}\sigma(n)$ (2)	$f_{dq}\bar{\sigma}(\Delta)$	$f_{SU(6)}\bar{\sigma}(\Delta)$	Ref.
3.9	130 ± 7 (3)	160 ± 15	195 ± 11	406 ± 22	54
4.0	117 ± 23	154 ± 15	176 ± 35	366 ± 72	55
5.0	73 ± 3	113 ± 11	109 ± 5	228 ± 10	56
5.45	56 ± 10	102 ± 8	85 ± 15	175 ± 31	57
8.0	37 ± 3	63 ± 5	56 ± 5	116 ± 9	58
11.7	25 ± 3 (4)	41 ± 3	38 ± 5	156 ± 21	59
13.1	15 ± 2	36 ± 2	22 ± 4	47 ± 7	60
16.0	18 ± 1 (5)	29 ± 2	28 ± 1	58 ± 2	61
			$f_{dq} \simeq 1.50$	$f_{SU(6)} = 25/8$	

(1) $\sigma(\pi^- p \to \pi^0\Delta^0)$ derived from isospin relation $\sigma(\pi^- p \to \pi^0\Delta^0) = \frac{1}{3}\sigma(\pi^+ p \to \pi^0\Delta^{++})$

(2) All $\pi^- p \to \pi^0 n$ cross sections are derived from the data of ref. [51] for $t < 1.5\ \text{GeV}^2$.

(3) forward hemisphere

(4) corrected for $t < 1.5\ \text{GeV}^2$

(5) The cross section for $t' > 1.0\ \text{GeV}^2$ is consistent with zero.

ment with SU(6) in all cases and very good agreement with the diquark model (especially for the narrow resonances). [*]

4. Conclusions

The (SU(6) - broken) diquark hypothesis is capable of explaining a large number of sum rule discrepancies between SU(6) and the experimental data for strangeness- and charge exchange two-body processes by introducing one extra parameter ($x=\sin^2\Gamma$). In this success two features of the diquark-model play a key role, namely

- the assumption of the diquark behaving as a spectator [69].
- the assumption that the sextet and triplet diquarks can take on unequal relative weights (=probabilities).

We have stated before that configuration mixing is not able to explain the quasi-two-body SU(6) violations. The reason essentially lies with the second of the above key assumptions [70]: The first assumption could (at least for strangeness-exchange reactions) be 'simulated' in configuration mixing by taking into account the symmetry breaking resulting from the mass-differences between the s and the u, d quarks. However there is no place inside configuration mixing for the second one because in this model the relative weight of the triplet and sextet diquarks is constrained to remain the one imposed by SU(6), i.e. equal to 1. Thus the only method left to modify the resulting sum rules is through the introduction of (strong) form factor effects. However with such an approach one could never explain the large effects

---

[*] Only in the ρ^0 case do we find a few points lying two or more standard deviations away from the diquark-model prediction. However, for all these points the disagreement with SU(6) is at least an order of magnitude larger. In addition, for broad resonances the cross sections certainly contain (non-detected) systematic effects.

Table IX

Test of diquark relation (13) versus SU(6) relation (14) for $\pi^- p \to \omega + (n,\Delta)$ and $x \simeq 0.64$ (in μb-units).

p_{lab} (GeV/c)	$\bar{\sigma}(\Delta) = \sigma(\Delta)$ (1)	$\bar{\sigma}(n) = \frac{p^*_\Delta}{p^*_n}\sigma(n)$ (2)	$f_{dq}\bar{\sigma}(\Delta)$	$f_{SU(6)}\bar{\sigma}(\Delta)$	Ref.
4.0	167 ± 50	276 ± 22	250 ± 75	522 ± 156	55
4.09	120 ± 10	262 ± 21	180 ± 15	375 ± 31	63
5.0	93 ± 3	164 ± 18	140 ± 5	291 ± 9	56,64
5.45	98 ± 6	135 ± 18	147 ± 9	306 ± 19	65
7.1	47 ± 1	71 ± 17	71 ± 2	147 ± 6	52
8.0	37 ± 3	54 ± 9	56 ± 5	175 ± 16	58
11.7	20 ± 4	22 ± 8	30 ± 6	63 ± 12	66
13.2	12 ± 1	17 ± 7	18 ± 2	38 ± 3	67
18.5	4 ± 1	7 ± 3	6 ± 2	13 ± 3	68
			$f_{dq} \simeq 1.50$	$f_{SU(6)} = 25/8$	

(1) $\sigma(\pi^- p \to \omega\Delta^0)$ derived from the isospin relation $\sigma(\pi^- p \to \omega\Delta^0) = \frac{1}{3}\sigma(\pi^+ p \to \omega\Delta^{++})$

(2) The $\pi^- p \to \omega n$ cross sections are derived from data on this reaction and on the reaction $\pi^+ n \to \omega\, p$ [49,72].

observed, mainly because the SU(6) violations are found to be present consistently at all forward t (see e.g. fig. 2).

The above findings raise the question of the validity of the diquark model outside the field of quasi-two-body reactions considered here. Indeed on the basis of only the analysis as made above it is quite possible that the baryon spends only a fraction of its time in a quark-diquark state but that peripheral quasi-two-body reactions are preferentially initiated from a baryon which is in this configuration. The first field one is immediately led to think about is the static properties of the baryons. In section III we will examine a particular static problem: the neutron charge radius.

III. NEUTRON CHARGE RADIUS IN THE SU(6)-BROKEN DIQUARK MODEL [*]

There are many static baryon properties which can in principle yield information as to the existence of diquarks as dynamical entities. The first one is of course the baryon spectroscopy itself; actually this formed the starting point of the diquark hypothesis. Other properties are the baryon electromagnetic mass splittings and their magnetic moments for which the predictions following from a SU(6) broken quark-diquark wave function had already been examined by Lichtenberg and coworkers [71]. The sum rules obtained were inconclusive in so far as they did not agree better - or worse - than the strict SU(6) ones. However it is questionable whether these early results constitute genuine diquark-model tests; all calculations involved were performed essentially keeping a three-quark basis and imposing a constant and equal gyromagnetic ratio for all quarks involved. The net effect is that the results obtained are merely those of an 'ordinary' SU(6)-broken model.

In this section we are going to examine the predictions from the broken quark-diquark model for the nucleon charge radii without imposing the above-mentioned simplifications. In particular we want to examine if this model yields a satisfactory interpretation of one of the long-standing problems of the symmetric quark model: the experimentally non-zero (negative) charge radius of the neutron.

Simple quark models [i.e. strict SU(6)] require the nucleon charge density to be proportional to its total charge, i.e. imply a zero neutron charge radius. Experiments indicate that the neutron has a charge radius which, though substantially smaller in magnitude than the proton radius, is still significantly differ-

---

[*] The material described in this part is based on a forthcoming paper "Diquark Clustering and the Neutron Charge Radius" with Z. Dziembowski and W.J. Metzger.

ent from zero and negative (see III.2). Especially the sign of this violation has eluded a description for many years.

Carlitz et al. have made the observation that diquark-clustering (if chosen appropriately) could explain the negative neutron charge radius [18]. Qualitatively, any model in which one assumes the proton and the neutron to possess a (ud) diquark core (charge: + 1/3), surrounded by a more distant positive u-quark for the proton and a negative d-quark for the neutron will yield a positive proton charge radius and a negative neutron charge radius. Carlitz et al. also showed this picture to hold up even quantitatively by relating both the neutron charge radius and the nucleon-delta mass difference to a spatial clustering caused by a OGE spin-spin interaction. In this approach one effectively assumes the existence of I=0 (S=0) triplet diquarks to explain both quantities.

The emphasis in our work will also be quantitative; we will relate the SU(6) violating quark-diquark structure needed to explain the negative neutron charge radius to the one required by the SU(6) violations in meson-baryon two-body reactions . As a result of this analysis information will be obtained on the confinement and the size of diquarks.

1. The Diquark Model Relations

In terms of a spatial charge distribution $\rho_N(\bar{r})$ the nucleon charge radius is defined by:

$$\langle r^2\rangle_N = \int d^3\bar{r}\; r^2\, \rho_N(\bar{r}) \tag{22}$$

Definition (22) requires that the distribution $\rho_N(\bar{r})$ is defined with respect to the center-of-charge of the nucleon. We will assume that this center coincides with the nucleon center-of-mass.

In the Breit frame the charge distribution $\rho_N(r)$ is related to its Fourier transform the electric form factor $G_{E,N}(q^2)$ through:

$$\rho_N(\bar{r}) = \int \frac{d^3\bar{q}}{(2\pi)^2} \cdot G_{E,N}(q^2).e^{i\bar{q}.\bar{r}} \tag{23}$$

where

$$G_{E,N}(q^2) = \langle N | \sum_i Q_i e^{-iq.r_i} | N \rangle \tag{24}$$

and Q_i is the charge operator for the i-th constituent. For the four-vector momentum transfer q^2 we use a metric such that $q^2 = +\bar{q}^2$. Substituting (24) into (23) one obtains:

$$\rho_N(\bar{r}) = \sum_i e_i f_i^N(\bar{r}) \tag{25}$$

where

$$f_i^N(\bar{r}) = \langle N | P_i.\delta(\bar{r} - \bar{r}) | N \rangle \tag{26}$$

with e_i the charge and P_i the projection operator of the i-th constituent.

Expression (25) clearly illustrates why strict SU(6) must lead to a zero neutron charge radius. In the symmetric quark model and for the case of the neutron, expression (25) reduces to:

$$\rho_n(r) = e_u \, .f_u^n(\bar{r}) + e_{d_1} f_{d_1}^n(\bar{r}) + e_{d_2} f_{d_2}^n(\bar{r})$$

Strict SU(6) implies a wave function which is totally symmetric in the three quarks, both in terms of the internal and the spatial variables, i.e. requires:

$$f_u^n(\bar{r}) = f_{d_1}^n(\bar{r}) = f_{d_2}^n(\bar{r})$$

Thus the spatial dependence factorizes and the remaining sum is identically zero.

Table X

$$f_u^p(\bar{r}) = f_d^n(\bar{r}) = 1/3\ \sin^2\Gamma . q_1(\bar{r}) + \cos^2\Gamma . q_0(\bar{r})$$

$$f_d^p(\bar{r}) = f_u^n(\bar{r}) = 2/3\ \sin^2\Gamma . q_1(\bar{r})$$

$$f_{s_1}^p(\bar{r}) = f_{s_3}^n(\bar{r}) = 2/3\ \sin^2\Gamma . D_1(\bar{r})$$

$$f_{s_2}^p(\bar{r}) = f_{s_2}^n(\bar{r}) = 1/3\ \sin^2\Gamma . D_1(\bar{r})$$

$$f_{t_1}^p(\bar{r}) = f_{t_1}^n(\bar{r}) = \cos^2\Gamma . D_0(\bar{r})$$

where

$$q_{0(1)}(\bar{r}) = \int d^3\bar{r}_D . d^3\bar{r}_q \ . \ \delta(\bar{r} - \bar{r}_q) |\phi_{t(s)}\ (\bar{r}_D, \bar{r}_q)|^2$$

$$D_{0(1)}(\bar{r}) = \int d^3\bar{r}_D . d^3\bar{r}_q \ . \ \delta(\bar{r} - \bar{r}_D) |\phi_{t(s)}\ (\bar{r}_D, \bar{r}_q)|^2$$

and (for k = 0,1)

$$\int d^3\bar{r} \ . \ q_k(\bar{r}) = \int d^3\bar{r} \ . \ D_k(\bar{r}) = 1$$

In the diquark–quark picture of the nucleon, expression (25) contains as many f_i^N-functions as there are different constituents, i.e. five for the proton (s_1, s_2, t_1, u and d) and again five for the neutron (s_2, s_3, t_1, u and d). These functions are given in table X. They are expressed via the quantities $q_0(\bar{r})$, $q_1(\bar{r})$ and $D_0(\bar{r})$, $D_1(\bar{r})$ in terms of the previously defined ϕ_t- and ϕ_s-functions. The normalization conditions of these quantities follow directly from the one for the ϕ-functions (see relation (3)). As we are from now on only dealing with diquarks in nucleons, we will switch to the labels 0 and 1 (instead of t and

s) to indicate those diquarks. Indeed as an inspection of table II easily shows, in the nucleon the t-diquark is an I=0 state only, the s-diquark an I=1 only.

Substituting (25) into (22) making use of the expressions of table X, and remembering that the diquarks carry the charges given in table I while u and d carry + 2/3 and - 1/3 respectively, leads to the following expressions for the nucleon charge radii:

$$\langle r^2\rangle_p = \sin^2\Gamma \langle r^2\rangle_{D_1} + \frac{\cos^2\Gamma}{3} [2\langle r^2\rangle_{q_0} + \langle r^2\rangle_{D_0}] \tag{27}$$

$$\langle r^2\rangle_n = \frac{\sin^2\Gamma}{3} [\langle r^2\rangle_{q_1} - \langle r^2\rangle_{D_1}] - \frac{\cos^2\Gamma}{3} [\langle r^2\rangle_{q_0} - \langle r^2\rangle_{D_0}] \tag{28}$$

where we have introduced the purely geometrical quantities (k= 0,1):

$$\langle r^2\rangle_{D_k} = \int d^3 \bar{r} . r^2 D_k(\bar{r}) \tag{29}$$

$$\langle r^2\rangle_{q_k} = \int d^3 \bar{r} . r^2 q_k(\bar{r}) \tag{30}$$

These $\langle r^2\rangle$-quantities measure the volumes in which the diquark and odd quark are confined. They are related to the transverse momenta (k_T) of those same entities by the Heisenberg relation.

Let us now relax our previous assumption that the diquark is point-like and accept the fact that it can have spatial structure. The $D_k(\bar{r})$-quantities are then in reality convolutions over the charge distributions of the diquarks themselves, and the $\langle r^2\rangle_{D_k}$-quantities defined in (29) become:

$$\langle r^2\rangle_{D_k} = \langle r'^2\rangle_{D_k} + \langle \zeta^2\rangle_{D_k} \tag{31}$$

where $\langle r^2 \rangle_{D_k}$ now measures the confinement of the diquark CM and $\langle \zeta^2 \rangle_{D_k}$ its size. Thus the diquark-quantities entering relations (27) and (28) are to be interpreted as sums of confinement and size contributions.

In the next section we will compare the expressions (27) and (28) with the experimental data.

2. Experimental Results

Measurements of the neutron charge radius, or alternatively, the neutron electric form factor, although complicated because of the well-known absence of pure neutron targets, have clearly shown these quantities to be significantly different from zero.

There are essentially two methods available to measure neutron form factors: elastic electron-deuterium scattering and scattering of thermal neutrons off electrons in heavy atoms. The former yields less precise values because of uncertainties resulting from the choice of the deuterium wave functions, relativistic corrections, etc.

From the Fourier-inverse of relation (23) it is straightforward to derive the expansions

$$G_{E,N}(q^2) = G_{E,N}(0)\ [1 - 1/6\ q^2 \langle r^2 \rangle_N + \ldots\ldots]$$

Thus we have the relation:

$$\langle r^2 \rangle_N = -\ 6/_{G_{E,N}(0)} \cdot \left. \frac{dG_{E,N}(q^2)}{dq^2} \right|_{q^2=0} \qquad (32)$$

Averaging the two most accurate experiments of the thermal neutron type [72] yields the result:

$$\left. \frac{dG_{E,N}(q^2)}{dq^2} \right|_{q^2=0} = (0.0195 \pm 0.0003)\ \text{fm}$$

or

$$\langle r^2 \rangle_n = (-\ 0.117 \pm 0.002)\ \mathrm{fm}^2 \tag{33}$$

The significance of this result is best appreciated by a comparison with the proton charge radius. The best empirical fits for the latter quantity, as derived from electron-proton elastic scattering [73], yield:

$$\langle r^2 \rangle_p = (+\ 0.846 \pm 0.055)\ \mathrm{fm}^2 \tag{34}$$

implying a (very significant) ratio:

$$\frac{\langle r^2 \rangle_n}{\langle r^2 \rangle_p} = -\ 0.138 \pm 0.009 \tag{35}$$

3. Comparison with the Model

Substituting the experimental values (33) and (34) in the L.H.S.'s of the diquark-quark model relations (27) and (28) leaves us with (only) two constraints between five (a priori unknown) parameters: (Γ, $\langle r^2 \rangle_{q_1}$, $\langle r^2 \rangle_{q_0}$, $\langle r^2 \rangle_{D_1}$ and $\langle r^2 \rangle_{D_0}$). Nevertheless the situation is not as unconstrained as these numbers suggest.

If diquark-clustering is a real substructure inside the nucleon and if diquarks are only lightly bound q-q states, (i.e. states significantly heavier than single quarks) it is natural to assume that:

$$0 \leqslant \langle r^2 \rangle_{D_1} \ll \langle r^2 \rangle_{q_0} \tag{36}$$

and

$$0 \leqslant \langle r^2 \rangle_{D_1} \ll \langle r^2 \rangle_{q_1} \tag{37}$$

It would then be the $\langle r^2 \rangle_{q_k}$ values which determine the size of the nucleons. Physically this makes it plausible that the $\langle r^2\rangle_{q_k}$ are quantities of the order of $\sim 1\mathrm{fm}^2$.

Leaving the exact values of $\langle r^2\rangle_{q_k}$ open (for the time being) but just assuming

$$\langle r^2\rangle_{q_0} \simeq \langle r^2\rangle_{q_1} \simeq R^2 \tag{38}$$

the expressions (27) and (28) become:

$$\langle r^2\rangle_p = \sin^2\Gamma \cdot \langle r^2\rangle_{D_1} + \frac{\cos^2\Gamma}{3} \cdot [2R^2 + \langle r^2\rangle_{D_0}] \tag{39}$$

$$\langle r^2\rangle_n = \frac{\sin^2\Gamma}{3} \cdot [R^2 - \langle r^2\rangle_{D_1}] - \frac{\cos^2\Gamma}{3} \cdot [R^2 - \langle r^2\rangle_{D_0}] \tag{40}$$

From the structure of (40) it is immediately clear that the non-zero value of $\langle r^2\rangle_n$ must arise from the SU(6)-violating situations that either $R^2 \neq \langle r^2\rangle_{D_0}$, or $R^2 \neq \langle r^2\rangle_{D_0}$ or both. The negative value of this quantity must arise from the dominance of the $\cos^2\Gamma$ -term. Let us - just as a starting point - try to make the $\langle r^2\rangle_n$ expression as negative as possible. This requires the additional assumptions:

$$\langle r^2\rangle_{D_1} \simeq R^2 \quad \text{and} \quad \langle r^2\rangle_{D_0} \simeq 0 \tag{41}$$

These assumptions also correspond to the most economical situation of just <u>one</u> SU(6) violating inequality causing a non-zero neutron charge radius.

In this approximation the charge radii become:

$$\langle r^2\rangle_p \simeq R^2\,(1 - 1/3\,\cos^2\Gamma) \tag{42}$$

$$\langle r^2\rangle_n \simeq - 1/3\,R^2 \cos^2\Gamma \tag{43}$$

The ratio of these two quantities is independent of R and given by

$$\frac{\langle r^2\rangle_n}{\langle r^2\rangle_p} \simeq \frac{\cos^2\Gamma}{\cos^2\Gamma - 3} \tag{44}$$

Substituting the experimental value for the L.H.S. (expression 35)) one calculates

$$\Gamma = (52.9 \pm 1.2)^0$$

We recall that in our application of the diquark model to quasi-two-body scattering a value of $\Gamma = (53.1 \pm 1.2)^0$ was found.

There are two elements which make this remarkable agreement seem fortuitous:

The first one has to do with the somewhat unnatural assumption (38) - an assumption which can only be approximately true, because it disagrees - if not in principle then at least in spirit - with the assumption that forms the basis of our SU(6) violating mechanisms, i.e. with the model that the odd-quark coupling to the diquark is different for the I=0 and for the I=1 diquark. A priori one should therefore expect that these two situations will also lead to different odd quark confinement quantities. However the fact that the $\Gamma = 53^0$ value is reached for the 'extreme' values of the quantities involved also creates an inverse relation. Assuming $\Gamma = 53^0$ and keeping the conditions (36) and (37) - but not imposing the conditions (38) and (41) - leads to equations which are only consistent and 'physical' (i.e. yield positive $\langle r^2\rangle$-values $< 1\ \mathrm{fm}^2$) only if relations (38) and (41) are satisfied (*). One can therefore, a posteriori, also look at

(*) To be more precise with $\Gamma = 53^0$, $\langle r^2\rangle_{D_1} \simeq R^2$ and assumptions (38), the equations (39) and (40) require $\langle r^2\rangle_{D_2} \simeq 0.07\ R^2$ at a 1 S.D.-level.

(38) as a physics result following from the value of Γ determined in another application of the diquark model.

The second objection could be the fact that the angle Γ as found in the quasi-two-body application of the diquark model was derived using wave functions with suppressed space parts (see section II). Including those space parts will lead to form factors dependent on the diquark isospin i.e. will in principle modify the Γ-value found. As stated before, it can however be shown that, at least in an impulse approximation (and for the reactions as we have considered them), the effect of including these form factors will be negligible [28].

4. Conclusions

We have shown that an SU(6) broken diquark-quark baryon wave function introduced to explain SU(6) violations in quasi-two-body reactions, also explains (both qualitatively and quantitatively) the negative charge radius of the neutron. Although our wave function still contains a preference for I=0 (ud) clustering, this result is obtained allowing both I=0 and I=1 clusters (i.e. without the drastic assumption of isospin scalar diquarks only).

As a byproduct of our analysis information was obtained on the quadratic sum of the diquark-CM's confinement and their sizes. This information - contained in relations (41) - should of course not be considered as much more than numerically formulated qualitative results; the approximations made do not warrant going further than this. In this sense we could then e.g. state that for the I=0 diquark the above sum covers only a small fraction of the nucleon volume, while for the I=1 diquark it essentially fills the entire nucleon volume. We can interpret these findings in terms of relation (31).

The large differences between the $\langle r^2 \rangle$-values of the D_1 and D_0 diquark could in principle be due either to a CM-confinement difference or to a size difference. A confinement difference is a priori inlikely. We have assumed all along that our diquarks obey SU(6); in the limit of exact SU(6) they should have equal mass. Classically (and also from relations (38)) we would then also expect approximately equal CM-confinements i.e.

$$\langle r'^2 \rangle_{D_1} \sim \langle r'^2 \rangle_{D_0} \sim 0 \ .$$

The fact that $\langle r^2 \rangle_{D_k}$ -differences are consequences of different (charge) sizes is a more likely explanation. In this case our results would imply a smaller and more tightly bound I=0 diquark and a larger, more loosely (or not) bound I=1 diquark. It is interesting to point out that the SU(6)-violating corrections due to OGE spin-spin interactions predict precisely an effect in this direction. A term of the type $\bar{s}_1 . \bar{s}_2$ leads to an attractive S=0 (I=0) force-component and a repulsive S=1 (I=1) one. One might be tempted to conclude that these same terms should also destroy the mass equality (i.e. make the I=0 diquark substantially lighter and the I=1 one substantially heavier). However the OGE-QCD terms - which affect the size of the diquark in first order, only modifies its mass in second order.

Are there alternative models to explain the (negative) neutron charge radius? Yes, configuration mixing does. In the beginning of section II we quoted a recent configuration mixing scheme where the mixing was introduced via the spin-spin term of the OGE [3]. As in the Carlitz approach (see above) such a scheme effectively introduces an attractive I=0 'diquark' and a repulsive I=1 one. We have introduced diquark states (both in the I=0 and I=1 state) bound primarily as a result of the attractive spin-independent part of the colour force, the spin-spin components of that same force acting as correction terms only. Configuration mixing of the type as sketched above leads to mixing of the {56} and {70}-plets and can thus explain the negative neutron charge radius. Therefore, while configuration mixing lacks the unequal-diquark-weight feature of the diquark model (which is essential for the interpretation of the SU(6)-violations in quasi-two-body reactions - see II.4), it does have enough properties in common with the latter model to provide a viable alternative explanation for the neutron charge radius.

Let us finally also mention that there exists an extensive literature on explaining the negative neutron charge radius in a more ad-hoc fashion, namely with models in which the quark transverse momentum (k_T) is postulated to depend on its own longitudinal variable x (= the Feynman-variable). [74], [75]. Of course, such models use a quark-parton model description of the nucleon. One of the consequences of such assumptions is that k_T becomes a non-factorizable variable. As shown by Seghal [74] and others this is a neccesary condition in order to explain a non-zero neutron radius. Apart from the fact that these early analyses suffered from some technical flaws pointed out by Gunion and Soper [76] they also required rather artificial $k_T(x)$-distributions, not supported by experiment. Our diquark approach is equivalent to using a flavour-dependent spatial distribution for the quarks in the nucleon or, alternatively, flavour dependent k_T -distributions. This works equally well as a mechanism to prevent k_T-factorization and has the advantage of being connected with the physics picture discussed before.

IV THE DIQUARK-PARTON MODEL

We are now going to 'translate' the diquark picture into a parton model description and see if it provides a consistent interpretation for the SU(6) violation observed in the $F_2^{en}(x)/F_2^{ep}(x)$-ratio. We first repeat some standard parton model formulae in order to show that strict SU(6) requires this ratio to be a constant equal or larger than 2/3, in disagreement with experiment. We then review the older diquark explanation for this discrepancy and show how this explanation disagrees with the diquark picture derived from the neutron charge radius. Finally we derive the diquark-parton formulae from the F_2^{eN} -functions and use these to propose an interpretation which is consistent with the neutron charge radius.

1. The Parton Model

In the parton model of the nucleon, we make the assumption that a very high momentum nucleon ($P_z \to \infty$) consists of massless, point-like objects, called partons. In such a frame each parton carries a longitudinal momentum xP and an effective mass m=xM (M = mass of the nucleon) [77].

In deep inelastic electron-nucleon scattering, the structure of the target enters through the well-known structure functions $F_2^{eN}(q^2, \nu)$, where q^2= the four-vector momentum transfer and ν^2= the rest-frame energy transfer. The scaling hypothesis - supported by experiment - implies that (at least in the limit $q^2 \to \infty$) the functions

$$F_2^{eN}(q^2,\nu) = F_2^{eN}(x_B) \tag{45}$$

where x_B is the dimensionless (finite) ratio $q^2/2M\nu$.

In the parton model we interpret the collision between the electron and the nucleon as a sum of (elastic) collisions between the electron and the partons. Elastic kinematics then gives that

$$q^2 = 2m\nu \tag{46}$$

Thus

$$x_B = \frac{q^2}{2M\nu} = \frac{2m\nu}{2M\nu} = \frac{m}{M} = x \tag{47}$$

i.e. x_B is equal to the momentum (c.q. mass fraction) number introduced before. In addition it can be shown, by a simple comparison of the formula for electromagnetic scattering of electrons on composite particle with the one on point-like particles, that:

$$\int \frac{F_2^{eN}(x)}{x} dx = \sum_i e_i^2 \qquad (48)$$

(e_i = charge of the ith parton) and thus that:

$$\frac{F_2^{eN}(x)}{x} = \sum_i e_i^2 \, q_i(x) \qquad (49)$$

where $q_i(x)$ stands for the x-space (number) density of the partons (and anti-partons) inside the nucleon and the sum runs over all partons contributing to the electromagnetic process considered.

In the standard quark model and for the proton and neutron respectively, the above relation can be explicitly written as

$$\frac{1}{x} F_2^{en}(x) = \frac{4}{9}[d(x) + \bar{d}(x)] + \frac{1}{9}[u(x) + \bar{u}(x)] + \frac{1}{9}[s(x) + \bar{s}(x)] \quad (50)$$

$$\frac{1}{x} F_2^{ep}(x) = \frac{4}{9}[u(x) + \bar{u}(x)] + \frac{1}{9}[d(x) + \bar{d}(x)] + \frac{1}{9}[s(x) + \bar{s}(x)] \quad (51)$$

where u, d, s, ($\bar{u}$, $\bar{d}$, $\bar{s}$) are the number densities for the quarks (c.q. the antiquarks) in the proton. We have neglected - for simplicity - the contributions of the c, b, t .... quarks. In the neutron formula (50) the isospin symmetries

$$u_n = d_p = d$$
$$d_n = u_p = u$$
$$s_n = s_p = s$$

have been used. There are constraints on the density functions arising from the nucleon quantum numbers and they lead to the sum rules:

$$\int [u(x) - u(x)]\, dx = 2$$

$$\int [d(x) - d(x)]\, dx = 1 \tag{52}$$

$$\int [s(x) - s(x)]\, dx = 0$$

In the region where the valence quarks dominate and the sea contributions are negligible (say x > 0.2) the expressions (50) and (51) yield the ratio

$$\frac{F_2^{en}(x)}{F_2^{ep}(x)} = \frac{\frac{4}{9}\, d(x) + \frac{1}{9}\, u(x)}{\frac{4}{9}\, u(x) + \frac{1}{9}\, d(x)} \tag{53}$$

SU(6) requires spatially fully symmetric wave functions, i.e. x-distributions for the u and d quark which, apart from the normalization, are identical. Hence

$$u(x) = 2d(x) \tag{54}$$

Substitution of this SU(6) requirement yields the SU(6) prediction mentioned before, namely that

$$\frac{F_2^{en}(x)}{F_2^{ep}(x)} = \text{constant} = 2/3 \tag{55}$$

Inclusion of sea contributions would require the constant in (55) to become > 2/3 (Note also that without imposing strict SU(6)-symmetry, expression (53) would give limiting values of 4 and 1/4).

Experimentally the ratio F_2^{en}/F_2^{ep} is not a constant > 2/3. Fig. 3 shows the most recent results [7]. It appears that this ratio varies from 0.9 to approx. 0.3 as x goes from 0.1 to 0.9. The violation of the SU(6) prediction is concentrated precisely

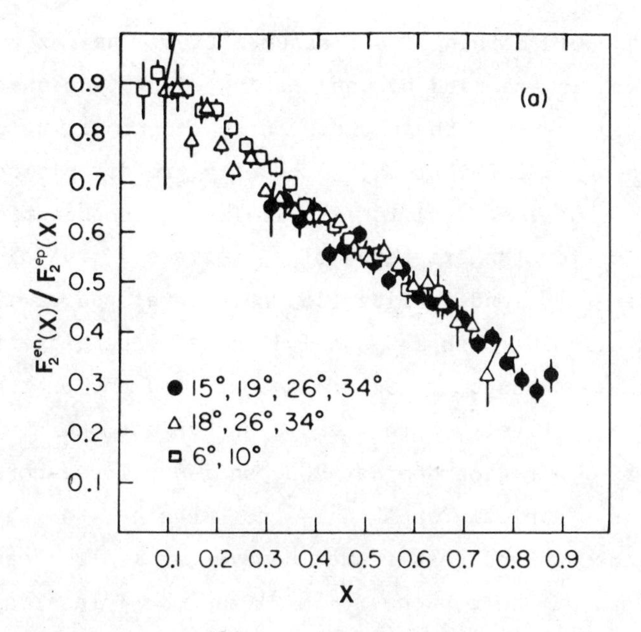

Fig. 3. Values of $F_2^{en}(x)/F_2^{ep}(x)$ obtained in the recent experiment of A. Bodek et al. (ref. 7).

in the region where the contributions from the sea are a priori expected to be most negligible. (*)

2. The Diquark Interpretation

Various explanations have been advanced to explain this SU(6)-violating $x \to 1$ limit. Feynman [13] has pointed out that a

---

(*) It has been noted that the data can be plotted in terms of variables which do make the F_2^{en}/F_2^{ep} -ratio constant (and approx. equal to 2/3) in the x-region from 0.2 to 0.8. [78] However this requires the introduction of new parameters different for the neutron the and proton and thus merely shifts the problem. Most of these improved variables do not have a clear theoretical significance.

limit of 1/4 would obtain if one assumes that - near $x = 1$ - all the momentum is carried by one leading quark (a u-quark in the proton and a d-quark in the neutron) with the remaining ud in an I=0 state near x=0. Relation (53) readily yields this prediction if one assumes $u(x) >> d(x)$ for $x \to 1$. This interpretation would yield evidence for the existence of an x-space I = 0 diquark at $x \simeq 0$. Close [14] and Carlitz [15] have developed similar ideas with models allowing both I=0 and I=1 cores. Starting from the proton and Δ transition form factors and the proton-Δ mass difference respectively, both authors arrive at the interpretation of an $x \simeq 0$ region dominated by an isoscalar x-core.

There is - strictly speaking - no meaning to the parton model in the nucleon rest-frame. However as x also measures the effective mass of the parton ($x = m/M$) and one 'classically' expects a parton to lie closer to the nucleon CM as it becomes heavier, x can also be interpreted as a variable measuring the distance from the nucleon CM. Thus $x \simeq 1$ would correspond to the nucleon center and $x \simeq 0$ to its periphery. In this sense the x-clusters mentioned above are equivalent to the spatial clusters of the rest-frame analysis. Using this picture one is then immediately confronted with a contradiction between the rest-frame consequence of the experimentally observed $F_2^{en}(x)/F_2^{ep}(x)$-ratio interpretation following Feynman, Close, Carlitz and the assumptions needed in this same frame to explain the neutron charge radius on the other [79]. Indeed, the neutron charge radius requires (in a three-quark as well as a quark-diquark description) a rest-frame core dominated by an I=0 (ud) configuration. The Feynman et al. interpretation of the $F_2^{en}(x)/F_2^{ep}(x)$ violation however implies this configuration to dominate near x=0 i.e. at the nucleon periphery instead of at the core! In addition a diquark dominating near x = 0 would also imply a diquark-mass smaller than for a single quark and thus also violate our intuitive mass-notions.

Actually the information contained in the data of fig. 3 implies a less far-reaching mechanism than the one invoked by Feynmann c.s. The data - which incidentically do not reach the 'one-quark-takes-all-x' limit - are sufficiently explained by the dominance of u(x) over d(x) 'at large x'. E.g. a value of $F_2^{en}/F_2^{ep} = 0.43$ can be obtained from (53) just by assuming that $u(x) \underset{-}{\sim} 5d(x)$. Such a F_2^{en}/F_2^{ep} -ratio is observed at $x \underset{-}{\sim} 0.7$ and does not require the remaining quarks to form an x = 0 'diquark' cluster. Of course the question then remains where this 'u(x) - over - d(x)' dominance originates from.

We will return to this problem, but let us first look at the $F_2^{eN}(x)$ expressions directly obtained in a (SU(6) broken) quark di uark picture (*). This implies introducing the x-behaviour of the diquarks as if they were pointlike objects of charge + 4/3 and + 1/3 respectively and going back to expression (49), now summing not only over single quarks but also over all possible diquarks. Obviously treating the diquarks as pointlike objects is a non-trivial simplification. This is especially true for the I=1 diquark for which the neutron charge radius interpretation has suggested a rather loose binding (if at all). However the neutron results should probably not be taken beyond the (qualitative) point that the I=1 diquark is a much looser structure than the I=0 one. In addition the fact whether the two quarks forming the diquark behave coherently or not is in last instance a matter of q^2 -region considered. Let us - for the moment - assume that a 'suitable' q^2 -region exists and come back to the criticism of this assumption further-on.

In evaluating (49) within the diquark-parton model all possible quark-diquark configurations should be taken with the rela-

---

(*) This was done for the first time by M.I. Pavkovic - see ref. 16 (The formulas in this publication contain several normalization errors however).

tive probabilities of the SU(6)-broken diquark wave functions (see table II). Neglecting all sea contributions one then obtains expression of the type:

$$\frac{1}{x} \cdot F_2^{en}(x) = \sum_i e_i^2 \; F_i^N(x) \qquad (56)$$

with the sum running over (five) quantities $f_i^N(x)$ completely analogous to those defined in table X for the nucleon charge radii calculations. They are nothing but the parton model (scaling region) equivalents of the $f_i^N(\bar{r})$. Explicitly one obtains the following expressions for the nucleons:

$$\frac{1}{x} \cdot F_2^{en}(x) = \sin^2\Gamma \cdot [\frac{3}{9} D_1(x) + \frac{3}{9} q_1(x)] + \cos^2\Gamma \cdot [\frac{1}{9} D_0(x) + \frac{1}{9} q_0(x)] \qquad (57)$$

$$\frac{1}{x} \cdot F_2^{ep}(x) = \sin^2\Gamma \cdot [\frac{11}{9} D_1(x) + \frac{2}{9} q_1(x)] + \cos^2\Gamma \cdot [\frac{1}{9} D_0(x) + \frac{1}{9} q_0(x)] \qquad (58)$$

The quantities $D_k(x)$ and $q_k(x)$ entering in the above expressions are again the parton model, scaling region, equivalents of the quantities $D_k(\bar{r})$, $q_k(\bar{r})$ introduced before. The $\sin^2\Gamma$-, $\cos^2\Gamma$-terms enter as a consequence of our allowing SU(6) violations. The other coefficients are easily reproduced by using the weight-factors of table X (or table II) and evaluating the weighted average of charges - squared of the diquarks or quarks concerned [*].

The scaling functions $D_k(x)$ and $q_k(x)$ must of course satisfy the normalization conditions

$$\int D_k(x) . dx = 1$$

---

[*] E.g. as can be read off from table II, the proton contains 2/3 of the time an I=1 diquark of the type s_1=uu (charge + 4/3) and 1/2 of the time an I=1 diquark of the type s_2=ud (charge + 1/3). The averaging yields: $2/3\ (+4/3)^2 + 1/3\ (+1/3)^2 = 11/9$ etc.

and $\qquad\qquad (k = 0,1) \qquad\qquad (59)$

$$\int q_k(x).dx = 1$$

In the deep-inelastic scattering process these functions correspond to the four contributions depicted in fig. 4.

It is easy to check that the relations (57) and (58) still contain the Feynman-Close-Carlitz limit as a special case; assuming the dominance of the I=0 configuration near x=0 and the dominance of the q_0-quark near x = 1 immediately reproduces the required factor 1/4. As explained above however, the nucleon rest-frame consequences of these assumptions are in disagreement with the quark-diquark properties required to interpret the neutron charge radius.

There is however another diquark-solution for the deep-inelastic structure function problem which does <u>not</u> suffer from this contradiction. From the same relation (57) and (58), but now using the assumption that for $x \to 1$ it are not the single quark densities which dominate but rather the diquark (both the I=1 and I=0) ones, one immediately finds

$$\frac{F_2^{en}(x)}{F_2^{ep}(x)} \underset{(x\to 1)}{\simeq} \frac{3\sin^2\Gamma.D_1(x) + \cos^2\Gamma.D_0(x)}{11\sin^2\Gamma.D_1(x) + \cos^2\Gamma.D_0(x)} \qquad (60)$$

Using the values of $\sin^2\Gamma$ and $\cos^2\Gamma$ for $\Gamma = 53^0$ and assuming that the $D_k(x)$-distributions are concentrated in the high x-region will imply a value for the F_2^{en}/F_2^{ep}-ratio - in that same high x-region - of approximately 0.30 (*). Obviously a diquark

---

(*) In ref. 80 an analogous claim is made (analyzing the data of ref. 81) assuming the large-x dominance of the I=0, S=0 diquark only. However such an assumption is not able to reproduce the experimentally observed (high x) F_2^{en}/F_2^{en} behaviour; as equation (60) easily shows, using only the I=0 diquark leads to the prediction $F_2^{en}/F_2^{ep} \simeq 1$.

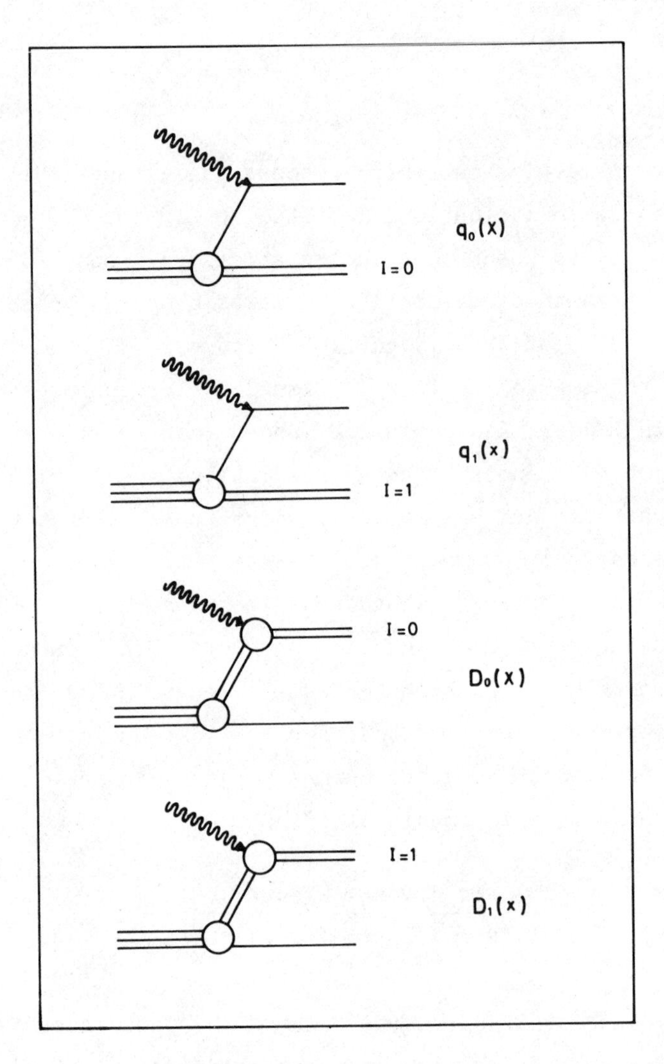

Fig. 4. Schematic representation of the (four) scaling components of the nucleon structure functions $F_2^{eN}(x)$ in the diquark model.

x-distribution peaking at large x does correspond to a rest-frame diquark contribution in the nucleon core and thus agrees with our charge radius finding in the nucleon rest-frame. (+)

(+) The above reasoning effectively corresponds to the (simplistic) approximation $D_1(x) = D_0(x) = \delta(1-x)$. It is of course possible to reproduce the data with more sophisticated $D_k(x)$ dependences.

Interesting questions remain however. The diquark picture should - at its best - only be valid in a limited region of momentum transfer, i.e. even if there exists genuine diquark substructure inside the baryons there should always be a high-momentum transfer region where the diquark again behaves like two separate (point-like) objects. Why is it then that the 'high-q^2' structure functions would (still) be interpretable via diquark-type partons? First of all it should probably be remembered that for the (x, q^2) regions presently explored (and used in the F_2^{en}/F_2^{ep} -ratio evaluation under discussion) the scaling violating contributions are not as small as generally assumed. Furthermore in discussing the effects of the diquark structure one should distinguish the effects which are due to the non-point-like nature of the diquark (and which give rise to scaling violations) and those that result from the clustering of its components itself (which do give a scaling contribution and remain present even after the diquark starts behaving as two separate quarks). The former indeed give effects which disappear as q^2 increases, while the latter keep influencing the structure functions even in the limit $q^2 \to \infty$ (*). Thus one expects that if a diquark prefers the high x-region, its (resolved) components will also have a tendency to populate this region. From the u, d content of the s_1, s_2 and t diquarks - and their weight in the broken SU(6) wave function - one then immediately predicts a high x-region u-over-d preference of approximately 2.5 to 1. This result must of course now be used in the expression appropriate for this single (resolved) quark regime - i.e. expression (53) - but, as explained above, it is exactly this kind of u(x) vs. d(x) be-

(*) Both effects are clearly seen and isolated in the quark-diquark parton model fits to the deep inelastic nucleon structure functions of ref. 81. The non-scaling terms are often associated with the scaling violations predicted by QCD. Coherent behaviour of two quarks (bound in a diquark) is an alternative - or complementary - interpretation.

haviour which this expression needs in order to reproduce the data. Thus the diquark explanation (based on expression (60) and in principle valid for a limited q^2-region only) 'continues' into a single quark interpretation (based on expression (53)) with essentially the same numerical consequences for the F_2^{en}/F_2^{ep}-ratio. The diquark picture keeps explaining the SU(6)-violating behaviour of the nucleon structure functions even after the scattering process starts to see the individual quarks forming the diquark, for the simple reason that these resolved single quarks keep an x-distribution related to the one of 'their' diquark.

There are other high-energy phenomena which are explained by the existence of a u-over-d dominance as $x \to 1$. One of the best-known interpretations of this type is the Ochs-Das-Hwa model for the inclusive beam fragmentation ratio $p \to \pi^+/p \to \pi^-$ [82]. However just as for the F_2^{en}-ratio explanation, a (high-x) $u(x)/d(x) \simeq 4$ or 5 would fit the data better than the 2.5 to 1 derived from the (resolved) diquark picture. It is now interesting to note that one of the criticism of the above picture could actually be connected with a mechanism increasing the u/d-ratio observed. Indeed although in the limit of $q^2 \to \infty$ both diquarks would be resolved, it is logical to expect the loosest one, i.e. the I=1 one, to 'yield' first. In this intermediate regime the uu-component of the I=1 diquarks would be an extra-source of high-x u-quarks. Actually there is interesting check on this assumption; if correct the $p \to \pi^+/p \to \pi^-$ ratio should eventually again decrease as $q^2 \to \infty$.

3. Conclusions

Extending the diquark model to a parton description i.e. introducing the x-behaviour of the diquarks as if they were point-like objects of charge + 4/3 and + 1/3 respectively, yields an interpretation of the well-known F_2^{eN} -SU(6) problem which is in qualitative agreement with the rest-frame interpretation of the neutron charge radius. The surprising result that the diquark assumption is able to explain such high q^2 data as those involved in the evaluation of the F_2^{en}/F_2^{ep} -ratio, can be (qualitatively) understood as resulting from the fact that the x-distribution of the two quarks forming the diquark is primarily an (approximate) transformation of the rest-frame mass-structure of the nucleon and does not depend on whether one sees the diquark as a whole or as two separate objects.

V. SUMMARY AND OUTLOOK

SU(6)-breaking coupled to a diquark-quark substructure hypothesis for the baryons, is capable of explaining a large variety of discrepancies between the experimental data and the predictions of the standard symmetric quark model. All this is accomplished through the introduction of one and only one new parameter, the SU(6)-breaking angle $\Gamma(\underline{\sim}53^0)$.

In particular this model explains:

(a) the SU(6) sum rule violations observed in the total and differential cross section, for a large body of quasi-two-body reactions (without destroying the known SU(6) successes!).

(b) the SU(6)-violating value of the charge radius of the neutron.

(c) the SU(6)-violating behaviour of the deep-inelastic neutron-proton structure functions.

In the field (a) the diquark hypothesis is coupled to a spectator assumption. In field (c) to the possibility of a parton model description. Applications (a) and (c) are dynamic: low- and high q^2 respectively. Application (b) is a static one.

Apart from the manifestations discussed in these lectures, there are many other indications of the existence of diquarks as dynamical entities. An (incomplete) list of topics where such indications have been found inc ıdes: the G_A/G_V-ratio in nucleon β-decay [83], hadronic weak decays [84], counting rules [85], the σ_L/σ_T-ratio in deep-inelastic electro-hadron production [80], polarized inclusive Λ-production [86], etc.

In addition, the diquark picture and its parameters should have measurable consequences for the dimensional scaling laws [87], the elastic p-p scattering [88], the elastic (and various transition) proton form factors [15] [87], the p_T-distribution of $(\mu^+\mu^-)$ pairs produced in π^-p vs. π^+p collisions [18] [76], the p_T-distribution of inclusive π^+ vs. π^- production both in deep-inelastic lepto- and hadro-production [18] [75] [76]. Of course in nearly all these fields the diquark picture will be subjected to the q^2-limitations mentioned before.

Acknowledgements

The work described here was done in close collaboration with Drs. Z. Dziembowski, W.J. Metzger and M. Zralek. Part I is completely based on ref. 10, published by M. Zralek et al.; Part II is based on a forthcoming paper with Drs. Z. Dziembowski and W.J. Metzger; I would like to thank these authors for allowing me to use the material contained in this paper prior to publication. In particular I would like to thank Dr. Z. Dziembowski for his continuous help in preparing this manuscript. Mrs. E. Dikmans-de Koning should be thanked for her patience in typing the numerous versions of this manuscript. This work was supported by the joint research programmes of the Dutch Organizations for Fundamental Research ZWO and FOM.

REFERENCES

[1] P.J. Litchfield, Proc. of Topical Conf. on Baryon Resonances, Oxford (1976), p. 339:

R.J. Cashmore, Proc. of the 19th International Conference on High Energy Physics, Tokyo (1978), p. 811.

[2] R. Horgan - Proc. of Topical Conference on Baryon Resonances Oxford (1976) p. 435.

[3] N. Isgur and G. Karl, Phys. Lett. 72B, 109, (1977), Phys. Lett. 74B, 353 (1978); Phys. Rev. D18, 4187 (1978); Phys. Rev. D19, 2653 (1979).

[4] A.J.G. Hey, Proc. of Topical Conf. on Baryon Resonances, Oxford (1976), p. 463.

[5] A. Bialas and K. Zalewski, Phys. Lett. 26B, 170 (1968), Nucl. Phys. B6 449, 463, 478 (1968).

[6] K. Bockmann et al., Phys. Lett. 28B 72 (1963);
M. Aderholz et al., Nucl. Phys. B8 503 (1968);
A. Kotanski and K. Zalewski, ibid 13B 119 (1969);
H. Friedman and R.R. Ross, Phys. Rev. Lett. 22, 152 (1969);
D. De Baere et al., Nuovo Cimento 56, 397 (1969);
H. Haber et al., Nucl. Phys. B17; 289 (1970).

[7] A. Bodek et al., Phys. Rev. D20, 1471 (1979).

[8] E. Hirsch et al., Phys. Let. 36B, 139 (1971).

[9] J.C. Kluyver et al., Nucl. Phys. B140, 141 (1978).

[10] M. Zralek, W.J. Metzger, R.T. Van de Walle, C. Dionisi, A. Gurtu, M. Mazzucato and B. Foster, Phys. Rev. D19, 820 (1978).

[11] M. Ida and R. Kobayashi, Prog. Theor. Phys. 36, 846 (1966).

[12] D.B. Lichtenberg and L.J. Tassie, Phys. Rev. 155, 1601 (1967).

[13] R.F. Feynman, "Photon-Hadron Interaction", Benjamin, New York (1972), p.150.

[14] F.E. Close, Phys. Lett. 43B, 422 (1973).

[15] R. Carlitz, Phys. Lett. 58B, 345 (1975); Phys. Rev. Lett. 36, 1001 (1976).

[16] M.I. Pavkovic, Phys. Rev. D13, 2128 (1976).

[17] H. M. Chan and H. Hogaason, Phys. Lett. 72B, 121 (1977); Phys. Lett. 72B, 400 (1978); Nucl. Phys. B136, 401 (1978).

[18] R.D. Carlitz, S.D. Ellis and R. Savit, Phys. Lett. 68B 443 (1977).

[19] G.P. Goldstein and J. Maharana; RHEL-preprint RL-79-048.

[20] K. Johnson and C. Thorn, Phys. Rev. D13, 1934 (1976); G. Preparata and K. Szego, Nuovo Cimento, 47A, 303 (1978).

[21] S.D. Drell, H. Guinn and M. Weinstein, SLAC-report (to be published).

[22] R.D. Carlitz and B.D. Creamer, Phys. Lett. 84B, (1979).

[23] D.B. Lichtenberg and R.J. Johnson, Hadronic Journal, 2, 1 (1978).

[24] See e.g. A. Le Yaouanc et al., Phys. Rev. D18, 1591 (1978) and references quoted therein.

[25] J.J.J. Kokkedee and L. Van Hove, Nuovo Cimento 42, 711 (1966).

[26] P.M. Fishbane, J.S. McCarthy, J.V. Noble and J.S. Trefil, Phys. Rev. D11, 1338 (1975).

[27] A. de Rujula, H. Georgi and S.L. Glashow, Phys. Rev. D12, 147 (1975).

[28] M. Zralek, University of Nijmegen, Internal Report HEN-190 , 1979 (unpublished).

[29] J.P. Ader et al., Nuovo Cimento 56A, 952 (1968).

[30] M. Zralek, W. Metzger and R.T. Van de Walle, Phys. Rev. D20, 344 (1979).

[31] H.J. Lipkin and F. Scheck, Phys. Rev. Lett. 18, 347 (1967).

[32] E. Hirsch, U. Karshon and H.J. Lipkin, Phys. Lett. 36B, 385 (1971).

[33] T. Hofmohl and M. Szeptycka, Nucl. Phys. B13, 53 (1969).

[34] S. Meshkov, G.A. Snow and G.B. Yodh, Phys. Rev. Lett. 12, 87 (1964).

[35] G.H. Trilling, Nucl. Phys. B40 13 (1972).

[36] J.C. Scheurer et al., Nucl. Phys. B33, 61 (1971).

[37] U. Karshon et al., Nucl. Phys. B29, 557 (1971).

[38] M. Aguilar-Benitez et al., Phys. Rev. D 6, 29 (1972).

[39] B. Chaurand et al., Nucl. Phys. B117, 1 (1976).
[40] S. Holmgren et al., Nucl. Phys. B119, 261 (1977);
G.G.G. Massaro et al., Phys. Lett. 66B, 385 (1977);
F. Marzano et al., Nucl. Phys. B123, 203 (1977).
[41] M. Aguilar-Benitez et al., Nucl. Phys. B124, 189 (1977).
[42] M.J. Losty et al., Nucl. Phys. B133, 38 (1978).
[43] D. Yaffe et al., Nucl. Phys. B75 , 365 (1974).
[44] D.J. Crennell et al., Phys. Rev. D6, 1220 (1972).
[45] H.G.J.M. Tiecke, thesis, University of Nijmegen, 1974 (unpublished);
W.J. Metzger, University of Nijmegen, Internal Report no. HEN-166, 1977 (unpublished).
[46] F. Marzano et al., Phys. Lett. 68B, 292 (1977);
J.J. Engelen (private communication).
[47] J. Gallivan et al., Nucl. Phys. B117, 260 (1976).
[48] M.G.D. Gilchriese et al., Phys. Rev. Lett. 40, 6 (1978).
[49] G.W. Brandenburg et al., Phys. Rev. D15, 617 (1977).
[50] K.J. Foley et al., Phys. Rev. D9, 42 (1974).
[51] J.A.J. Matthews et al., Phys. Rev. Lett. 26, 400 (1971).
[52] S.U. Chung et al., Phys. Rev. D12, 693 (1975).
[53] P. Sonderegger et al., Phys. Lett. 20, 75 (1966).
[54] B. Haber et al., Phys. Rev. D11, 495 (1975.
[55] M. Aderholz et al., Phys. Rev. 138, B897 (1965);
[56] D.J. Schotanus et al., Nucl. Phys. B22, 45 (1970).
[57] I.J. Bloodwordt et al., Nucl. Phys. B81, 231 (1974).
[58] M. Aderholz et al., Nucl. Phys. B8, 45 (1968).
[59] D. Evans et al., Nuovo Cimento 16A, 299 (1973).
[60] J.H. Scharenguivel et al., Nucl. Phys. B36, 363 (1972).
[61] R. Honecker et al., Nucl. Phys. B131, 189 (1977).
[62] G.C. Benson et al., Phys. Rev. Lett. 22, 1074 (1969);
M.J. Emms et al., Nucl. Phys. B98, 1 (1975);
G.S. Abrams et al., Phys. Rev. Lett. 23, 673 (1969);

L.E. Holloway et al., Phys. Rev. D8, 2814 (1973);
N. Armenise et al., Nuovo Cimento 65A, 637 (1970);
M.S. Farber et al., Nucl. Phys. B29, 237 (1971);
J.C. Anderson et al., Phys. Lett. 45B, 165 (1973);
K. Paler et al., Lett. Nuovo Cimento 4, 745 (1972);
V.N. Bolotov et al., Yad. Fiz. 21, 316 (1975), Sov. J. Nucl. Phys. 21, 166 (1975).

[63] D.G. Fong et al., Phys. Rev. D9, 3015 (1974).

[64] C.L. Pols et al., Nucl. Phys. B25, 109 (1970).

[65] I.J. Bloodworth et., Nucl. Phys. B35, 79 (1971).

[66] D. Evans et al., Nucl. Phys. B51, 205 (1973).

[67] J.A. Gaidos et al., Nucl. Phys. B72, 253 (1974).

[68] N.N. Biswas et al., Phys. Rev. D2, 2529 (1970).

[69] M. Zralek, University of Nijmegen, Internal Report HEN-172 (1978) (unpublished)

[70] Z. Dziembowski, W. Metzger and M. Zralek, University of Nijmegen, Internal Report HEN-191, 1979 (unpublished).

[71] J. Carroll, D.B. Lichtenberg and J. Franklin, Phys. Rev. 174 1681 (1968);
D.B. Lichtenberg, Phys. Rev. 178, 2197 (1968).

[72] V.E. Krohn and G.R. Ringo, Phys. Rev. D9, 1305 (1973);
L. Koester et al., Phys. Rev. Lett. 36, 1201 (1976).

[73] F. Borkowski, P. Peuser, G.G. Simon, V.H. Walther and R.D. Wensling, Nucl. Phys. A222 269 (1974).

[74] L.M. Seghal, Phys. Lett. 53, 106 (174).

[75] F.E. Close, F. Halzen, D.M. Scott, Phys. Lett. 68B, 447 (1977).

[76] J.F. Gunion and D.E. Soper, Phys. Lett. 73B. 189 (1978).

[77] For details and more complete derivations - up to formulae (50) and (51) - see F.E. Close - "An Introduction to Quark and Partons" Academic Press (1979) - Chapter 9, 10, 11.

[78] W.B. Atwood, SLAC-PUB 185 (1975).

[79] A. Le Yaouanc, L. Oliver, O. Pene and J.C. Raynal, Phys. Rev. D15, 844 (1977) and Phys. Rev. D18, 1733 (1978).

[80] L.F. Abbot, E.L. Berger, R. Blankenbecler and G.L. Kane, Phys. Lett. 88B, 157 (1979).

[81] I. Schmidt and R. Blankenbecler, Phys. Rev. D16, 1318 (1977).

[82] K.P. Das and R.C. Hwa, Phys. Lett. 68B, 459 1977;
W. Ochs, Nucl. Phys. B118, 397 (1977).

[83] N. Isgur, Acta Phys. Pol. B8, 1081 (1977).

[84] T. Hayashi, T. Karino and T. Yanaguda, Progr. Theor. Phys. 60 1066 (1978).

[85] S. Brodsky - "Color Symmetry and quark confinement" Moriond Meeting 1977. (Ed. Tran Thanh Van).

[86] B. Andersson, G. Gustafson and G. Ingelman, Phys. Lett. 85B, 417 (1979).

[87] F. Gunion, Phys. Rev. D10, 242 (1974).

[88] See e.g. P. Collins and F. Gault, Phys. Lett. 73B, 330 (1978).

DISCUSSIONS

CHAIRMAN: Prof. R.T. Van de Walle

Scientific Secretaries: A. Hornoes and A. Kreimer

DISCUSSION

- GINSPARG:

If the separation of the quark from the diquark requires an angular momentum, why should the results apply to the lowest lying states?

- VAN DE WALLE:

There is no angular momentum introduced. The separation between the diquark and the quark is due to the fact that quarks cluster as a result of the color forces. Remember that the expectation value of $\langle \bar{3} | \lambda_1 \lambda_2 | \bar{3} \rangle$ (λ_1, λ_2 = Gell-Mann matrices for color SU(3)) is negative in the one gluon approximation, i.e. produces q-q binding.

- VANNUCCI:

I wonder about the Q^2 dependence of the quark-diquark structure. Deep inelastic phenomena give evidence for 3 quarks without clustering.

- VAN DE WALLE:

The quark-diquark structure for the baryons is indeed expected to be a Q^2 dependent phenomena. At high Q^2 the baryon would again look like 3 quarks without clustering, and only for relatively low Q^2 processes (which incidentally the two-body processes discussed turn out to be) would it show diquark substructure. Moreover, even at low Q^2 there might be signs of a "direct" 3 quark state. The diquark v.s. symmetric baryon wave function is not expected to be strictly either/or. It is possible that the baryon spends a fraction of its time in a 3 quark configuration and the remaining one in a quark-diquark configuration, and that the peripheral two-body processes are preferentially initiated from baryons in the diquark-quark mode. However, baryon spectroscopy suggests (via the dominance of the minimal spectrum) that the baryon spends most of its life as a diquark-quark state.

- JAFFE:

I would like to ask you why you believe that SU(6) has anything to do with the real world.

- VAN DE WALLE:

Because experimental data belong to the real world, and SU(6) successfully describes and parametrizes a lot of that data. That same reason makes it interesting to try to modify it (slightly) if there exist systematics in the patterns with which it is violated.

- JAFFE:

But does it make sense to modify a model which we know to be wrong ab initio, in order to explain small deviations from it? After all, the fact that SU(6) works so well is a mystery.

- VAN DE WALLE:

All that we have been trying to do is to modify this mystery slightly (with one parameter and guided by the possibility of quark clustering) so as to make the agreement even better and the mystery even deeper.

- JAFFE:

It is a good parametrization of our ignorance, I will grant you that.

- VAN DE WALLE:

Parametrization is an important starting point in physics. Remember the Balmer lines!

- JAFFE:

I believe there is a piece of good experimental evidence for the existence of a $56,1^-$ multiplet (outside the so-called minimal spectrum) which is expected in relativistic quark models, namely, the Δ (1920), $\frac{5}{2}^-$ state. Reinders has done some mass fits on this state, and might want to comment on it.

- REINDERS:

Dalitz, Horgan and I have tried to assign the Δ (1920) to the $56,1^-$ multiplet of the N=3 excitation level (not the N=1 level). In the symmetric SU(6)⊗O(3) quark model we derived a sum rule which connects this state with known states of lower multiplets, and this predicts the $\Delta\, D_{35}$ at 2080 MeV. So the symmetric quark model does not easily accomodate this state. There is however, also no evidence for a minimal multiplet structure, where at the N=3 level one has only two multiplets (viz. the $70,3^-_3$ and $70,1^-_3$). The last one does not contain a J=5/2 Δ state, so one is forced to assign the Δ (1920) to the 70, 3^-_3 which would be very low in this case. One would expect the $70,1^-_3$ to be even lower and therefore essentially degenerate with the $70,1^-_1$, which is very improbable in my opinion.

- VAN DE WALLE:

I cannot see that the present classification of the Δ (1920) $5/2^-$ as a $56,1^-$ instead of $70,3^-$ is compelling. There are other states in the 2 GeV mass region asking for $70,3^-$ classification. Furthermore, let me again repeat that the diquark-quark model picture vs. the symmetric 3 quark model does not have to be strictly either/or. States outside the minimal spectrum could be manifestations of the fraction of the time the baryon spends in the symmetric 3 quark configuration. For our picture to hold, all that is needed is that the baryon is a diquark-quark structure <u>most</u> of the time, and that would not be excluded by the existence of one state (or multiplet) outside the minimal spectrum.

FAREWELL TO INSTANTONS

Adrian Patrascioiu

Department of Physics
University of Arizona
Tucson, Arizona 85721

1. INTRODUCTION

At the present time the instanton is being used in QCD to explain color confinement, dynamical symmetry breaking with fermion mass generation, and to resolve the U(1)-problem (Callan et al., 1978; Carlitz and Creamer, 1979). In my lecture I would like to show that for some technical reasons it is doubtful that the instanton accomplishes these feats and that it is even possibly irrelevant to the structure of QCD altogether.

The objects of my discussion will be the determinants associated with the quantum fluctuations of given fields in the background potential of the instanton field. I will show that for massless fields:

(a) The ratio of the instanton and the free field determinant depends upon the infrared cutoff if the computation is carried out in flat space R^n or upon the ultraviolet cutoff if the computation is carried out over the sphere S^n.

(b) The value of the determinant depends upon the boundary conditions used to define the operator.

The first point is important in understanding the role of the instanton in, for example, pure QCD_4. The second one has to do with the treatment of the fermion, which I believe was not properly done.

The organization of the lecture is as follows: in Section 2, I illustrate the type of difficulties I will be discussing by studying quantum determinants in one dimension. The main advantages of working in one dimension are the lack of infinities and the fact that for some boundary conditions there is a closed formula (Van Vleck's trick) for computing determinants. In Section 3 I will discuss QCD_4 and show that the quantum determinants relevant are in fact infrared divergent. In Section 4 I touch upon the problem of fermions and illustrate in QED_2 why I believe that fermion determinants have been calculated improperly. Some conclusions are stated last.

2. QUANTUM DETERMINANTS FOR MASSLESS FIELDS IN ONE DIMENSION

The problem is to compute Det M where

$$M = - \partial_t^2 + V(t) \tag{2.1}$$

and the boundary conditions imposed at T_1 and T_2 are such that M is a hermitian operator (for a more complete treatment of quantum determinants in one dimension, see Patrascioiu, 1979a). Please notice that for massless problems $V(t) \to 0$ for $|t| \to \infty$.

Dependence of the Determinant upon the Boundary Conditions

First I will list some boundary conditions which render M hermitian. They are:

$$\text{(a)} \quad \phi_n(T_1) = \phi_n(T_2) = 0 \tag{2.2}$$

$$\text{(b)} \quad \phi_n(t) = \phi_n(t + T_2 - T_1) \tag{2.3}$$

$$\text{(c)} \quad \phi_n(t) = -\phi_n(t + T_2 - T_1) \tag{2.4}$$

These are the correct boundary conditions for computing the boson kernel, the trace of the evolution operator for bosons and fermions, respectively. Only boundary condition (2.3) takes a simple form on the sphere; indeed, for $-T_1 = T_2 \to \infty$, (2.3) selects smooth functions over the sphere.

To show the importance of the boundary conditions, consider a harmonic oscillator

$$M = -\partial_t^2 + \omega^2 \tag{2.5}$$

It is well known (Patrascioiu, 1979a) that on the interval 0 to T

$$\text{with boundary condition (2.3), Det } M = 2 \sinh \frac{\omega T}{2} \tag{2.6}$$

$$\text{with boundary condition (2.4), Det } M = 2 \cosh \frac{\omega T}{2} \tag{2.7}$$

For $\omega T >> 1$ (2.6) and (2.7) become equal; this is due to the fact that $\omega \neq 0$ (massive field). For $\omega \to 0$, (2.6) gives

$$\text{Det } M \to 0 \tag{2.8}$$

while (2.7) gives

$$\text{Det } M \to 1 \tag{2.9}$$

Thus for massless fields the boundary conditions are important.

Infrared Divergence in R^n, Ultraviolet Divergence in S^n

The operator I consider is

$$M(g) = -\partial_t^{\ 2} + \frac{g(g+1)}{1+t^2} \tag{2.10}$$

(This choice is motivated by the operators which appear in QCD -- see Section 3). It is obvious by direct inspection that M(g) and M(o) have the same short distance behavior, but differ in their behavior for large $|t|$ (notice that for $|t| \to \infty$, $\partial_t^{\ 2} f \sim \frac{1}{t^2} f$).

If one chooses the boundary condition $\phi_n(0) = \phi_n(T) = 0$, it is extremely easy to compute Det M (see Appendix) and find that indeed

$$\frac{\text{Det } M(g)}{\text{Det } M(o)} \propto T^g , \tag{2.11}$$

that is, that the ratio is infrared divergent.

Now projecting $R^1 \to S^1$ with the map

$$\begin{aligned} \phi &= 2 \tan^{-1} t \\ \hat{\Phi}(\phi) &= \left(\frac{1}{1+t^2}\right)^{\frac{1}{2}} \Phi(t) \end{aligned} \tag{2.12}$$

one finds that M(g) gets mapped into

$$M_c(g) = -\partial_\phi^{\ 2} - \frac{1}{4} + \frac{g(g+1)}{4 \cos^2 \frac{\phi}{2}} \tag{2.13}$$

For $\phi = \pi$, $M_c(g)$ is singular while $M_c(o)$ is not; hence, Det $M_c(g)$/ Det $M_c(o)$ must be ultraviolet divergent if $\phi = \pi$ is included in the domain over which the computation is done. This is borne out by direct computation since one finds

$$\frac{\text{Det } M_c(g)}{\text{Det } M_c(o)} = \prod_{n=0}^{\infty} \frac{(n + \frac{g+1}{2})\ (n + \frac{g+3}{2})}{(n+\frac{1}{2})\ (n+\frac{3}{2})} \tag{2.14}$$

and the product is obviously (ultraviolet) divergent. Thus, in going from R^1 to S^1, infrared divergences become ultraviolet ones. That this transmutation of infrared divergences into ultraviolet ones is in no way peculiar to one dimension can be seen by studying QED_2 (Patrascioiu 1979a).

3. INFRARED DIVERGENCES IN QCD_4

The Gaussian approximation of the path integral in QCD_4 involves: (a) finding a classical solution, and (b) taking into account quantum fluctuations around it, that is, computing the quantum determinant. 't Hooft (1977) pointed out that if the boundary conditions are imposed on a sphere of radius R centered about the instanton, then the relevant operator is

$$M_{J,L} = -\partial_r^2 - \frac{3}{r}\partial_r + \frac{4\vec{L}^2}{r^2} + \frac{4(\vec{J}^2 - \vec{L}^2)}{1 + r^2} + \frac{4\vec{T}^2}{(1 + r^2)^2} \tag{3.1}$$

Here $\vec{J} = \vec{L} + \vec{T}$, $\vec{T}$ is the isospin and $\vec{L}$ the orbital angular momentum. For future reference I notice that

$$\begin{aligned} \vec{L}^2 &= l(l + 1), \quad l = 0, \tfrac{1}{2}, 1, \ldots \\ \vec{J}^2 &= j(j + 1) \end{aligned} \tag{3.2}$$

and that at given (l,j) the multiplicity is $(2l + 1)(2j + 1)$. One needs

$$\frac{\text{Det } M}{\text{Det } M^o} = \prod_{j,l} \left(\frac{\text{Det } M_{J,L}}{\text{Det } M^o_{J,L}}\right)^{(2l + 1)(2j + 1)} \tag{3.3}$$

where

$$M^o_{J,L} = -\partial_r^2 - \frac{3}{r}\partial_r + \frac{4\vec{L}^2}{r^2} \tag{3.4}$$

To avoid cumbersome notation, let me define the following ratios

$$D = \frac{\text{Det } M}{\text{Det } M_o} \tag{3.5}$$

$$D_{J,L} = \frac{\text{Det } M_{J,L}}{\text{Det } M^o_{J,L}} \tag{3.6}$$

Computation of $D_{J,L}$

By direct inspection of (3.1) and (3.4) one sees (Section 2) that both Det $M_{J,L}$ and Det $M^{o}_{J,L}$ must depend upon both the ultraviolet and the infrared cutoff. However, also by inspection, $D_{J,L}$ should be independent of the ultraviolet cutoff, but dependent upon the infrared cutoff R. This is borne out by a direct computation. The substitution $t = 1/r^2$ puts the operators in a form suitable for the use of Van Vleck's formula and one obtains (Patrascioiu 1978) for $T^2 = 3/4$:

$$D_{J,L} = R^{2(j-1)} \frac{\Gamma(2j+1)\,\Gamma(2l+2)}{\Gamma(j+l+\frac{5}{2})\,\Gamma(j+l+\frac{1}{2})} \tag{3.7}$$

It is instructive to see how this answer compares with the one obtained by 't Hooft (1977). First recall that for algebraic reasons 't Hooft chooses to compute $D_{J,L}$ as

$$D_{J,L} = \frac{\text{Det}\,\frac{1}{4}\,(1+r^2)\,M_{J,L}\,(1+r^2)}{\text{Det}\,\frac{1}{4}\,(1+r^2)\,M^{o}_{J,L}\,(1+r^2)} \tag{3.8}$$

One can easily show that this is equivalent to computing $D_{J,L}$ on the sphere instead of on the line (Patrascioiu 1979b). Thus he finds a discrete spectrum even for $R \to \infty$. He then argues that for $R \gg 1$ one has

$$D_{J,L} = \prod_{n\geq 0}^{\infty} \frac{(n+l+j+\frac{1}{2})\,(n+l+j+\frac{5}{2})}{(n+2l+1)\,(n+2l+2)} \tag{3.9}$$

Now this product is independent of the infrared cutoff R and is ultraviolet divergent. Introducing an ultraviolet cutoff Λ (in n) one can easily verify that in (3.9)

$$D_{J,L} \propto \Lambda^{2(j-1)} \tag{3.10}$$

The moral of this exercise is twofold:

(a) 't Hooft eigenvalues are not (strictly) correct.

(b) If one uses 't Hooft's naive eigenvalues by arguing that $R \gg 1$, one picks up fake ultraviolet divergencies which are in effect infrared divergences (Patrascioiu 1979a).

The first point has been investigated by Nienast and Stack (1979) who found that, even for $R \gg 1$, 't Hooft's eigenvalues in

(3.9) are valid only for $n \leq O(R)$ while, for $n \geq O(R)$, the eigenvalues change in such a way that the infinite product in (3.9) is ultraviolet convergent and (3.7) is reproduced.

The second point leads one to suspect that the use of 't Hooft's eigenvalues to calculate D may yield an answer with ultraviolet divergences more severe than one would expect in perturbation theory. That this is probably the case will be shown next when I discuss the question of regularization.

Ultraviolet Divergence of D and Its Regularization

In perturbation theory, D would be given by the sum of all 1-loop diagrams with external field insertions. In spite of the fact that the external field is nonsingular, the singular behavior of the free field propagator makes the diagrams with up to four legs ultraviolet divergent.* This is the first point I would like to make: genuine ultraviolet divergences have to do with 1 (not j). Please see M^o in (3.4) and recall that the propagator is the inverse of M^o.

To cure these ultraviolet divergences I propose using the Pauli-Villars method. It consists in replacing the free field propagator $\Delta(x - y)$ with a regulated propagator $\Delta_{reg}(x - y)$ as follows:

$$\Delta_{reg}(x - y) \equiv \Delta(x - y) - \Delta_M(x - y) \simeq \begin{cases} 0 \text{ for } |x - y| < \frac{1}{M} \\ \Delta(x - y) \text{ for } \\ |x - y| > \frac{1}{M} \end{cases} \tag{3.11}$$

Here $\Delta_M(x - y)$ is the free field propagator for a field of mass M. This is the second point I would like to make: only if $M >> 1$ is the subtraction in (3.11) local. Needless to say, if the subtraction is not local, neither is the counterterm.

Let me now make contact with 't Hooft's (1977) computation of D. He regularizes using an x-dependent mass:

$$M(x) = \frac{M}{1 + r^2} \tag{3.12}$$

and takes first $R \to \infty$, then $M \to \infty$. Therefore, for all points x such that $r^2 >> M$, the subtraction field is practically massless.

*If one uses a singular potential, such as the instanton field in the singular gauge, one encounters additional ultraviolet singularities; they are different from the divergences usually encountered in field theory.

This is not a local subtraction (see 3.11); hence, the counterterm that goes with it cannot be the local one he uses (his Eq. 7.4).

Computation of D

From (3.7) using the appropriate multiplicities one has:

$$D = \prod_{j,l} D_{J,L} = \tag{3.13}$$

$$R^{\sum_{j,l} 2(j - l)(2l + 1)(2j + 1)} \prod_{j,l} \frac{\Gamma(2j + 1)\ \Gamma(2l + 2)}{\Gamma(j + l + \frac{5}{2})\ \Gamma(j + l + \frac{1}{2})}$$

Please notice that if one were to use 't Hooft's space dependent mass Pauli-Villars regulator with $R^2 >> M$, then the dependence upon the infrared cutoff R in (3.13) would disappear. As stated previously, that would not be a local subtraction, so one must either take $M >> R^2$ or simply use ordinary Pauli-Villars regularization with constant mass. Unfortunately, in either case the algebra for determining the eigenvalues becomes untractable. But I have argued above (Section 2) that the ultraviolet divergences have to do with l, so presumably the Pauli-Villars regulator would make the sum in l convergent. Thus I propose to calculate

$$S(\alpha) = \sum_{j,l} 2(j - l)\ (2l + 1)\ (2j + 1)\ e^{-\alpha l} \tag{3.14}$$

for $\alpha \to 0$. Some trivial algebra yields

$$S(\alpha) = \frac{4}{\alpha} + \frac{5}{6} + O(\alpha) \tag{3.15}$$

For $\alpha \to 0$, the divergent piece in $S(\alpha)$ would be cancelled by the regulator leaving:

$$D \propto R^{\frac{5}{6}} \tag{3.16}$$

This is the main result about QCD_4 showing that for $R \to \infty$, $D^{-1/2} \infty R^{-5/12}$. The result implies that in spite of the existence of the instanton, there may be no tunneling between different classical vacua in QCD_4. In obtaining this answer, it was crucial that the cutoff was in l. In fact any cutoff symmetric in l and j would make the power of R in (3.13) vanish. I tried to emphasize why I think that a proper regularization would be equivalent to a cutoff in l. I may add that the result in (3.16) is consistent with $D \geq 1$, as required by a theorem proved recently (Hogreve et al., 1978).

4. FERMION DETERMINANTS

Fermions obey Fermi-Dirac statistics. This requires certain modifications of the path integral if one wants to describe fermions. For example, the integration is performed over anti-commuting c-number fields. Another modification concerns the boundary conditions to be used in computing the fermion determinant. It is well known that the correct boundary conditions for computing the trace of the evolution operator for fermions include antiperiodicity in time (Dashen et al., 1975; Gildener and Patrascioiu, 1977). (Please note that for large Euclidean time computing the trace is as good as computing the vacuum to vacuum amplitude.) A common attitude has been to say that, for large Euclidean times, the boundary conditions used do not matter. Actually, Eqs. (2.8) and (2.9) show that, for massless fields, the boundary conditions make a difference even for $T \to \infty$.

The moral of this discussion is that, for massless fermions, trying to solve the problem on the sphere S^n instead of in R^n would present no advantage, since the boundary condition would be guaranteed to spoil any spherical symmetry the problem may have. The stress here is on the physics, not the mathematics: the physically correct definition of the fermion determinant does not select smooth functions over the sphere.

QED_2 offers a nice example of the importance of the boundary conditions. Indeed, the determinant of $i\not\partial - e\not A$ can be computed to all orders in perturbation theory in two dimensions. One can then compute the determinant through the eigenvalue problem. One easily finds that boson boundary conditions (zero field) give the wrong answer (Patrascioiu 1979b). The culprits are the zero modes, which are in fact the really boundary condition-sensitive eigenvalues (a zero mode produces the difference between (2.8) and (2.9) too).

5. CONCLUSIONS

I have presented some evidence that in QCD_4 the ratio

$$\frac{\text{Det } M^o}{\text{Det } M} \propto R^{-5/6} \text{ for } R \to \infty. \tag{5.1}$$

My findings are preliminary in that I cannot at the present time find a regularization scheme for which, at the same time, I can compute the determinants and the counterterms. But experience with models in one dimension and QED_2 (Patrascioiu 1979b) makes me believe that (5.1) is actually true.

Assuming it true, what implications does such a result have?

Since it is an infrared divergence one is encountering, one of these alternatives could be the case:

(a) There is no tunneling in QCD^4 and the theory can be defined around the old, naive vacuum.

(b) There is tunneling and the instanton is important, but the present weak coupling scheme is unsatisfactory (Patrascioiu 1978). That this is not just idle speculation can be seen in one dimension with $V(x) = \lambda(x^2 - a^2)^4$.

The second conclusion of my lecture is that massless fermions have been mistreated since care has not been taken to choose the physically correct definition of the fermion operator. Thus, irrespective of any infrared divergences, one cannot believe the results about fermion mass generation in the presence of the instanton.

APPENDIX: VAN VLECK'S FORMULA

Let

$$M(\alpha) = -\partial_t^{\,2} + \alpha V(t) \tag{A.1}$$

and compute Det M with boundary condition $\phi_n(0) = \phi_n(T) = 0$. One can prove (Levit and Smilansky, 1976) that

$$\frac{\text{Det } M(\alpha)}{\text{Det } M(o)} = \frac{N(T,\alpha)}{N(T,o)} \tag{A.2}$$

where $N(t,\alpha)$ is defined by

$$\begin{aligned} &M(\alpha)\ N(t,\alpha) = 0 \\ &N(o,\alpha) = 0 \\ &N(o,\alpha) = 1 \end{aligned} \tag{A.3}$$

REFERENCES

Callan, C., Gross, D., and Dashen, R., 1978, Phys. Rev., D17:2717.
Carlitz, R. and Creamer, D., 1979, Phys. Lett., 84B:215.
Dashen, R., Hasslacher, B., and Neveu, A., 1975, Phys. Rev., D12: 2443.
Gildener, E. and Patrascioiu, A., 1977, Phys. Rev., D16:1802.
Hogreve, H., Schrader, R., and Leiler, R., 1978, Nucl. Phys., B142:525.

Levit, S and Smilansky, U., 1976, Weizmann Institute Report WIS-76/3-Ph.
Neinast, R. and Stack, J., 1979, Phys. Rev., D19:1214.
Patrascioiu, A., 1978, Phys. Rev., D17:2764.
Patrascioiu, A., 1979a, Phys. Rev., D19:3800.
Patrascioiu, A., 1979b, Phys. Rev., D20:491.
't Hooft, G., 1977, Phys. Rev., D14:3432.

CONSTRAINTS ON LOW ENERGY COUPLING CONSTANTS IN GRAND UNIFIED THEORIES

R. Petronzio

CERN

Geneva, Switzerland

The standard description of "low energy" interactions is made in terms of a gauge theory for the electroweak and the strong interactions based on the group $SU(3)_{colour} \otimes [SU(2) \otimes U(1)]_{e-w}$. All basic gauge fields are coupled to massless fermions and those mediating the electroweak interactions are also coupled to a set of scalar fields. The instability of the vacuum of the scalar sector generates a spontaneous symmetry breaking of the electroweak interactions and provides masses for the fermions and the residual scalar fields (Higgs's).

If all these interactions are indeed fundamental, it is conceivable that they will unify at a very large energy scale denoted by M_U to some universal interactions[1,2]. If, on the contrary, they represent just a "low energy" limit of some more fundamental force, the above picture may still be valid provided the "super interactions" become effective at energies larger than M_U.

In such a picture, the values of coupling constants at low energies cannot in general be arbitrary because either they have to meet precise values at the unification or they have to respect some bounds if the interactions must not become strong before the scale M_U. The analysis and the definition of such bounds, which can be easily transferred into limits on fermion and Higgs masses, is the subject of this discussion.

I will limit myself to the standard model[3] with only one Higgs doublet

$$\phi = \begin{pmatrix} \frac{\phi_1 + i\phi_2}{\sqrt{2}} \\ \frac{\phi_3 + i\phi_4}{\sqrt{2}} \end{pmatrix} \equiv \begin{pmatrix} \phi_+ \\ \phi_0 \end{pmatrix}$$

Then, the coupling constants which are involved are those for:

i) the strong interaction: α_s

ii) the electroweak SU(2): α_W

iii) the electroweak U(1): α'

iv) the quartic scalar interaction: λ

v) the Yukawa interaction of the scalars with the fermions: D, U, L.

The last coupling constants are in fact matrices in the generation space coupled as follows:

$$\begin{aligned} \mathcal{L} &= \phi_0[\bar{d}_L D d_R + \bar{u}_R U^+ u_L + \bar{\ell}_L L \ell_R] + \\ &+ \phi^+[\bar{u}_L D d_R - \bar{u}_R U^+ d_L + \bar{\nu}_L L \ell_R] \end{aligned} \tag{1}$$

where d, u, ℓ are vectors in the generation space:

$$u = \begin{pmatrix} u \\ c \\ t \\ \vdots \end{pmatrix} \qquad d = \begin{pmatrix} d \\ \lambda \\ b \\ \vdots \end{pmatrix}$$

The scalar potential contains an imaginary mass which generates the spontaneous symmetry breaking characterized by a non-vanishing vacuum expectation value of the scalar field: $\langle\phi\rangle_0 = \langle\phi_0\rangle_0 = \eta \sim 176$ GeV. The scale η fixes the "low energy" scale of our future considerations.

I will first consider the restrictions on the gauge couplings. The basic ingredient is the renormalization group invariance which allows to relate the values of the coupling constant renormalized at η with those at M_U.

The general relation is expressed by the renormalization group equation

$$\frac{d\alpha}{dt} = \beta(\alpha) = \beta_1 \alpha^2 + \beta_2 \alpha^3 + \dots \tag{2}$$

and at one loop level takes the form:

$$\frac{1}{\alpha(\eta)} - \frac{1}{\alpha(M_U)} = \beta_1 \ln\left[\frac{M_U^2}{\eta^2}\right] \tag{3}$$

Neglecting Higgs contributions, one has for the gauge couplings[4]:

$$\frac{1}{\alpha} = \frac{8}{3}\left[\frac{L_U N}{3\pi} + \frac{1}{\alpha_0}\right] - \frac{L_U}{3\pi}\left(\frac{11}{2}\right)$$

$$\frac{1}{\alpha_W} = \left[\frac{L_U N}{3\pi} + \frac{1}{\alpha_0}\right] - \frac{L_U}{3\pi}\left(\frac{11}{2}\right) \tag{4}$$

$$\frac{1}{\alpha_S} = \left[\frac{L_U N}{3\pi} + \frac{1}{\alpha_0}\right] - \frac{L_U}{3\pi}\left(\frac{33}{4}\right)$$

where $L_U \equiv \ln (M_U^2/\eta^2)$, α_0 is the value of the unique gauge coupling of the unifying gauge group [SU(5) [1], for example] and N is the number of generations. For simplicity I have introduced the fine structure constant of QED instead of the α' of the unbroken U(1) gauge group.

By substituting, as an input, the values of α and α_s (~ 0.1) into Eqs. (4) one gets the following relations:

$$\sin^2\theta_W = \frac{\alpha}{\alpha_W} = \frac{1}{6} + \frac{5\alpha}{9\alpha_S} \sim .2$$

$$L_U = \frac{2\pi}{11}\left[\frac{1}{\alpha} - \frac{8}{3\alpha_S}\right] \simeq 63 \tag{5}$$

The first equation gives us the low energy value of the Glashow-Weinberg-Salam angle and the second tells us the scale at which the superunification occurs.

The determination of the value of α_s depends explicitly upon N

$$\frac{1}{\alpha_0} \simeq 6.7\,[9.7 - N] \tag{6}$$

Equation (6) can be justified only if α_0 is not too large: this can be translated into a limit on N: $N \lesssim 9$. Within existing superunification schemes large values of α_0 would produce undesirable effects such as an unacceptably fast proton decay.

The hypothesis that gauge interactions unify to some explicit super gauge force, gives us a determination of the θ_{GWS} angle very close to the experimental values, but makes the values of the low energy coupling constants critically dependent upon those at M_U. There exist however, a possible "anti-unifying" scenario where low energy values of gauge couplings are independent from those at high energy. This may happen if all the theories under consideration are infrared (and not asymptotically) free at energies of the order of few low energy units (η) [5], so that their coupling constants grow with increasing the energy. By looking at Eqs. (4) one can see that they contain an implicit bound on α, α_s, α_w coming from the condition $1/\alpha_0 \geq 0$. If the value of α_0 is sufficiently larger than the corresponding low energy one, the values of α, α_s, α_w will be close to those which saturate the bound and numerically independent from $1/\alpha_0$ In particular, unification values different for the various gauge couplings would still lead to the same set of low energy couplings. The discussion based on a single one-loop expansion of the β function is certainly not adequate in a strong coupling regime. However, the value of α_0 has a rapid variation only in a region very close to M_U and numerical calculations based on Eqs. (4) can be regarded as reasonable estimates. The saturation conditions read:

$$\frac{1}{\alpha} \geq \frac{L_U}{3\pi}\left(\frac{8}{3}N - \frac{11}{2}\right) \tag{7}$$

$$\frac{\alpha L_U}{3\pi}\left(N - \frac{11}{2}\right) \leq \frac{\alpha}{\alpha_W} \left(\equiv \sin^2\theta_W\right) \leq \left[1 - \frac{\alpha L_U}{3\pi}\frac{5}{3}N\right] \tag{8}$$

In such a picture the value of the unification mass has to be given; with some amount of prejudice, it can be set equal to the Planck mass ($L_U \sim 77$). From the saturation of both Eqs. (7) and (8) one then gets:

$$\sin^2\theta_W \sim .18$$
$$\alpha_e \sim \frac{1}{137}$$
$$N \sim 8$$

With the same values of L_U and N, one gets for α_s:

$$\alpha_s(\eta) \simeq 0.1 \quad ^{*)}$$

The above results are obtained in the approximation where the N generations of fermions appear at sufficiently low energy to influence the behaviour of the renormalized coupling constants all

---

*) The two-loop β function expression has been used in this case, due to the critical value of N.

the way from M_U down to η. The discussion on bounds on fermion masses will in fact confirm this hypothesis. In this "anti-unifying" picture one finds a possible explanation of why coloured interactions become strong in the low energy region. There the heaviest fermion generations decouple from the interactions, and the self gluon coupling starts to dominate the behaviour of the renormalized charge. The interactions become definitely strong when only few active generations are left.

UPPER BOUNDS ON YUKAWA AND QUARTIC HIGGS COUPLINGS

I consider first the Yukawa coupling of the Higgs doublet to the fermion fields. For simplicity I will restrict the analysis to the case where there is only one heavy fermion (of type U) while all the other masses are light compared to the mass scale η. This is a realistic description of the case N = 3 where the yet undiscovered top quark would be the only sizeably heavy fermion.

The heavy quark Yukawa coupling is denoted by h and the relation with corresponding fermion mass is

$$M_F = h(\eta^2)\cdot\eta \tag{9}$$

The renormalization scale dependence of h is regulated by the equation:

$$8\pi^2 \frac{dh^2}{dt} = \beta[h, \alpha_s, \alpha_w, \alpha] \tag{10}$$

The β function depends also on other coupling constants; at the one-loop level and in the approximation $\alpha_s \gg \alpha_w$, α, it is given by

$$\beta[h, \alpha_s] = [9/4\, h^2 - 16\pi\alpha_s]\, h^2 \tag{11}$$

Its sign depends on the condition

$$h^2 \gtrless \frac{64}{9}\pi\alpha_s \tag{12}$$

When $\beta < 0$, the Yukawa coupling is driven to zero by the strong interactions. The point $h_c^2 = 64/9\ \pi\ \alpha_s$ is ultraviolet unstable: values of h larger than h_c^2 will certainly lead to a strong interactions regime at sufficiently high energies. By demanding that this should not happen before a high, but finite energy M_U, one gets an absolute upper bound on h not so far from $h = h_c \sim 250$ GeV. The numerical values for this bound if $M_U \simeq 10^{15}-10^{16}$ GeV are:

$$M_F \leq 200\ \text{GeV} \qquad (N=3)$$
$$M_F \leq 250\ \text{GeV} \qquad (N=8) \tag{13}$$

would the hypothesis of only one heavy fermion be released, similar limits would hold for the sum of fermion masses. In any case all fermion generations are expected to be accessible at energies of order η.

The upper bounds on quartic Higgs self coupling are again regulated by the corresponding β function entering in the renormalization group[6]:

$$32\pi^2 \frac{d\lambda}{dt} = \beta(\lambda, h, g', g) \tag{14}$$

where $g' \equiv \sqrt{4\pi\alpha'}$; $g = \sqrt{4\pi\alpha_W}$

$$\beta(\lambda, h, g, g') = 4\lambda^2 + 12\lambda h^2 - 3\lambda(3g^2 + g'^2) - 36h^4 + 9/4\,[2g^4 + (g^2 + g'^2)^2] \tag{15}$$

For λ sufficiently large the β function is always positive and an upper bound is obtained along the same path as in previous sections.

The sign of β is always positive in the case of light fermions ($h \simeq 0$): when fermions are heavy it becomes negative for small values of λ and an infrared stable point appears, λ_c.

The derivative of $\beta(\lambda, h)$ with respect to λ calculated at the infrared stable point increases with h: as a consequence the upper limit on low energy value of λ becomes closer and closer to the fixed point value λ_c. We will see in the next section that if $\beta(\lambda) < 0$ for $\lambda = 0$ there exist a lower bound on λ which gets very close to λ_c, implying a strong correlation, in this region, between Higgs and heavy fermion masses.

LOWER BOUND ON THE HIGGS MASS

Lower bounds on Higgs self couplings are obtained by requiring that the classical minimum of the scalar potential remains an absolute one even in the presence of quantum corrections. This stability condition must be valid both in the region $\phi > \eta$ and $\phi < \eta$.

The analysis is based on the notion of effective potential. I will recall only the essential steps which lead to its definition; for an exhaustive discussion, see, for example, the Coleman lectures in 1973 [7].

The connected Green's functions can be obtained by the generating functional

$$e^{iW(J)} \equiv Z(J) = \langle 0|0\rangle_J = \int d[\phi] \exp\left\{\int \mathcal{L}(x)d^4x + J(x)\cdot \phi(x)d^4x\right\}$$

where J(x) is an external classical field. The following relation holds:

$$\frac{\partial W(J)}{\partial J} \equiv \phi_c = \frac{\langle 0|\phi|0\rangle_J}{\langle 0|0\rangle_J}$$

which provides an implicit relation $J = J(\phi_c)$. By a Legendre transformation one introduces

$$\Gamma(\phi_c) = W(J) - \int J(x)\,\phi_c(x)\,d^4x$$

The quantity $\Gamma(\phi_c)$ satisfies

$$\frac{\partial \Gamma}{\partial \phi_c} = -J \tag{16}$$

In the quantum theory the problem of the spontaneous symmetry breaking is reconducted to the analysis of solution of Eq. (16) when $J = 0$ for values of ϕ_c different from zero. The part of $\Gamma(\phi_c)$ which survives by taking only field configurations which are constant in space-time is called the effective potential; at tree level it coincides with the potential present in the Lagrangian. Radiative corrections introduce ultraviolet divergences which are cured by a renormalization procedure with subtractions defined at some value M of the field. Therefore

$$V(\phi) = V(\phi, M)$$

For large values of ϕ ($\gg \eta$) one can drop the quadratic term and retain only the quartic:

$$V(\phi) \simeq V^{IV}(\phi/M)\,\phi^4$$

The choice of the subtraction field strength M is arbitrary: the equivalence of different choices is expressed, as usual, by the renormalization group equation[8]

$$\left[-\frac{\partial}{\partial \ln M^2} + \beta\frac{\partial}{\partial \lambda} + \beta_{g_i}\frac{\partial}{\partial g_i} + \beta_h\frac{\partial}{\partial h} + \beta_g\frac{\partial}{\partial g} + 4\gamma\right]V^{IV}(t,\lambda,h,g,g') = 0 \tag{17}$$

where $t \equiv \ln \phi^2/M^2$ and γ is a term related to external legs renormalization. The solution is of the form:

$$V^{IV}(t,\lambda,h,g,g') = V^{IV}(0,\tilde{\lambda}[t,\lambda,h,g,g'])\cdot \cdot\exp\int_0^t dt'\, 4\gamma[\tilde{\lambda}(t,\ldots)] \tag{18}$$

where $V^{IV}(0,\lambda) \equiv \lambda$

and $\tilde{\lambda}$ satisfies Eq. (14).

Therefore, the condition $V(\phi) \geq 0$ for $\phi > \eta$ is satisfied if

$$\lambda[\phi/M,\ldots] \geq 0 \quad \text{for} \quad \phi > \eta \tag{19}$$

going back to Eq. (15) one can see that the β function at $\lambda = 0$ becomes negative for sufficiently large values of h. It follows that those values of λ which lie at very high energies between zero and the infrared stable point λ_c are driven to zero at small energies by radiative corrections. Actually, this happens only if the available "time" ($t = \ln \phi^2/M^2$) is arbitrarily large. In a finite amount of time $[t_{max} = \ln M_U^2/\eta^2]$ $\lambda(\eta)$ will stay positive if $\lambda(M_U)$ is close enough to λ_c.

The condition $\beta(0) \geq 0$ can be translated into an inequality between fermion and vector boson masses:

$$12\, M_F^4 \leq 3\,(2M_W^4 + M_Z^4) \tag{20}$$

The above condition is not an absolute bound on fermion masses[9]: heavier fermions are compatible with heavy Higgs (λ close to λ_c). As anticipated at the end of the last section, heavy fermion and Higgs masses are then strongly correlated. In Fig. 1 the curve denoted by b represents the lower bound under consideration. The point M_F^* denotes the value of M_F which saturates the condition (20).

If $\beta(\lambda)|_{\lambda=0} > 0$, λ will stay positive at all energies; however, a new type of instability may occur in this case: radiative corrections can dig a deeper minimum for $\phi < \eta$ [10,11].

To analyze this possibility one has to write the complete form of effective potential in the one-loop approximation:

$$V(\phi,\sigma) = -\frac{\mu^2\phi^2}{2} + \frac{\lambda}{4!}\phi^4 + A\phi^2 + B\phi^4 + \frac{\bar{\beta}(\lambda,\ldots)}{4!}\phi^4 \ln[\phi^2/\sigma^2] \tag{21}$$

where $\bar{\beta}(\lambda, \ldots)$ is the β function of the four-point Green function of the field ϕ truncated of the external legs renormalization.

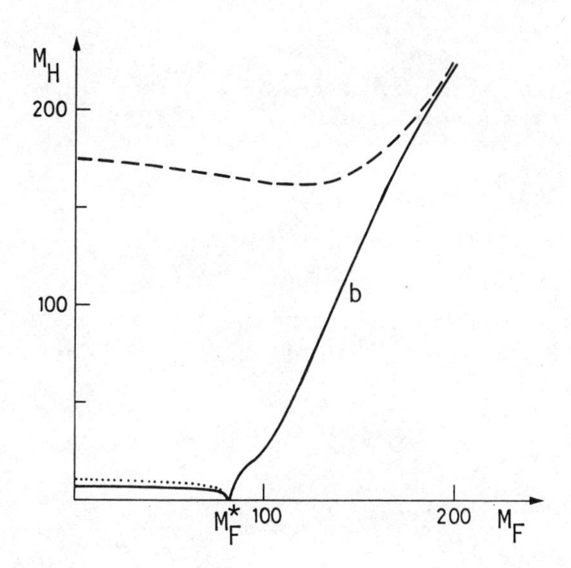

Fig. 1 Bounds on the mass of the Higgs boson (M_H) as a function of the top quark mass (M_F) in the case of three generations. We have taken $\sin^2\theta_W = 0.2$. The dashed line and the full line represent the upper and the lower bound, respectively. The dotted line is the prediction of the massless theory. The curves end in correspondence to the upper bound on M_F, Eq. (13).

The constants A and B have to be fixed by specifying the subtraction conditions at $\phi^2 = \sigma^2$.

A convenient choice is that which preserves for the renormalized μ and λ the same relations as the tree approximation[11]:

$$\begin{aligned} &i) \quad \left.\frac{\partial V}{\partial \phi}\right|_{\phi^2=\sigma^2 \equiv 6\mu^2/\lambda} = 0 \\ &ii) \quad \left.\frac{\partial^2 V}{\partial \phi^2}\right|_{\phi^2=\sigma^2} = \frac{\lambda\sigma^2}{3} \end{aligned} \tag{22}$$

With this particular renormalization prescription one gets:

$$V(\phi,\sigma) = \frac{\lambda\phi^4}{4!} - \frac{1}{2}\mu^2\phi^2 + \frac{\bar{\beta}(\lambda,\ldots)}{4!}\left[\phi^4 \ln \phi^2/\sigma^2 - \frac{3}{2}\phi^4 + 2\sigma^2\phi^2\right] \tag{23}$$

The instability is certainly absent if $\partial^2 V/\partial\phi^2|_{\phi^2=0} < 0$ (the origin is a maximum); this happens if:

$$\left.\frac{\partial^2 V}{\partial \phi^2}\right|_{\phi^2=0} = -\lambda + \bar{\beta}(\lambda,\ldots) < 0 \tag{24}$$

If $\lambda \leq \bar{\beta}(\lambda)$ the extremum in the origin is a minimum and a bound is obtained from the condition that it should not be deeper than the minimum in σ:

$$V(\sigma,\sigma) \leq V(0,\sigma)$$

or

$$V(\sigma,\sigma) \leq 0$$

This implies the condition:

$$\frac{\bar{\beta}(\lambda,\ldots)}{2} \leq \lambda \tag{25}$$

Instabilities occur when $\lambda \sim O(g^4)$; one can therefore neglect $O(\lambda^2)$ terms in the above equation, which then reads:

$$\frac{\beta(0)}{2} \leq \lambda \tag{26}$$

Alternatively it may be written as

$$M_H^2 \geqslant \frac{3}{32\pi^2\eta^2} \left\{ [2M_W^4 + M_Z^4] - 4M_F^4 \right\} \tag{27}$$

This is the lower bound on Higgs mass preventing vacuum instabilities for $\phi < \eta$.

The above condition subsists if fermions are not too heavy so to make $\beta(0) < 0$; in this case however, one recovers the lower bound related to instabilities for $\phi > \eta$.

It has been speculated[12] that there may exist some symmetry principle which forbids the appearance of a quadratic term in the potential. In such a picture the Higgs mass would be fixed by the condition

$$\left.\frac{\partial^2 V}{\partial \phi^2}\right|_{\phi^2=0} = 0 = -\lambda + \bar{\beta}(\lambda) = -\lambda + \beta(0) + O(\lambda^2)$$

Recalling Eq. (26) one sees that such a mass M_H^* is simply a factor $\sqrt{2}$ larger than the lower bound of Eq. (27). With three generations and a mass for the top quark of few tens of GeV, one gets:

$$M_H^* \sim 10 \text{ GeV}$$

In Fig. 1 [13] is summarized the situation concerning the bound on M_H as a function of the heaviest fermion mass M_F. The dotted curve represents the value of $M_H = M_H^*$.

The bounds which have been obtained are rather severe as a consequence of the stability conditions imposed on such a large energy scale. The resulting picture is that of a big desert extending from η up to M_U energy scales. While a violation of lower bounds could still be obtained by introducing new Higgs multiplets, a breaking of the upper bounds would mean that the enormous desert is actually populated by new types of interactions.

REFERENCES

1. H. Georgi and S.L. Glashow, Phys. Rev. Letters 32, 438 (1974).
2. J.C. Pati and A. Salam, Phys. Rev. D8, 1240 (1973) and Phys. Rev. D10, 275 (1974).
3. S.L. Glashow, Nuclear Phys. 22, 579 (1961);
   S. Weinberg, Phys. Rev. Letters 19, 1264 (1967);
   A. Salam, Elementary Particle Theory, Ed. N. Svartholm (Almquist and Wiksell, Stockholm, 1968), p. 387.
4. H. Georgi, H.R. Quinn and S. Weinberg, Phys. Rev. Letters 33, 451 (1974).
5. L. Maiani, G. Parisi and R. Petronzio, Nuclear Phys. B136, 115 (1978).
   G. Parisi, Phys. Rev. D11, 909 (1975).
6. T.P. Cheng, E. Eichten and L.F. Li, Phys. Rev. D9, 2259 (1975).
7. S. Coleman, *in* Laws of hadronic matter, 1973 International School of Subnuclear Physics (ed. A. Zichichi), (Academic Press, 1975
8. S. Coleman and E. Weinberg, Phys. Rev. D7, 1888 (1973).
9. N.V. Krasnikov, Soviet J. of Nucl. Phys. (October, 1978), Russian edition.
   Pham Quang Hung, Phys. Rev. Letters 42, 873 (1979).
   H.D. Politzer and S. Wolfram, Caltech preprint, CALT 68-691 (1978
10. S. Weinberg, Phys. Rev. Letters 36, 294 (1976).
11. A.D. Linde, JETP Letters 23, 64 (1976);
12. S. Weinberg, Phys. Letters 82B, 387 (1979).
    J. Ellis, M.K. Gaillard, D.V. Nanopoulos and C.T. Sachrajda, Phys Letters 83B, 339 (1979).
13. N. Cabibbo, L. Maiani, G. Parisi and R. Petronzio, Nucl. Phys. B158, 295 (1979).

CAN ONE TELL QCD FROM A HOLE IN THE GROUND?

(A DRAMA IN FIVE ACTS)

A. De Rújula, J. Ellis, R. Petronzio,
G. Preparata and W. Scott

CERN
1211 Geneva 23

THE PLOT

A new religion (*Quod Cern Demonstraturum*) attempts to impose itself upon an imaginary world. Its opponents and defenders struggle to decide whether QCD describes their reality, whether one can prove that it so does, and whether its much publicized miracles are fake.

THE CHARACTERS

<u>The Ayatellis</u> - A prophet of QCD. As if by divine inspiration, he knows the (ultimate) truth at any point in time.

<u>Biscotte</u> - A sorcerer's apprentice. Deus ex 400 GeV machina who performs the prodigies that prove the prophet's most recent truth.

<u>De Oracle</u> - Interpreter of the dogma and arbiter of the tournaments. Preaches to the masses and predicts the past.

<u>Giuliano Bruno</u> A heretic. Criticizes QCD from without and harasses its blind followers.

<u>Pestilonzio</u> - An infidel. Devil's advocate who attempts to undermine QCD from within.

ACT I - THE DAWN OF PREHISTORY

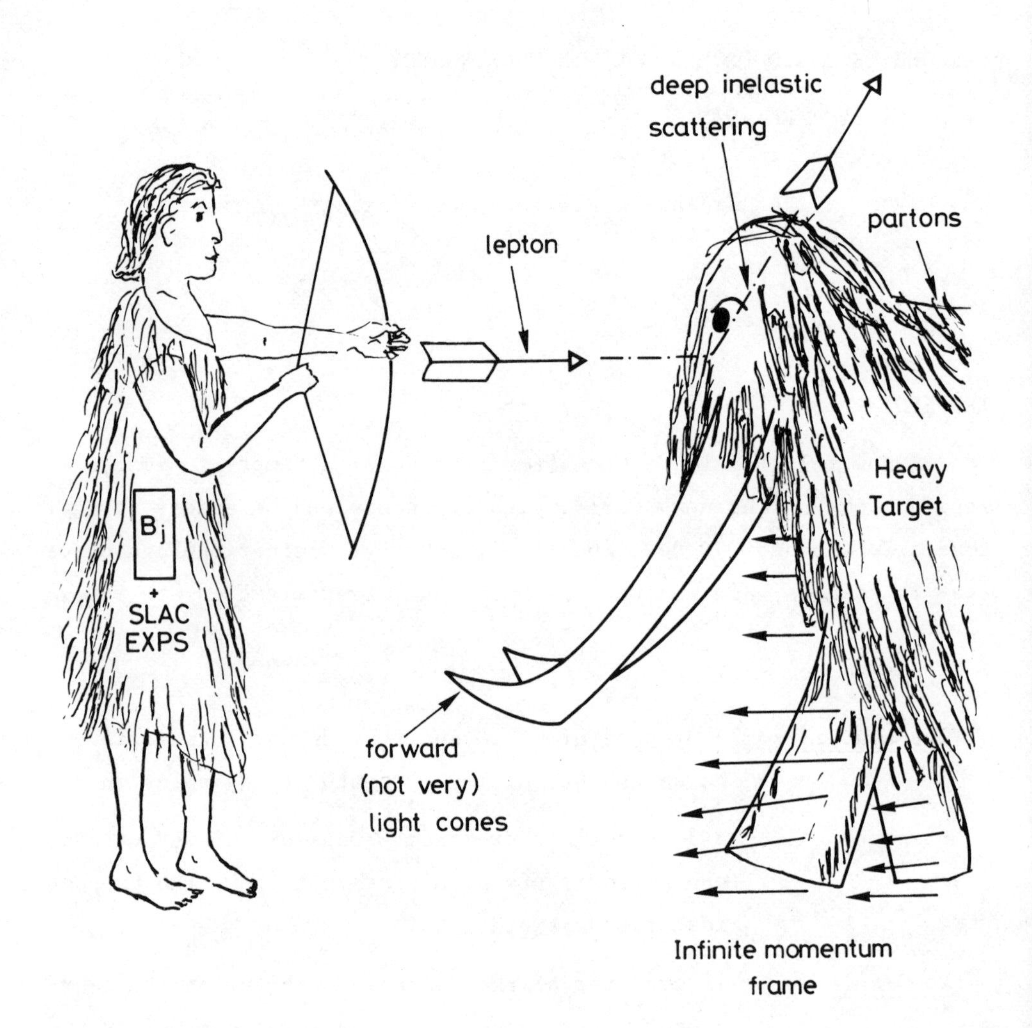

|*The characters enter as they are announced. The* AYATELLIS *wears a coat of many colours and a peculiar hat. He carries (and rings) a small bell, and bears a rolled up prayer-mat. He sits cross-legged on his mat.* BISCOTTE *carries a large brief-case containing much computer output, transparencies and a hand calculator. He takes out his output, sits behind it and opens it up.* DE ORACLE *wears what appears to be a torn sheet. He seats himself behind a large sign "Theatrical Division (Delphi)".* GIULIANO BRUNO *wears a conical hat labelled the "light-cone". It has "future" written on the front and "past" written on the back. He sits next to* BISCOTTE. PESTILONZIO *has horns on his head and wears a black cape from which emerges a long black tail. He carries a multicoloured pitchfork. He sits between* BRUNO *and* DE ORACLE.|

ACT I - |*Announced by a transparency*|

THE DAWN OF PREHISTORY - 1967

|AYATELLIS *rises from his mat as if inspired, ringing his bell. He places a transparency triumphantly on the projector.*|

AYETELLIS: Behold! A new prophet has arisen in the West, and has revealed unto us a new deep and inelastic truth[1]. It is vouchsafed unto us that at short distances protons will be seen to have a point-like structure which will first be made manifest in lepton-hadron scattering at large momentum transfers. As every first-year novice knows, the cross-sections for these processes illustrated in Fig. 1 are proportional to

$$\sum_X |\langle X|J|p\rangle|^2 \qquad (1.1)$$

where X is an inelastic hadronic final state and the momentum transfer is q. As every second-year novice knows, the cross-section (1) can be rewritten using the miracle of momentum conservation in the form

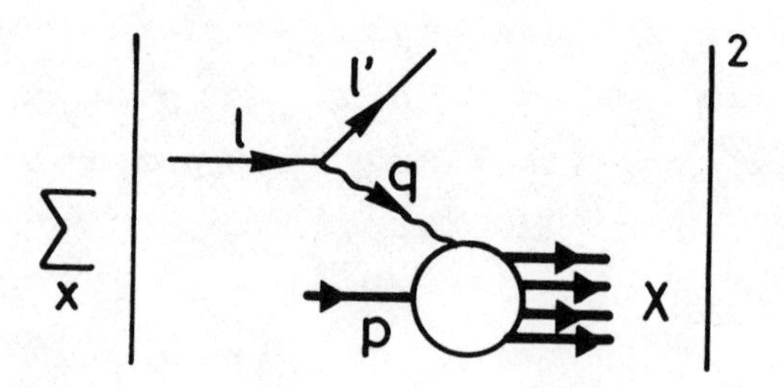

Fig. 1 The deep inelastic scattering cross-section.

$$\int d^4y\, e^{iq\cdot y} \langle p | [J^\dagger(y), J(0)] | p \rangle \tag{1.2}$$

If we take account of the spin of the currents, the cross-section (2) can be decomposed into dimensionless structure functions: two in the case of eN or μN scattering

$$F_{1,2}^{eN,\mu N}(x, Q^2) \quad : \quad Q^2 \equiv -q^2 \qquad x \equiv \frac{Q^2}{2m_N \nu} = \frac{-q^2}{2p\cdot q} \tag{1.3a}$$

and three in the case of νN or $\bar{\nu}$N scattering

$$F_{1,2,3}^{\nu N, \bar{\nu} N}(x, Q^2) \tag{1.3b}$$

The great prophet of the West has meditated for many days in the wilderness of Palo Alto on the arcane mysteries of current algebra commutators[1]. He concludes that these studies lead infallibly to the prediction that the deep inelastic structure functions (1.3) should <u>scale</u> at large Q^2:

$$F_{1,2,3}(X, Q^2) \xrightarrow[Q^2 \to \infty]{} F_{1,2,3}(X) \qquad (1.4)$$

with no asymptotic dependence on Q^2.

I now call on our worthy experimental colleague to bear witness to this <u>deep</u> and <u>inelastic</u> miracle.

<u>BISCOTTE</u>: [*approaches absent-mindedly as if from a different drama*] During the past few years the MIT-SLAC collaboration[2] has performed several experiments on inelastic electron scattering using the spectrometer facility at SLAC. Here in Fig. 2 is a plot of the structure functions νW_2 (that's F_2) plotted versus q^2 for fixed ω (that's fixed x). For $x = 0.25$ and $q^2 > 1\ \text{GeV}^2$ there is no dependence on q^2 consistent with the scaling hypothesis.

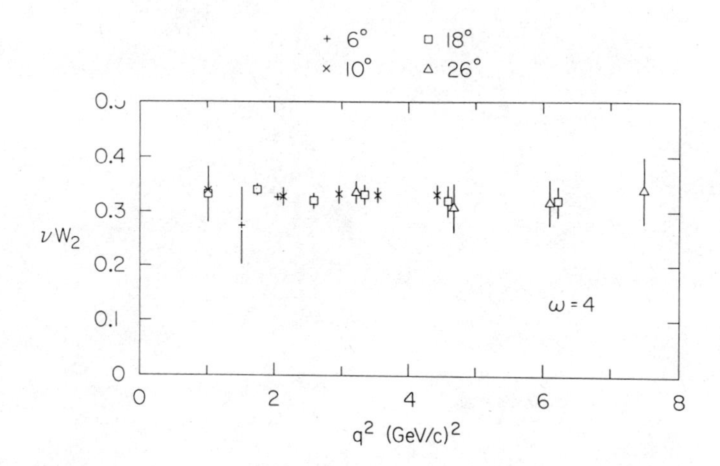

Fig. 2 Data[2] on the deep inelastic structure functions in e-N scattering from SLAC, which appear to exhibit scaling.

AYATELLIS: [*rings bell*] Indeed, just and merciful is the Lord, and glorious are the miracles vouchsafed to the faithful! It is a corollary of this beautifully confirmed miracle of scaling that the total neutrino-nucleon cross-section should be proportional to the centre-of-mass $(\text{energy})^2$ s, since at large Q^2 no other dimensionful parameter may appear in the cross-section besides G_F^2:

$$\sigma_{\text{total}}(\nu + N \to \mu^- + X) \propto G_F^2\, s \tag{1.5}$$

Since $s \simeq 2m_N E_\nu$ it follows that the total νN cross-section should rise linearly with the neutrino beam energy E_ν, and I now call on our worthy experimental colleague to bear witness to this deep and inelastic truth.

BISCOTTE: From the CERN 1.2 m heavy liquid bubble chamber[3] we have in Fig. 3 a plot of the neutrino total cross-section as a function of neutrino energy. The data are consistent with a linear increase as predicted by the scaling hypothesis.

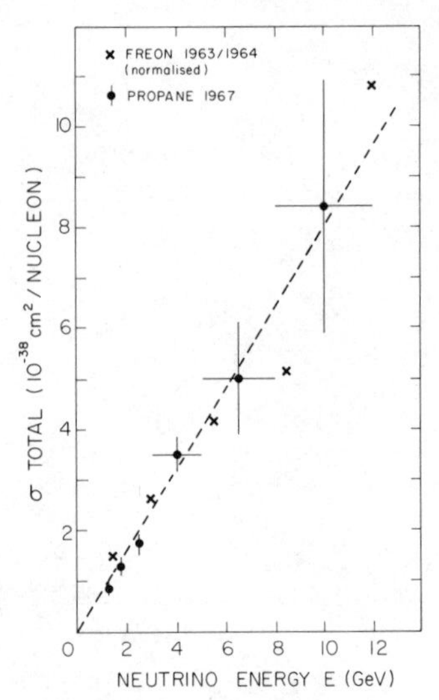

Fig. 3 Data[3] on the ν-N scattering total cross-section from CERN which appear to exhibit the linear rise expected on the basis of scaling.

[AYATELLIS *rings bell*]

Now what is all this stuff about scaling?

AYATELLIS: Please let me explain to you the basis for this deep and inelastic truth. You will recall the exponential factor $e^{iq\cdot y}$ in the formula (1.2). By the principle of complementarity handed down from our quantum mechanical forefathers we know that large momenta $q^2 \to \infty$ are associated with short distances

$$y^2 \to 0 : \text{ the light cone} \quad (1.6)$$

Contributions to the integral (1.2) where $y^2 \neq O(1/q^2)$ as $q^2 \to \infty$ would be cancelled by the rapidly oscillating exponential factor. The basis for scaling is the revelation that at short distances (large momenta) the strong interactions should vanish. One way to visualize this miracle is to observe that when one observes in a deep inelastic collision the hadron target on a time scale $\ll$ the typical hadronic time scale of 10^{-23} seconds, then the hadron will seem to be "frozen" - the impulse approximation. The hadron will then appear at these short times and distances to consist of a bundle of independent, non-interacting, point-like constituents, as in Fig. 4, referred to by the prophets[4] of the West as "PARTONS".

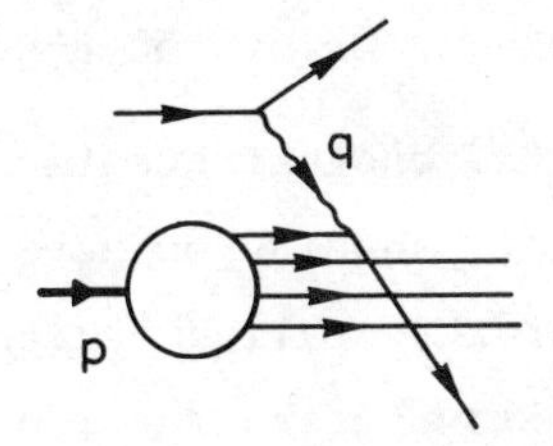

Fig. 4 At high q^2, the nucleon is revealed[4] as a collection of non-interacting constituents, the partons.

The answer to your question about the basis for the miracle of scaling is that

"Strong interactions are sōlved - they do not exist"

(at least at short distances).

BISCOTTE: I am very sorry I still don't understand.

DE ORACLE: [*rises spectrally and starts to pontificate arrogantly*] Elementary, my dear Biscotte!

A structure function depends on two variables $F = F(x,Q^2)$. Let F, by convention, be chosen to be dimensionless. Let the strong interactions, by decree, be switched off. There is then only one mass scale in the problem: m, the target mass. Thus, $F = F(x,Q^2/m^2)$, on dimensional grounds. Let Q^2/m^2 tend to infinity. In this limit F may do either of three things:

$$\begin{aligned} &i)\ F \to 0 \\ &ii)\ F \to f(x) \\ &iii)\ F \to \infty \end{aligned} \tag{1.7}$$

Examples of the three kinds of behaviour are found in Nature. The ratio $\sigma_L/\sigma_T \equiv (2xF_1 - F_2)/2xF_1$ behaves as (i). Examples of a nontrivial scaling limit (ii) are F_1, F_2 and F_3. An example of (iii) is σ_T/σ_L. To summarize, in "free" field theory (no strong interactions) the scaling behaviour of suitably defined structure functions is just the consequence of the absence of a mass scale.

But there is, Biscotte, another question you may have asked, and a very good one at that, should you have asked it. ¿Is there a simple interpretation of the scaling variable x? The answer is yes, and I predict the following explanation will become useful later[4].

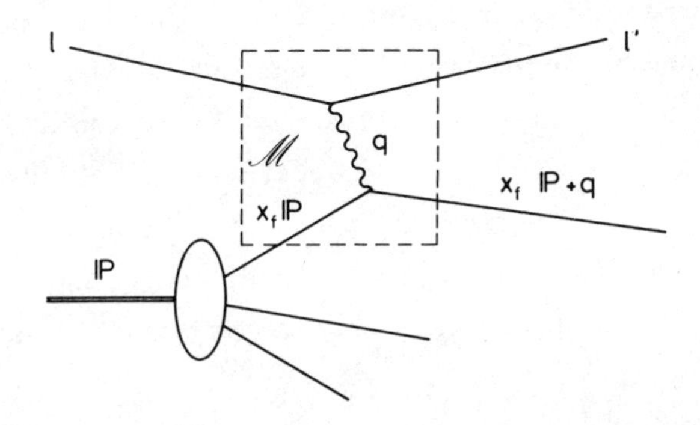

Fig. 5 The kinematics of deep inelastic scattering.

Let x be defined as usual $x = -q^2/2\mathbb{P}\cdot q$, in terms of variables that refer to the scattered lepton. Let the lepton scatter on a hypothetical nucleon constituent as in Fig. 5. Let masses be neglected and let x_f be the fraction of target four-momentum carried by the struck quark (quark: a lapsus linguae due to any vision into the future). Now, a deep statement:

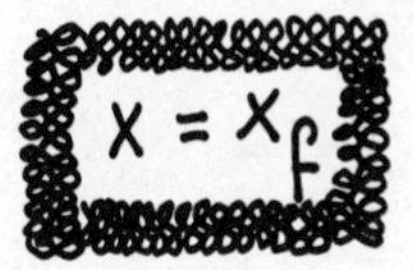

The proof is simple. The hadron inclusive cross-section is of the form

$$\frac{d\sigma}{d\ell'^3} = \int dx_f \, F(x_f) \, |\mathcal{M}|^2 \delta_+(p'^2 - \mu^2) \qquad (1.8)$$

where $F(x_f)$ is the probability for the constituent to have a fraction x_f of the target four-momentum, and we are integrating over all possible fractions. $F(x_f)$ reflects the "Fermi motion" of proton constituents and the corresponding bound state dynamics. The quantity $\mathcal{M}$ is the scattering amplitude of the lepton and the hadron constituent, as in Fig. 5. For a point-like constituent $\mathcal{M}$ is trivial. The delta function $\delta_+(p'^2 - \mu^2)$ is a left over of the (inclusive) phase space summation over all outgoing recoil momenta $p' = x_f \mathbb{P} + q$, and μ is the constituent mass. Massage this δ function a little:

$$\delta(p'^2 - \mu^2) = \delta([x_f \mathbb{P} + q]^2 - \mu^2)$$

$$= \frac{1}{2\mathbb{P}\cdot q}\,\delta\left(x_f - x + O\left(\mu^2/q^2\right)\right) \simeq \frac{1}{2\mathbb{P}\cdot q}\,\delta(x_f - x) \qquad (1.9)$$

Not only do I conclude $x = x_f$ (*Quod Constituibat Demonstraturum*) but also

$$d\sigma/d\ell'^3 \sim F(x) \qquad (1.10)$$

that is, the observed x distribution is a direct measure of the constituent's "Fermi motion".

But I see Giuliano jump up and down, I'd better run off the stage and let him display his heresies.

[*The heretic springs up from his seat as if inflamed and initiates a heated harangue*]

GIULIANO BRUNO: "*In principio erat conum lucis*"[5]

In the Beginning there was the Light Cone[6]; this is the reason why things look so very much simpler at large momenta. Strong interactions do exist, otherwise it is we who would not exist.

But the physics at Light Cone distances does not care about them. But let me try to shed some light on some obscure points of the Light Cone.

Every time the natural philosophers in the Far West[2] fire their electron gun, they make a measurement of an arcane quantity:

$$[J_\mu(x), J_\nu(y)] \tag{1.11}$$

the commutator of two currents for distances $x_\mu - y_\mu$ which all lie on the Light Cone: $(x - y)^2 = 0$.

Gell-person[7] and the Current-Algebra forefathers have taught us that when you are at the tip of the Light-Cone $[x_\mu - y_\mu = 0]$ the current-commutator is given by the same expression that you would find in free field theory. They showed us that Nature, on the tip of the Light Cone, can only read the "Free Field Book".

But how about the whole Light Cone? After many penitences and prayers with a fellow monk, in our monastery on the East Side of New York[6], we were able to generalize the teachings of the Current Algebra forefathers to the whole Light Cone, and we wrote the formula

$$[J_\mu(x), J_\nu(y)] \xrightarrow[(x-y)^2 \to 0]{} \frac{1}{(x-y)^2} \sum_n O^{(n)}_{\mu\nu,\alpha_1\ldots\alpha_n}\left(\frac{x+y}{2}\right)(x-y)^{\alpha_1} \cdots (x-y)^{\alpha_n} \tag{1.12}$$

It is in the singularity $1/(x - y)^2$ that Nature behaves as if there were no interactions; that is where scaling comes from.

Thus it is on the Light Cone, and only there, that our theological dream, that things become simple and calculable, becomes a reality. But, beware: do not leave the Light Cone, or you will soon fall into a mess. This is what the sect of the "Partonists", those religious fanatics who blindly follow the doctrine of Feyn-person[4], do not comprehend, and they have strange fantasies about a world populated by unlikely, point-like creatures, the partons.

To them and to everybody here let me recall the conclusions of the Edict of the 1971 Coral Gables Council[8]: << *Extra conum lucis nulla salus* >>. There is no salvation outside the Light Cone.

<u>AYATELLIS</u>: [*rises as if inspired*] Giuliano's Light Cone is indeed a most revealing illumination. But it seems to me that there is no contradiction between the Light Cone and the parton language which the heretic abhors. It is indeed revealed unto us by another prophet of the West[9] that

"Nature reads Free Field Theory Books"

The operators appearing in Giuliano's Light Cone expansion will be revealed to behave exactly <u>as if</u> they were made out of free quark fields. Thus the operators $O^{(n)}_{\mu_1 \ldots \mu_n}(0)$ will have the same algebraic properties as

$$O^{(n)}_{\mu_1 \cdots \mu_n}(0) = \frac{1}{n}\left[\bar{q}(0)\gamma_{\mu_1}\overleftrightarrow{\partial}_{\mu_2}\cdots\overleftrightarrow{\partial}_{\mu_n}q(0) + \text{permutations}\right] \tag{1.13}$$

and all the c number singular functions appearing in the operator product expansion[10] will take their canonical power-law form. This revelation will then enable all the predictions of the parton model to be recovered from the light-cone expansion, with the added miracle that it will be seen that <u>partons</u> are <u>quarks</u>. An example[11] of the miraculous predictive power of this <u>deep</u> and <u>inelastic</u> insight is the following ratio between eN and νN structure functions:

$$\frac{18}{5}\left(F_2^{ep} + F_2^{en}\right) = F_2^{\nu N} \tag{1.14}$$

where the constant of proportionality comes from dividing by the average of the quark $(\text{charge})^2$:

$$\frac{1}{2}\left(\frac{4}{9} + \frac{1}{9}\right) = \frac{1}{2}\left(\frac{5}{9}\right) = \frac{5}{18} \tag{1.15}$$

For the benefit of our worthy experimental colleague

$$\frac{18}{5} = 3.6 \tag{1.16}$$

and I now call on him to bear witness to the deep and inelastic truth (1.14).

BISCOTTE: The plot in Fig. 6 shows the structure function $F_2^{\nu N}(x)$ from the Gargamelle neutrino antineutrino experiments at CERN[12] plotted for events in the scaling region $Q^2 > 1$ GeV^2, $W^2 > 4$ GeV^2. The solid curve is the structure function F_2^{ed} measured at SLAC multiplied by the factor 18/5 to take account of quark charges. The broken curve shows the effect of Fermi motion and measurement errors in the Gargamelle neutrino experiment.

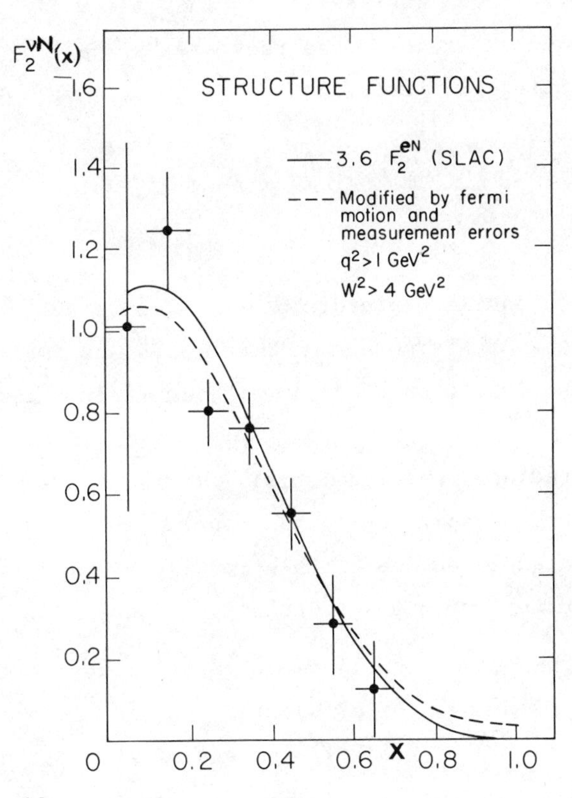

Fig. 6 Data[12] on $F_2^{\nu N}$ and F_2^{eD} compared: their ratio is consistent with the 18/5 = 3.6 expected in the quark-parton model.

Clearly the electron data and the neutrino data are in very good agreement provided that one takes account of the fractional quark charges.

Incidentally, did you know that Eq. (1.16) is due to Llewellyn[11] Smith?

PESTILONZIO: [*shaking with some effort a big volume with one hand, the pitchfork with the other, the horns with the head, and the tail with the (deleted)*]

Wait, wait! Books on field theory have a second volume, on Interactions!

In any interacting and renormalizable field theory known to date, scaling violations must exist[13]. They may well be tiny, i.e., logarithmic, but also in this case they are "always" expected to grow with increasing Q^2 according to:

$$\frac{d\,F(x,Q^2)}{d\ln Q^2} \simeq (\ln Q^2)^P \; ; \; P \geq 0 \tag{1.17}$$

You can hardly reconcile a picture of strong interactions where scaling violations are large at small energies and small at large energies with the existence of an underlying field theory.

DE ORACLE: Ladies, gentlemen et al., it is now late '67, and time to offer some recapitulation and conclusions.

We have contemplated the birth of a wild prophecy: asymptotic scaling of inclusive structure functions.

We have witnessed an unbelievable result: scaling is observed experimentally. A minor goof of Nature: scaling is precocious (true to $\sim 20\%$ at $Q^2 > 2\ \mathrm{GeV}^2$) instead of asymptotic (i.e., never true).

We have seen two characteristically unsuccessful theoretical approaches emerge around these questions:

i) The formal Operator Product Expansion, that emphasizes the asymptotic nature of the scaling predictions;
ii) A more "physical" parton model, that is in trouble marrying the strong bound state dynamics with the negligible strong interactions required to observe a "point-like" scaling behaviour.

Moreover, Pestilonzio reminds us that in no known field theory do the interactions die away as Q^2 increases, such as to approach scaling limits. There appears to be only one possible conclusion:

FIELD THEORY IS DEAD †

Some in the Wild West do indeed stick to this conclusion.

But I am ready to offer two optimistic oracular predictions:
i) Scaling will remain with us as a crucial property of Nature, and we shall make fundamental progress in understanding it. Progress is needed for, after all, a physicist is like a bicyclist: if she ever stops, she collapses;
ii) Field theory, like the characters in comic strips, never dies...

GIULIANO BRUNO: ...or never reaches maturity!

DE ORACLE: The East Coast field theory crusaders shall revenge.

ACT II - THE COMING OF GAUGE

vacuum polarization
soft gluon emission
4g vertex
tH
3g vertex
$q\bar{q}g$ vertices

ACT II - THE COMING OF GAUGE

AYATELLIS: [*rises as if inspired, as usual.*] Oh ye of little faith: it is premature to conclude that field theory is dead. It is merely in need of reincarnation. And behold, the Lord is just and merciful, a new prophet has arisen in Holland[14] who teaches us how this miracle may be performed. He teaches to us the way to calculate reliably with gauge theories.

Local gauge invariance is the property that you can make on space-time dependent phase transformation on complex fields in the theory, e.g., for fermions

$$\psi(x) \to \exp\left[ie_\psi \Lambda(x)\right]\psi(x) \tag{2.1}$$

which leaves the theory invariant. In order for the fermion kinetic energy to be invariant under such a transformation (2.1) it is necessary that there be a vector field $A_\mu(x)$ coupled to the fermion:

$$e_\psi \bar{\psi}(x)\gamma_\mu \psi(x) A^\mu(x) \tag{2.2}$$

which has the transformation property

$$A_\mu(x) \to A_\mu(x) + \partial_\mu \Lambda(x) \tag{2.3}$$

when the gauge transformation (2.1) is made. An example of such a gauge theory is QED, with $\psi(x)$ an electron field, $A_\mu(x)$ the photon field. The phase transformation $\Lambda(x)$ has no internal symmetry group properties, and this simplest type of gauge theory such as QED is called Abelian.

Distinguished patriarchs of generations past instructed us in the art of calculating with Abelian gauge theories: the beardless prophet of the tulip fields tells us how to calculate with non-Abelian gauge theories. I prophesy unto you that great indeed will be the popularity of these non-Abelian theories. Let me tell you why. If you look in the Oxford English Dictionary[15] you will find that the Abelians were a sixth century sect in North Africa who practiced chastity after marriage. It is clear why Abelian theories have died out, and why non-Abelian theories are certain to be much more popular.

In non-Abelian theories, the fermions $\psi(x)$ and gauge transformation $\Lambda(x)$ acquire internal symmetry indices:

$$\psi_i(x) \to \exp\left[ig\Lambda^a(x)\lambda^a\right]_{ij}\psi_j(x) \tag{2.4}$$

Again, to ensure the invariance of the fermion kinetic energy, a gauge field $A_\mu^a(x)$ must be introduced, with the transformation law[16]

$$A_\mu^a(x) \to A_\mu^a(x) + \partial_\mu\Lambda^a(x) + f_{abc}\Lambda^b(x)A_\mu^c(x) \tag{2.5}$$

where the f_{abc} are the structure constants of the internal symmetry group. The kinetic energy term of the gauge fields now requires the gauge fields to play with themselves via trilinear and quadrilinear terms (another reason for the popularity[17] of non-Abelian theories):

$$-\frac{1}{4}F_{\mu\nu}^a(x)F_{\mu\nu}^a(x) : F_{\mu\nu}^a(x) \equiv \partial_\mu A_\nu^a(x) - \partial_\nu A_\mu^a(x) + g f_{abc} A_\mu^b(x) A_\nu^c(x) \tag{2.6}$$

in order to ensure invariance under the non-Abelian gauge transformations (2.4) and (2.5).

Groups of crusaders in the United States have calculated[18] with these theories, and preach unto us that the strong interactions are described by just such a non-Abelian gauge theory. It is to act upon the colour degree of freedom of quarks as the index i in Eq. (2.4). Just as the photon is massless in QED, so also the gauge vector bosons of this theory - the gluons - are pronounced to be massless. The interactions by which they play with themselves are believed to confine quarks and gluons in some mysterious way which the Good Lord will reveal unto us in his own good time. This non-Abelian theory of the strong interactions shall be known as QCD:

Quod Cern Demonstraturum

Pestilonzio objected in the first act that scaling was impossible in field theory because of the logarithms encountered in perturbation theory:

$$g + g^3 \ln Q^2 + g^5 \ln^2 Q^2 + \cdots \tag{2.7}$$

The beardless prophet of the tulip fields and gallant crusaders of the American East Coast have vouchsafed unto us a resolution of this problem that is unique to non-Abelian gauge theories: they are asymptotically free[18]. This means that if we compute by summing the logarithms of (2.7) the effective Q^2 dependent coupling

$$\alpha_s(Q^2) \equiv \frac{g^2(Q^2)}{4\pi} \tag{2.8}$$

then (miracle of miracles, the Lord is indeed just and merciful) if falls to zero at large Q^2, according to the following pregnant formula

$$\alpha_s(Q^2) \underset{Q^2\to\infty}{\approx} \frac{12\pi}{(33-2f)\ln Q^2/\Lambda^2} \tag{2.9}$$

In Eq. (2.9), f is the number of quark flavours with masses $\ll Q$, and Λ is an unknown parameter characterizing the absolute scale of the strong interactions.

The miracle of asymptotic freedom provides the faithful with a qualitative understanding of the parton model, but there are some doctrinal subtleties which mean that scaling in deep inelastic scattering is not exact in QCD. The problem is that the effective coupling (2.9) decreases only very slowly as $Q^2 \to \infty$, and some of its effect lingers on in the Nirvana of high Q^2. To see this one must sum all the logarithms of perturbation theory. For a structure function the leading logarithms are

$$F_0 + F_1 g^2 \ln Q^2 + F_2 g^4 \ln^2 Q^2 + \dots \tag{2.10}$$

while there are also non-leading logarithms

$$G_1 g^2 + G_2 g^4 \ln Q^2 + \dots \tag{2.11}$$

and so on. We may sum the leading logarithms by using the operator product expansion[10] of Act I, and the renormalization group of Stueckelberg and Peterpersonn[13]. One finds results which are most simply expressed for the moments of deep inelastic structure functions

$$F_i(n, Q^2) \equiv \int_0^1 dx\, x^{n-1} F_i(x, Q^2) \tag{2.12}$$

and the simplest results of all apply to non-singlet combinations of structure functions such as F_2^{ep-en} or $F_3^{\nu N}$. For these QCD makes the miraculous predictions[19]

$$F_i(n, Q^2) = F_i(n, Q_0^2) \left[\frac{\ln Q^2/\Lambda^2}{\ln Q_0^2/\Lambda^2} \right]^{\frac{4}{33-2f}\left[-1 + \frac{2}{n(n+1)} - 4\sum_{j=2}^{n} \frac{1}{j}\right]} \tag{2.13}$$

Notice that the QCD prediction is for the Q^2 dependence starting from an a priori unknown initial condition at $Q^2 = Q_0^2$. The exponents in the $\ln Q^2/\Lambda^2$ dependence are familiarly known as the anomalous dimensions d_n [20].

On the basis of Eq. (2.13) I prophesy unto you a miraculous corollary. It is clear from the form of the moments (2.12) that higher moments $(n \to \infty)$ weight preferentially X closer to 1. It is also clear from (2.13) that the anomalous dimensions d_N increase logarithmically as $n \to \infty$. Therefore the structure function at large X must fall as $Q^2 \to \infty$. Indeed, the rate of fall of the structure functions $F_i(X,Q^2)$ should be faster and faster at larger and larger X [21].

I now call on our worthy experimental colleague to bear witness to this deep and inelastic miracle.

BISCOTTE: There is new evidence from a muon scattering experiment[22] at the Fermi National Accelerator Laboratory which shows evidence for deviations from scaling. This plot (Fig. 7) shows the ratio of the measured structure function to the scaling structure function plotted versus q^2.

The trend of the data is that the structure function increases with increasing Q^2 at small x and decreases with increasing Q^2 at large x. Now that you mention it, the SLAC data[23] show the same effect (Fig. 7b). The point is, you see, that although scaling holds perfectly at $x = 0.25$, if you go to other x values, you can see a definite Q^2 dependence. In fact, every man and his dog[24] (Fig. 8) now sees scaling deviations. In each case the trend of the data is that the structure function increases with increasing Q^2 at small x and decreases with increasing Q^2 at large x. Are you saying that you predicted all this?

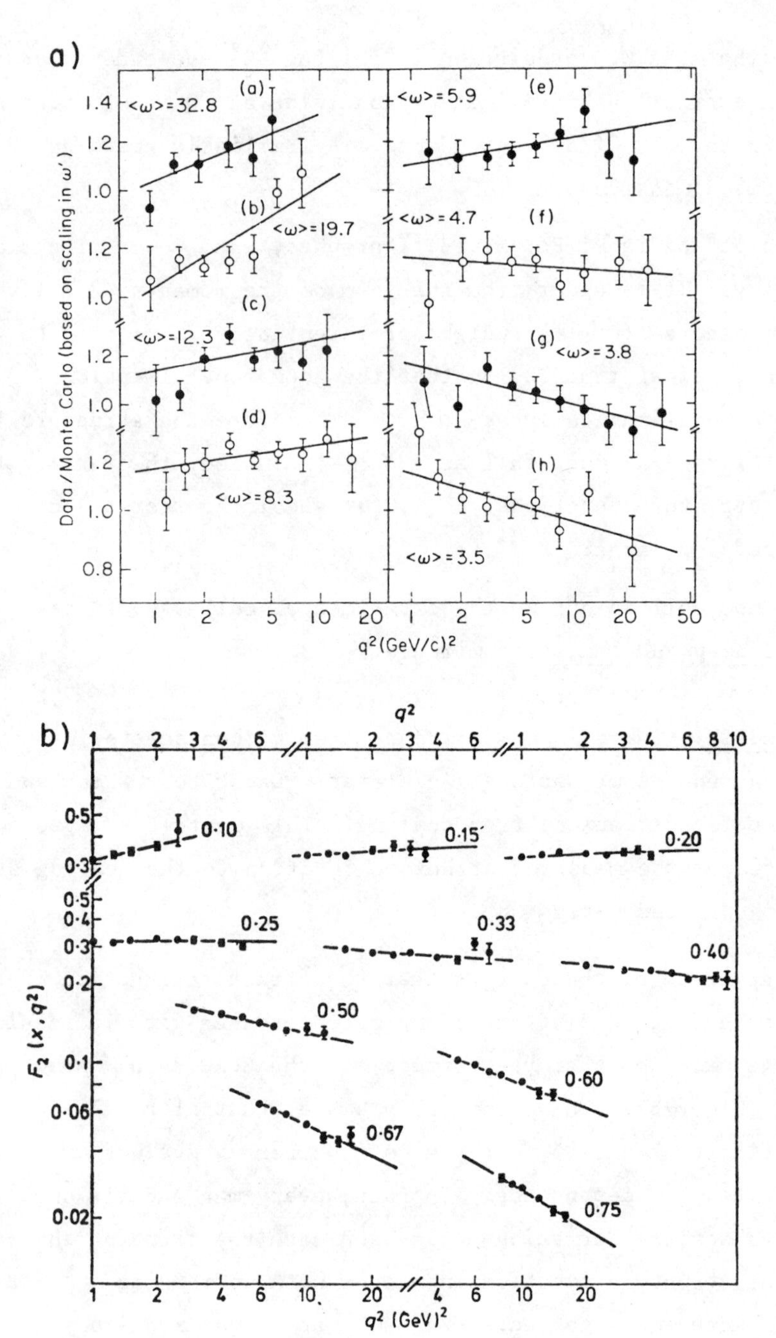

Fig. 7 Data[22,23] on μ-N and e-N scattering which appear to exhibit scaling violations.

AYATELLIS: [*rings bell*] Yes indeed, the Lord is just and merciful. See, here in Fig. 8 are curves[25] for the Q^2 dependence calculated on the basis of the leading logarithmic QCD predictions (2.13). See how beautifully they fit the experimental data. Thus are the faithful rewarded. QCD is proved in the leading log approximation!

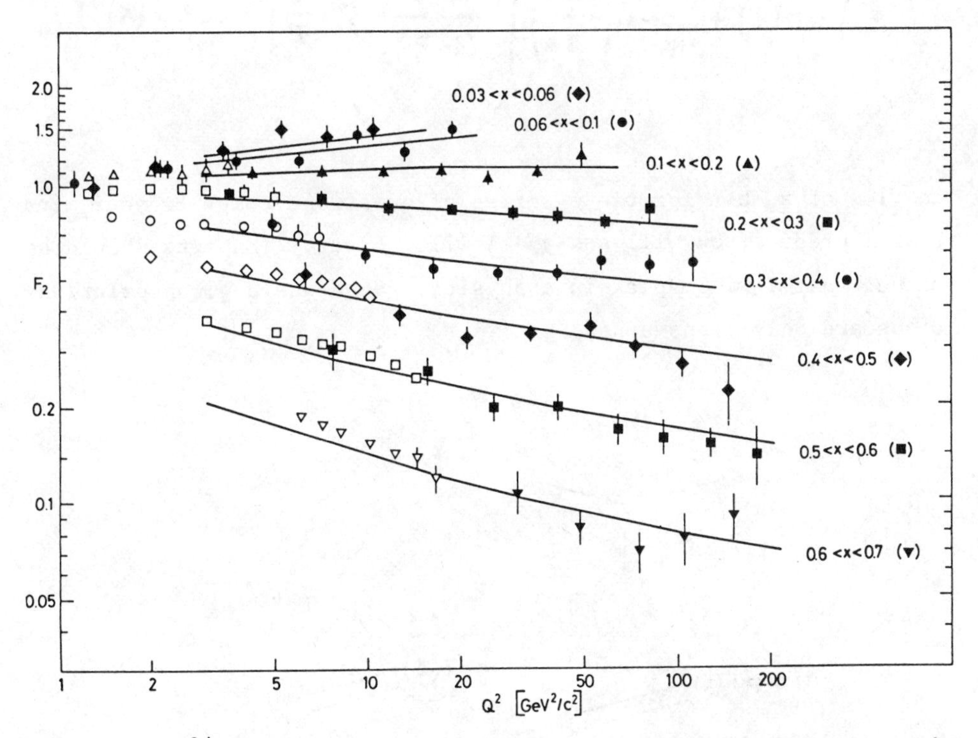

Fig. 8 Data[24] on ν-N scattering from the CDHS Collaboration which appear to exhibit scaling violations. The solid curves are predictions of QCD [AYATELLIS *rings bell*].

BISCOTTE: What does he mean by the "leading log approximation"?

DE ORACLE: I wish you had not asked that question. In a minute, you too will wish you had not asked that question.

Remember that we had obtained scaling upon neglect of quark interactions at short distances (Fig. 5, redrawn in Fig. 9a). The next step is to include quark interactions in QCD perturbation theory. An example of a first order correction is the gluon brems strahlung diagram of Fig. 9b. These corrections led to deviations from scaling. To lowest order in α_s, the prediction for a structure function F becomes[26]:

$$F(x,q^2) = F(x) + \frac{\alpha_s}{2\pi} \ln\left(-\frac{q^2}{\mu^2}\right) \int_x^1 \frac{dz}{(1-z)_+} F\left(\frac{x}{z}\right) + \frac{\alpha_s}{2\pi} g(x) + \cdots \tag{2.14}$$

In view of such a formula a series of questions comes to your mind that I predict you will ask and I will answer. (Experts should be warned that I will speak in a physical gauge where gluon polarizations are only transverse.)

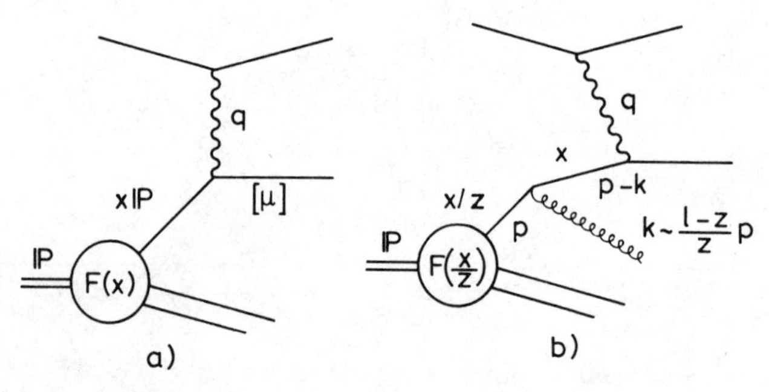

Fig. 9 First order corrections to the naive quark-parton model from gluon bremsstrahlung in QCD.

BISCOTTE: O.K. Why $\ln(-q^2/\mu^2)$?

DE ORACLE: The probability of radiation from a relativistic coloured quark of mass μ increases logarithmically with the momentum scale in the process. Remember that we are computing a hadron-inclusive observable. Thus, in the diagrams of Fig. 9, we are summing over the phase space of all outgoing hadrons (quarks and gluons). The second diagram has an extra gluon of momentum k and an extra integration over d^3k. Relative to the first diagram there is also an extra quark propagator $1/[(p-k)^2-\mu^2]$. Let $p=(E,\vec{p})$ be the quark four-momentum, $\beta=|\vec{p}|/E$ its velocity, and let θ be the angle between $\vec{p}$ and the gluon momentum $\vec{k}$. Also $p^2\simeq\mu^2$, $k^2=0$. Believe it or not, you are now ready to contemplate the birth of the famous logs: first notice

$$\frac{1}{(p-k)^2-\mu^2}\sim\frac{1}{-2p\cdot k}=\frac{-1}{2Ek(1-\beta\cos\theta)} \tag{2.15}$$

$$d^3k=k^2\,dk\,d\cos\theta\,d\varphi$$

Among the phase space integrations I will stumble upon

$$\int_{-1}^{1}\frac{d\cos\theta}{1-\beta\cos\theta}=\ln\frac{1-\beta}{1+\beta}=\ln\frac{1-\beta^2}{(1+\beta)^2}\simeq\ln\frac{1-\beta^2}{4}=\ln\frac{E^2-p^2}{4E^2}=-\ln\frac{4E^2}{\mu^2} \tag{2.16}$$

that is, the famous log!

BISCOTTE: I thought you were after $\ln(-q^2)$, not $\ln 4E^2$?

DE ORACLE: Peanuts. $\ln E^2$ and $\ln(-q^2)$ differ by the log of some dimensionless scaling variable. Logarithms that do not grow with momentum scale I dumped into the unspecified $O(\alpha_s/\pi)$ term of Eq. (2.14). The leading log approximation consists in keeping only the logs that grow with the momentum scale.

BISCOTTE: Why $dz/(1-z)_+$ in Eq. (2.14)?

DE ORACLE: The quantity z is defined in Fig. 9b, where the quark fraction of four-momentum degrades from x/z, to x. At this point the quark is hit by the lepton probe and contributes to the structure function at $-q^2/2\mathbb{P}\cdot q = x$. To answer your question I would have to put together the rest of the phase space integrations over the outgoing gluon momentum. But you know the answer since

$$\frac{dz}{1-z} \propto \frac{dk}{k} \tag{2.17}$$

and this the spectrum of bremsstrahlung radiation[26] you learned in school. The mysterious + sign in $(1-z)_+$ refers to the contribution of virtual diagrams that I did not draw. They subtract the (infrared) singularity at $z = 1$.

BISCOTTE: Why $F(x/z)$ in Eq. (2.14)?

DE ORACLE: The quarks seen by the probe at a fraction of target four-momentum x may, as in Fig. 9b, be those that had a higher fraction x/z $[x < z < 1]$ and descended to fraction x by gluon bremsstrahlung. $F(x/z)$ is the probability that the quark originally had the corresponding fraction of momentum.

BISCOTTE: Equation (2.14) still does not look like the Ayatellis formula Eq. (2.13) for the evolution of structure function moments Moreover, it contains the unknown quark mass μ. What's going on?

<u>DE ORACLE</u>: Patiente. Equation (2.14) is not yet cast in a useful form. Suppose we went to higher orders of perturbation theory, as in Fig. 10 where more gluons have been emitted and they are seen to loop now and then into quarks or more gluons. For N such gluon emissions we get an overall radiation probability $[\ln(-q^2/\mu^2)]^N$. The effect of loops can be reabsorbed into a running coupling constant $\alpha_s \sim 1/b \ln(-q^2/\Lambda^2)$ at each vertex. Thus the actual perturbation theory parameter is

$$\left(\alpha_s \ln \frac{-q^2}{\mu^2}\right)^N = \left(\frac{1}{b \ln \frac{-q^2}{\Lambda^2}} \ln \frac{-q^2}{\mu^2}\right)^N \tag{2.18}$$

which has two bad features. First, it depends on quark masses. The hint on how to get rid of these is already given by the first order formula Eq. (2.14): compare two momentum scales q_1^2 and q_2^2 and exploit a fascinating property of logs,

$$\ln \frac{-q_1^2}{\mu^2} - \ln \frac{-q_2^2}{\mu^2} = \ln \frac{q_1^2}{q_2^2} \tag{2.19}$$

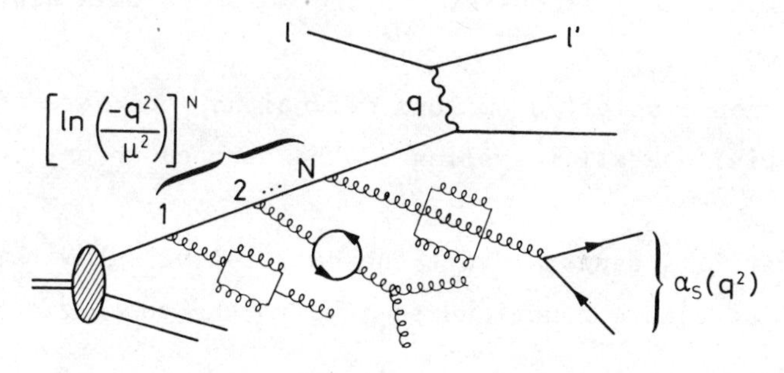

Fig. 10 Higher order corrections to the naive quark-parton model in QCD.

The second bad feature of an expansion in $\alpha_s \ln Q^2$ is that it is quantity of $O(1)$. One is forced to sum perturbation theory to all orders in $\alpha_s \ln Q^2$ before it makes numerical sense. The result of this exercise, which I spare you, can be most simply cast as the Ayatellis' moment formula Eq. (2.13).

Now I can answer your original question. The moment Eq. (2.13) is dubbed a "leading log" formula in the sense that we have taken into account all terms of $O(\alpha_s \ln Q^2)^N$ and neglected $O(\alpha_s)$ corrections to the result (the sub-leading terms of $O[\alpha_s^N (\ln Q^2)^{N-1}]$. Incidentally, the mythical "anomalous dimensions" d_n in Eq. (2.13) are nothing[26] but z^n moments of the bremsstrahlung spectrum $dz/(1 - z)_+$.

BISCOTTE: Thank you very much, everything is now as clear as mud!

GIULIANO BRUNO: Mad!

PESTILONZIO: [*with a devilish smile*] Wait, wait!
ADMITTEDLY, the property of asymptopic freedom of QCD allows you to apply to strong interactions the techniques of perturbation theory (in some cases)
BUT, the prediction that structure functions shrink to small x as Q^2 increases is a general consequence[27] of momentum migration through bremsstrahlung processes. By no means is it a distinctive feature of QCD.
IN FACT, the non-singlet "anomalous dimensions" d_n are identical (up to a trivial overall rescaling of the charge) in any vector theory.
THUS, the Ayatellis cannot invoke the qualitative behaviour of Q^2 evolution of structure functions to jump to the conclusion that QCD is vindicated.
HOWEVER, your findings have a positive social impact, as they suggest the following strategy:

INFALLIBLE PROGRAM TO JUSTIFY YOUR SALARY

1. Take a QED calculation (RUSSIAN[28], if possible) at random
2. Change $\alpha \rightarrow \alpha_s(Q^2)$
3. Change QED into QCD in title*
4. Publish
5. Go to 1.

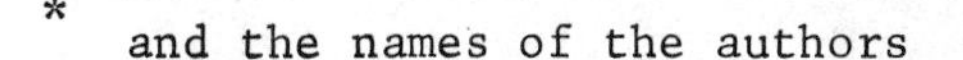

* and the names of the authors

GIULIANO BRUNO: As the ancients used to repeat: *"Natura non facit saltus"*, Nature does not make jumps. But you go on jumping all the time from quarks and gluons to the real hadronic world.

You need more than a concoction of quarks, gluons and leading logs[29] to jump out of the Light Cone into the dark labyrinths of the hadronic world.

You seem to enjoy the taste of your "asymptotic freedom" but you do not realize that it is nothing but hallucination, a mirage lingering upon that hopeless, desert world, that is the perturbative world.

By feverishly wandering in that world, picking up diagram after diagram, colliding with log after log, how can you make it less of a desert? How can you fail to see creatures like quarks and gluons which nobody in his right senses[30], like our fellow experimenters[31] have ever seen?

What happens to that wonderful world which has been revealed to us in the last three decades? That rich and strange world which is populated by hundreds of different particles, fat and thin, big and small[32], light and heavy?

What happens then to all the mass scales that come with these wonderful creatures? to the fine structure effects which never cease to excite our wonder?

To these disturbing questions, rather than wake up, you believe that you can settle all the questions by uttering the magic formula

"CONFINEMENT"

A very big step from Science back to Magic!

To this I will answer, like that great lawyer from Frascati: "*Quousque tandem ingenua QCD abutere patientiae nostrae?*" Until when will naive QCD abuse our patience?

DE ORACLE: Not even alchemists dreamed of the dimensional transmutation that underlies asymptotic freedom, a worthier trick than turning paper dollars into gold. The dimensionless coupling constant $g^2 \sim \alpha_s \sim 1/\ln Q^2/\Lambda^2$ is not a constant, and is determined in terms of a dimensionful dynamical parameter Λ. Weak strong interactions at short distances and the corollary of scaling are made compatible with strong strong interactions at larger distances. Moreover, we have seen the pattern of small deviations from scaling predicted by asymptotically free QCD be faithfully reproduced by our experimental friends (*Lapsus linguae*, by Nature). Thus, this seems like a good place to conclude that we have understood, and to quit.

Unfortunately, we are reminded by Pestilonzio that the observed pattern of deviations from scaling would also be the expected one in any other (non-QCD) field theory. But, and this cannot be overemphasized, only non-Abelian gauge theories like QCD are asymptotically free and allow us to do a consistent and predictive perturbation theory for the strong interactions.

More worrisome is a fact that Giuliano Bruno will not allow us to forget for a second: confinement has been put in by hand, perturbation theory by itself would imply that quarks and gluons get out.

Let me conclude with some oracular predictions:

i) Confinement shall remain for years to come the main source of controversy and/or the scorn of heretics. We are nowhere near having established that QCD confines. (A property of a theory is said to be fully established the day that mathematical physicists start to try to prove it.)

ii) The leading log approximation shall be used and indeed, abused, to the point that it will become the misleading log approximation.

iii) Miracles that do not concern us today (neutral currents and charm) will convert the masses to the gauge cult. Theorists and experimentalists alike will rush after QCD, like ambulance chasing drivers.

iv) Time to stop for a few moments, since moments are to become the language of many a future development.

ACT III - THE QUEST FOR THE WHOLLY SCALING VARIABLE

ACT III - THE QUEST FOR THE WHOLLY SCALING VARIABLE[33]

AYATELLIS: In the last act, Giuliano Bruno reproached the preachers of QCD because they were trying to apply the asymptotic predictions (2.13) in a range of finite Q^2. From complementarity, finite Q^2 means finite $y^2 < 0$, inside the light-cone and not on it. The heretic correctly emphasizes that to make predictions inside the light-cone we need to understand the finite mass-scales, possibly non-perturbative, which are associated with physical hadrons as opposed to the confined quarks and gluons. But behold, the new religion of QCD is so powerful and all-embracing[34] that it enables us to venture inside the light-cone and take into account many aspects of the hadronic mass-scales. It is revealed unto us that these mass corrections are generally of order $(1/Q^2)^n$, and may either be computed, or can be argued to be small in the Q^2 range of interest.

The mass-scales which one can imagine as sources of $(1/Q^2)^n$ corrections to scaling are:

- the target nucleon mass $m_N \longrightarrow O(m_N^2/Q^2)^n$
- the quark masses $m_q \longrightarrow O(m_q^2/Q^2)^n$
- higher twist / non-perturbative effects $\longrightarrow O(p^2/Q^2)^n$

where p is of the same order as a typical hadronic $p_T \sim 300$ MeV. Let us meditate in turn on each of these effects.

Target nucleon mass

At high Q^2 the moments $F_i(N,Q^2)$ (2.12) project out the contributions to the structure functions of operators $O_{\mu_1 \ldots \mu_n}$ of definite spin n. Field theory predictions, e.g., for anomalous dimensions, apply to operators of definite spin. However, at finite Q^2 the moments (2.12) combine together the contributions of operators of all spins $\leq n$ and are hence very complicated from a field theory point of view. Nevertheless, the Good Lord has revealed to us through the Austrian prophet Nachtperson[35] how one may redefine the moments at finite Q^2 so as to continue the projection on to definite spin and take exact account of the finite nucleon target mass. The solution is to use Gegenbauer polynomials and O(4).

[*Flourishes hands and rings bell.*] Let there be O(4)! A quick back-of-the-envelope calculation reveals the appropriate subasymptotic form for the F_3 moments: [DE ORACLE *produces large (1 m by 1 m 50cm) envelope from behind his "Theatrical Division - Delphi" sign. On the front it is addressed to him, on the back it reads*[36]:

$$F_3(n,Q^2) = \int_0^1 \frac{dx}{x^3}\, \xi^{n+1} F_3(x,Q^2)\left[\frac{1+(n+1)\sqrt{1+4m_N^2x^2/Q^2}}{n+2}\right] \tag{3.1}$$

AYATELLIS *rings bell and continues preaching.*] See how elegantly the m_N^2/Q^2 effects are taken exactly into account by the O(4) analysis. Notice how many of the effects arise from the introduction[37] of the

$$\text{\underline{Wholly Scaling Variable}} \qquad \xi \equiv \frac{2x}{1+\sqrt{1+4m_N^2x^2/Q^2}} \tag{3.2}$$

in fact, if one forgets the complication due to spin, then the corrections to scaling from target nucleon mass effects amount to replacing scaling in x by scaling in ξ:

$$F(x) \rightarrow F(\xi) \tag{3.3}$$

Quark masses

These are not important when only light (u, d, s) quarks are involved, as in eN or μN scattering. For the chiral symmetric dogma asserts that the effective quark masses at large Q^2 are:

$$m_{u,d} = O(10)\ \text{MeV}$$
$$m_s = O(200)\ \text{MeV} \tag{3.4}$$

These are clearly much less than the nucleon mass, and we are therefore confident that the effects of

$$O\left(m_q^2/Q^2\right)^n \ll O\left(m_N^2/Q^2\right)^n \tag{3.5}$$

and that quark mass effects are negligible by comparison, the corrections to scaling coming from the wholly scaling variable ξ.

Higher twist and non-perturbative effects

These reflect the fact that the target is not a free quark, but is instead a nucleon which is a bag of partons with radius $R \sim 1$ fermi. The partons inside this bag will typically be off-shell, and have finite transverse momenta, on a scale

$$\langle p \rangle = O\left(\frac{1}{R}\right) = O(300)\ \text{MeV} \tag{3.6}$$

We expect, therefore, that all the many different corrections of this type to scaling will be of the order

$$O\left(p^2/Q^2\right)^n = O\left(\frac{0.1\ \text{GeV}^2}{Q^2}\right)^n \tag{3.7}$$

Examples of diagrams which may give corrections of this type are shown in Fig. 11. Figures 11a and b are examples of higher twist effects, while Fig. 11c features the mythical instanton, the simplest possible non-perturbative correction to the naive predictions of perturbative QCD. All these effects are dogmatically asserted to

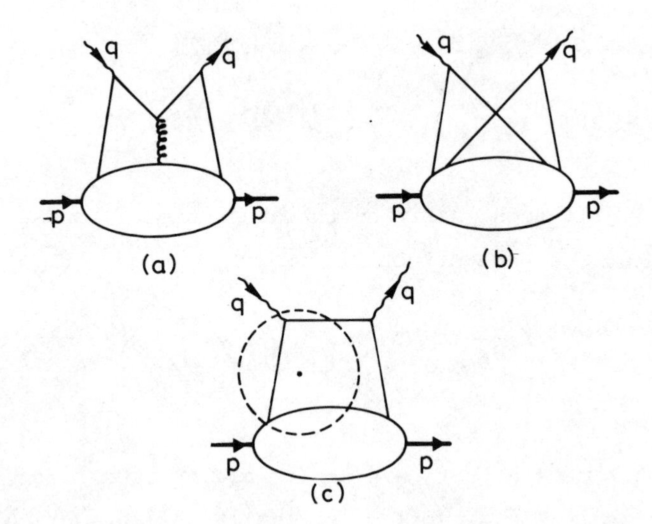

Fig. 11 Some examples of $(1/Q^2)^n$ corrections to deep inelastic structure functions: (a) and (b) are higher twist diagrams, while in (c) lurks the devilish instanton.

be much smaller than $(m_N^2/Q^2)^n$ effects because $\langle p \rangle \ll m_N$.

Combining all these considerations, we see that the dominant subasymptotic corrections due to hadronic mass-scales to the asymptotic perturbative QCD predictions come from ξ scaling and O(4).

BISCOTTE: Hold it a minute! What is this variable ξ and why does he keep saying "oh four" for?

DE ORACLE: Elementary, my dear colleague. You may have noticed that our prophet is British. Thus, he starts all sentence with "Oh!". Upon landing in our imaginary world he said "Oh, four!", referring to his awe at three space and one time dimensions, many more than what most of his abstract and heavenly colleagues could deal with.

BISCOTTE: Oh!, I see, and ξ?

DE ORACLE: The ξ variable embodies trivial kinematical effects of a non-negligible target mass m. Let me abuse the parton model to make this abundantly unclear. Recall Fig. 5 and the words around it. I will repeat the calculation of the inclusive cross-section $d\sigma/dl'^3$ keeping the terms of $O(m^2/Q^2)$ in which we are interested and dropping quark mass effects of $O(\mu^2/Q^2)$. All that changes is the famous δ function of Eq. (1.9): $\delta(p'^2 - \mu^2) \simeq \delta(p'^2)$, with $p' = x_f \mathbb{P} + q$. To halfwit:

$$\delta(p'^2) = \delta\left([x_f \mathbb{P} + q]^2\right) =$$

$$\delta\left[x_f^2 m^2 + 2x_f \mathbb{P}.q + q^2\right] = \frac{1}{2\mathbb{P}.q}\,\delta(x_f - \xi) \tag{3.8}$$

where ξ is the sensible root of the second order equation in x_f:

$$\xi \equiv 2x \Big/ \left(1 + \sqrt{1 + 4m^2x^2/-q^2}\right) \qquad \text{(cf. 3.2)}$$
$$x \equiv -q^2/2\mathbb{P}.q$$

Insert $\delta(x_f - \xi)$ into the expression for the lepton scattering cross-section to obtain:

$$d\sigma/dl'^3 \propto F(\xi) \tag{3.9}$$

By the by, notice our inconsistencies, E.G., $q^2(\text{me}) = -Q^2$ (others). This and many other inconsistencies, we hope, will keep our audience on their toes.

The conclusion is: if quarks are point-like (their short distance interactions are neglected) and the target mass corrections are kept (this could be done more correctly than I did), inclusive lepton scattering scales as a function of ξ. A corollary is that interaction effects should be investigated as deviations from exact ξ scaling (rather than exact x scaling).

BISCOTTE: This transparency (Fig. 12) shows the data for the electron proton structure function measured at SLAC[23] plotted versus Q^2 for fixed x ranges. Perhaps all the scaling deviations at fixed x reflect exact scaling in ξ?

AYATELLIS: No, I am afraid not, my worthy experimental colleague. If we compare the deviations seen in the data in Fig. 12 with curves calculated[38] from scaling in ξ (with the appropriate kinematic corrections to take account of spin) we see that they have insufficient violation of scaling to fit the experimental data. We have

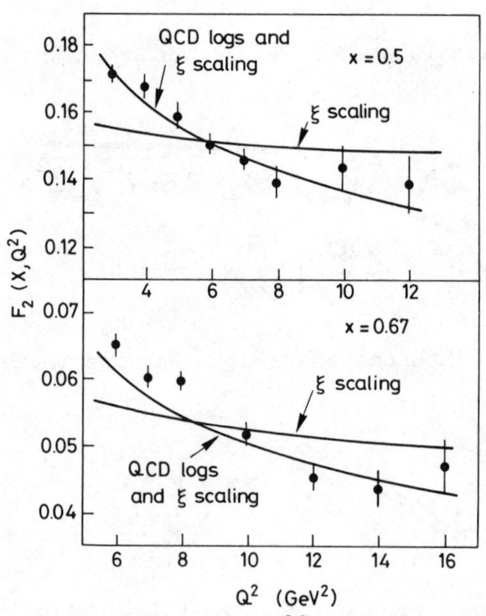

Fig. 12 SLAC ep scattering data[23] compared with curves based on ξ scaling alone[38], and on ξ scaling combined with the QCD perturbative logarithms[25].

already seen that quark mass and higher twist effects are surely negligible. We therefore conclude that something else is needed to fix the experimental data, and this can only be the logarithms of QCD. Indeed, if we compare the data with predictions[25] coming from QCD with both the ξ scaling and the QCD logarithms, then we see that they fit the data very well [*rings bell vigorously*] Indeed, the Lord is just and merciful, and great are the miracles with which He (or She) rewards our faith. QCD is still proved!

PESTILONZIO: [*threatening*] You keep repeating always the same litany: QCD is right!

CHORUS: *ORA PRO NOBIS*

PESTILONZIO: Higher twists are negligible!

CHORUS: *ORA PRO NOBIS*

PESTILONZIO: ξ scaling is relevant!

CHORUS: *ORA PRO NOBIS*

PESTILONZIO: But let's check if it is really true.

A particular type of higher twist effect which is easy to estimate is that related to the existence of a primordial $p_T \neq 0$ in the target. Its presence will affect the definition of the wholly scaling variable ξ leading to a new one, ξ_{p_T}. To first order in p_T^2/Q^2 it can be written as:

$$\xi_{p_\perp} \approx \xi\left(1 + a\frac{p_\perp^2}{Q^2}\right) \tag{3.10}$$

Scaling is now predicted to occur in a new variable, ξ_{p_T}; assuming the structure functions to behave near $\xi \sim 1$ as $(1 - \xi)^A$, one gets a simple relation between $F(\xi_{p_T})$ and $F(\xi)$

$$F(\xi_{p_\perp}) \cong F(\xi)\left[1 \pm O\left(\frac{p_\perp^2}{Q^2}\frac{1}{1-\xi}\right)\right] \tag{3.11}$$

The kinematic background to the genuine dynamical effects is therefore modified by a correction factor. Let me superimpose on your curve (Fig. 13) the uncertainty coming from this "higher twist" effect for $p_T^2 \simeq 2\text{GeV}^2$. When these corrections are accounted for you can hardly conclude - at low Q^2 - you've seen the logs.

BISCOTTE: What is a higher twist?

DE ORACLE: Elementary, my dear colleague. In the real world out there [*points to the audience, realizes his mistake, points in the direction of some different high energy physics lab*] when the data do not quite agree with the prophet's latest prejudice, some feel forced to reanalyze them just once more, just to give them a final little massage. This is called giving the data a higher twist.

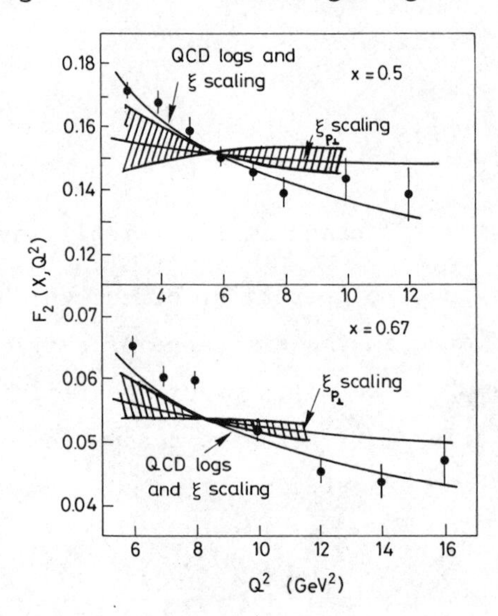

Fig. 13 The region of uncertainty due to some typical higher twist effects related to an intrinsic p_T^2 of quarks of ~ 0.15 GeV is superimposed on the SLAC ep scattering data of Fig. 12.

In our imaginary world, however, the meaning of twist is more straight. Let us rewrite the QCD prediction for a structure function in a somewhat symbolic form:

$$F(\xi,Q^2) = f(\xi) * (\ln Q^2 \text{ correcs.}) * \left(1 + O\left(\frac{M^2}{Q^2}\right) + O\left(\frac{M^2}{Q^2}\right)^2 + \cdots\right) \quad (3.12)$$

The first term on the right-hand side is the exact scaling result, the $(\ln Q^2)$ term refers to perturbative interaction corrections (the formula down to this point is called the "twist two" approximation). But this is not all, There are corrections that vanish as $1/Q^2$, called "twist four"; $1/Q^4$, called "twist six", etc. An example of a specific and easily calculable higher twist correction was already mentioned by Pestilonzio: the effect of a non-zero transverse motion of the hadron constituents $\langle p_T \rangle \neq 0$. Pestilonzio argued that if this effect is not forgotten, structure functions should scale (up to logs) in the variable $\tilde{\xi}(p_T) = \xi(1 + \langle p_T^2 \rangle / Q^2)$. Let me take this for granted and Liz expand for small $\langle p_T^2 \rangle / Q^2$:

$$F \cong F(\tilde{\xi}(p_T)) \simeq F(\xi) + \frac{\langle p_T^2 \rangle}{Q^2} \xi F'(\xi) + \cdots \quad (3.13)$$

Suppose, for the sake of definiteness, that $F(\xi) \simeq (1 - \xi)^A$. Then

$$F \simeq (1-\xi)^A \left[1 - \frac{\langle p_T^2 \rangle}{Q^2} A \frac{\xi}{1-\xi} + \cdots\right] \quad (3.14)$$

The message is clear: no matter how large Q^2 is, one can always consider a ξ sufficiently close to 1 for higher twists not to be negligible. This message can be easily translated into moment language. Again suppose that the structure function scales in $\tilde{\xi}(p_T)$, and take the usual ξ moments of it:

$$\int \xi^n F(\tilde{\xi})\, d\xi = \int \xi^n \left[F(\xi) + \frac{\langle p_T^2 \rangle}{Q^2} \xi F'(\xi) + \cdots \right] \tag{3.15}$$

To proceed, recall the integration theorem by the mathematician Parts (who was also a private in the British army) to get

$$\int \xi^n F(\tilde{\xi})\, d\xi \simeq F_n \left[1 - \frac{\langle p_T^2 \rangle}{Q^2} (n+1) + \cdots \right] \tag{3.16}$$

In moment language: no matter how high Q^2 is, there is always a sufficiently large n moment to take of the structure function; that the higher twist effects will be non-negligible. QCD (*Quasi Completely Disgusting*)

GIULIANO BRUNO: "*Errare humanum est, perseverare diabolicum*" To err is human, to persevere in error is devilish. You keep making the same mistakes all the time. Perturbative QCD formulae can be at best applied to the asymptotic regions, when $Q^2 \to \infty$ and the physics happens on the Light Cone. But what do you do instead? You apply them to values of Q^2 which are so low that one can only get but a glimpse of the Light Cone. Don't you know that particles exist with large spins and masses as large as your values of Q^2? No perturbative world can host such creatures. Thus p_T cut-offs, higher twists and similar gadgets belong to a world which does not, cannot communicate with the naive, perturbative QCD world.

There is no map that you can draw to describe the features of the world, where ξ scaling, among other things, holds. Before doing this you must have a grasp of this world that you simply do not possess[39].

In the deep south of Italy, with a small group of fellow heretics[40] we are able to draw such maps, and in due time I will show how well does it reproduce the known features of these fascinating places.

But before that, let me repeat to you the statement I heard from De Oracle. *Abusus non tollit usum*" You cannot legitimate your use of naive QCD by abusing it!

But now, why don't you Biscotte, reveal how shameless are the blind prophets of QCD by showing us the SLAC data at low Q^2, and how they totally disagree with the predictions of perturbative QCD!

BISCOTTE: Figure 14 shows the structure function F_2^{ep} measured at SLAC[23] plotted versus ξ for small fixed values of Q^2. The bumps are the nucleon resonances and the arrow indicates the position of the elastic scattering peak.

AYATELLIS: [*rings bell*] Indeed, another glorious miracle for QCD! These data are in beautiful agreement with perturbative QCD. These theological scholars [*indicates* PESTILONZIO *and* DE ORACLE] have shown us that the higher twist effects are non-uniform in N and Q^2. They showed that the relative higher twist effects

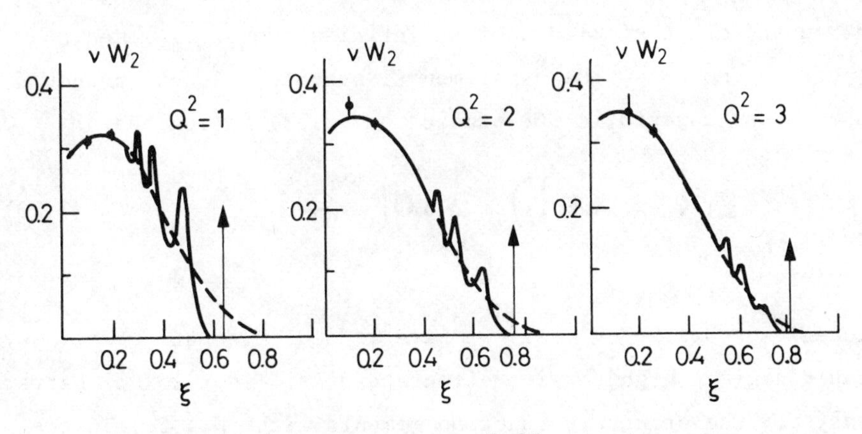

Fig. 14 SLAC ep scattering data[2,23] at low values of Q^2.

$$\frac{\Delta F_i(N,Q^2)}{F_i(N,Q^2)} \approx \frac{aN}{Q^2} \tag{3.17}$$

where $a = O(p^2) \approx 0.1\ \mathrm{GeV}^2$. It is clear that the non-uniformity implies that at any fixed Q^2, the higher twist correction will only be negligible for low N moments:

$$\frac{\Delta F_i(N,Q^2)}{F_i(N,Q^2)} \ll 1 \Leftrightarrow \frac{aN}{Q^2} \ll 1 \tag{3.18}$$

This corresponds to the restriction

$$N \ll \frac{Q^2}{a} \tag{3.19}$$

on the moments for which the higher twist effects may be neglected, and we can reliably use the predictions of perturbative QCD[41].

Since at fixed Q^2 we may only use the first N moments consistent with (3.19), it follows that the asymptotic QCD perturbation theory predictions (2.13) will only work in an average sense. After all, exact reconstruction of the structure functions $F_i(x,Q^2)$ would require exact knowledge of an infinite number of moments. With just N moments, the experimental structure function should average to the asymptotic QCD prediction (2.13) in regions

$$\Delta x = O\left(\frac{1}{N}\right) \gg O\left(a/Q^2\right) \tag{3.20}$$

If indeed $a = O(0.1\ \mathrm{GeV}^2)$ as we have argued, then at $Q^2 \sim 2\ \mathrm{GeV}^2$ one could neglect higher twist effects and still get 25% or better accuracy for the structure function moments with $N \lesssim 5$. In that case the data should average out to the QCD predictions in bins $\Delta x = O(0.2)$. Figure 15 shows that the data do indeed[41] average to

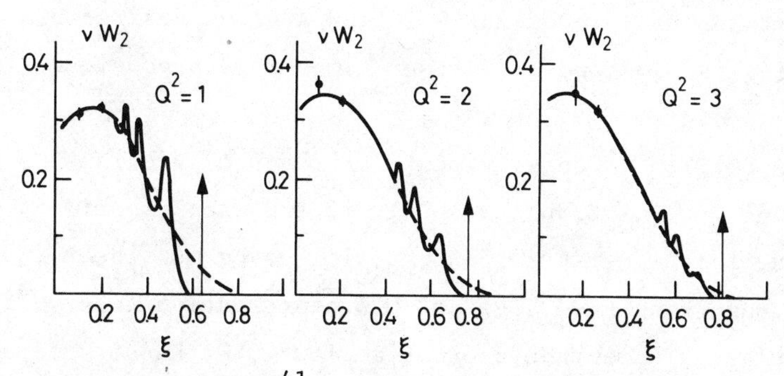

Fig. 15 The extrapolation[41] back to low Q^2 of the asymptotic QCD logarithmic scaling violations. The AYATELLIS claims that these average very reasonably the resonance bumps in the data.

the asymptotic QCD prediction, indicating that the first few moments should indeed agree with the miraculous predictions (2.13) of QCD perturbation theory.

May I remind our worthy experimental colleague that the structure function moments are indeed the most simple, direct and reliable test of the QCD miracles, and ask him to show the moments (even down to $Q^2 = 1\ \text{GeV}^2$ for $N \leq 5$, let's be brave) so that we may check the $(\ln Q^2/\Lambda^2)^{-1/d_N}$ behaviour predicted by QCD?

DE ORACLE: Before we follow your enthusiastic momentum into a momentous discussion of moments, I would stop momentarily for some conclusions.

i) The variable ξ correctly and consistently takes care of target mass effects. Perturbative and higher twist corrections are deviations from exact ξ scaling.

ii) ξ scaling is a correct zeroth order approximation, but is it useful? After all, the difference between exact ξ scaling and exact x scaling is nominally a higher twist effect: $O(m^2/Q^2)$, with m the target mass. Yes, ξ is useful states the QCD party line. Effects other than proton mass effects are of order M^2/Q^2, with $M^2 \sim \langle p_T^2 \rangle$, m_q^2, R_p^{-2}, Λ^2,... all of which are ~ 4 times smaller than the (proton mass)^2. For a given desired precision ξ scaling is expected to be four times more precocious (in Q^2) than x scaling.

iii) The Devil's advocate reminds us that the previous paragraph is full of wishful thoughts. Only if and when experiment measures the size of a higher twist effect will we be confident that we may also be observing the "logs" of QCD perturbation theory.

iv) The heretic reminds us that at low Q^2 the structure function consists of a bunch of resonances, looking nothing like a scaling curve. Baloney, says the believer, take off your glasses so as to see fuzzy in x resolution and you will see the QCD truth (in its present incomplete disguise).

I dare to predict that experimentalists will dare take moments of their data. This, since non-perturbative, confinement and higher twist effects will keep on haunting us, will clarify the situation only in a sense. Each one of you will tighten his anchorage to his respective prejudices.

ACT IV - ARE MOMENTS MAGIC ?

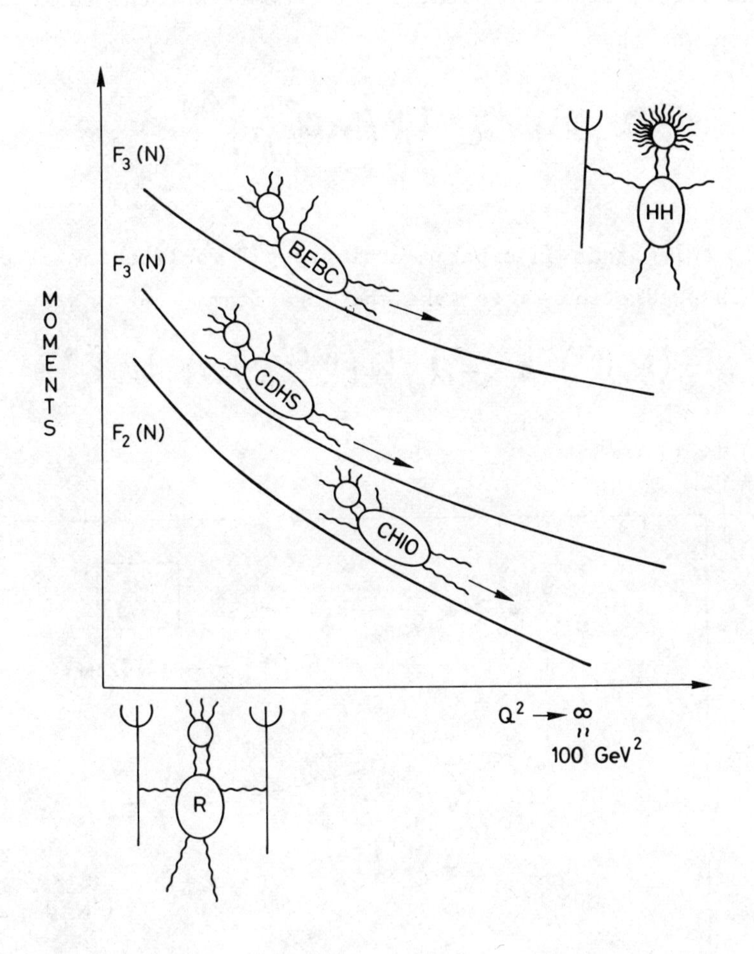

ACT IV - ARE MOMENTS MAGIC?

[BISCOTTE *puts on* BEBC-GGM *moments* (Fig. 16) *and clears throat modestly*]

AYATELLIS: Let us take a few moments together to meditate on the significance of these data. We remember that the basic QCD prediction was that

$$F_i(N,Q^2) \propto \bar{F}_i^N \left(\ln Q^2/\Lambda^2\right)^{-d_N} \tag{4.1}$$

The first thing that an experimental group[42] would think of doing to check this prediction is to take the logarithm

$$\ln F_i(N,Q^2) \approx -d_N \ln(\ln Q^2/\Lambda^2) + \ln \bar{F}_i^N \tag{4.2}$$

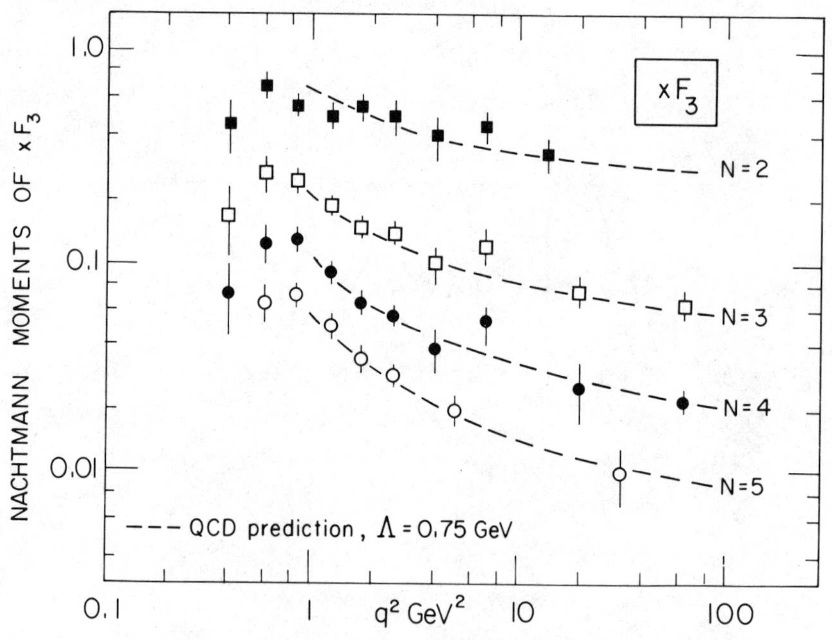

Fig. 16 Moments[12] of the deep inelastic structure function $F_3^\nu(x,Q^2)$ as extracted from BEBC and Gargamelle data.

It is apparent from (4.2) that if one plots against each other $\ln F_i(N,Q^2)$ and $\ln F_i(N',Q^2)$ one should get straight lines with slopes $d_N/d_{N'}$, the ratios of anomalous dimensions[19,20]. I now call on our worthy experimental colleague to bear witness to this miracle.

[BISCOTTE *shows ln ln plots* (Fig. 17)]

AYATELLIS:]*rings bell; everybody whistles the tune of the 1950's pop song "Magic Moments"*] Glorious are the miracles vouchsafed unto the faithful! Indeed the data do lie on straight lines, but so would anything including Ms Claudia Cardinale. But if she were plotted, her straight lines would not have the slopes predicted by QCD. Lo and behold, we see that the slopes do agree with those predicted by QCD. Any other vector gluon theory with small coupling would also predict the same slopes, since the anomalous dimensions are basically just the moments of the bremsstrahlung spectrum. But in lowest order scalar gluons[43] would predict different slopes

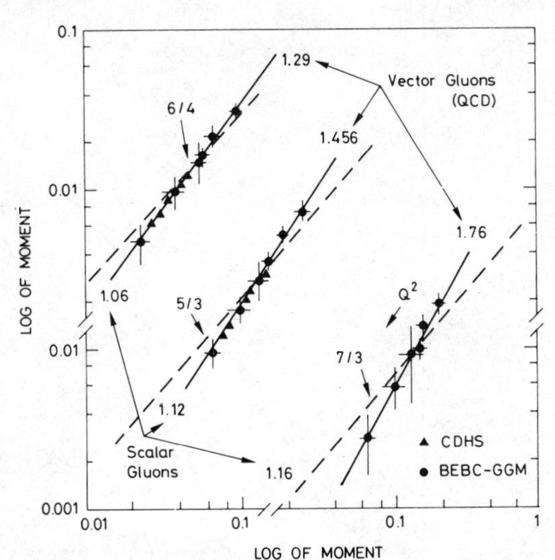

Fig. 17 Log-log plots of moments of $F_3(x,Q^2)$ from BEBC-GGM[42] and CDHS[24] data. The solid lines have the slopes prophesied by vector gluon theories such as QCD, the dashed lines are the heretical predictions of scalar gluon theories in lowest order[43,44].

$$\frac{d_N}{d_{N'}} = \left[\frac{1-\frac{2}{N(N+1)}}{1-\frac{2}{N'(N'+1)}}\right] \quad (4.3)$$

and we see that they do not agree with the data. QCD is proved! [*rings bell, much whistling of "Magic Moments"*]

There is a second check of the formula (4.1) which an experimental group[42] would think of doing. They would take the powers $[F_i(N,Q^2)]^{-1/d_N}$ which according to QCD (4.1) are predicted to rise linearly with $\ln Q^2$:

$$[F_i(N,Q^2)]^{-1/d_N} \propto \ln Q^2/\Lambda^2 \quad (4.4)$$

The plots should rise linearly on a logarithmic scale, and I now call on our worthy experimental colleague to bear witness to this miracle. [BISCOTTE *shows* F_N^{-1/d_N} *plots* (Fig. 18)]

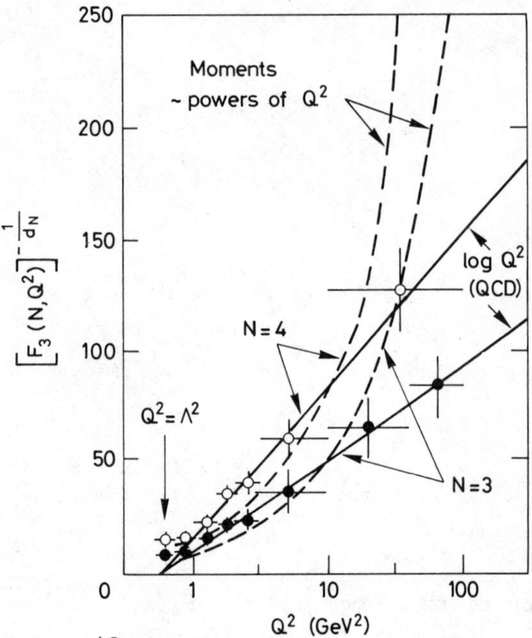

Fig. 18 The moments[42] of $F_3(x,Q^2)$ raised to the powers $-1/d_N$. QCD predicts a linear rise in $\log Q^2$, while other theories in general[44] predict a power rise in Q^2.

See, indeed the experimental data do rise linearly, consistently with (4.4), and behold[44], it is impossible to fit the data on the moments for $Q^2 > 1\ \text{GeV}^2$ with a power-law behaviour of the type

$$\left[F_i(N,Q^2)\right]^{-1/d_N} \propto \left(Q^2/Q_0^2\right)^p \qquad (4.5)$$

as would be expected in a non-asymptotically free QCD theory with a fixed point coupling $g^* \neq 0$. Indeed, the Lord is just and merciful. QCD is proved! [*Much bell-ringing and whistling*]

VOICE FROM THE AUDIENCE: [*A beautiful feminine voice, but the authors being non-sexists*[45] *they can't give further details*]

"Telegram, a telegram just arrived!"

ALL: Is it for me!? Is it from Sweden!?

THREE VOICES SOUNDING LIKE GLASHOW, SALAM AND WEINBERG:

No! No! No! It is for the Ayatellis and from a certain Rabbi Haim Hairy[46].

ALL: [*Various mumblings of unfulfilled expectations*]

DE ORACLE: Do not open it, I clairvoyantly see what it says. It says, [*proceeds with a strong Eastern accent*] "Dear respected Ayatellis", the $\ln F_n$ versus $\ln F_m$ plot used to "measure" anomalous dimensions is MEANINGLESS. I can prove on general grounds that the data must lie close to the QCD predictions. Consider the structure function $xF_3(x,Q^2)$. Let its form at some $Q^2 = Q_0^2$, be approximated by

$$xF_3(x,Q_0^2) \simeq x^{\frac{1}{2}}(1-x)^3 \qquad (4.6)$$

where $1/2[3]$ is a Regge pole (counting rule) guess (prejudice). As thou knowest, in field theories the bulk of xF_3 must move toward smaller x as Q^2 increases, as a consequence of the increasing bremsstrahlung of whatever quanta the theory bremsstrahls. Let me incorporate this truth into a general ansatz:

$$xF_3(x,Q^2) = \mathcal{N}(Q^2,Q_0^2)\, x^{1/2(1+f'(Q^2-Q_0^2))} (1-x)^{3(1+g'(Q^2-Q_0^2))} \tag{4.7}$$

where $\mathcal{N}$ is a normalization factor and the new exponents represent smooth departures from $1/2$ and 3. Take x moments of this ansatz and plot $\ln F_n = (\text{slope } [n,m]) * \ln F_m$. The slope $[n,m]$ factors have upper and lower bounds corresponding to $f' = 0$ and $g' = 0$, respectively. One may compare them with the QCD predictions, for which slope $[n,m] = d_n/d_m$. Take $n = 5$, $m = 3$ for instance. The result is

$$1.71 \text{ (upper bound)} \geq 1.46 \text{ (QCD)} \geq 1.18 \text{ (lower bound)} \tag{4.8}$$

The same pattern recurs for other (n,m) choices. Moreover, the experimental error bars are of the order of magnitude of the difference between my upper and my lower bound.

In plain English, dear Ayatellis, this means: the alleged tests of QCD are not tests of QCD but, at most, tests of the existence of some kind of strong interactions. Destructively yours, R

<u>AYATELLIS:</u> Are moments bound to fit QCD? Please permit me to enter into a polite doctrinal dispute with my eminent rabbinical colleague. It seems to me that his distinguished bounds cannot be exactly correct. A scalar gluon theory[43] in the lowest order predicts 1.12 for the slope of $\ln F_3(5)$ versus $\ln F_3(3)$. This lies slightly outside the "bounds" of my eminent rabbinical colleague, which are

1.71 to 1.18 for this particular combination of moments. A scalar gluon theory should be perfectly respectable so why does it lie outside the bounds of my eminent rabbinical colleague? In fact, we know that all field theories predict a form of evolution of $F_3(x,Q^2)$ which must be more complicated than the $x^{1/2f}(1 - x)^{3g}$ form that he assumed. For it has been shown by distinguished theologians[47] that while field theories do predict a rising power of $(1 - x)$ as $x \to 1$, corresponding to $g' > 0$ in the notation of my eminent rabbinical colleague, nevertheless they predict a <u>fixed power</u> as $x \to 0$, corresponding to $f' = 0$. If the form of $F_3(x,Q^2)$ used by my eminent rabbinical colleague were exactly correct, then all field theories should predict 1.71 for the slope of $\ln F_3(5)$ versus $\ln F_3(3)$, whereas in fact we know that QCD predicts 1.46 and the scalar gluon theory predicts 1.12. Thus, I would humbly submit to my eminent rabbinical colleague that the form of the structure function must be more complicated than he had assumed, and that a measure of the error made in using this simple[48] form to bound may be

$$O(1.71 - 1.46) \text{ or } O(1.71 - 1.12) \approx O(0.5) \tag{4.9}$$

In fact, an appropriate measure of the significance of these log moment slope plots may be the difference between the vector gluon prediction of 1.46 and the scalar gluon prediction of 1.12. We see in Fig. 19 that the data on Nachtpersonn moments - particularly those from the BEBC-GGM[42] collaboration which have smaller error bars because of their longer lever arm in Q^2 - seem to favour vector gluons and QCD over scalar gluons. Furthermore, a modest reduction in the error bars would provide a convincing discrimination between the two classes of theories. QCD lives! [*rings bell, much whistling*]

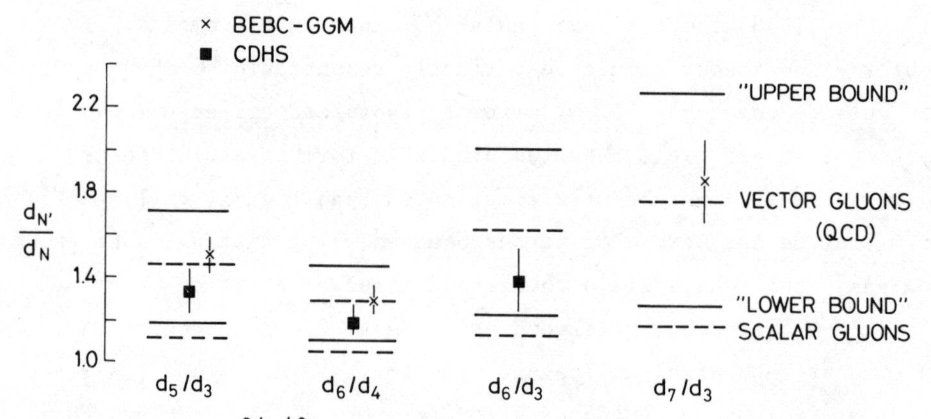

Figure 19 Values[24,42] of the slopes on log-moment plots (cf. Fig. 17) compared with the upper and lower "bounds" of the Rabbi Haim Hairy[46] and the lowest order predictions of vector and scalar gluon theories.

SAME DELICATE FEMININE VOICE: "Telegram! a new telegram just arrived!"

ALL, VOICE, ETC.: Is it for me!? Is it from Sweden!?

FEMININE VOICE: The telegram is from the northernmost confines of civilization and...

DE ORACLE: [*shutting his eyes tight*] Stop! My mind will read it. [*proceeds with a strong accent from somewhere in North Europe*]

"Dear Herr Professor Doctor Thingymachaw Ayatellis! The Gargamelle-BEBC QCD analysis is meaninglesss! No prudent scientist would use data for $Q^2 <$ fünf GeV^2! For $Q^2 >$ fünf GeV^2, the data cannot distinguish a QCD log from the Q^2 power law behaviour of fixed point theories! Foldingly, spindlingly and mutilatingly yours! DESY!!!

GIULIANO BRUNO: Long live the Reformists!, !!!.

DE ORACLE: For those of you who do not understand the unknown acronym "DESY" I will translate it. It stands for: "*Diese Ellises of Schweiz are Ydiots!*"

AYATELLIS: Please permit me to enter into a polite doctrinal dispute with the eminent Lutheran pastor[49]. Why does he only use data with $Q^2 > 5\ GeV^2$? It is not surprising that he cannot discriminate between a $\ln Q^2/\Lambda^2$ and a $(Q^2/Q_0^2)^p$ dependence if he restricts himself to $Q^2 > 5\ GeV^2 >> \Lambda^2$, the typical hadronic scale. The BEBC-GGM analysis used data down to $Q^2 = 1$ or $2\ GeV^2$. We saw in the previous act that this is perfectly reasonable as long as one restricts oneself to low values of N. Perturbative QCD is powerful enough to justify its applicability in this range of Q^2. We recall that higher twist and other $O(1/Q^2)$ effects are expected to be

$$O\left(\frac{p^2}{Q^2}\right)N \approx \frac{0.1N}{Q^2} \tag{4.10}$$

for the Nth moment, and this is small already for $Q^2 \gtrsim 1$ or $2\ GeV^2$ for $N = 3,4,5$. In any case, we have seen that the data show no signs of higher twist effects in the range of Q^2 down to $1\ GeV^2$. Let me urge my eminent Lutheran colleague to be brave and try applying QCD perturbation theory down to $Q^2 \sim 1\ GeV^2$: it works! In other words, there is a moral: don't get your knickers in a twist about higher twist! QCD lives [*rings bell, much whistling*]

PESTILONZIO: [*impetuous*] Wait! Wait! If the Ayatellis blows his own trumpets, I will blow my higher twist trumpets (Fig. 20)! They are superimposed to lowest twist predictions, according to the following formula:

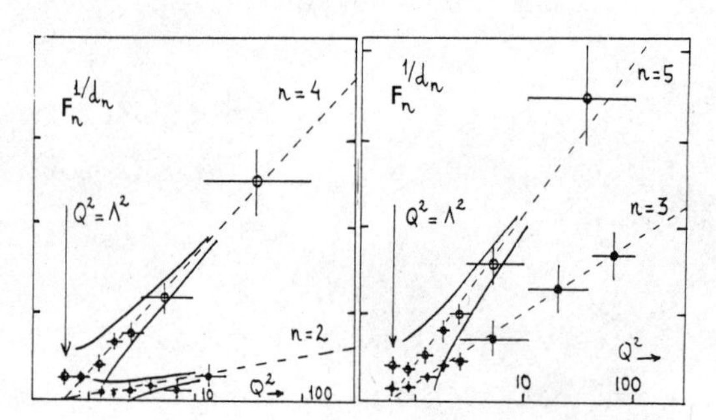

Fig. 20 Higher twist trumpets against QCD predictions for massaged moments of xF_3 structure function of Fig. 18.

$$F_n(Q^2, p_\perp^2) \simeq F_n(Q^2)\left[1 \pm \frac{\langle p_\perp^2 \rangle}{Q^2}(n+1) + \ldots\right] \tag{4.11}$$

with $p_T^2 \sim 0.2\ \text{GeV}^2$.

No conclusion can be reached with low Q^2 data if you do not fit simultaneously the value of the strong coupling constant scale Λ and the unknown magnitude of higher twist effects.

AYATELLIS: Well, what you say may be true, though there are still no signs in the data that higher twist effects are significant for $Q^2 > 1\ \text{GeV}^2$. Nevertheless, to sate your appetite for a pound of flesh, let us proceed to higher Q^2, where the higher twist effects will have died away, and perturbative QCD will be revealed in all its puritan glory!

PESTILONZIO: [*happy*] If power-like corrections may be tamed by going to higher Q^2, still other measurable corrections are expected to survive at higher energies. The expression for moments of structure functions beyond leading log can be written as:

$$F_n(Q^2) = F_n(Q^2)_{\text{Leading Log}} \cdot [1 + C_n \alpha_s(Q^2) + \dots] \tag{4.12}$$

The size of the corrections is fully calculable[50] within renormalization improved perturbation theory. Do the experimental data agree with QCD predictions?

AYATELLIS: Do not be disheartened, oh ye of little faith. There is a simple trick which enables us to exorcize the devil of these higher orders without too much trouble. The corrections take the form

$$F_i(N, Q^2) \approx (\ln Q^2/\Lambda^2)^{-d_N} \left[1 + \frac{A_N}{\ln Q^2/\Lambda^2} + \frac{B \ln\ln Q^2/\Lambda^2}{\ln Q^2/\Lambda^2} + \dots\right] \tag{4.13}$$

and we see that they all vanish asymptotically, with the dotted terms suppressed by $O(1/\ln Q^2)^2$ relative to the leading term. Since the $\ln\ln Q^2/\Lambda^2$ term is reasonably constant in the present range of Q^2 we can absorb the A_N and B correction terms into a redefinition of the Λ parameter for each different value of N:

$$F_i(N, Q^2) \approx (\ln Q^2/\Lambda_N^2)^{-d_N} \left[1 + O\left(\frac{1}{\ln Q^2}\right)^2\right] \tag{4.14}$$

Exhausting calculations by perservering bands of monks[50] in the Calvinist city of Geneva and elsewhere reveals that the second order QCD predictions are that the Λ_N should increase with increasing N. The manner of this increase is fixed, but the absolute values of the Λ_N cannot be predicted by QCD.

I now call on our worthy experimental colleague to bear witness to this miracle of the increasing Λ_N.

BISCOTTE: Figure 21 shows the value of Λ obtained from fits to the Q^2 dependence of the moments of xF_3 from the CERN neutrino experiments[24,42]. Note that in the BEBC analysis elastic and quasi-elastic contributions to the structure functions have been included explicitly while in the CDHS analysis the structure functions have been extrapolated for $x > 0.7$ assuming a $(1 - x)^3$

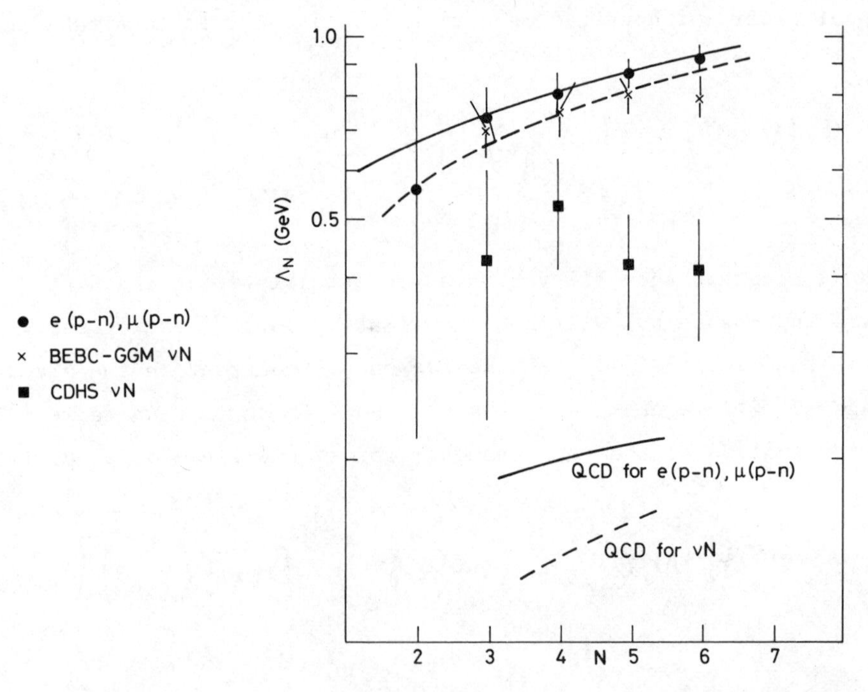

Fig. 21 Values of Λ_N extracted from eN and νN experiments, compared with the N dependence expected from second order QCD calculations[51]. The absolute vertical scale of the theoretical curves is arbitrary, but their slopes and the distance between them are fixed.

form. On the other hand, electromagnetic radiative corrections which have been accounted for in the CDHS analysis have been ignored in the present BEBC-GGM analysis. Also shown are the values of Λ obtained from fits[52] to the moments of the structure function $F_2^{ep} - F_2^{eN}$ measured in electron and muon scattering experiments. Note that the various data sets extend over different ranges of Q^2.

AYATELLIS: Behold, just and merciful is the Lord. QCD is proved yet again. [*rings bell, muffled whistling*] The experimental data in Fig. 21 are clearly not in disagreement with the Λ_N dependence expected[51] from second order QCD. There are problems with the difference between the BEBC-GGM and CDHS values of Λ_N being different, but much of this arises from differences in the methods of analysis. Indeed the (ep - en) and (μp - μn) show tantalizing signs of agreement with the expected trend of increasing Λ_N. QCD lives! [*rings bell*]

PESTILONZIO: [*disgusted*] Two remarks: First, everybody who has worked with Deuterium knows that - contrary to what chemistry manuals say - Deuterium STINKS! You can smell it mainly in the large x region, crucial to the Λ_N analysis. At high x the ill-understood highe moentum components of deuteron wave functions play an important role. I would like the xF_3 data - besides those on $F_2^{ep} - F_2^{en}$ - to agree with theoretical predictions*. Second, in absorbing the $O(\alpha_s)$ corrections into a N dependence of the scale Λ you have carefully hidden that the corrections themselves are sometimes 50% of the "leading" result. Why am I to believe first or second order perturbation theory when it converges so slowly?

GIULIANO BRUNO: At last "*In datis veritas*"; the truth shines in the data. You can all see that the naive QCD parametrization does not work! And this is not an experimental pitfall; CDHS and

* and among themselves!

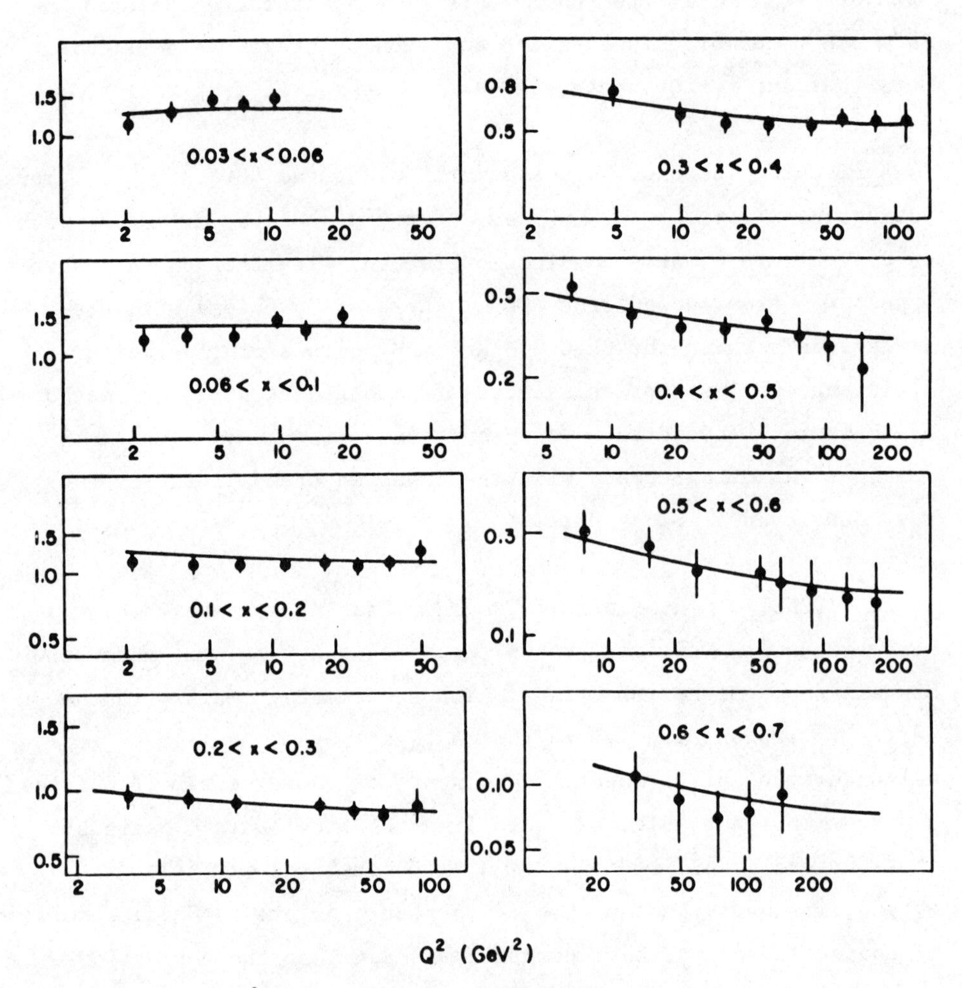

Fig. 22 $F_3(x,Q^2)$ versus CDHS. A vindication of BRUNO and his band of heretics[40].

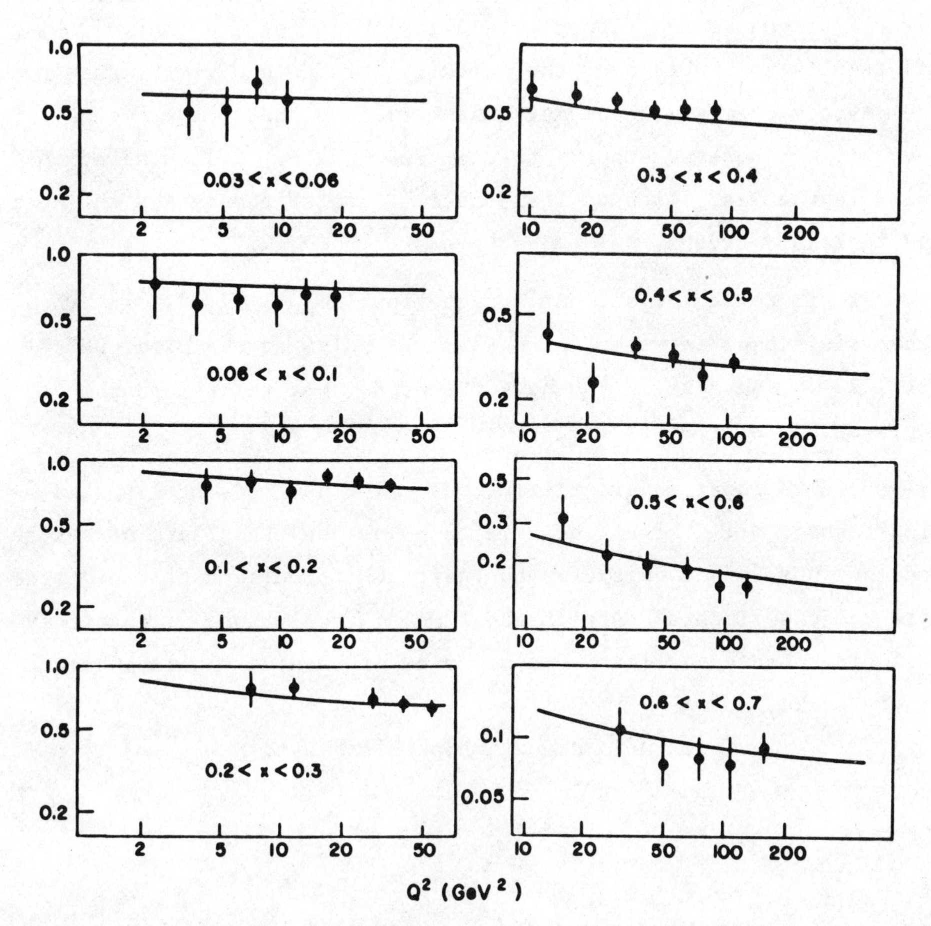

Fig. 23 $xF_3(x,Q^2)$ versus CDHS. Another vindication of BRUNO and his band of heretics[40].

BEBC-GGM data have different intervals in Q^2 and they agree when they overlap.

Thus Nature, through Biscotte data reject the false prophecies of the naive QCD priests. The simple truth that we do not live in a world of quarks and gluons (the dilapidated world of naive QCD) has finally triumphed over the perturbative QCD clockwork, showing the poverty of its mechanisms. And more than that, experiments have given beautiful support to the ideas that my fellow heretics and I in the Deep South of Italy have[40] used to comprehend the deep and inelastic truths.

In Figure 22 you see our expectations for $F_2(x,Q^2)$ in νN scattering compared with CDHS data. xF_3 also agrees with CDHS (Fig. 23)! And finally you have the proof that there is nothing magic about the moments (Fig. 24)

But I harbour no great illusions! The new Aristotelians, I am almost sure, will for a long time to come, deny the facts of Nature and continue to add epicycles to their QCD formulae until they agree with the experimental data of the time. My only hope is that they will discontinue the loathsome practice of burning their opponents at the stake!

"*Spes ultima dea*", hope dies last.

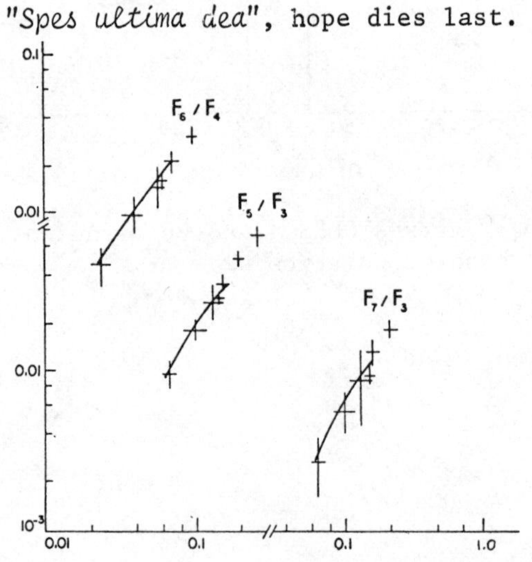

Fig. 24 Nothing magic about moments, according to BRUNO[40].

DE ORACLE: Indeed they will not burn them at the stake; that is, as long as they keep on producing theories of fire sausages. But, time for some conclusions.

Moments are not magic, just tricky. But they are a simple language to show the limitations of theory and to define the goals of experiment.

Theory is limited by our inability to give reliable estimates of higher twist effects. No doubt twist four is a simpler problem to solve than confinement. Thought and model estimates would be highly welcome.

Calculations to higher orders of perturbation theory than the $O(\alpha_s)$ corrections we discussed, present a formidable computational problem. Moreover, they are likely not to change the qualitative behaviour of the predictions, much as the $O(\alpha_s)$ corrections did not change the qualitative behaviour of the leading log predictions. In this sense, $O(\alpha_s)$ or $O(\alpha_s^2)$ corrections are very hard to measure in a single type of experiment. Notice that this is the only sense in which anything makes physical sense. Should deep inelastic scattering remain the most precise observable compatible with QCD, it may be regarded as the place where α_s (or the parameter Λ in a fixed convention) is best measured. Then one may move to other experiments (e.g., hadron annihilation into lepton pairs) to try and observe $O(\alpha_s)$ corrections. These corrections to the "leading"comparison of experiments are large in an absolute and convention independent sense. Large means easy to measure.

The recommendations to experiment were probably clear in our discussion: do not ignore higher twists, measure them at relatively low Q^2 or relatively high moments. Then move to higher Q^2 or smaller moments, knowing the size of what you neglect. Try to fit to perturbative logs whatever you see there. Then tell the world. By the way, hire a theorist for every dozen or so experimentalists to help you with your QCD radiative corrections (for the smaller QED ones, this is not unheard of).

Quantitative QCD is in its infancy, and work is needed to distinguish it from sausages. Though, as an Oracle, I know which way the future will move, I fear the reaction of my co-actors, and I invite them to offer their own conclusions.

CONCLUSIONS

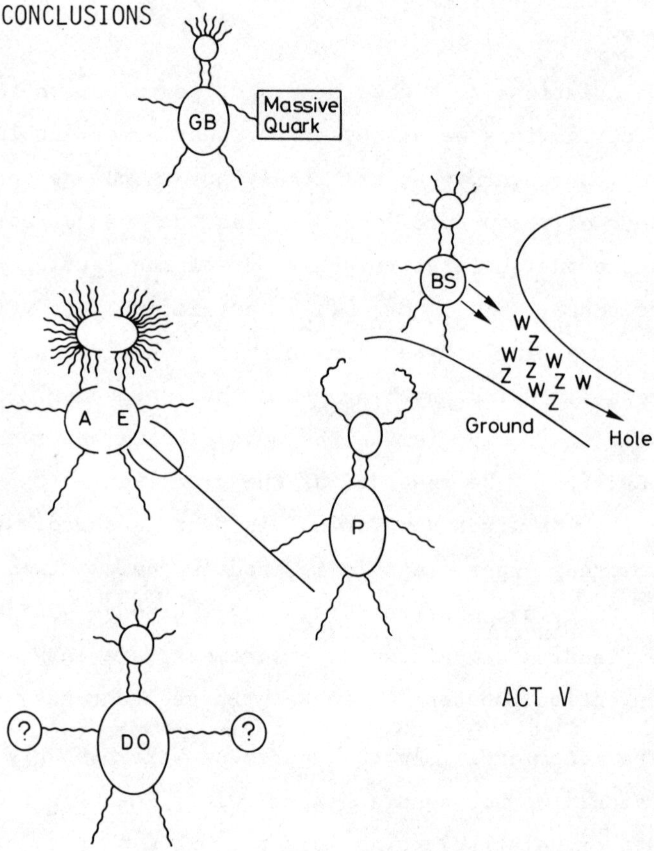

GB
Massive Quark
BS
W Z W Z W Z Z W W Z
Ground
Hole
A E
P
? DO ?

ACT V

ACT V - CONCLUSIONS

GIULIANO BRUNO: As the Roman General who brought the "*Pax Romana*" to France used to say: "*Libenter homines id quod volunt credunt*"; men rather believe what they strongly desire. So it will take some time to realize finally that

† PERTURBATIVE QCD IS DEAD †

So my dear friends, start looking for real alternatives; look for heresies! One is already there; it is now strong and healthy. Study it and try to reject it. For such is the only practicable road to find out whether the QCD Lagrangian is the Lagrangian of the real world.

BISCOTTE: Well, it seems to me that all this analysis of deep inelastic scattering data is a fiasco. When I got into the neutrino business I wanted to study weak interactions. Maybe now I will go and hide myself in a hole in the ground and look for them there.

AYATELLIS: The conclusion is clear. QCD is obviously correct and its miracles have been convincingly proven by the experimental data (or if not, it soon will be).

Under these circumstances, it no longer requires a tremendous act of faith to believe in QCD, and I must go seek some new more far-out cult to believe in. I have resolved that I will go and work on grand unified theories of the strong, weak and electromagnetic interactions. These theories clarify and create many amusing new mysteries. One of the miracles they predict is the ultimate instability of matter: protons should have a finite lifetime of $10^{30 \pm 2}$ years. This raises many interesting theological problems on which I shall meditate. For example:

IS GOD MADE OF PROTONS?

DE ORACLE: The Ayatellis hopes She is, so that he may replace Her on Her throne.

PESTILONZIO: [*diabolic and emphatic*] We have seen that the conclusions of QCD prophets are - at least - premature. Thus I will not move to the contemplation of

GOD'S ULTIMATE TRUTH
**rand *nified *heories*

(it would be a bit inconsistent with my role). I will rather keep wandering through hell and wondering about the diabolic mechanism that makes *Quod Confinement Dictat* consistent with simple (minded?) perturbative predictions.

DE ORACLE: We have discussed only a limited set of observables: inclusive structure functions. They are so far the most quantitative tests of QCD, though much less quantitative than we would like them to be. We have repeatedly emphasized that "QCD is only predictive for the first few moments..."; we meant "...of structure functions". QCD is a theory and will not die like all other "theories" of the strong interactions did. It will do it differently.

Consensus on a "correct" theory cannot arise just from the analysis of structure functions. It can arise, if at all, from the overall picture. It is fair to say that no theory other than QCD scores as many stars in its confrontation with the properties of hadrons [*flashes Michelin Guide to Hadrons so fast that nobody has time to digest it*].

Notice that not even the heretic criticized the beautiful

but the often blind industry of exploiting it via a perturbation theory that does not lead to confinement. While QCD is not fully understood nor established, alternatives are healthy. The orthodoxy is so generally followed that we do not have alternative theories to convert the comparisons to experiment into a healthier game. In this, the heretic could not be more right.

I predict that, to improve upon QCD inspired experiments, many a well-meaning experimentalist will dream of a bigger and better apparatus. This comes as no surprise since, if Sigmund Freud was right, most of you [*points at the audience*] share that dream.

And now, ladies and gentlemen, we are ready for the conclusive 600M Swiss Franc question:

CAN ONE TELL QCD FROM A HOLE IN THE GROUND?

A tentative answer is reflected in the Figure below:

D I S C U S S I O N S

CHAIRMAN: Prof. A. De Rújula

Scientific Secretaries: M.B. Gavela, A. Nicolaidis and F. Rapuano

DISCUSSION 1

- GUENIN:

What is exactly proven in the affirmation that gauge group plus renormalizability uniquely defines the interaction?

- DE RUJULA:

Gauge invariance and renormalizability severely restrict the possible choices of interaction Lagrangians. For example:

$$\frac{g}{m} \bar{\Psi} \sigma_{\mu\nu} \partial^{\mu} \psi A^{\nu}$$

is gauge invariant but not renormalizable. The list I gave of renormalizable polynomial four-dimensional flat space theories is complete. The proof is done by exhaustion.

- KLEINERT:

So your only arbitrariness in your gauge theories is the choice of not gauging the baryon number.

- DE RUJULA:

The choice of $SU(3)_c$ is the simplest one but not the only possible one to make, with just the information that baryons are made of 3 quarks. Once you choose the local gauge group and the representation to which quarks belong there is no extra arbitrariness.

- KAUL:

Why not SO(3) as the underlying color gauge group instead of SU(3)?

- DE RUJULA:

That is a very good question.

- ZICHICHI:

Asymptotic freedom is lost for more than 2 flavors.

- DE RUJULA

That is a very good answer.

- JAFFE:

SO(3) is not an acceptable color group because all the representations are self conjugate. Therefore $Q - Q$ and $Q - \overline{Q}$ interactions are identical. If $Q - \overline{Q}$ bind to form mesons, then $Q - Q$ bind to form fractional baryon number hadrons with the same masses as ordinary mesons.

- DE RUJULA:

This is what I will write down as my answer.

- KLEINERT:

What are the experimental limits on the size of a coupling to baryon number?

- DE RUJULA:

The best limits on such a long range force come from the Eötvos experiment. It is done, say, for atoms containing different number of neutrons and the same number of protons and electrons. The limits are extremely stringent: the strength of the corresponding charge is many orders of magnitude smaller than the electric charge.

- GINSPARG:

Depending on your criterion for "naturality" you can choose the coupling to baryon number as small as you like and in particular small enough to be in accord with Eötvos experiment, so I don't agree with your objection. $U(3) \simeq SU(3) \otimes U(1)$ so the coupling constants are in principle independent.

- DE RUJULA:

My argument only applies to the choice of U(3) as gauge group, not $SU(3) \otimes U(1)$. Should the group be U(3), the audience and I would interact strongly. There is experimental proof of the contrary.

- MIKENBERG:

SU(3) à la Gell-Mann happens in the QCD Lagrangian just by chance. Is that also true for isospin, and if so, why does it work so well?

- DE RUJULA:

It works so well because the mass difference of the light quarks happens to be so small on the hadronic scale.

- SISKIND:

If the underlying gauge group is not the simple choice SU(3) but rather a more complicated one (e.g. 10 colors), what would be the experimental evidence?

- DE RUJULA:

The prediction for $\pi^0 \rightarrow \gamma\gamma$ would be wrong by a large amount (a factor of $10^2/3^2$ for your choice). Another constraint is

$$R = \frac{(e^+e^- \rightarrow \text{hadrons})}{(e^+e^- \rightarrow \mu^+\mu^-)}$$

which is predicted to be proportional to the number of quark colors.

- CORBO':

You need 3 colors to get the right factor for $\pi^0 \rightarrow \gamma\gamma$. But with respect to what? How do you define the decay constant f_{π_0}?

- DE RUJULA:

If I remember correctly, the prediction for $\pi^0 \rightarrow \gamma\gamma$ is normalized to the observed charged pion decay rate. PCAC and very light quarks are extra necessary assumptions, as well as an understanding of the Steinberger-Adler-Bell-Jackiw anomaly.

- PREPARATA:

I do not believe that theoretically the $\pi^0 \rightarrow \gamma\gamma$ has been shown beyond reasonable doubt to be related to the Adler anomaly. The problem is that the relation of the amplitude which is given by the Adler anomaly and the physical $\pi^0 \rightarrow \gamma\gamma$ amplitude is a very difficult one to be established (see my 1972 Erice Lectures). Thus I think that the symmetry of the baryon wave function is a much better case for having 3 colors.

- OLIENSIS:

Why don't non-perturbative effects spoil the derivation of the axial anomaly and $\pi^{\circ} \to \gamma\gamma$?

- DE RUJULA:

Assume they don't. $\pi^{\circ} \to \gamma\gamma$ works, so this is an indication that the derivation (true, for once, to all orders in perturbation theory), can be trusted.

- SCHMIDT:

If one is dissatisfied with the $\pi^{\circ} \to \gamma\gamma$ argument in favour of 3 colors and thus SU(3), but approves for a solution of the spin statistics dilemma, is it not possible to use parastatitistics, or is this the same as SU(3) color?

- DE RUJULA:

I think that parastatistic is just another name for "hidden" non-gauged color. For $\pi^{\circ} \to \gamma\gamma$, R, and Drell-Yan, the parastatistic approach would reproduce the standard color-counting results.

- VANNUCCI:

The argument for color coming from the Drell-Yan cross-section is not established experimentally. Because of uncertainties in the sea distribution in the case of pp reactions there is still room for an extra factor of 3.

- DE RUJULA:

True: of all these numbers ($\pi^{\circ} \to \gamma\gamma$, R, D-Y), the weakest is the one concerning Drell-Yan. Moreover, there are big next to leading QCD corrections to the naïve Frell-Yan calculations of the order of 50-100%.

- PREPARATA:

You should not talk about Drell-Yan but about muon pairs, because it is muon pair production and not Drell-Yan that have been measured.

- DE RUJULA:

A Drell-Yan pair has been measured to weight an estimated 150 KG.

- PETRONZIO:

Transverse momentum distribution of muon pairs produced in hadron-hadron collision, can provide also a good test of QCD. For example, the average values of $P_{\perp}^{2}$ is expected to scale - just by naïve dimensional analysis - with the dimuon invariant mass squared Q^{2} according to

$$\langle p_{\perp}^{2} \rangle \simeq \alpha_{s}(Q^{2})\, Q^{2} f(\tau) + \text{const.}$$

where $\tau \equiv Q^{2}/s$, s being the total c.m. energy squared; the constant takes care of a non-perturbative component of $\langle P_{\perp}^{2} \rangle$. The function $f(\tau)$ can be calculated in perturbative QCD. The results are in agreement with experimental data (see for example the rapporteur talk by G. Altarelli at the EPS conference, Genève 1979).

- PATON:

The spin-statistics problem can be solved by introducing a discrete permutation symmetry rather than a continuous symmetry.

- DE RUJULA:

Yes, but it would not afford the aesthetic pleasure of being able to gauge the symmetry.

- HANSSON:

Is there any argument excluding the existence of new fermions belonging to 6 or 8 or higher representations of $SU(3)_c$?

- DE RUJULA:

Not in principle. However, the quarks in ψ/J and Υ are not "quixes" (that is quarks belonging to a 6 representation), nor queights, nor quens. The way to test this is to consider

$$\frac{\Gamma(\psi \to 3\text{ gluons} \to \text{hadrons})}{\Gamma(\psi \to e^+e^-)} \sim$$

If constituent quarks are traded for quixes, say, this quantity acquires an extra factor 7^2 which would destroy the present reasonable agreement between experiment and the model predictions.

- MARTIN:

I want to make an advertising statement that I shall show in my lectures, that there is experimental evidence that the forces between quarks, including the b quark, are flavor independent. Hence the b quark is a quark like the others.

- KLEINERT:

The existence of an additional 6 of quarks is desirable to make supermultiplets fit existing fundamental particles.

- DE RUJULA:

Even in super-schemes I do not think that you are forced to introduce these representations; but the confidence level for this answer is not higher than 25%.

- ANDERSON:

Could or will you or someone explain in more detail what the potential is of dimuon production by hadrons for the verification of QCD?

- DE RUJULA:

I will not lecture on that. The experimental situation is not yet entirely satisfactory. In the case of proton beams, although we have the structure functions as measured in deep inelastic scattering, the antiquark content of the nucleon is poorly known (there are, for instance, significant radiative corrections with relatively large uncertainties). For pion beams, the structure function is that measured in the same experiment so we have no check. It is in antiproton-nucleon experiments that we can make neater tests of QCD; this is a valence-valence quark fusion and the valence quark distribution are best known.

- PREPARATA:

After having reminded people that the application of naïve QCD to μ-pairs production is on a much stakier footing than the deep inelastic scattering, I would like to point out that the experimental observed independence of $\langle P_{\perp}^2 \rangle$ from Q^2 at fixed s is something very difficult to understand within the naïve framework. Why? Because if $\langle P_{\perp}^2 \rangle$ depends on s and not on Q^2, the only variable which tells us when we are probing short distances, this means that this effect must be a consequence of the structure of hadronic interactions, irrespective of the distances at which we probe them. But then I cannot understand how the same results can emerge from two types of dynamics which are alleged to be very different: the soft hadronic interactions and the hard quark-parton naïve QCD behaviour.

- PETRONZIO:

I'd like to stress again that one can calculate the corrections to the estimate for the total Drell-Yan cross section which are obtained by using quark and antiquark distribution functions extracted from deep inelastic electron and $\nu\ (\bar{\nu})$ experiments. They predict an experimental rate which is larger than what one would expect from the lowest order approximation: experimental data indicate a deviation from the lowest order prediction in the expected direction and of the right order of magnitude.

DISCUSSION 2

- PREPARATA:

The solution

$$\alpha_s(Q^2) = \frac{\alpha_s(\mu^2)}{1 + b \ln \frac{Q^2}{\mu^2}}$$

is only a particular solution of differential equations. This comes from a special way of writing the β function.

- COLEMAN:

Yes, it is in the leading log approximation. It is a unique solution for small values of the running coupling constant. Indeed, to my knowledge, it is only the renormalization group equations that enable us to justify the summation of leading logarithms.

- CHRISTOS:

I would just like you to repeat your arguments of the last ten minutes of your lecture. In particular, I am interested in the physical significance of what you showed - also your b, which is defined from $\beta(g) = -bg^3 + O(g^5)$.

- DE RUJULA:

As I stated

$$\alpha_s(Q^2) = \frac{\alpha_s(\mu^2)}{1 + b\alpha_s(\mu^2) \ln Q^2/\mu^2}$$

where $b = \frac{33 - 2n}{12\pi}$, $n =$ number of flavors. It is just a mathematical identity to equate this to $\alpha_s(Q^2) \equiv (b \ln Q^2/\Lambda^2)^{-1}$

this will be so if

$$\Lambda^2 = \mu^2 \exp\left\{-1/b\,\alpha_s(\mu^2)\right\} \tag{1}$$

Now, had we chosen another renormalization point $\tilde{\mu}^2$, we may wonder whether the corresponding Λ

$$\tilde{\Lambda}^2 = \tilde{\mu}^2 \exp\left\{-1/b\,\alpha_s(\tilde{\mu}^2)\right\}$$

is such that $\Lambda = \tilde{\Lambda}$,so that the physical quantity Λ is, as it should be, renormalization point independent.
Equate Λ to $\tilde{\Lambda}$ and you will obtain:

$$\alpha_s(\tilde{\mu}) = \frac{\alpha_s(\mu)}{1 + b\alpha_s(\mu)\ln \tilde{\mu}^2/\mu^2}$$

which is our starting equation. The Lord is merciful.
What this tells us is that QCD is a one parameter theory even though we seem to have the two parameters, $\alpha_s(\mu)$ and μ . This is because these two parameters occur only through the combination (1). This formula determines the coupling constant in the $O(\alpha_s)$ approximation. Since we cannot predict everything at this time, we don't know what the actual value of Λ is. Saying that this parameter is 500 MeV, is not unlike saying that in QED the coupling constant, at zero momentum transfer, is $\alpha = 1/137$.

- COLEMAN:

When you first think of QCD with all quark masses set equal to zero you think of a one parameter family of theories. The only parameter is the coupling constant g. As Alvaro explained, however, defining the renormalization coupling constant is a tricky business. Because if you define it in a naïve way, on the mass-shell, which in a fully massless theory means all external momenta equal to zero, you find yourself sitting on π particle thresholds and then you might find everything diverging or vanishing for purely kinematical reasons. Thus, it is desirable to define g away from thresholds at some non-zero mass. Therefore, it seems that we have two parameters, the coupling constant and the mass scale at which it is defined. However, one might wish to choose another mass scale, still describing the same theory, QCD. We can draw curves in the g, M plane

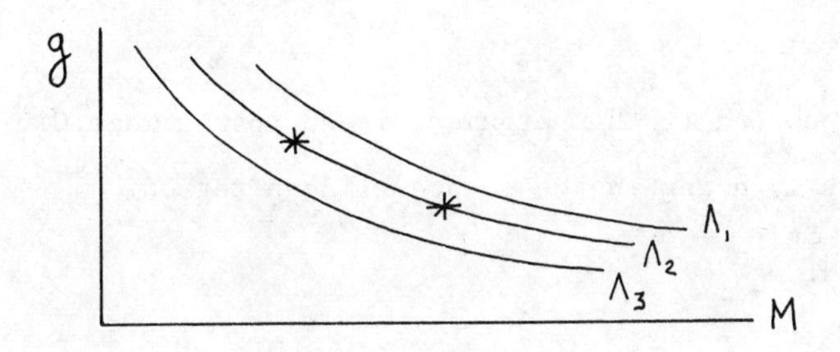

such that any two points on the same curve represent the same theory. Also in the same region where g is small, one can compute what these curves are like, by means of ordinary perturbation theory.

- GINSPARG:

I'm confused about this insistence on the deep-euclidean regime. We know this was required in the original Callan-Symanzik equations to eliminate the inhomogeneous term, but subsequently mass-dependent renormalization group schemes have been developed and shown to agree with the original Callan-Symanzik equations to lowest order.

- COLEMAN:

There are several ways of approaching this problem: you could start up with a theory with masses and go to sufficiently high momentum where the masses can be ignored; then you replace the theory with a fully massless one, and then you analyze the asymptotic behaviour of the massless theory.

Equivalently, you can choose not to separate the problem into two parts that way and just study the behaviour of the massive theory and go to the region where the masses can be ignored without bothering to replace it with a massless theory. It's only a matter of taste.

- GINSPARG:

I'm concerned if this argument is an answer when Giulano worries about actually being in the deep-euclidean region.

- COLEMAN:

In e^-N or νN collision we are in the euclidean region. Whether deep or shallow is another question. All the momenta are not in the euclidean region, but we are studying the behaviour of the Wilson coefficient function of the particular term in the operator product expansion and the argument of that Wilson coefficient function is in the deep euclidean region.

- PREPARATA:

I would like to stress that the light cone expansion, which is so dear to my heart, is a very strong extra assumption to be added to the renormalization group analysis that you have carried out so lucidly in your lectures here in Erice a few years ago. Your analysis is in the deep euclidean region and you cannot get

out of it without some other means. But nobody has ever been able to prove the light cone expansion beyond perturbation theory, and I have the strong suspicion, even though I cannot prove it, that in a confined theory things are much more subtle than in a perturbation theory.

- PATRASCIOIU:

If you look at a $\lambda \phi^4$ zero mass theory but you put a ϕ^3 term, is it there still dimensional transmutation?

- COLEMAN:

Well, dimensional transmutation per se is a phenomenon that takes place only in purely massless theories; you must throw away the ϕ^3 term; that is dimensionful, like a mass term.

- SZWED:

In the simplest QCD diagram, there are at least three masses involved. Which one plays the role of Q^2 in $\alpha_s(Q^2)$?

- COLEMAN:

The correct way of calculating the process is using the operator product expansion and the renormalization group equations. There, looking at the moments of the structure functions, factorisation occurs and the only mass which is involved in the part we calculate is Q^2. If you look at this graph

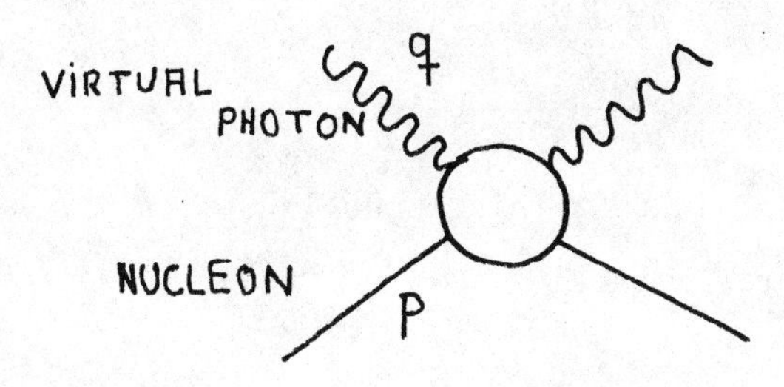

before I do the momentum analysis, there are two four-momenta, q and P: the momentum of the virtual photon and the momentum of the nucleon. When I do the momentum analysis I can separate the graph like this

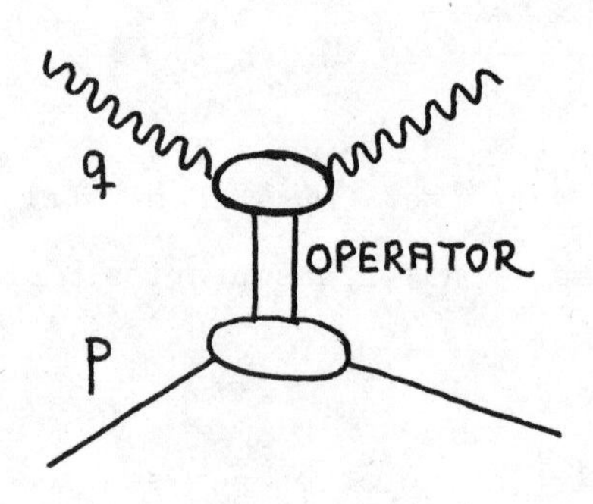

and break it in the product of two parts, the upper one only depends on the running coupling constant, while the lower one stays as an unknown coefficient in both the theoretical and experimental analysis.

- CORBO':

Is there some symmetry principle that assures you that a theory like $\lambda\phi^4$ remains massless?

- COLEMAN:

No, there is not such a symmetry principle, but this is a case I can consider; I can consider the case of $\lambda\phi^4$ with vanishing meson renormalized masses.

- CORBO':

Could it happen that this theory acquires a mass when I sum up higher order diagrams?

- COLEMAN:

This could happen. This is what we believe to happen in QCD.

- OLIENSIS:

What is a leading logarithm?

- DE RUJULA:

When you compute something in a renormalizable four dimensional field theory, not only you do have powers of the coupling constant, but often also powers of the logarithm of the momentum scale at which you are working. A leading log is defined as the biggest power of the logarithm that occurs at a given order of perturbation theory. Sometimes it is possible, and the renormalization group does it elegantly for you, to sum perturbation theory to all orders in the approximation that you only keep this dominating log terms.

- OLIENSIS:

Are there processes where you can do a leading log resummation but don't have a renormalization group equation? How do experiments confirm the predictions of these techniques?

- DE RUJULA:

Yes. Sometimes the two ways work, sometimes leading logs are the only known way of deriving some results. The Q^2 behaviour of fragmentation functions in processes like $e^- p \to e^- \pi X$ was first attacked by leading logs and could have been an answer to your question, but the results have been recently justified, I believe, in the formal approach by A. Müller.

- COLEMAN:

For example, QCD at small Q^2 cannot be treated with leading logs.

- PATON:

What is the justification for assuming that the $\alpha_s(Q^2)$ in deep inelastic processes is related to the α_s that occurs in gluon decay of charmonium? Even forgetting that Q^2 is no longer euclidean, and bound state troubles, what is the Q^2 of α_s in the gluon decay calculation of quarkonium?

- DE RUJULA:

The question is whether you can factorize your ignorance of the bound state problem and get rid of it by taking ratios of observables, to be then consistently predicted in perturbation theory. First, one can ask the corresponding question for ortho-positronium. It can be shown to next to trivial order that the bound state decay is proportional to

$$|\psi(0)|^2 \cdot (\sigma_{e^+e^- \to \text{photons}})$$

where $\psi(0)$ is the wave function at the origin and σ is the threshold annihilation cross section to first non trivial order in perturbation theory. The same factorization has been shown to occur for paracharmonium to first order of perturbation theory, the order beyond

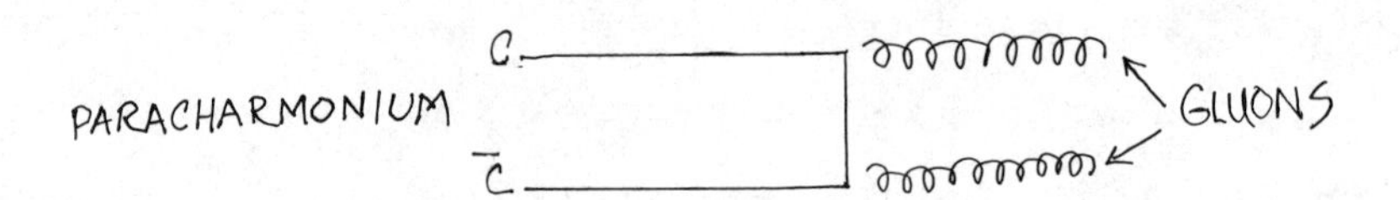

Nobody knows if factorization occurs to all orders in perturbation theory, even in QED. If it does, ratios like $\psi \to$ hadrons/$\psi \to e^+e^-$ are computable in perturbation theory and the bound state dynamics drops from them. Only one mass scale occurs in the calculation of these inclusive observables: M_ψ.

- PREPARATA:

And what is the mass scale of the final gluons?

- DE RUJULA:

Zero!

- KLEINERT:

So, you are not in the deep euclidean region.

- DE RUJULA:

No. Then I am now doing perturbation theory, not the formal approach, but I know that I am computing infrared-safe, threshold singularity-safe observables.

- JAFFE:

In paracharmonium annihilation into two gluons beyond leading order in QCD, we might expect the appearance of large logarithms of the form log $(m_c^2 R^2)$ where m_c is the quark mass and R is the bound state radius. So I understand your answer to Paton as saying such terms do not appear?

- DE RUJULA:

Yes. To the order to which it has been studied the $J^P = 1^-, 0^-$ annihilation is local, and no such terms arise.

- PREPARATA:

In J/ψ decay into three gluons, you are testing the wave function at around a distance of the order of the ρ compton wave length, so you are not at very short distance. But, of course, the usual objection to naïve QCD makes the situation even worse.

- DE RUJULA:

Not true, the annihilation of a heavy quark of mass M take place at distances of order $\sim 1/M$, at least in a non-relativistic context.

- KLEINERT:

What coupling constant do you use for this zero mass gluon?

- DE RUJULA:

Let me answer the question in a somewhat simpler process. Consider $e^+e^- \to q\bar{q}$, whose squared amplitude is

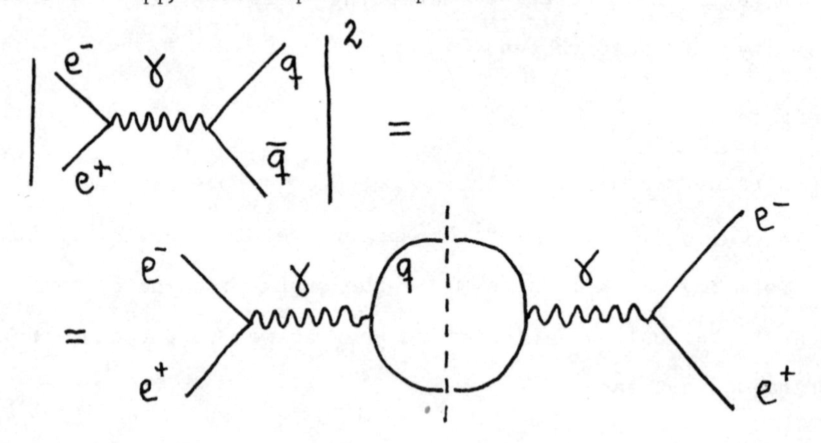

the next order in perturbation theory is

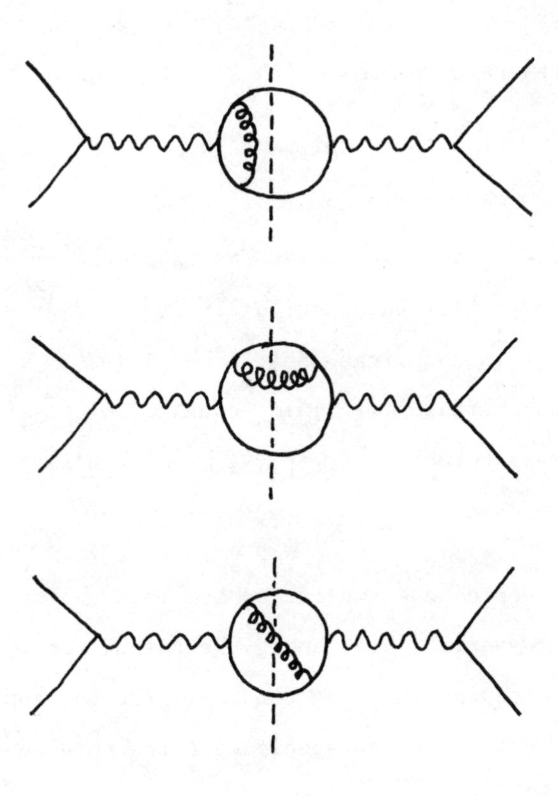

But the infrared divergencies cancel between these diagrams (Kinoshita, Lee, Nauemberg theorem), so you are no to be worried about the fact that small momenta appear. What you are dealing with in this inclusive calculation are momenta of order Q, the only scale in the infrared finite cross section $\sigma_{tot}(Q)$. The zero mass of quarks and gluons, the small momentum scale, has vanished from the answer.

DISCUSSION 3

- SZWED:

Is the list of renormalizable theories given during the lecture complete? Is there a theorem which says that these are the only ones?

- COLEMAN:

This is the list of known renormalizable theories:

1) Polynomial interactions in $P(\phi)$, degree ≤ 4.
2) Yukawa couplings $\bar{\psi}(a + b\gamma_5)\psi\phi$.
3) Gauge fields minimally coupled to conserved currents.
4) Mass terms: ϕ^2, $\bar{\psi}\psi$, $A_\mu A^\mu$ for U(1) gauge fields only.

- GINSPARG:

In fact, people have considered the most general theories and found that a convergent high energy behaviour requires very delicate relationships between coupling constants for the appropriate cancellations. The theories so specified correspond identically to gauge theories with names given by spontaneous symmetry breaking.

- COLEMAN:

They did not use renormalizability. This is the work of Norton, Cornwall and Tiktopoulos and also Llewellyn-Smith. They demand that tree graphs are well behaved in a certain sense when all external momenta get large simultaneously. It was not clear to what extent this criterion is equivalent to renormalizability (it is obeyed by all known renormalizable theories).

- ADKINS:

Has Hepp's theorem been formulated in a way that is general

enough to cover all types of renormalizable theories and all methods of regularization?

- COLEMAN:

I believe it has, but it is a question you should ask an expert. Hepp's original theorem was quite restricted; it required a very strong regularization and subtractions at zero external momenta. Since then, the machine has been improved considerably, and I understand that it works for other regularizations (including dimensional regularization) and for subtractions at arbitrary momenta.

- HANSSON:

How do you decide, for an actual physical process, what Q^2 to put into $\alpha_s(Q^2)$? More specifically, is it only in the ψ or Υ decays that you take $Q^2 = M^2_{resonance}$?

- COLEMAN:

In principle, you can choose any Q^2. However, some choices are better for practical calculations.

- DE RUJULA:

In the above process

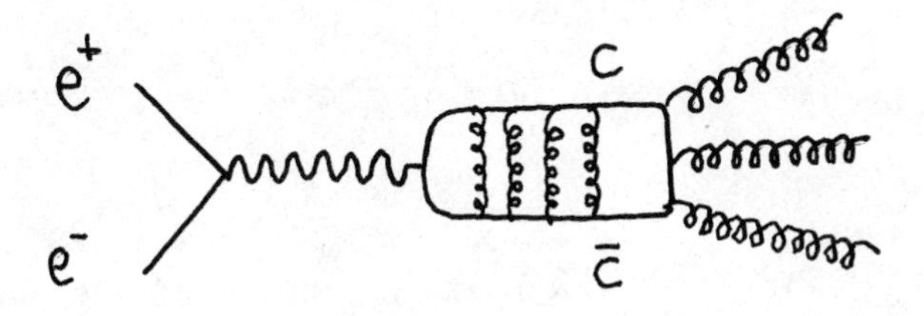

what you are actually computing is

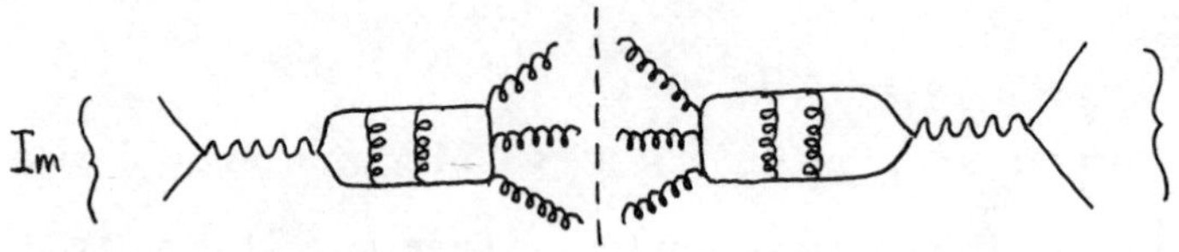

This is a infrared-safe process, because of the cancellation order by order of mass singularities, and that means that there is just a mass-scale, the Q^2 of the photon.

By using $\alpha_s(Q^2)$ we take into account and tame the effects of higher order graphs at the leading log level.

The exact meaning and value of the Q^2 coefficient (whether one means $2Q^2$ or $Q^2/2$, say) can only be ascertained after one has calculated the next to leading log corrections.

- JAFFE:

Do you mean simply that Kinoshita's theorem applies to charmonium decays?

- DE RUJULA:

No. The theorem is not proven in general in QCD. It is proven for some processes to some orders of perturbation theory.

- MARTIN:

This question is directed to Sidney Coleman: everybody, including you, likes to predict masses or mass differences. However, we are satisfied when we know that a theory (take QCD) is renormalizable; i.e. has a priori arbitrary masses.

- COLEMAN:

What is not fixed by the theory are the electric charge, the parameters of the Higgs system (including its Yukawa couplings), the Weinberg angle, and the mass scale of strong interactions. But, in principle, we can predict hadron mass scale ratios in the limit of vanishing "bare quark masses" (more strictly, vanishing Higgs Yukawa couplings).

- PREPARATA:

I have doubts on the way of calculating mass differences. People compute the contribution of one gluon exchange, but why not two or more gluon exchanges? The strong coupling constant should be big because the momentum transfer is very small.

I also want to remind you that, apart from Q^2, there is another mass scale, that is the heavy quark mass (charm in the case of ψ) that should not be neglected.

- DE RUJULA:

The effect of a non-zero quark mass, in a particular renormalization scheme, on the coupling constant α_s is explicitly known. Well below the threshold for the production of these quarks, their effects can be neglected in agreement with the Appelquist-Carrazzone theorem. Very far above threshold, the coupling constant is what is supposed to be for a given number of massless quarks. However, as one goes, for example, from the mass of charmonium to the mass of bottonium, the mass of the charmed quark becomes relevant steeping up logarithmically in its effect on the coupling constant.

- GINSPARG:

We can make a stronger statement. If one uses a dimensional regularization scheme, the differential equation for the coupling constant, a dimensionless parameter, does not involve dimensionful parameters like masses.

- COLEMAN:

The question is whether you can use renormalization group techniques near threshold where there are two mass parameters.

- PREPARATA:

I think we cannot. The Λ you take in the Callan-Symanzik formalism has to be much bigger than all masses. In all these calculations you are sitting on something almost at rest.

- LOW:

You rest on the smallness of the coupling constant. In QED, for the decay of positronium virtual electron-positron pairs are created and they are important. Similarly, if you have here creation of charmed particles, I do not see how you can ignore them.

- DE RUJULA:

I am saying that if you do define the physical coupling constant in terms of the ratio R

$$R = 3 \Sigma_i Q_i^2 \left(1 + \frac{\alpha_s}{\pi}\right)$$

smoothed over the energy resolution $\langle R^{exp} \rangle = \int_0^\infty \frac{R \, dQ'^2}{(Q'^2 - Q^2)^2 + \Gamma^2}$

bigger than the resonance width, then you know how the coupling constant α_s, in this definition, varies as you cross thresholds of quark masses.

- LOW:

The fact that the formula exist is clear. The problem is whether or not there are regions where the calculation is only correct when the coupling constant is intrinsically weak.

- DE RUJULA:

First, you must convince yourself in the case of timelike observables (it is often not a proof, but at most a very reasonable argument), that there are observables that you can calculate in perturbation theory provided that the coupling constant is really small.

Would you continuosly be in the deep euclidean region, I do believe that you would exactly know what you are doing with this only assumption.

In the timelike region you need more assumptions. For example, the ratio

$$\frac{(e^+e^- \to \text{hadrons})}{(e^+e^- \to \mu^+\mu^-)}$$

is given in QCD at lowest order by

$$R = 3\sum_i Q_i^2 \sqrt{1 - \frac{4m_i^2}{Q^2}}\ \theta\left(\sqrt{Q^2} - 2m_i\right)$$

where the square root is a threshold factor, Q being the mass of the photon. You have something like

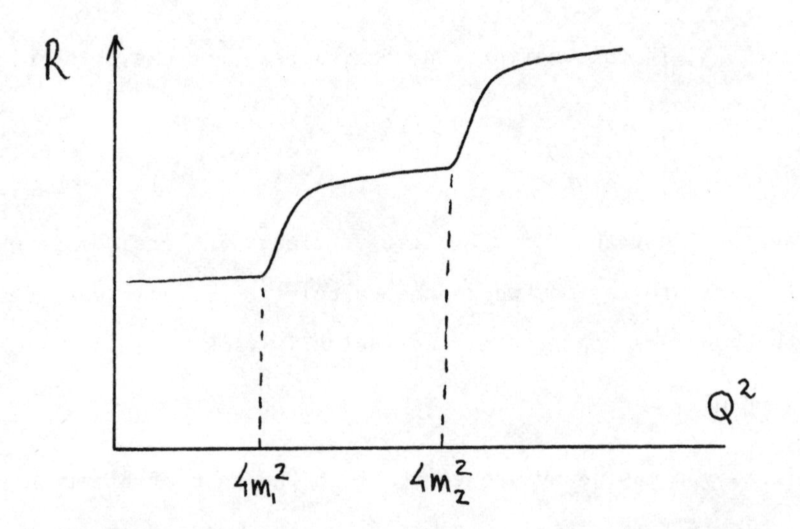

It is well known that the experimental result is different because of the existence of high peak resonances (ψ , Υ , ...). For example, at the ψ mass the calculated ratio is off by a factor of several thousand.

Should we be in QED, we should understand this fact by solving the bound state problem: close to a threshold, there are coulombic singularities that do bind the outgoing particles

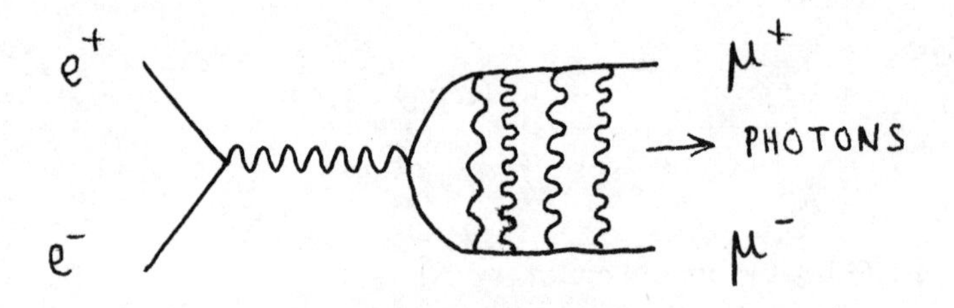

These corrections turn out to be, near the quark threshold, of the type

$$1 + \alpha/v$$

v being the quark relativistic velocity

$$v = \sqrt{1 - \frac{4 m_q^2}{Q^2}}$$

All orders in $1/v$ must be summed to obtain the Coulombic bound states.

In QCD you do not know how to solve the bound state problem. But perturbation theory gives a hint on how to avoid this problem. This can be done by moving from the real axis by a small amount $i\Gamma$ in Q^2. Thus

$$v = \sqrt{1 - \frac{4 m_q^2}{Q^2 + i\Gamma}}$$

and one avoids getting close to the region sensitive to bound state singularities. Fortunately, there are dispersion relations that allow you to extend your information from the data to the complex plane. If one believes that the danger of bound state singularities can be avoided in QCD as one knows it can in QED, one is in good shape: there is a "truce line" in the complex Q^2 plane, not far from the timelike axis, where one can compare perturbation theory with "smoothed" data, analitically continued with help of a dispersion relation. The "smoothed" data and QCD perturbation theory are in very satisfactory agreement.

About the problem of mass differences I would like to make a general remark. The contribution that De Rujula and the naïve QCD theorists take is of a "hyperfine" type, i.e. for baryons

The reason to do this is obviously the success of this analysis in, for example, atomic physics. But, in order to understand its physics, we can adopt a different representation of the self-mass of a given system, e.g. hydrogen, through dispersion relations:

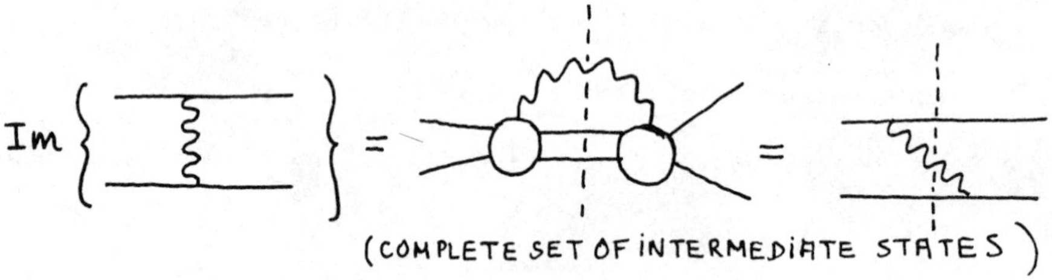

Im { } = =

(COMPLETE SET OF INTERMEDIATE STATES)

$$\Pi(E) = \int \frac{dE'}{E'-E} \; \mathrm{Im}\,\Pi(E')$$

$$\mathrm{Im}\,\Pi(E) = \frac{\Gamma}{2}\,, \quad \mathrm{Re}\,\Pi(E) = -\Delta M$$

Γ is the width of the state

ΔM is the mass of the state

For e.m. interactions, everything is perfectly well understood, but for gluons, due to their non-existence as physical states, the whole picture does not make any sense, unless we introduce into this calculational scheme some major theoretical modifications.

What makes sense, on the other hand,is to keep to the quantum mechanical self-mass effects and to try to compute the real parts of the self energy functions. This has been done by G. Fogli and myself and by N. Tömquist and M. Roos for mesons, with extremely good results. In order to understand qualitatively why higher spin states have higher mass, we must realize:

(i) That the sign of ΔM turns always out to be negative for low mass states.

(ii) For higher spin states the size of ΔM is smaller, because of, among other factors, the statistical factors.

Finally, for the Λ°-Σ mass difference if we look at the diagrams

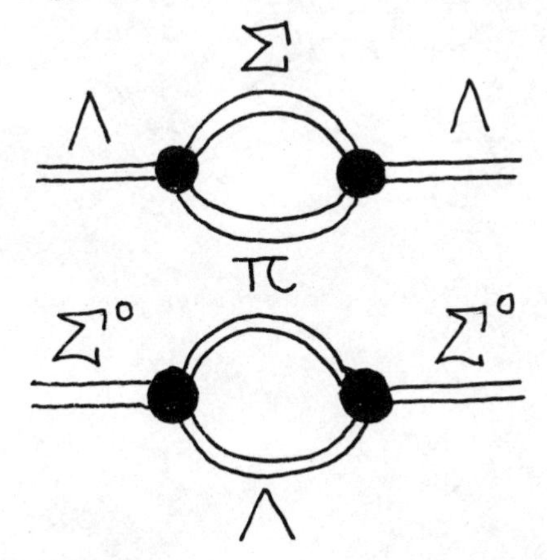

again the isospin factor makes the Λ contribution in absolute value bigger than the Σ° contribution.

DISCUSSION 4

- OLIENSIS:

Could you explain the work developing a renormalization group equation for the wave function?

- DE RUJULA:

Recently, some people have proven that one can derive in QCD the power behaviour of the elastic form factor for a bound state of a certain number of quarks. The naïve expectation has been known for many years. It states that assuming the wave function of the bound state, say the proton, is soft (it does not have large components of order Q^2), one needs to exchange at least two hard quanta (one in the case of $Q\bar{Q}$ mesons) in order to be able to recombine the same object in the final state.

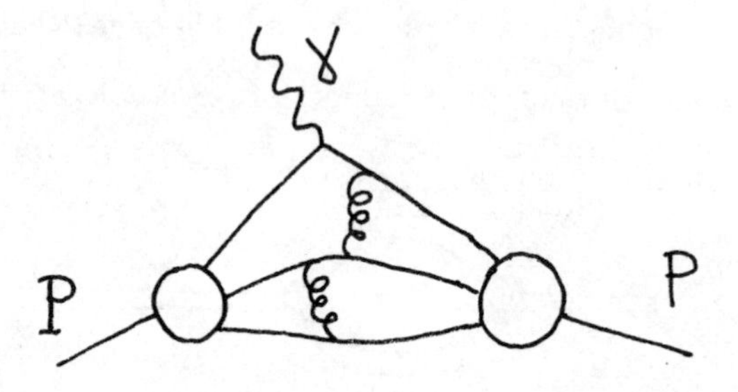

By simple power counting from propagators one obtains

$$F_P \sim \frac{1}{(Q^2)^2}$$

Brodsky, Lepage and others have recently shown that in QCD the pion form factor behaves as

$$F_\pi \sim \frac{\Sigma_n c_n (\ln Q^2/\Lambda^2)^{d_n}}{Q^2}$$

where d_n are calculable anomalous dimensions.

This is a weaker statement than the corresponding structure function Q^2-evolution statement, in that the value of the form factor at a given Q^2 is not enough to determine the unknown coefficients. The reason for that is that this is not an incoherent process as the structure function one is supposed to be. You have to know a lot about the overlap of the wave functions with themselves after each quanta has been emitted or absorbed. This requires a deep knowledge of the high momentum components of the Bethe-Salpeter equation for the bound state. Perturbation theory can only help you in the hard part, that is, where all the quanta are far from their mass shell. It has been shown that, first, the hard wave function is perturbatively calculable in terms of the soft one and, second, that this knowledge is enough to predict the evolution of a form factor up to the unknown C_n coefficients in the above formula. To summarize: the impact of all I quoted is that the power behaviour of a form factor is calculable in QCD, up to logarithms whose power behaviour is also known.

- ROMANA:

Is there a way for deriving the X distribution of the sea quarks?

- DE RUJULA:

Some people do that starting from the valence quark distribution that is known to behave not unlike $\sqrt{X}(1 - X)^3$ and by assuming that sea quarks are generated by radiative processes from valence quarks

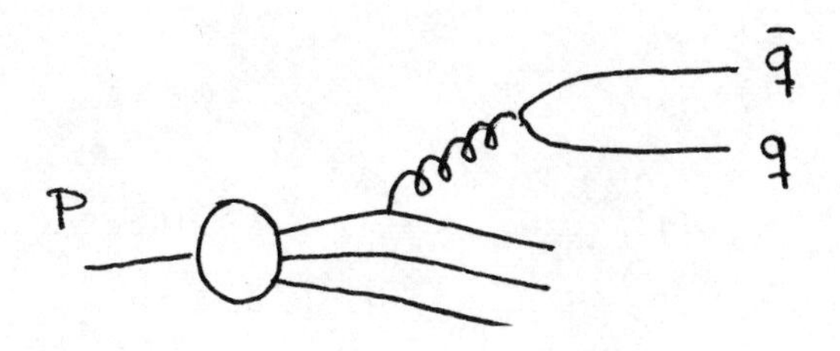

This gives results for the X distribution of the sea behaving like $(1 - X)^7$. All of this is based on the assumption that a certain

starting point Q_o^2 where you can suppose the hadron to be made of valence quarks only, is big enough for perturbation theory to make sense. That does not seem to me a realistic assumption.

- RAPUANO:

Is a relation like the Drell-Yan still valid for the elastic form factors that in QCD acquire a logarithmic dependence on Q^2?

- DE RUJULA:

Yes, several recent papers prove that a Drell-Yan West relation follows from QCD. I think this is not to be taken too seriously for technical reasons. The point is that it is not always true that the argument of the running coupling constant is Q^2. You are, as you go near $X = 1$ constrained in the phase space, so that the relevant momentum is not Q^2 but something like $Q^2(1 - X)$ and therefore I think that a perturbative calculation is not easy to defend because you are forced to configurations where momenta are indeed small.

- DAUM:

Is it right that the main experiment in QCD is that showing the three gluon vertex?

- DE RUJULA:

Yes.

- DAUM:

Is there a theoretician having an idea on how to measure that vertex?

- DE RUJULA:

The cleanest way of doing it is tough: consider toponium decay into 3 gluons where the three gluons should be hard enough for the analysis of the decay plane to be clean. Suppose also that you have a high number of events; then you could measure the T-odd observable

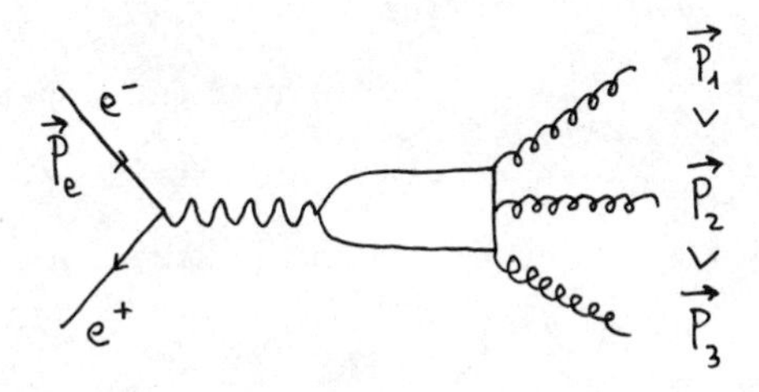

p_1 and p_2 being the hardest and next to hardest gluon momenta; p_e being the electron momentum.

This observable changes sign when you change the sign of all three momenta.

There is a theorem that states that a T-odd observable is proportional, in perturbation theory, to the absortive part of the final state interaction.

The lowest order graphs that give a non-zero contribution to that observable are gluon exchange and the four gluon vertex:

The average value of that observable is proportional to:

$$\langle (\vec{p}_1 \times \vec{p}_2) \cdot \vec{p}_e \rangle \propto$$

This ratio is given by some number times $\alpha_s(Q^2)$ where the α_s refers to the gluon couplings because nothing appears in the ratio but the 3 and 4 gluon interactions. This is, in principle, the cleanest way I know of testing the nonabelian nature of color charge.

- DAUM:

You have also proposed the energy flow analysis and we have done it at DESY and found it is not a good way to do it. We can separate between 3 gluon Monte Carlo events and two jet events but we cannot separate from phase space.

- DE RUJULA:

Being an optimist I think that your own interpretation of your own phase space model is pessimistic. First, it does not fit the thrust distribution and to compare momentum flow data at fixed thrust with the phase space Monte Carlo you have to "renormalize" each thrust bin to the correct number of events. Second, the large thrust (very jetty) events are predicted by the phase space Monte Carlo to have <u>more</u> of a three-lobed shape than either the data or the three gluon model (into which the number three is built in). I find the tendency of your phase space Monte Carlo to produce a strong "threeness" very suspicious. I have my theory of what is going on, that I will not put in writing because no prudent person should simply criticize someone else's Monte Carlo.

- CHRISTOS:

I would like to ask you about a paper by E. Witten concerning the charmed quark contribution to the proton structure functions. With a heavy quark flowing around we have two large mass scales, Q^2 and $4m_c^2$. Now E. Witten calculated the charm quark contribution using the product expansion which amounts to only considering diagrams like

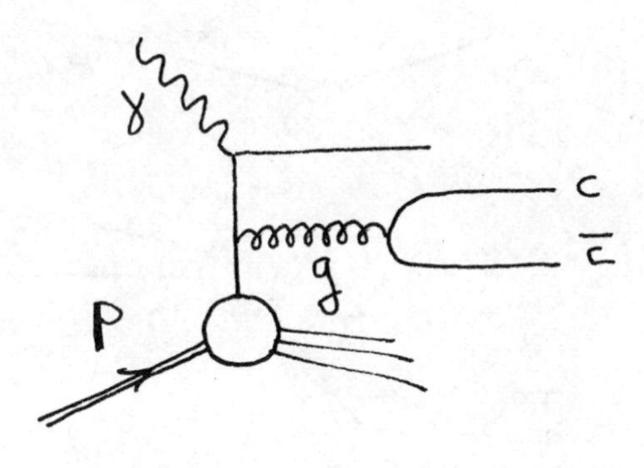

Why does not include diagrams like

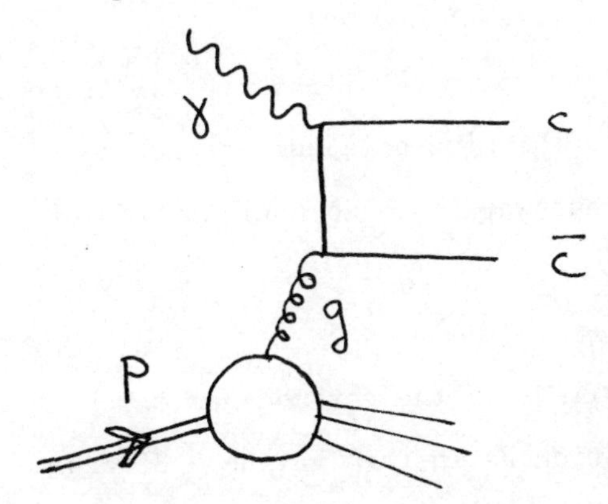

- DE RUJULA:

Perhaps because this is not a contribution to the structure function defined as the scattering of the photon on the heavy quark. The paper you refer to I read more than a couple of years ago. My memory is not good enough to remember its details.

- PAUSS:

How can one measure the structure function of the pion?

- DE RUJULA:

There are at least two ways of doing it. One of them is

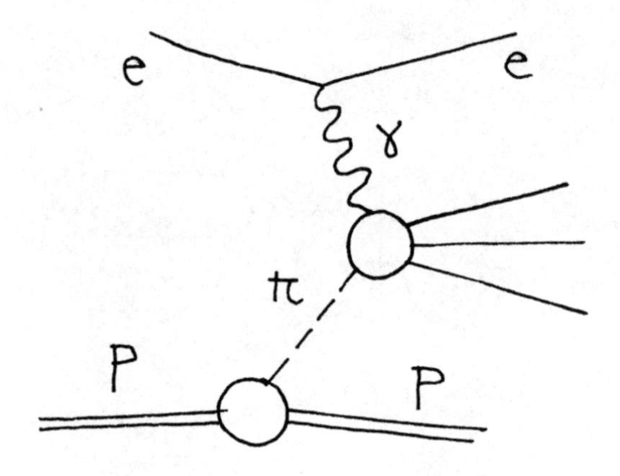

One observes the proton recoiling with small momentum, so that we get close to the pion pole and extract a $(t - m_\pi^2)^{-1}$ contribution. The other way is to do annihilation of pions on protons.

$$\pi^- p \rightarrow \mu^+ \mu^- X$$

If you believe in the chromodynamic way of computing, this is $Q\bar{Q}$ annihilation as in the "valence" diagram.

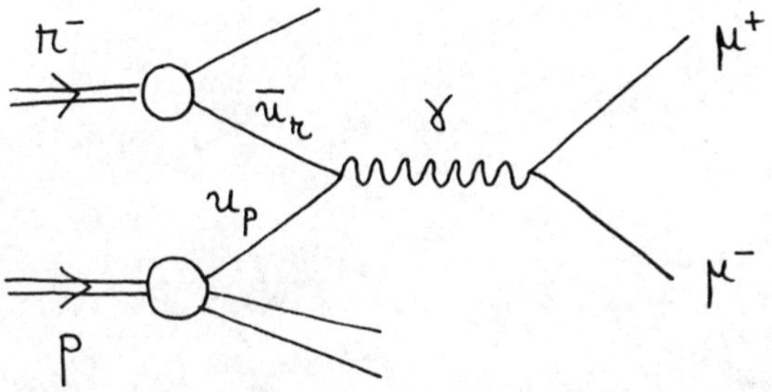

This gives you non linear tests of the $\bar{U}_\pi$ times U_p quark distributions. However, the QCD corrections to order α_s (to the comparison of annihilation and deep inelastic scattering) are very large and it is necessary to include them in order to extract the pion structure function. On the other hand, these $O(\alpha_s)$ corrections are very large and thus very interesting and presumably not so hard to measure.

- SISKIND:

Given that toponium has not been found and e^+e^- machines have low luminosity, what is a possible experiment to perform?

- DE RUJULA:

The comparison of the fragmentation of $J/\psi \longrightarrow$ pions and $\Upsilon \longrightarrow$ pions. Diagrams where the triple vertex of gluons is present are quite important. I think the best thing we can do at this moment is to compare the shapes of these two fragmentation functions in a model with/without color-charged gluons.

- HORNOES:

Can you give a simple explanation of what a moment of a structure function is?

- DE RUJULA:

No.

REFERENCES

1. J.D. Bjorken, Phys. Rev. 179:1547 (1969).
2. E.D. Bloom et al., SLAC-PUB 796 (1970) unpublished.
3. I. Budagov et al., Phys. Lett. 30B:364 (1969).
4. R.P. Feynman, Phys. Rev. Lett. 23:1415 (1969);
J.D. Bjorken and E.A. Paschos, Phys. Rev. 158:1975 (1969);
R.P. Feynman, "Photon Hadron Interactions", Benjamin, New York (1972).
5. Genesis, νW_2, Q^2, X (4004 B.C.).
6. B.L. Ioffe, Phys. Lett. 30B:123 (1969);
R.A. Brandt and G. Preparata, Nucl. Phys. B27:541 (1971);
Y. Frishman- Ann. Phys. (New York) 66:373 (1971).
7. M. Gell-Mann, Phys. Rev. 125:1067 (1962); Physics 1:63 (1964).
8. "Broken Scale Invariance and the Light Cone", Tracts in Mathematics and the Natural Sciences, ed. M. Dal Cin, G.J. Iverson and A. Perlmutter, Gordon and Breach, New York (1971).
9. H. Fritzsch and M. Gell-Mann, in Ref. 8.
10. K.G. Wilson, Phys. Rev. 179:1499 (1969).
11. C.H. Llewellyn Smith (sic), Phys. Rep. 3C:261 (1972).
12. H. Deden et al., Nucl. Phys. B85:269 (1975).
13. E. Stueckelberg and A. Peterman, Helv. Phys. Acta 5:499 (1953);
M. Gell-Mann and F.E. Low, Phys. Rev. 95:1300 (1954);
C.G. Callan, Phys. Rev. D2:1541 (1970);
K. Symanzik, Comm. Math. Phys. 18:227 (1970).
14. G. 't Hooft, Nucl. Phys. B33:173 (1971); B35:167 (1971).
15. Oxford English Dictionary, Oxford University Press, Oxford (1961) p. 15.
16. C.N. Yang and R. Mills, Phys. Rev. 96:191 (1954 A.D.).
17. J. Ellis, "Savage Sinners", Midwood, Norwalk (1977) any page.
18. H.D. Politzer, Phys. Rev. Lett. 30:1346 (1973);
D.J. Gross and F.A. Wilczek, Phys. Rev. Lett. 30:1343 (1973);
G. 't Hooft, unpublished.
19. D.J. Gross and F.A. Wilczek, Phys. Rev. D8:3633 (1973) and D9:980 (1974);
H. Georgi and H.D. Politzer, Phys. Rev. D9:416 (1974).
20. Guiness Book of World Records, N.and R. McWhirter, Bantam, New York (1977) p.11.
21. G. Parisi, Phys. Lett. 43B:207 (1973).
22. C. Chang et al., Phys. Rev. Lett. 35:901 (1975).
23. E.M. Riordan et al., SLAC-PUB-1634 (1975).
24. J.G.H. de Groot et al., Zeit. Phys. C1:143 (1979); Phys. Lett. 82B:292, 456 (1979).
25. A.J. Buras and K.J.F. Gaemers, Nucl. Phys. B132:249 (1978).
26. C. Weizsäcker and E.J. Williams, Zeit. Phys. 88:612 (1834);
L.N. Lipatov, Sov. J. Nucl. Phys. 20:181 (1974);
G. Altarelli and G. Parisi, Nucl. Phys. B126:298 (1977).

27. A.M. Polyakov, Proc. 1975 Int. Symp. on Lepton and Photon Interactions at High Energies, Stanford, ed. W.T. Kirk, SLAC, Stanford (1975) p. 855 and references therein;
J. Kogut and L. Susskind, Phys. Rev. D9:697, 3391 (1974).
28. V.N. Gribov and L.N. Lipatov, Sov. J. Nucl. Phys. 15:438 (1972).
29. J. Child, "A French Chef Cookbook," Bantam, New York (1971) p. 37.
30 A. De Rújula, R.C. Giles and R.L. Jaffe, Phys. Rev. D17:285 (1978).
31. L.W. Jones, Rev. Mod. Phys. 49:717 (1977);
See however,
G.S. LaRue, W.M. Fairbank and A.F. Hebard, Phys. Lett. 38:1011 (1977);
However, see however,
R.N. Boyd et al., Phys. Rev. Lett. 43:1288 (1979);
and CERN experiment WA44 (unpublished).
32. "Guinness Book of World Records" (loc. cit. Ref. 20) p. 169; See also, Particle Data Group, Phys. Lett. 75B:1 (1978).
33. M. Python, Grail Pictures, Inc. (1973).
34. M. West, private communication.
35. O. Nachtmann, Nucl. Phys. B63:237 (1973); B78:455 (1974).
36. S. Wandzura, Nucl. Phys. B112:412 (1977).
37. H. Georgi and H.D. Politzer, Phys. Rev. 37(E):68 (1976); Phys. Rev. D14:1829 (1976).
38. R. Barbieri, H. Ellis, M.K. Gaillard and G.G. Ross, Nucl. Phys. B117:50 (1976).
39. G. Preparata, "Lepton-Hadron Structure" ed. A. Zichichi, Academic Press, New York (1975) p. 54.
40. P. Castorina, G. Nardulli and G. Preparata, CERN preprint TH. 2670 (1979).
41. A. De Rújula, H. Georgi and H.D. Politzer, Ann. Phys. (New York) 103:315 (1977).
42. P.C. Bosetti et al., Nucl. Phys. B142:1 (1978).
43. D. Bailin and A. Love, Nucl. Phys. B75:159 (1974);
M. Glück and E. Reya, Phys. Rev. D16:3242 (1977).
44. J. Ellis, "Current Trends in the Theory of Fields", ed. J.E. Lannutti and P.K. Williams AIP, New York (1978) p. 81.
45. M.K. Gaillard, private communication.
46. H. Harari, Nucl. Phys. B161:55 (1979).
47. A. De Rújula, H. Georgi and H.D. Politzer, Phys. Rev. D10:1649 (1974).
48. C. DeLieto, Imperial College preprint ICTP-79-23 (1979).
49. E. Reya, Phys. Lett. 84B:445 (1979).
50. E.G. Floratos, D.A. Ross and C.T. Sachrajda, Nucl. Phys. B192:493 (1979) and references therein and thereout.
51. A. Para and C.T. Sachrajda, CERN preprint TH. 2702 (1979).
52. T.W. Quirk et al., Oxford University preprint (1979).

FATE OF FALSE VACUA - FIELD THEORY APPLIED TO SUPERFLOW

H.Kleinert

Institut für Theoretische Physik

Freie Universität Berlin, 1000 Berlin 33

ABSTRACT

The quantized states of superflow in a circular tube form a great number of metastable "false vacua". At the level of the field equations of motion there is absolute topological stability. Fluctuations cause decay with "critical bubbles" of large energies acting as a necessary trigger, just as in the evaporation process of a superheated liquid. The similarity and differences with respect to the infinite set of gauge field vacua are discussed.

I Introduction

During the last few years, two aspects of gauge theories have found special attention in elementary particle physics:

1) The existence of infinitely many vacua which differ by their topological properties.
2) The communication among the vacua via solutions of the field equations when continued to imaginary "time".

   The so obtained euclidean form of 3+1 dimensional field theory is formally equal to statistical mechanics in 4 spatial dimensions with the coupling constant g^2 playing the role of a temperature.

In this lecture I would like to show that a very similar situation is responsible for an important set of physical phenomena: The existence of superflow and its extremely slow decay in superconductors and superliquid ^{3}He. Certainly, since these systems are of a statistical nature as they are, no analytic continuation is necessary to imaginary time, the inverse temperature $\tau=1/T$ playing this rôle from the beginning. In superconductors, the theory has been developed a long time ago[1)] but has recently found an essential improvement[2)]. In ^{3}He, on the other hand, whose superliquid transition was discovered seven years ago[3)], the more complicated interplay of dynamics[4)] and topology has been understood only during the last year.

Common to both systems is the formation of a condensate of Cooper pairs: In a superconductor, these consist of two electrons (on the surface of the Fermi sea) in an s-wave. The attraction is caused by phonon exchange and is very weak. Physically, the exchange diagram accounts for the accumulation of positive ions along the path of the electron which acts as an attractive potential wake[+)]. The binding energy determines the critical temperature at which the pairs break up, due to thermal collisions. It can be found as

$$T_c = \mu\, e^{-1/g^2} \approx 1\ {}^{\circ}K \tag{1}$$

Here μ is the ultraviolet cutoff for the phonon spectrum and g^2 is the strength of the potential wake. It turns out that all low-energy properties of the super-conductor can be described by using only this single energy parameter T_c (apart from the actual temperature, T employing natural units $\hbar=1$, v_F = Fermi velocity = 1,2m = 1). Thus systems with many different μ and g. are identical superconductors (see Fig.1). This is quite analogous to the existence of a dimensionally transmuted coupling constant in gauge theory. There an artificial mass parameter μ is needed to define a coupling strength of the mass-

---

+) I thank V.Weisskopf for an illuminating discussion of this process.

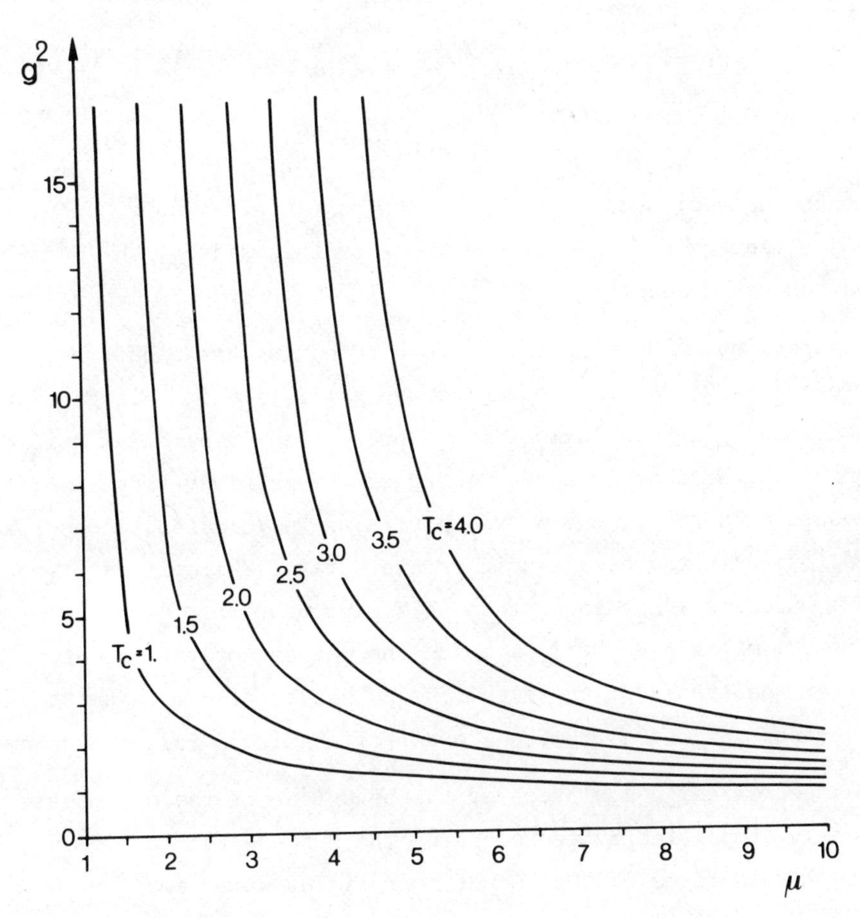

1. The curves of identical theories in g^2,μ space. The renormalization group determines the reparametrization of any fixed theory along the corresponding curve.

less theory but all physical quantities depend only on the combination

$$\Lambda^2 = \mu^2 \ e^{-1/g^2(\mu^2)} \tag{2}$$

The critical temperature gives directly the size of Cooper pairs via+)

+) This relation holds only close to T_c. Trivial numerical factors are left out, for simplicity.

$$\xi(T) = \frac{1}{T_c}\left(1-\frac{T}{T_c}\right)^{-1/2} \approx 1000\ \mathring{A}\left(1-\frac{T}{T_c}\right)^{-1/2} \tag{3}$$

The presence of such large bound states causes the superconductor to be coherent over this distance $\xi(T)$. For this reason, $\xi(T)$ is called the coherence length.

In superliquid ^3He, the interatomic potential has a hard repulsive core for $r < 2.7\ \mathring{A}$. In the degenerate Fermi liquid, this gives rise to strong spin correlation with a preference of parallel spin configurations. Because of antisymmetry of the pair wave function, this amounts to a repulsion in even partial waves. Indeed, Cooper pairs are formed in the p-wave spin triplet state. The critical temperature is a thousand times lower than in superconductors: $T_c = .27^o mk$ at p=35 bar . Since the masses of the ^3He atoms are larger than those of the electrons by about the same amount, the coherence length has the same order of magnitude in both systems.

The theoretical description of the behaviour of the condensate is greatly simplified by reexpressing the fundamental euclidean action directly in terms of the Cooper pair fields which are

$$\sigma(x) = \psi_e(x)\ \psi_e(x) \tag{4}$$

$$\sigma_{ai}(x) = \psi_{^3He}(x)\ \sigma_a\ \partial_i\ \psi_{^3He}(x) \tag{5}$$

respectively. Such a change of field variables can easily be performed in a path integral formulation[3)] in which the partition function of the system reads

$$Z = \sum_{\psi_e} e^{-A[\psi_e]} \quad \text{or} \quad \sum_{\psi_{^3He}} e^{-A[\psi_{^3He}]} \tag{6}$$

By going from integration variables ψ to σ one can immediately find the alternative form[6)]

$$Z = \sum_{\sigma} e^{-A[\sigma]} \qquad (7)$$

where σ is the Cooper pair field (4) or (5).

The new action is very complicated. For temperatures close to T_c, however, it can be expanded in powers of the field σ and its derivatives. For static fields

$$-A[\sigma] \equiv F/T = \frac{1}{T} \int d^3x \, f$$

$$= \frac{1}{T} \int d^3x \left[(-\log \frac{\mu}{T} + \frac{1}{g^2})|\sigma|^2 + \frac{1}{2T_c^2} |\sigma|^4 + \frac{1}{T_c^2} |\partial\sigma|^2 + \ldots \right] \qquad (8)$$

where the dots denote the higher powers of σ and of their derivatives, each accompanied by an additional factor $1/T_c$.

In ^{3}He one has to take care of all different contractions among the spatial and spin indices i and a, respectively+). This generates 3 derivative terms

$$f_{der} = \frac{1}{T_c^2} (\partial_i \sigma^+_{aj} \partial_i \sigma_{aj} + \partial_i \sigma^+_{aj} \partial_j \sigma_{ai} + \partial_i \sigma^+_{ai} \partial_j \sigma_{aj}) \qquad (9)$$

and five quartic potential terms

$$f_{pot} = \frac{1}{T_c^2} \{ |tr(\sigma^T\sigma)|^2 + \beta_2 [tr(\sigma^+\sigma)]^2 + \beta_3 \, tr[(\sigma^+\sigma)(\sigma^+\sigma)^*] + \beta_4 \, tr(\sigma^+\sigma)^2 + \beta_5 \, tr[(\sigma\sigma^+)(\sigma\sigma^+)^*] \} \qquad (10)$$

+) They contract separately, i.e. spin with spin, orbital with orbital indices, such that the theory is $SO(3)_{spin} \times SO(3)_{orbit}$ invariant.

It may be worth remarking that the transition from fields ψ_e or $\psi_{^3He}$ to that of the Cooper pair (4) or (5) is completely analogous to going from the fields ψ of the massive Thirring model to the scalar field ϕ defined by

$$\partial_\mu \phi \equiv \varepsilon_{\mu\nu} \bar{\psi} \gamma^\nu \psi \tag{11}$$

of the Sine-Gordon theory - the only difference being that in more than two dimensions the exact pair action is non-local and very complicated. However, if the system is studied close to the critical temperature and only with respect to its low-energy properties, the expansion of the free energy up to forth order in the field and up to second order in the derivatives contains all relevant information.

We shall now demonstrate how the field theory (8), and its extension (9),(10) in the case of ^{3}He, is capable of accounting for the properties of superflow.

II Superconductor

First we observe that with the critical temperature (1) the mass term in the action (8) can be written as

$$- \log \frac{T_c}{T} |\sigma|^2 \approx - (1- \frac{T}{T_c})|\sigma|^2 \tag{12}$$

It has the wrong sign for $T<T_c$ such that the field has no stable minimum at $\sigma=o$ but oscillates around a new place

$$|\sigma|^2 = T_c^2 (1- \frac{T}{T_c}) + \sigma(1-\frac{T}{T_c}) \tag{13}$$

Thus if T lies sufficiently close to T_c, the higher power of contribute less and less and may be neglected. It is useful to take the factor $T_c(1- \frac{T}{T_c})^{1/2}$ out of the field σ and write the renormalized free energy as

$$f = - |\sigma|^2 + \frac{1}{2} |\sigma|^4 + |\partial\sigma|^2 \tag{14}$$

where we have made use of the coherence length (3) to introduce a dimensionless space variable and dropped an overall energy density factor proportional to $(1-\frac{T}{T_c})^2\ T_c^2$. The minimum of f lies now at $|\sigma|=1$ where it has the value

$$f = -1/2$$

This negative energy accounts for the binding of the Cooper pairs in the condensate and is therefore called condensation energy. In terms of (14), the partition function in equilibrium can be written as

$$Z = \sum_{\sigma} e^{-\frac{1}{T}\int d^3x\, f} \tag{15}$$

Let us now consider the flow properties of the system. Certainly, there is a divergenceless current

$$j(x) = \frac{1}{2}\ \psi^{\dagger}(x)\partial_x\ \psi(x) \tag{16}$$

associated with the transport of particle number. The important question to be understood is: How can this current become super?

In order to see this let us set up a current in a long circular wire. If the thickness is chosen much smaller than the coherence length, transverse variations of σ are strongly suppressed with respect to longitudinal ones by the Boltzmann factor and the system depends only on the coordinate along the wire. If the cross section of the wire is absorbed in the definition of the temperature, we may simply study the partition function (15) for a one-dimensional problem along the z-axis. The field may be decomposed in polar coordinates

$$\psi(z) = \rho(z)e^{i\gamma(z)} \tag{17}$$

such that the free energy

$$f = -\rho^2 + \frac{1}{2}\rho^4 + \rho_z^{\ 2} + \rho^2\ \gamma_z^{\ 2} \tag{18}$$

leads to field equations

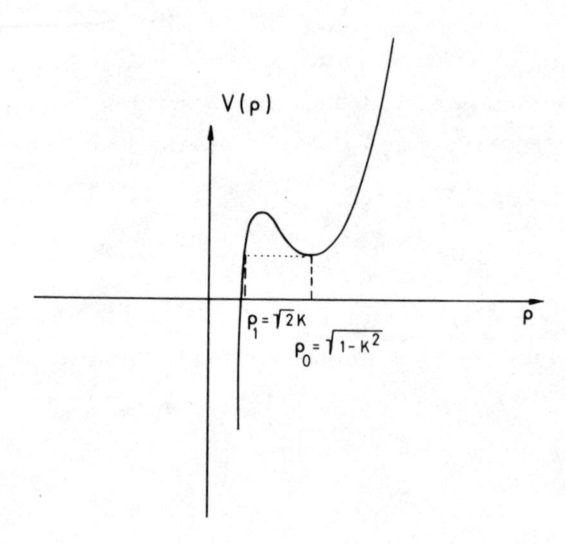

2. The potential $V(\rho)= -\rho^2+\rho^4/2-j^2/\rho^2$ showing the barrier to the left of ρ_0 to be penetrated if the supercurrent is to relax.

$$j = \rho^2 \gamma_z = \text{const.} \tag{19}$$

$$\rho_{zz} = -\rho + \rho^3 + \frac{j^2}{\rho^3} \tag{20}$$

The latter corresponds to the mechanical motion of a mass point in the potential, turned upsidedown (see Fig.2),

$$V(\rho) \equiv -\rho^2 + \frac{1}{2}\rho^4 - \frac{j^2}{\rho^2} \tag{21}$$

if z is considered as a "time". Obviously, there is a stationary solution

$$\gamma(z) = kz \quad .$$
$$\rho(z) \equiv \rho_0 = \sqrt{1-k^2} \tag{22}$$

Since the wire is closed, the phase $\gamma(z)$ has to be periodic over the length L and must be quantized according to

$$k_n = \frac{2\pi}{L} n \tag{23}$$

The corresponding energy density is (see Fig.3)

$$f(k_n) = -\frac{1}{2}(1-k_n^2)^2 \tag{24}$$

with a current

$$j(k_n) = \rho_o^2 (1-\rho_o^2) = k_n (1-k_n^2) \tag{25}$$

Notice that this current is bounded by

$$|j| < j_c \equiv \frac{2}{3\sqrt{3}} \tag{26}$$

No solution of the field equations can support a larger current than given by this critical value.

We shall now demonstrate that all states of current j_n smaller than j_c are infact "super" in the sense of having an extremely long lifetime (in practice ranging from hours to years). In the sense of field theory each state k_n can be considered as a "false vacuum" which eventually will decay to the true vacuum, the state of no current[7].

In order to understand this enormous stability of the states we notice that the temperature is very small such that the temperature fluctuations leave ρ very close to ρ_o. We can thus picture the field configuration as a spiral of radius ρ_o wound around the wire with the azimuthal angle representing the phase $\gamma(z) = k_n z$ (see Fig.4).

If the temperature is zero, ρ is frozen at ρ_o and the winding number is absolutely stable on topological grounds. The current runs on forever. The "false vacuum" has an infinite lifetime.

In order that the current may relax by one unit it is necessary that at some place thermal fluctuations carry $\rho(z)$ to zero. There the phase becomes undefined and may slip by 2π.

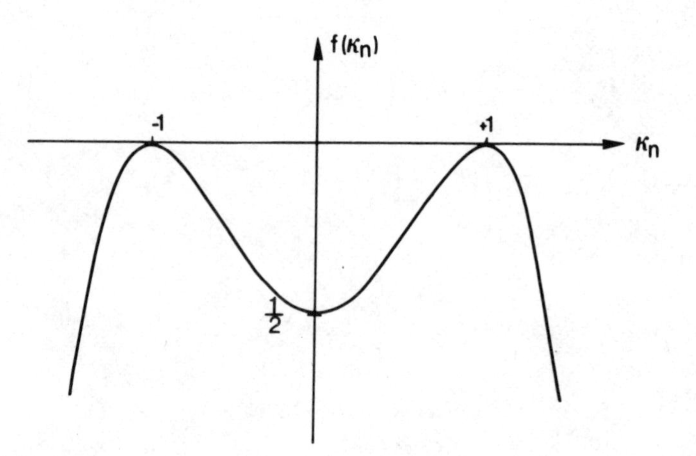

3. The condensation energy as a function of the velocity parameter $k_n = \frac{2\pi}{L} n$.

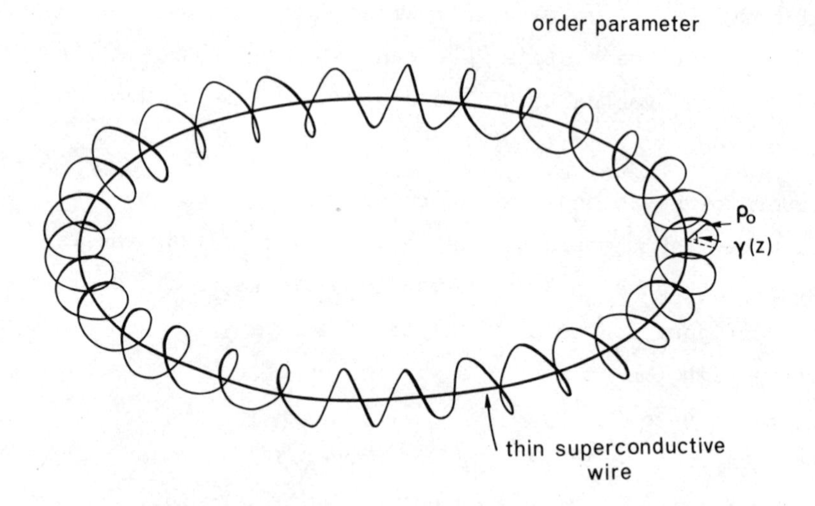

4. If ρ is frozen at ρ_0, the field configuration may be pictures as a spiral of radius ρ_0 with pitch $\frac{\partial\gamma(z)}{\partial z} = \frac{2\pi}{L} n$. The supercurrent is absolutely stable since the winding number n is locked topologically.

Now, since $T << 1$, such phase slips are extremely rare. In order that an excursion of $\rho(z)$ to $\rho=o$ has an appreciable measure in the functional sum (7) one must first look for solutions of the equations of motion[+)] which carry $\rho(z)$ as closly as possible to zero. From our experience with mechanics we are used to imagining the motion of a mass point in the potential $-V(\rho)$. It is easily realized that there is, in fact, a solution which carries $\rho(z)$ from ρ_o at $z=-\infty$ up the curve $V(\rho)$ to $\rho=2K^2$ and back to ρ_o at $z=\infty$ (see Fig.5). Explicitly :

$$\rho_b(z) = 1-k^2 - \frac{\omega^2}{2}/ch^2 \frac{\omega}{2}(z-z_o) \tag{27}$$

with energy

$$F_b = \int dz\, f(\rho_b) = \frac{4}{3}\omega = \frac{4}{3}\sqrt{2(1-3k^2)} \tag{28}$$

where ω is the curvature of $V(\rho)$ close to ρ_o

$$V(\rho) \approx \omega^2 (\rho-\rho_o)^2 + \ldots \tag{29}$$

The solution reaches the point of smallest ρ at z_o :

$$\rho(z_o) = 2k^2 \tag{30}$$

This is still non-zero and does not permit a phase slip. We shall now see, however, that quadratic fluctuations around this solution are sufficient to reduce the current. Let us insert a small deviation

$$\rho(z) = \rho_b(z) + \delta\rho(z) \tag{31}$$

into the free energy. Respecting the equation of motion for ρ_b , the lowest variation of F is of second order

$$\delta^2 F = \int dz\, \delta\rho\, (-\partial_z^{\,2} + V''(\rho))\delta\rho \tag{32}$$

But this expression is not positive definite. This can be seen

---

+) This follows from the method of steepest descent generalized from integrals to path integrals.

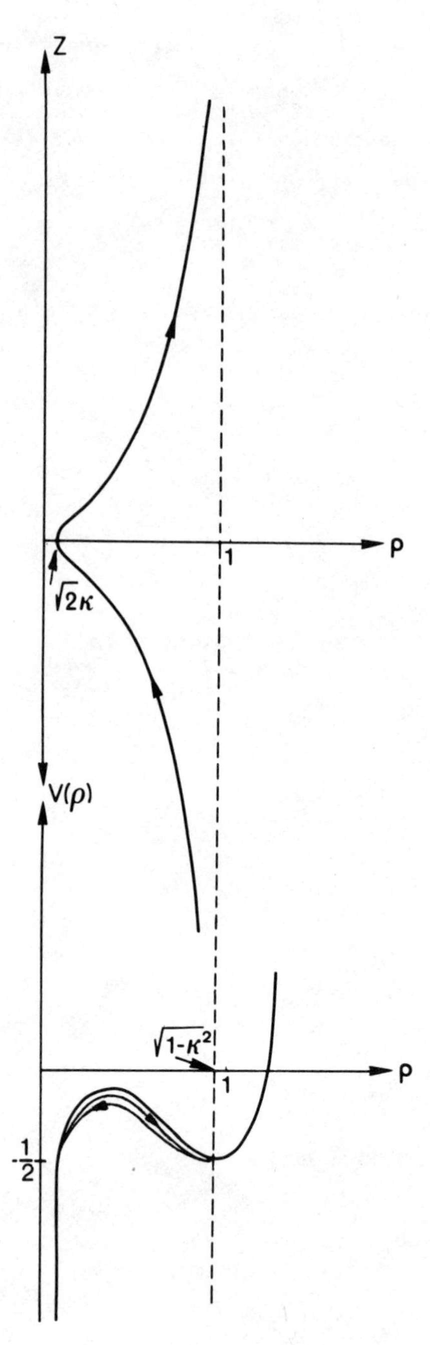

5. An extremal excursion ("critical bubble") corresponds to a mass point sitting at ρ_0, rolling under the influence of negative gravity up the hill unto the point $\rho=2k^2$, and returning back to ρ_0, the variable z playing the role of a time variable.

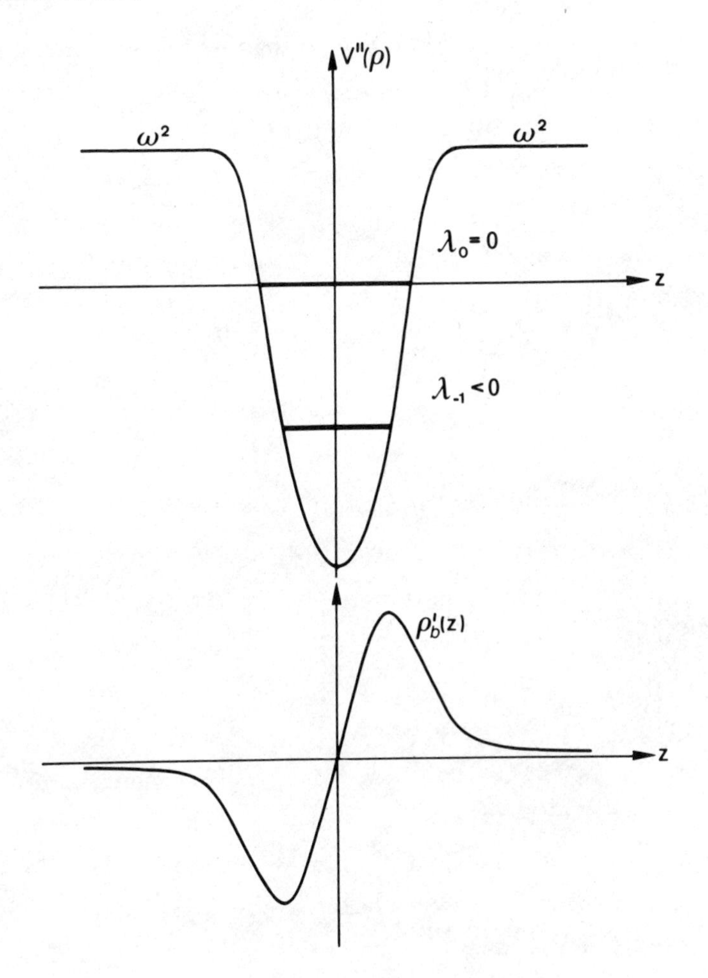

6. The infinitesimal translation of the "critical bubble, ρ_b' is an antisymmetric wave function of zero energy. Hence there must be a negative-energy bound state.

either by explicitly solving the eigenvalue problem

$$(-\partial_z^{\,2}+V''(\rho_b))\psi_n(z)= (-\partial_z^{\,2}-1+3\rho_b^{\,2}-3\frac{j^2}{\rho_b^4})\psi_n(z)=\lambda_n\psi(z) \tag{33}$$

or by the following much simpler reasoning : Equ.(33) is a Schrödinger equation in a potential which asymptotically is a constant ($=\omega^2$). For z approaching z_o , the potential becomes smaller due to the curvature of V(ρ) decreasing (see Fig.6). At $z=z_o$ it has a minimum at a negative value. The small vibrations are unstable if there is a bound state with a negative eigenvalue $\lambda_{-1}<o$. Its existence can be argued by noticing first that there is certainly a zero eigenvalue due to translational invariance: Since the wire is very long, the solution $\rho_b(z)$ which has its closest approach to the origin at z_o exists for any finite z_o ($<<L$) . Thus a small translation of z_o is certainly a fluctuation which does not change the energy such that

$$\psi_o = [\rho_b(z+\delta z_o)-\rho_b(z)] / \delta z_o = \rho_b'(z) \tag{34}$$

must solve the fluctuation problem (33) with $\lambda_o=o$. This can indeed be verified by an explicit calculation: Since

$$-\rho_{b\ zz} + V'(\rho_b) = o \tag{35}$$

one has

$$-\partial_z^2 \rho_b' + V''(\rho)\rho_b' = o \tag{36}$$

such that (33) is fulfilled. Now this zero frequency solution has an important property: Since $\rho_b(z)$ is an even function in $z - z_o$, $\rho_b'(z)$ is odd and has a node at z_o . Therefore it cannot be the ground state of the Schrödinger problem and there must be another lower lying state, i.e. with $\lambda_{-1}<\lambda_o= o$. The vibration problem is therefore unstable [7].

The whole process is quite analogous to the nucleation of the vapour phase in a superheated liquid [7] . There bubbles have to form containing vapour inside. The extremal size (critical bubble) is determined by the balance of surface tension and pressure (solution of equations of motion of surface). The bubbles are unstable against radial fluctuations: expansion leads to the transformation of the whole liquid into vapour (the volume energy wins), contraction leads to the return to the liquid phase (the surface energy wins). Associated with this is a negative eigenvalue of the corresponding differential equation [7]. Because of the complete analogy one may call the distorted solution $\rho_b(z)$ a "critical bubble" whose decay mediates the phase slip.

For the precise calculation of the decay rate we refer to Ref.2).

Let us summarize the physical situation: Thermal fluctuations lead to the presence of "critical bubbles" along the uniform spiral. Their presence is very rare due to the small Boltzmann factor+)

$$e^{-\frac{1}{T}\frac{4}{3}\omega} \ll 1 \tag{37}$$

The bubbles cause distortions of the spiral bringing the size of the field close to zero somewhere along the wire. The quadratic fluctuations around the bubbles do reach the zero field point and allow the phase to slip. The current relaxes by one unit of 2π and the spiral returns to the uniform configuration with one winding number less (and the bubble having disappeared)

What are the parallels with gauge theories? The quantum number $k = \partial_z \gamma$ characterizing the superliquid velocity of the "false vacua" can be written as

$$k = e^{i\gamma(z)}\, i\partial_z\, e^{-i\gamma(z)} = U^{-1}(z)\, i\partial_z\, U(z) \tag{38}$$

Thus the circular superliquid velocity corresponds to a vector potential for a pure gauge transformation. In the superliquid, the energy depends on k like

$$f = -\frac{1}{2} + k^2 + O(k^2) \tag{39}$$

In a gauge theory it does not, due to gauge invariance. Notice that if the gauge field had a mass term

$$f = A_z^2$$

then also its energy would have a k^2 dependence and the infinitely many vacua of different winding number would no longer be degenerate.

+) A cross section factor of the wire is absorbed into T, for simplicity, together with the previous factors $(1-\frac{T}{T_c})^2\, T_c^{\,2}$ when going from (8) to (14).

III Superliquid ^{3}He

Consider now the topologically more interesting system of superliquid ^{3}He. The discussion of the minima of the energy (10) is not as trivial as in the case of the superconductor (12). Here at least 11 local extrema are known with the two lowest ones being present in the laboratory, depending on pressure and temperature. They are called B and A phase(for a phase diagram see Ref.3 or 6). If the size of the field σ_{ai} is frozen, they may be parametrized as

$$B: \quad \sigma_{ai} = \rho_o R_{ai}(z)\, e^{i\phi(z)} \tag{40}$$

$$A: \quad \sigma_{ai} = \rho_o \ell_a(z)(\phi_i^{(1)} + i\phi_i^{(2)})(z) \tag{41}$$

where $R_{ai}(z)$ is an arbitrary rotation matrix. $\underset{\sim}{\phi}^{(1)} \perp \underset{\sim}{\phi}^{(2)}$ are unit vectors characterizing the plane in which the Cooper pairs move and

$$\underset{\sim}{\ell} = \underset{\sim}{\phi}^{(1)} \times \underset{\sim}{\phi}^{(2)} \tag{42}$$

points in the direction of the arbital angular momentum of the Cooper pairs.+)

We can now immediately see that the B phase has, in a long circular tube, superflow properties very similar to those of the thin superconductor. For ρ_o frozen, a uniform flow is given by

$$\phi(z) = k_n z \, , \; k_n = \frac{2\pi}{L} n \tag{42}$$

and relaxation can occur only by fluctuations of ρ_o to the origin. For this the formation of energetic "critical ρ bubbles" is needed causing a high stability of the superflow states k_n (the false vacua). The situation is quite different in the A phase. Let ρ_o

---

+) In writing this form in the A phase we have taken into account the hyperfine interaction which forces the spin direction into the orbital plane (see Refs. 3 and 6).

be frozen and consider a uniform superflow along the z axis with $\underset{\sim}{\ell}||z$. A Galileian factor e^{ikz} is equivalent to rotating $\underset{\sim}{\phi}^{(1)} \perp \underset{\sim}{\phi}^{(2)}$ around $\underset{\sim}{\ell}$ as one proceeds along the tube. The configuration many again be visualized by the end point of $\underset{\sim}{\phi}^{(1)}$ (say) forming a spiral of unit radius around the tube.

So far everything looks the same as in the previous two cases. The important difference, however, is that we have chosen $\underset{\sim}{\ell}$ to be parallel to the z axis. If the direction of $\underset{\sim}{\ell}$ is allowed to vary, it is quite easy to see that the spiral can be deformed until it is a straight line for $n \doteq$ even, or only one winding is left for $n =$ odd. The proof for this is identical to the standard way of showing the doubly connectedness of the rotation group SO(3). The positions of the dreibein $\underset{\sim}{\phi}^{(1)}, \underset{\sim}{\phi}^{(2)}, \underset{\sim}{\ell}$ may be parametrized by a rotation matrix which is necessary to transform it to a certain fixed configuration (say $\underset{\sim}{\phi}^{(1)}, \underset{\sim}{\phi}^{(2)}, \underset{\sim}{\ell}$ in x,y,z direction). This in turn may be written as

$$e^{i\theta \underset{\sim}{n} \underset{\sim}{L}} \tag{43}$$

where $\underset{\sim}{n}$ is the axis and θ the angle of rotation. The vectors $\theta\,\underset{\sim}{n}$ are lying in a sphere of radius π with diametrially opposite surface points identified.

The position of the dreibein along the circular tube may be drawn as a closed path in this parameter space. It is easily seen that there are two topologically inequivalent classes depending on whether there is an even or odd number of jumps between diametrially opposite points (see Fig.7). But the state of superflow which was set up in the beginning ($\underset{\sim}{\ell}||z$ and $\underset{\sim}{\phi}^{(1)}$ describing a spiral) corresponds exactly to a straight line path from the south to the north pole, jumping to the south and continuing again to the north pole and so on n-times. The continuous deformability of the path to either of the two fundamental ones (which is the point at the origin or a single straight line from south to north pole) is equivalent to the continuous reduction of superflow to no or one unit of flux. In either case the flow would not be "super" at all. Thus, contrary to superconductors

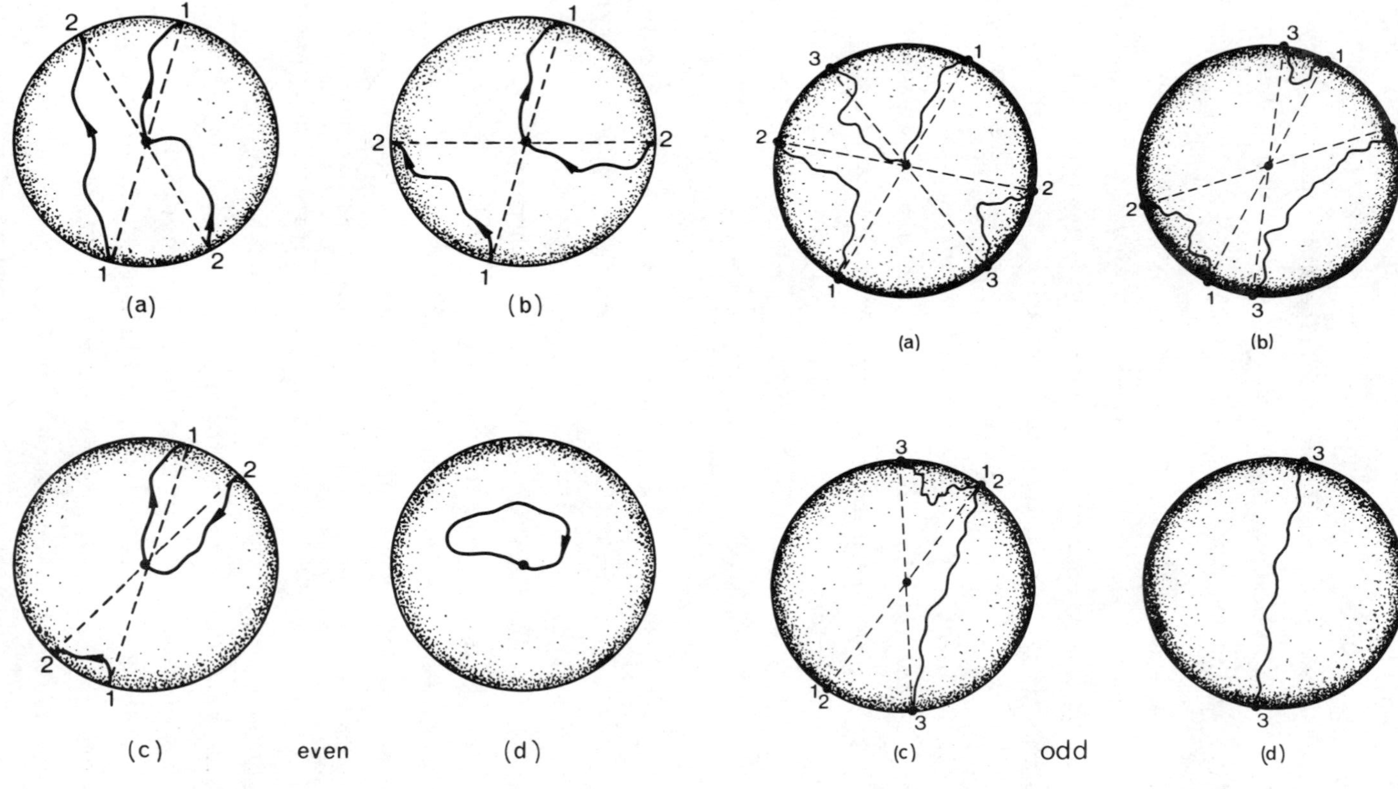

Fig. 7. States of superflow correspond to a closed contour in the SO(3) parameter sphere. For an even number of flux quanta the contour can be deformed continuously into a point which is a state of no current. States with an odd quantum number can be contracted until there is only one diametral line from south to north pole (say) corresponding to a single unit of flux.

and ^{3}He B, the freezing of ρ_o does not topologically stabilize the flow. There is, however, another energy barrier, which still allows superflow to occur: It lies in the space of directions of the dreibein. By looking at the derivative terms of the free energy one can see that a superflow $\partial \sigma_{ai} \propto v_s \sigma_{ai}$ causes an attraction to the ℓ vector of the form

$$f_{der} \propto -(\ell \cdot v_s)^2$$

Thus there is, in fact, a potential barrier against the deformation described before: In the idealized case that the vector ℓ is tightly locked to the forward direction, the spiral $\phi^{(1)}$ around the tube becomes topologically stable. Decay can occur by the formation of a "bubble" in directional space (say the angle between ℓ and v_s). Again, a very small Boltzmann factor causes a high stability of the different current states. Notice that now the superflow is related to

$$A_z = R\partial_z R^{-1} \tag{45}$$

where R is the rotation matrix (43). The different "false vacua" correspond, just as before, to states of pure gauges, now in a non-abelian version. Again, the bending energies are equivalent to inserting a mass term in a gauge theory destroying the degeneracy of the infinitly many vacua of different winding number.

Finally, let us mention one more interesting aspect of the topology of superflow in ^{3}He-A : If a magnetic field is turned on parallel to the z axis it attracts the spins of the Cooper pairs. Since these lie in the orbital plane there is a force pulling ℓ out of the direction of flow. It turns out that for H larger than a certain critical value there is a new equilibrium position in which ℓ forms a fixed non-zero angle with the current. As a consequence, there is an additional topologically stable quantum number: The number by which ℓ winds around the z axis. Since the other number, by which $\phi^{(1)}$ winds around the direction of flow, is completely independent of this, there is a doubly infinite set of false vacua characterized by two macroscopic quantum numbers n_1, n_2. ^{3}He-A has become a double superliquid [5]. What is the physical nature of this second superflow? Since the Cooper pairs are formed in a p-wave there are two mechanisms of matter flow: The first consist of the Cooper pairs themselves moving. The second is observed even if all Cooper pairs are at rest. If their orbital planes are not all parallel but have $\nabla \times \ell = 0$, the circular orbits of neighbouring pairs do no longer cancel each other and a transport of neighbouring pairs is observed. This is quite analogous to the source of the magnetic field in the presence of matter

$$\nabla \times B = j + \nabla \times M$$

Even with no current flowing a magnetic field is generated if the magnetization has a curl due to the non-cancellation of the microscopic current loops.

IV Conclusion

At very low temperatures, many-body systems such as superconductors and superliquid ^{3}He can be described by a simple field theory. Many of the properties of the vacuum which have recently come under investigation in the field theory of elementary particles are present in these systems and carry responsibility for the phenomenon of superflow.

REFERENCES

1. J.S.Langer and V.Ambegaokar, Phys.Rev. 164,498 (1967)
   D.E. Mc Cumber and B.I.Halperin, Phys.Rev. 131, 1054 (1970)

2. H.Duru, H.Kleinert and N.Ünal, Berlin preprint, Oct.1979

3. For a review and references see
   H.Kleinert, Fortschr.Physik 26,565-671 (1978) and
   Collective Field Theory of Superliquid ^{3}He, Berlin Preprint, Sept.1978.

4. H.Kleinert, Y.R.Lin-Liu and K.Maki, Suppl.J.de Phys. 6, C6-59 (1978) (paper presented at the Low Temperature Conference LT15) Phys.Letters 70A,27 (1979)
A.Fetter, Suppl.J.de Phys. 6, C6-46 (1978) Phys.Rev.Lett.40, 1656 (1978).

5. H.Kleinert, Phys.Letters 71A, 66 (1979).

6. This is discussed in my 1978 Erice lectures to appear in "The New Aspects of Subnuclear Physics, ed. by A.Zichichi, Plenum Press 1978.

7. J.S.Langer, Am.Phys.41,108 (1967)
P.Frampton, Phys.Rev.Letters 37, 1378 (1976) and Phys.Rev.D15, 2924 (1977).
S.Coleman, Phys.Rev. D15, 2929 (1977) and Erice Lectures 1977.

MY LIFE AS A PHYSICIST

H.B.G. Casimir

Eindhoven (The Netherlands)

"But you were going to be an architect". Those were the first words Ehrenfest[1)] spoke to me when I went to see him before the summer holidays in 1926. I had recently passed my final school-examination, intended to study theoretical physics and wanted some advice. Since I have no gift for drawing and since my faculties for visualizing and remembering complicated structures in space are no more than average, this struck me as a curious remark. But Ehrenfest was firmly convinced that any kind of talent should manifest itself at an early age. He has occasionally visited our home and had apparently been impressed by the zeal and concentration with which I built towers and castles with wooden blocks. (If all kids that play with wooden blocks were to become architects there would be a surplus of architects, I'm afraid). He next asked me how I had fared at school and what I had read and studied beyond the school curriculum. That was next to nothing. I had had a very good physics teacher, a former student of H.A. Lorentz, and what he taught us had impressed me, that was

1) Paul Ehrenfest, b. Vienna 1880, d. Amsterdam 1933, held the chair of theoretical physics at Leiden from 1912 until his death. He is known for his contributions to statistical mechanics and quantumtheory, but his influence as a teacher, critic and elucidator has been greater than that of his original contributions.

all. On the whole I had been rather lazy, working just hard enough to get high marks in all subjects, which was not very hard. And so Ehrenfest took a rather dim view of the situation. The fact that I had not had difficulties with any scholastic subject must have impressed him unfavourably if at all. But he was very kind, told me to go and see Uhlenbeck and Goudsmit, who where at that time, like myself, living in the Hague, and advised me to study some calculus during the summer holidays. Anyway, I had first to pass the "candidaatsexamen"; the choice between mathematics, experimental physics and theoretical physics had to be made afterwards, so we could wait and see. It was obvious however, that he did not think I would make it as a theoretical physicist.

The curriculum at Leiden was in those days not very heavy. Mathematics was traditional and somewhat old-fashioned: Calculus, analytical geometry and some algebra. But it was very competently taught and provided a sound background. Physics was rather old-fashioned too and the teaching left something to be desired. However, Uhlenbeck, who was Ehrenfest's assistant during my first year at Leiden - he and Goudsmit both got their Ph.D. on July 7, 1927 and then went to Ann Arbor, Michigan - ran an unofficial seminar in which we studied a recently appeared book by Frank and Jordan "Anregung von Quantensprüngen durch Stösse" (Excitation of quantum jumps by collisions); there we really learnt something about modern physics. I had to present the sections dealing with collisions of the second kind and apparently I did rather well; Uhlenbeck reported to Ehrenfest and as a result I was invited to the famous Colloquium. There we heard about the latest developments, for Ehrenfest kept in touch with all the leading figures of those days and many came to lecture at Leiden. I remember that Pauli came and lectured about his non-relativistic spin theory and that Ehrenfest was surprised that an equation could be invariant under rotation although it could not be written in tensor form. Pauli replied somewhat vaguely "Weyl has told me that there exist a kind of Dublett-Grössen". Dirac spent quite some time at Leiden, working on his book, and I heard him lecture on the first chapters before they were published.

I duly passed my "candidaats" in June 1928, spent one or two months in Göttingen - mainly learning German - and in September I seriously began to study quantum mechanics. By that time the theory was well-established (Dirac's relativistic equation was published early in 1928) and useful textbooks began to appear. Ehrenfest understood at once that group theory provided the clue to understanding the semi quantitative rules of the vector model on the basis of quantum mechanics. He invited Wigner to Leiden. That was not entirely successful. The trouble was that Wigner was too polite to be clear. It is obvious that if someone lecturing to theoretical physicists would spend considerable time explaining how to solve a simple quadratic equation then the audience would feel insulted. To Wigner many things appeared to be just as simple, so he carefully avoided explaining them. (After all, according to a slightly apocryphical legend, he himself had learnt group theory from Johny von Neumann on one rainy Sunday afternoon.) Heitler was more helpful: his ideas about what was trivial and what was not were closer to ours. I also remember a brilliant lecture by Van der Waerden. But to me the great revelation was Weyl's book, Gruppentheorie und Quantenmechanik. I still think it is a wonderful book, and the little I know about group theory and part of what I know about quantum mechanics I learnt there, directly or indirectly.

In the spring of 1929 Ehrenfest took me with him to a conference at Bohr's institute. For a young physicist in those days that was something like "coming out" for a young lady in high society. As a matter of fact this meeting became for me the beginning of a new phase of my education: Bohr asked me to stay at Copenhagen, provided me with a stipend and for the next two years I spent more than half of the time at Copenhagen.

During that meeting Oudsmit, who had come over from Ann Arbor, asked me whether I could calculate the hyperfine interaction in an s-state. That was a famous problem. If you write this interaction as the interaction of a nuclear magnetic dipole and a distribution of electronic magnetic dipole moments, which is a logical thing to do on the basis of Pauli's theory, then the integral diverges for an s-state, where the density is not zero at the origin. Now I had been deeply impressed by the fact that in Dirac's theory the magnetic field of an electron results from a current and as you may recall, the current distribution in the fundamental state of the hydrogen atom is particularly beautiful. It can be described by saying that the charge contained inside a spherical shell rotates as a rigid body around the z-axis, but the angular velocity of successive shells is such that the linear velocity at the equator is always the same, namely
it was therefore very easy to calculate the magnetic field at the centre.
The result is by now well known:

$$H = \frac{8\pi}{3} \cdot \frac{e\hbar}{2mc} \cdot |\psi(0)|^2$$

Goudsmit then provided some estimates for $|\psi(0)|^2$ for sodium and other simple atoms and everything seemed to fit nicely.
I suggested that we should publish a note together but Goudsmit was more generous and said I should publish alone. So I wrote a manuscript - it must have been a pretty clumsy one - and sent it to Goudsmit. For a long time I heard nothing and by the time I received an answer with suggestions for some improvements, Fermi's famous paper had appeared.

In retrospect it seems rather curious that no one had thought of introducing a spin current in the framework of Pauli's theory. In classical electrodynamics one can write

$$i = c \operatorname{curl} M$$

that is the meaning of the magnetization. Why should one not do the same for an atom and write

$$i = \frac{e\hbar}{m} \operatorname{curl} (\psi^* \vec{s} \psi)$$

Of course I was somewhat disappointed that my first attempt to do some work of my own came to nought, but this was not the end of my occupation with hyperfinestructure. In the summer of 1932 I spent some time at Lise Meitner's institute at Berlin. By the way I vividly remember how particles were counted there by Geiger counters connected to a string electrometer. One peered through the eyepiece and every jerk of the string was ticked off by hand. The electronic age in counting had at that time hardly started and anyway it was England rather than Germany that was leading in that respect. While at Berlin I also visited Schüler who was then doing spectroscopical work in a small lab connected to the astronomical observatory. He showed me some of his results and this led to a short paper on configuration interactions in hyperfinestructure. Also, when Schüler later found strong evidence for the influence of a nuclear quadrupolemoment, he informed me at once and I worked out the theory, first in a preliminary note and then in a fairly lengthy paper, published as a prize essay by "Teylers Tweede Genootschap"-an old-fashioned Dutch institution- in which I gave a fairly thorough analysis of the electromagnetic interaction of electrons and nuclei. That paper may have been of some use in the further development of the subject. It was later reprinted by Freeman & Co as a paperback. I returned once more to the theory of hyperfinestructure in 1942 when I estimated together with a student the influence of a nuclear magnetic octupole. It was far too small to be observed by the methods of those days.

Back to Copenhagen. During my stay I often acted as a kind of secretary to Bohr, who liked to have someone listening and taking notes while he was trying to formulate his thoughts. For him writing an article was not just formulating things he knew already, it was a way of thinking. The story is told that he once asked Dirac to help him, but that did not work out. Dirac's characteristic advice "one should never begin a sentence before knowing the end" was not acceptable to Bohr. Since sentences came very slowly, shorthand was not required, but one had to get accustomed to Bohr's voice, which was rather indistinct and soft, whether he was speaking Danish, English, or German.

It is impossible to convey by some short remarks the depth of Bohr's influence on my contemporaries and myself. Ehrenfest taught me the importance of clear, crisp formulation and was a master at finding simple examples illustrating the gist of a physical theory. Later Pauli forced me not to shun elaborate mathematical analysis, but Bohr was both more profound and more practical. He had an excellent grasp of orders of magnitude; in that respect he resembled the creative engineers I have known. Afterall, as a young man, he had done beautiful experiments on surface tension and had built with his own hands most of the apparatus.

The following anecdote may illustrate this point. Close to Bohr's institute there is a body of water - I hesitate whether I should call it a lake or a pond - about three kilometers long and between 150 and 200 meters wide, the Sortedamsø. It is crossed by several bridges. One day Bohr took me on a walk along that lake and over one of the bridges. "Look", he said, "I'll show you a curious resonance phenomenon". The parapet of that bridge was built the following way. Stone pillars, about four feet high and ten feet apart were linked near their top by a stout iron bar or probably it was a tube, let into the stone. Halfway between two pillars an iron ring was anchored in the stonework of the bridge, and two heavy chains, one on each side, were suspended between shackles welded to the topbar close to the stone pillars and that ring. Bohr grasped one chain, near the top bar, and set it swinging and to my surprise the chain at the other end of the top bar began swinging too. "Isn't that a remarkable example of resonance?" Bohr said. I was much impressed, but suddenly Bohr began to laugh. Of course resonance was quite out of the question, the coupling forces were extremely small and the oscillations were strongly damped. What happened was that Bohr while moving the chain was rotating the top bar, which was let into but not fastened into the stone pillars, and in that way he moved both chains simultaneously. I was crestfallen that I had shown so little practical sense, but Bohr consoled me, saying that Heisenberg had also been taken in; he had even given a whole lecture on resonance.

That bridge, known in Bohr's institute as the resonance bridge, figures also in another story. In the late summer of 1933 I was again at Copenhagen for a conference. I had recently married and my wife had come with me. One evening we were sitting with a small group of people and someone suggested that we should go for a walk over the resonance bridge. This somehow led to a wager between Placzek and myself that I would swim across. The stake, 50 Danish crowns, was considerable, considering our respective financial positions. I later heard that Placzek, himself a bachelor at the time, had based his bet on a theory of marriage: he had expected that my young bride at the last moment would hold me back. Little did he know my wife. It was not much of a swim, but it is a curious feeling to swim more or less near the center of a town, accompanied by an escort of surpised ducks and even more so to have to walk home to your hotel with the water dripping from your clothes. A few years ago, when visiting Copenhagen, I pilgrimed to that bridge, but it had been widened and the parapet was no longer there.

The 1933 meeting was the last time I saw Landau; I don't think he ever came to western Europe again. One evening all of us were received at Bohr's home. Vicky Weisskopf, Otto Frisch and Hans Kopfermann performed some music and Landau, who had no ear for music at all, was pulling faces and making rather a nuisance of himself in a somewhat childish way. Afterwards Dirac walked up to him and said "if you don't like music, why don't you leave the room" to which Landau at once replied "it's the fault of Mrs Casimir. She isn't interested in music either and I proposed that we should leave the room together. Why didn't she come along?" To this Dirac replied politely "I suppose she preferred listening to the music to going out of the room with you". Landau, for once, had no repartee.

This was also the last time I saw Ehrenfest: he returned to Holland and, on the 25th of September, he took his own life.

But let me tell a bit more about my early days at Copenhagen. Gamow was there part of the time. He was a bit of a playboy putting much energy and ingenuity into rather innocent practical jokes, but he also wrote his book on Atomic Nuclei & Radioactivity. Landau was there too. His was one of the quickest minds I have ever come across. His physical reactions however were slow. From that point of view he was just the opposite of Bohr, who was a slow, but incredibly persevering thinker, but whose mechanical reactions were quick. (Much later I chanced upon an interview with Harald Bohr, Niels' younger brother, the mathematician who in his youth had been the star of the Danish national soccer team. He said about Niels as a goal-keeper "he was in a way excellent, but he could never make up his mind whether he should stay in his goal or run out".) Landau could be extremely aggressive. If he disliked somebody - and he could take sudden and unmotivated dislikes - then he could tease and badger in a very nasty way, but he could also be a very good friend. He was in those days a convinced revolutionary, anti-bourgeois, anti-

capitalist, but he was no Marxist. Theoretical Marxism and dialectic materialism he considered to be sheer nonsense, like any other political theory, like any branch of philosophy.
He once gave a lecture on scientific life in Russia to a student's society. During the discussion that followed someone asked about the freedom of teaching. He answered in the following way (this was before the Lysenko controversy): "One has to distinguish between meaningful and meaningless branches of learning. Meaningful, like mathematics, physics, chemistry, biology and so on and meaningless, like theology, the whole of philosophy, sociology and the rest. Now the situation is simple. For the meaningful disciplines there is complete freedom. I must admit that for the meaningless disciplines there is a preference for certain schools of thought, but after all it is of no importance whether one prefers one kind of nonsense to the other. (Ob man den einen oder den andere Quatsch bevorzgut)." No wonder he later got into some difficulty with the Stalin regime.

While at Copenhagen I also started work on my Ph.D. thesis. It dealt with the quantum mechanics of rotating solid bodies and was inspired by a short note of O.Klein, who had shown that the angular momenta in a rotating frame of reference obey commutation rules with inverted signs. The thesis contained no applications to molecular spectra: it treated the kinematics of rotation, some special aspect of the representations of the rotation group and so on. It also contained what became later known as "Casimir operator", a quadratic form in the infinitesimal operators of a semi-simple Lie group that commutes with all the infinitesimal operators. Such operators are now playing a certain role in field theory. They have also been of some importance in pure mathematics, where they have been applied in the proof of complete reducibility of reducible representations. (First in a joint paper by Van der Waerden and myself, written during my stay at Zürich, later - more elegantly - by R. Brauer).

I obtained my Ph.D. at Leiden in November 1931 and stayed at Leiden as assistant to Ehrenfest. It was not a very productive winter. Ehrenfest was rather depressed and some work I tried to do on internal conversion and on the photoelectric effect did not succeed too well. My stay at Berlin in the summer of 1932 provided some new inspiration and, while there, I received a letter from Pauli asking me to become his assistant.

To work with Pauli was a new experience. If you define a genius as someone who is able to create things that are to begin with beyond his own understanding, then he may not have been a genius, but his intellectual powers were tremendous. He was also known as a merciless critic who did not mince his words. I found him on the whole kind and considerate, and although he did not underestimate his own abilities he was as critical of himself as of others. My first task as an assistant took a curious and for me rather fortunate turn. The young French theoretician Jacques Solomon had been at Zürich the preceding term. (I had met him briefly at Copenhagen a few years earlier; I remember that Landau did not like him at all, probably because Solomon was an "orthodox" communist. Solomon was later killed by the Germans.) He had written a paper in which he had added some details, especially an energy-momentum tensor, to Einstein's recent five-dimensional theory and he had sent Pauli the second proof of this paper, with a letter in which he stated that he would feel more happy if the paper would be published under the names of Pauli and Solomon. Pauli thought this over and told me that he had not only suggested the problem to Solomon but that he had also indicated how to tackle it. So he felt that a joint publication would be appropriate, but if the paper was to bear also his name he wanted to be sure that it was correct. So he instructed me to check the formulae. I was in those days reasonably good at juggling with tensors and the task was not a difficult one. To my dismay I found that most formulae, at least the essential ones, were wrong. I told Pauli, who at once cabled to Paris that publication should be postponed. Soon afterwards there came a letter from Jean Langevin, then editor of the Journal de Physique, with apologies: the second proof had not been returned in time, there had been only very few corrections in the first proof and so the article had already been printed and circulated, with all the errors and with Pauli's name. Pauli was less angry than he might have been, but he resolutely tackled the problem. He redid the calculations himself - and, to my satisfaction confirmed my results - and wrote a masterly introduction to an entirely new version. This new article was sent to Solomon who translated it into French (but Pauli told him not to change one single formula). Then a curious thing happened. Pauli came into my room and looked worried, almost embarassed. "Look here" he said "you found that mistake; perhaps you should be co-author or at least I should put in an acknowledgement. But this is an awkward situation and I should not like to harm Solomon's future. How do you feel about it?". I answered that I did not think I deserved a special acknowledgement. It might have been different if I had found the errors at my own initiative. But Pauli had told me to check the formulae, which was simply a routine business. The great man really looked relieved. This little incident shows that Pauli was, his sharp tongue notwithstanding, very conscientious in dealing with the work of others. It also helped to establish friendly relations between master and assistant. And I feel quite happy that I have made at least one contribution to general relativity, and that a strictly anonymous one.

Pauli, in those days, was not a very happy man. His first marriage had broken up after a few months and he had not yet met his second wife. Also, he had put a lot of work into has article on quantum mechanics in the Handbuch der Physik, but he did not quite see where to go from there. He did not like solid state physics and some of the other straightforward applications of the new theories. And I do not think he really believed in Einstein's approach to unified field theories. One of his consolations was his car. He had had some difficulty to pass the driving test and he was not a good driver, but he managed, and as far as I know accidents were limited to some dents and scratches incurred while leaving or entering his garage. Sometimes he would take me out in the evening and we would have a quiet meal at some country inn. His car also kept him from having a glass too many, which was just as well.

To the last mentioned point, however, there occurred one notable exception. That was on the occasion of the spring meeting of the Swiss Physical Society at Luzern. Pauli had driven us - that is David Inglis, Homi Bhabha, I believe also Felix Bloch and myself from Zürich to Luzern in the morning and apart from Pauli's slightly disconcerting habit of saying from time to time "Ich fahre ziemlich gut (I'm driving rather well)", a statement he underlined by turning around to his passengers and by releasing his hold on the wheel, nothing untoward happened. In the evening Pauli was drinking fruit juice with a wry face. Suddenly he changed his mind and ordered a whisky-soda. That was alright, but when he had ordered a second one and showed no sign of wanting to stop there we became really worried. So we made a plan: we would offer him more drinks and then Inglis would drive us home. The first part of the operation succeeded, but, when we suggested that Inglis would drive, Pauli refused. He was going to drive and as far as he was concerned we could come along or stay at Luzern. By then the last train to Zürich had left, and anyway we did not want to let Pauli go all by himself. So we went, with Inglis sitting beside Pauli, ready to grasp the wheel in an emergency; we had been joined by Elsasser, who had missed the last train and was sitting on the floor of the car. Pauli sounded his horn several times, hit one curb, swerved to the other side of the street where he hit the other curb and then managed to find his bearings and get going. It was a memorable trip.

Pauli would still from time to time say "Ich fahre ziemlich gut" but when the car went screeching around curves Inglis would say sternly "Das heisst nicht gut fahren (that is not called good driving)", which somehow had a sobering effect. Once a rising moon came just over the crest of a hill and Pauli started to swear at the driver who did not dim his headlights.
Once Pauli said, "Here, I know a short cut," and suddenly turned into an unpaved track. It came to an end at the wagon shed of a farm and after some angry comments about people who, overnight, put wagonsheds across his short cut Pauli turned around and went back to the main road. But that was all, we came safely home.

I mentioned Bhabha. He had come to Zürich after R.H. Fowler had written a letter of recommendation .... of a kind. Fowler realized that Bhabha was very gifted, but he also thought him rather opinionated and unruly, so he felt he needed a strong hand hand. "You can be as brutal as you like" he wrote. This Pauli enjoyed immensely. He showed me the letter and repeated over and over again - that was one of his habits, if a certain short sentence struck him he would repeat it several times, tasting and retasting it, so to say - "I can be as brutal as I like." I wonder whether Fowler was subtle enough to understand that this letter was the best way to make Pauli friendly disposed towards Bhabha. But it did work out that way and they became good friends, although it must be admitted that Bhabha sometimes went to Wentzel rather than to Pauli when he wanted to discuss an entirely new idea.

As to my own work, I mentioned already the work I did in mathematics. I also wrote a short paper on radiation damping, explaining a paradox that had been discussed at Copenhagen.

And I tried to calculate Compton scattering by bound electrons. Today I do not feel sure about the result, but my paper has been useful to others because it introduced a projection operator that simplified summing over states of positive energy only. I should have liked to stay one more year at Zürich and Pauli would have been willing to keep me, but Ehrenfest urged me to come back to Leiden. Was he already planning his own suicide and did he want to make sure that during the time that would elapse before a successor could take over there would at least be someone who could keep things going? In September 1933 I settled down at Leiden, where I stayed until 1942. Kramers came to Leiden as Ehrenfest's successor in the course of 1934. He became a close personal friend, he taught me many a useful mathematical tricks and we had many talks about physics, but we never did any major research together.

I have spoken already about my work on hyperfinestructure. Gradually, however, I became more interested in the work that was going on in the low temperature Laboratories and in 1936 I changed my position for one in the Kamerlingh Onnes Labs. Low temperature physics was then as it is now - a fascinating subject. Liquid helium was in those days only available in very few laboratories and among these the Leiden Labs were the biggest by far. And once you had liquid helium you could do interesting experiments with quite simple apparatus. Even a rather clumsy theoretician like myself could build part of what was needed with his own hands. And I had very pleasant relations with my colleagues at Oxford and Cambridge. Let me mention some of the subjects I worked on.

After the discovery of Meissner (B=0 in the superconducting state), Gorter and I worked out the thermodynamics of superconductors, which is fairly elementary, but notoriously tricky. Heisenberg himself once made a serious error. We also introduced a two fluid model that provided a useful approximate description until it was superseded by the BCS theory. I devised a simple method to determine the change with temperature of the penetration depth of the magnetic field into a superconductor. I bungled the experiments but the method later became useful in more skillful hands. Together with Dupré I worked out a simple thermodynamic theory of paramagnetic relaxation. The basic idea, according to which the spinsystem can have a well-defined temperature that is different from the lattice temperature, has turned out to be quite useful, also in recent work on nuclear spin magnetism. My own experiments dealt mainly with adiabatic demagnetization. I gave some lectures on the subject at Cambridge (England) and that led to the publication of a Cambridge Tract on Magnetism and very low Temperatures. It was later reprinted as a Dover publication. Consideration of the symmetry relations valid for the magnetoresistance of single crystals of bismuth led me to a careful study of Onsagers symmetry relations for irreversible processes.

By that time, 1942, The Netherlands had been occupied by the Germans, Leiden University had been closed, and when the Philips Company offered me a job in a quiet corner of the research laboratories I was glad to accept. It was there that I realized that Onsager's theory could also be applied to electrical networks, and immediately after the war I published some papers in which I tried on the one hand to clarify some aspects of Onsager's derivation and on the other hand to widen its field of application.

After the war I was offered a co-directorate of the Philips research labs and I decided to stay at Eindhoven. At first I still found time for some research of my own. Together with Polder I worked out the influence of retardation on London - Van der Waals forces. Somewhat later I derived a formula predicting a universal attraction between perfectly conducting plates. This formula has since been derived in a number of different ways, has been generalized, first by Lifshitz and most recently by Schwinger, and has been confirmed experimentally. To my knowledge this attraction is the only macroscopic quantum-effect that is entirely determined by Planck's constant and the velocity of light and does not depend on another atomic constant like the charge of the electron.

Gradually my managerial duties increased. I became a member of the executive board of the Philips Company, I had also to supervise work in research laboratories outside the Netherlands and it was already quite a job to take an intelligent interest in the multitude of projects that were going on. I did from time to time publish a paper. Sometimes I tried to clarify some aspects of existing theories, sometimes I worked out some special problem in electrodynamics, but that was all.

A few years before his death Pauli paid a visit to Holland. One evening we were sitting together with a small group of people and someone asked me if I had not had a difficult time of it as a young assistant to Pauli. Pauli looked at me expectantly and so I felt I had to make up some kind of story. I said "Not really, Pauli had then just bought his car and there was a kind of tacit understanding between us that as long as I would refrain from making comments on his driving he would not say anything about my physics. Now I won't brag about my physics, but I think that in those days it was somewhat better than Pauli's driving." Everyone laughed but Pauli had the last word: "Maybe it was like that. To-day I don't drive any longer, and you, Herr Direktor, don't do physics anymore. Die Sache stimmt noch immer (Things still tally)."

In a way Pauli was right, and so the story of my life as a physicist ends here. But the things I learnt as a physicist stood me in good stead throughout the activities of my later years.

CLOSING CEREMONY

The Closing Ceremony took place on Friday, 10 August 1979. The Director of the School presented the prizes and scholarships as specified below.

PRIZES AND SCHOLARSHIPS

- Prize for Best Student - awarded to Douglas FONG
  Massachusetts Institute of Technology, Cambridge, USA.

Eleven scholarhips were open for competition among the participants. They were awarded as follow:

- Patrick M.S. Blackett Scholarship - awarded to Paul GINSPARG
  Cornell University, Ithaca, USA.
- James Chadwick Scholarship - awarded to M. Belen GAVELA LEGAZPI
  Université Paris XI, Orsay, France.
- Amos-de Shalit Scholarship - awarded to François VANNUCCI
  LAPP, Annecy-le-Vieux, France.
- Gunnar Källen Scholarship - awarded to Giora MIKENBERG
  DESY, Hambourg, FRG.

- André Lagarrigue Scholarship - awarded to George CHRISTOS
  University of Oxford, UK.

- Giulio Racah Scholarship - awarded to Hans HANSSON
  Chalmers Tekniska Högskola, Göteborg, Sweden.

- Giorgio Ghigo Scholarship - awarded to Irene CAPRINI
  Institute of Physics and Nuclear Engineering, Bucharest, Rumani

- Enrico Persico Scholarhip - awarded to Michael P. SCHMIDT
  Yale University, New Haven, USA.

- Peter Preiswerk Scholarship - awarded to Gianpiero PAFFUTI
  University of Pisa, Italy.

- Gianni Quareni Scholarship - awarded to John OLIENSIS
  Princeton University, USA.

- Antonio Stanghellini Scholarship - awarded to In-Gyu KOH
  Sogang University, Seoul, Korea.

- Honorary Mentions for the Best Scientific Secretaries
  ex-aequo awarded to

  - M. Belen GAVELA LEGAZPI
    Université Paris XI, Orsay, France.

  - Argyris NICOLAIDIS
    Collège de France, Paris, France.

  - Federico RAPUANO
    University of Rome, Italy.

- The following participants gave their collaboration in the scientific secretarial work:

- Gregory ADKINS
- Leonard ANDERSON
- Irene CAPRINI
- Guido CORBO'
- Douglas FONG
- M. Belen GAVELA LEGAZPI
- Paul GINSPARG
- Arne HORNOES
- In-Gyu KOH
- Arthur KREYMER
- Holger LIERL
- Jnanadeva MAHARANA
- Giora MIKENBERG
- Argyris NICOLAIDIS
- John OLIENSIS
- Lazlo PALLA
- Felicitas PAUSS
- Ashok RAINA
- Federico RAPUANO
- Michael P. SCHMIDT
- Tokuzo SHIMADA
- Thom STERLING
- Jevzy SZVED
- François VANNUCCI

P A R T I C I P A N T S

Gregory ADKINS	University of California Physics Department UCLA LOS ANGELES, CA 90024, USA
Elisabetta ALBINI	Istituto di Fisica dell'Università Via Celoria, 16 20133 MILANO, Italy
Leonard ANDERSON	Laboratoire de Physique Nucléaire des Hautes Energies Ecole Polytechnique 91128 PALAISEAU, France
Mauro ANSELMINO	Istituto di Fisica dell'Università Corso M. D'Azeglio, 46 10125 TORINO, Italy
Shoyo AOYAMA	Istituto di Fisica dell'Università Via F. Marzolo, 8 35100 PADOVA, Italy
Jean Jacques AUBERT	Laboratoire de Physique des Particules IUT(IN2P3) Chemin de Bellevue B.P. 909 74019 ANNECY-LE-VIEUX, France
Sergio BERTOLUCCI	Laboratori Nazionali di Frascati INFN 00044 FRASCATI, Italy

Daniel BERTRAND
Inter-University Institute for
High Energies (ULB-VUB)
Boulevard du Triomphe, C.P. 230
1050 BRUXELLES, Belgium

Irene CAPRINI
Institute of Physics and Nuclear
Engineering
Department of Theoretical Physics
P.O. Box 5206
BUCHAREST, Romania

Hendrik B.G. CASIMIR
De Zegge, 7
HEEZE near Eindhoven
N-BR, The Netherlands

Leonardo CASTELLANI
Istituto di Fisica Teorica
Largo E. Fermi, 2
50125 FIRENZE, Italy

Jasbinder S. CHIMA
Imperial College of Science and
Technology
The Blackett Laboratory
Prince Consort Road
LONDON SW7 2BZ, UK

George CHRISTOS
University of Oxford
Department of Theoretical Physics
1 Keble Road
OXFORD OX1 3NP, UK

Antonio CODINO
Laboratori Nazionali di Frascati
INFN
00044 FRASCATI, Italy

Richard N. COLEMAN
University of Rochester
Department of Physics and Astronomy
ROCHESTER, NY 14627, USA

Sidney COLEMAN
Harvard University
Department of Physics
Lyman Laboratory of Physics
CAMBRIDGE, MA 02138, USA

Robert COQUEREAUX	Centre de Physique Théorique CNRS - Section 2 Luminy, Case 907 13288 MARSEILLE Cedex 2, France
Guido CORBO'	Istituto di Fisica dell'Università Piazzale delle Scienze, 5 00185 ROMA, Italy
Luigi DADDA	Politecnico di Milano Piazza Leonardo da Vinci, 32 20123 MILANO, Italy
Hans Jürgen DAUM	University of Wuppertal Fachbereich 8 Gausstrasse 20 5600 WUPPERTAL, FRG
Johannes G.H. DE GROOT	CERN, EP Division 1211 GENEVE 23, Switzerland
Alvaro DE RUJULA	CERN, TH Division 1211 GENEVE 23, Switzerland
Sergio DI LIBERTO	Istituto di Fisica dell'Università Piazzale delle Scienze, 5 00185 ROMA, Italy
Friedrich DYDAK	Institut für Hochenergiephysik Universität Heidelberg Albert-Ueberle-Str. 2 6900 HEIDELBERG 1, FRG
John C. ECCLES	"Ca' a la Gra'" 6611 CONTRA (Locarno), Switzerland
Johnathan R. ELLIS	CERN, TH Division 1211 GENEVE 23, Switzerland
Douglas FONG	Massachusetts Institute of Technology Laboratory for Nuclear Science CAMBRIDGE, MA 02139, USA

Masaki FUKUSHIMA
Massachusetts Institute of Technology
Laboratory for Nuclear Science
CAMBRIDGE, MA 02139, USA

José V. GARCIA ESTEVE
Universitad de Zaragoza
Facultad de Ciencias
Departamento de Fisica Atomica y Nuclear
ZARAGOZA, Spain

Maria Belen GAVELA LEGAZPI
Université Paris XI
Laboratoire de Physique Théorique et Particules Elémentaires
Bâtiment 211
91405 ORSAY, France

Jos GERIS
Katholieke Universiteit Leuven
Instituut voor Theoretische Fysika
Departement Natuurkunde
Celestijnenlaan 200B
3030 LEUVEN, Belgium

Gianrossano GIANNINI
Istituto Nazionale di Fisica Nucleare
Via Livornese, 582/a
56010 S.PIERO A GRADO, (Pisa), Italy

Paul GINSPARG
Cornell University
Laboratory of Nuclear Studies
ITHACA, NY 14853, USA

Marcel GUENIN
Université de Genève
Département de Physique Théorique
32 Boulevard d'Yvoy
1211 GENEVE 4, Switzerland

Hans HANSSON
Chalmers Tekniska Högskola
Department of Physics
Fack
41296 GÖTEBORG, Sweden

Arne HORNOES	University of Bergen Department of Physics Allégt. 55 5014 BERGEN, Norway
Maurizio IORI	Département de Physique des Particules Elémentaires CEN, Saclay, B.P. 2 91190 GIF-SUR-YVETTE, France
Robert L. JAFFE	Massachusetts Institute of Technology CAMBRIDGE, MA 02139, USA
Romesh KAUL	Freie Universität Berlin Institut für Theoretische Physik Arnimallee 3 1000 BERLIN 33, FRG
Hagen KLEINERT	Freie Universität Berlin Institut für Theoretische Physik Arnimallee 3 1000 BERLIN 33, FRG
In-Gyu KOH	Sogang University Department of Physics SEOUL, Korea
Bernd KOPPITZ	DESY/F14 Notkestrasse 85 2000 HAMBURG 52, FRG
Arthur KREYMER	Indiana University Department of Physics BLOOMINGTON, IN 47401, USA
Holger LIERL	University of Dortmund Postfach 500500 4600 DORTMUND 50, FRG
Erich LOHRMANN	DESY Notkestrasse 85 2000 HAMBURG 52, FRG

William LOUIS	Rutherford Laboratory Chilton DIDCOT, Oxon OX11, OQX, UK
Francis E. LOW	Massachusetts Institute of Technology CAMBRIDGE, MA 02139, USA
Dieter LÜKE	DESY Notkestrasse 85 2000 HAMBURG 52, FRG
Jnanadeva MAHARANA	Rutherford Laboratory Chilton DIDCOT, Oxon OX11, OQX, UK
André MARTIN	CERN, TH Division 1211 GENEVE 23, Switzerland
Juan MATEOS	Universitad de Salamanca Facultad de Ciencias Departamento de Fisica Teorica SALAMANCA, Spain
Clara MATTEUZZI	CERN, EP Division 1211 GENEVE, Switzerland
Giora MIKENBERG	DESY Notkestrasse 85 2000 HAMBURG 52, FRG
Alain MILSZTAJN	Département de Physique des Particules Elémentaires CEN, Saclay, B.P. 2 91190 GIF-SUR-YVETTE, France
Marcella MORICCA	Istituto di Fisica dell'Università Piazzale delle Scienze, 5 00185 ROMA, Italy
Dusan NESIC'	Faculty of Natural and Mathematical Sciences Dept. of Physics and Meteorology P.O. Box 708 11011 BELGRADE, Yugoslavia

Argyris NICOLAIDIS
Laboratoire de Physique
Corpuscolaire
Collège de France
11, Place Marcelin-Berthelot
75231 PARIS Cedex 05, France

John OLIENSIS
Princeton University
Physics Department
P.O. Box 708
PRINCETON, NY 08540, USA

Gianpiero PAFFUTI
Istituto di Fisica dell'Università
Piazza Torricelli, 2
56100 PISA, Italy

Lazlo PALLA
Eötvös University
Institute for Theoretical Physics
Puskin U. 5-7
1088 BUDAPEST, Hungary

Jack E. PATON
University of Oxford
Department of Theoretical Physics
1 Keble Road
OXFORD OX1 3NP, UK

Adrian PATRASCIOIU
University of Arizona
Department of Physics
TUCSON, AZ 85281, USA

Felicitas PAUSS
Max-Planck-Institut für Physik
und Astrophysik
Föhringer Ring 6
8000 MUNCHEN 40, FRG

R. Sebastian PEASE
Culham Laboratory
U.K. Atomic
Energy Authority
ABINGDON, Oxfordshire OE14 3DB, UK

José A. PENARROCHA GANTES
Universitad de Valencia
Departamento de Fisica Teorica
VALENCIA 10, Spain

Roberto PETRONZIO	CERN, TH Division 1211 GENEVE 23, Switzerland
Mario POSOCCO	Istituto di Fisica dell'Università Via Marzolo, 8 35100 PADOVA, Italy
Giulano PREPARATA	CERN, TH Division 1211 GENEVE 23, Switzerland
James PROUDFOOT	Rutherford Laboratory Chilton DIDCOT, Oxon OX11 0QX, UK
Stefano RAGAZZI	Istituto di Fisica dell'Università Via Celoria, 16 20133 MILANO, Italy
Ashok K. RAINA	Université de Lausanne Institut de Physique Théorique Faculté des Sciences Dorigny 1015 LAUSANNE, Switzerland
Federico RAPUANO	Istituto di Fisica dell'Università Piazzale delle Scienze, 5 00185 ROMA, Italy
Hans REINDERS	University College London Department of Physics Gower Street LONDON, WC1E 6BT, UK
Albert ROMANA	Laboratoire de Physique Nucléaire des Hautes Energies Ecole Polytechnique 91128 PALAISEAU, France
Mario RONCADELLI	Istituto di Fisica dell'Università Via A. Bassi, 6 27100 PAVIA, Italy

David L. RUBIN
University of Michigan
The Harrison M. Randall Laboratory of Physics
ANN ARBOUR, MI 48109, USA

William SCOTT
CERN, EP Division
1211 GENEVE 23, Switzerland

Michael P. SCHMIDT
Yale University
Department of Physics
Sloane Laboratory
217 Prospect Str.
NEW HAVEN, CT 06520, USA

Bernhard SCHOLZ
Siegen University
Physics Department
Adolf-Reichwein-Str.
Fachbereich 7
5900 SIEGEN 21, FRG

Gordon SEMENOFF
University of Alberta
Department of Physics
EDMONTON, Alberta T6G 2J1
Canada

Tokuzo SHIMADA
Max-Planck-Institut für Physik und Astrophysik
Föhringer Ring 6
8000 MUNCHEN 40, FRG

Eric SISKIND
CALTECH - 256-48
High Energy Physics
PASADENA, CA 91125, USA

Thom STERLING
University of Michigan
The Harrison M. Randall Laboratory of Physics
ANN ARBOUR, MI 48109, USA

James L. STONE
CERN, EP Division
1211 GENEVE 23, Switzerland

Jevzy SZWED	Max-Planck-Institut für Physik und Astrophysik Föhringer Ring 6 8000 MUNCHEN 40, FRG
Jean TEILLAC	Commissariat à l'Energie Atomique 29/31 rue de la Fédération 75752 PARIS Cedex XV, France
Sam C.C. TING	Massachusetts Institute of Technology Department of Physics CAMBRIDGE, MA 02139, USA
Ludovico TORTORA	Istituto Superiore di Sanità Viale Regina Elena, 299 00161 ROMA, Italy
Remigius T. VAN DE WALLE	Fysisch Laboratorium Toernooiveld 6525 ED NIJMEGEN, The Netherlands
François VANNUCCI	Laboratoire de Physique des Particules Elémentaires B.P. 909 74019 ANNECY-LE-VIEUX
Luis VAZQUEZ MARTINEZ	Departemento de Metodos Matematicos de Fisica Facultad de Ciencias Fisicas Universitad Complutense de Madrid MADRID 3, Spain
Achille VENAGLIONI	Istituto di Fisica dell'Università Via A. Bassi, 6 27100 PAVIA, Italy
Victor F. WEISSKOPF	Massachusetts Institute of Technology Department of Physics CAMBRIDGE, MA 02139, USA

Gunter WIRTHUMER

Institut für Theoretische Physik II
Tech. Universität Wien
Karlsplatz 13
1040 WIEN, Austria

INDEX